Polymer Reference Book

T.R. Crompton

Rapra Technology Limited

Shawbury, Shrewsbury, Shropshire, SY4 4NR, United Kingdom
Telephone: +44 (0)1939 250383 Fax: +44 (0)1939 251118
http://www.rapra.net

Chemistry Library

First Published in 2006 by

Rapra Technology Limited

Shawbury, Shrewsbury, Shropshire, SY4 4NR, UK

©2006, Rapra Technology Limited

A catalogue record for this book is available from the British Library.

Every effort has been made to contact copyright holders of any material reproduced within the text and the authors and publishers apologize if any have been overlooked.

ISBN: 1-85957-526-9

Typeset by Rapra Technology Limited
Printed and bound by Lightning Source UK Limited

Contents

Preface

The aim of this book is to familiarise the reader with all aspects of the techniques used in the examination of polymers, covering chemical, physiochemical and purely physical methods of examination.

It is the purpose of this book to describe the types of techniques now available to the polymer chemist and technician, and to discuss their capabilities, limitations and applications. All types of modern instrumentation are covered including those used in general quality control, research analysis, process monitoring and the mechanical, electrical, thermal and optical characteristics. Aspects such as automated analysis and computerised control of instruments are covered.

The book is intended, therefore, for all staff, whether direct or peripheral, who are concerned with the instrumentation and methodology in the polymer laboratory including laboratory designers, engineers and chemists, and those concerned with the implementation of analytical specifications and process control limits. Practically all of the major newer instrumental techniques and many of the older classical techniques have been used to examine polymers and all of these are discussed in detail.

The book covers not only instrumentation for the determination of metals, non metals, functional groups, polymer structural analysis and end-groups in the main types of polymers now in use commercially, but also the analysis of minor non-polymeric components of the polymer formulation, whether they be deliberately added, such as processing additives, or whether they occur adventitiously, such as residual volatiles and monomers and water. Fingerprinting techniques for the rapid identification of polymers and methods for the examination of polymer surfaces and polymer defects are also discussed.

Additionally, other features of polymer characterisation are discussed such as the determination of molecular weight, polymer fractionation techniques, chemical and thermal stability, resin cure, oxidative stability, photopolymers, glass and other transitions, crystallinity, viscoelasticity, rheological properties, thermal properties, flammability testing, particle size analysis and the measurement of the mechanical, electrical and optical properties of polymers.

1

Elemental analysis is now a well accepted method for the analysis of polymers and copolymers and their additive systems. Also, as so many different types of polymers are now used commercially it is advisable when attempting to identify a polymer and its additive system to classify it by first carrying out an elemental analysis for metallic and non-metallic elements. Instrumentation discussed for carrying out these analyses include; atomic absorption spectrometry including its graphite furnace, atom trapping and vapour generation modifications, Zeeman atomic absorption spectrometry, inductively coupled plasma atomic emission spectrometry, visible ultraviolet spectroscopy, polarography and voltammetry, ion chromatography, x-ray fluorescence spectroscopy and neutron activation analysis for metals (Chapter 1) and furnace combustion methods, oxygen flask combustion and x-ray fluorescence spectroscopy for non-metallic elements (Chapter 2).

Equally important, particularly in the case of copolymers, is the determination of functional groups. Frequently it is necessary to obtain a detailed picture of the microstructure of a polymer involving the determination of minor concentrations of functional and end-groups. Instrumentation discussed for carrying out this type of analysis (Chapter 3) includes, infrared and near infrared spectroscopy, Fourier transform infrared spectroscopy, NMR and PMR spectroscopy, and complimentary techniques such as reaction gas chromatography, pyrolysis gas chromatography mass spectrometry and pyrolysis gas chromatography - Fourier transform infrared spectroscopy. Recent work on the combination of pyrolysis with mass spectrometry and infrared spectroscopy is discussed in detail as well as work on new techniques for polymer characterisation such as time of flight secondary ion mass spectrometry (SIMS), x-ray photoelectron spectroscopy (XPS), tandem mass spectrometry (MS/MS), Fourier transform ion cyclotron mass spectrometry, matrix assisted laser description/ionisation mass spectrometry (MALDI/MS), radiofrequency glow discharge mass spectrometry, microthermal analysis, atomic force microscopy and scanning electron microscopy (SCM) - energy dispersive analysis using x-rays (EDAX).

These powerful new techniques are throwing considerable light on polymer structure.

Impurity inclusions and surface defects are a cause of many difficulties to the polymer producer and user. Equipment used for studying these phenomena discussed in Chapter 4 include electron microprobe x-ray emission/spectroscopy, NMR micro-imaging, various forms of surface infrared spectroscopy, e.g., diffusion reflection FTIR, ATR, also photoacoustic spectroscopy and x-ray diffraction - infrared microscopy of individual polymer fibres. Newer techniques such as scanning electron microscopy (SECM), transmission electron microscopy, time of flight secondary ion mass spectrometry (TOFSIMS), laser induced photoelectron ionisation with laser desorption, atomic force microscopy and microthermal analysis are discussed.

Instrumentation for the determination of non-polymeric volatiles and water in polymers are discussed in Chapter 5 and include gas chromatography, high performance liquid chromatography, polarography, headspace analysis, headspace analysis - gas chromatography - mass spectrometry and purge and trap analysis.

If a simple qualitative identification of a plastic is all that is required then fingerprinting techniques discussed in Chapter 6 may suffice. Fingerprinting instrumentation discussed include glass transition, pyrolysis techniques, infrared spectroscopy, pyrolysis - Fourier transform infrared spectroscopy, Raman spectroscopy and radio frequency slow discharge mass spectrometry.

Instrumentation for the determination of organic additive mixtures in polymers discussed in Chapter 7 fall into two categories; (a) direct methods such as Fourier transform infrared spectroscopy, ultraviolet spectroscopy, luminescence and fluorescence spectroscopy, NMR spectroscopy, mass spectrometry and polarography; or (b) techniques based on a preliminary separation of individual additives from each other in an extract of the polymer and subsequent determination of individual separated compounds. Separation instrumentation discussed includes high performance liquid chromatography, gas chromatography, ion chromatography, supercritical fluid chromatography and thin-layer chromatography, also complementary techniques such as high performance liquid chromatography combined with mass spectrometry or infrared spectroscopy. Newer developments in additive identification include pyrolysis - gas chromatography mass spectrometry, x-ray photoelectron spectrometry (XPS) and secondary ion mass spectrometry (SIMS) and x-ray fluorescence spectrometry.

Physical methods of examination of polymers are discussed in Chapters 8-19. These include polymer fractionation and molecular weight determination (Chapter 8) involving instrumentation for high performance gel permeation chromatography (ion exclusion chromatography), supercritical fluid chromatography, gas chromatography, thin-layer chromatography, ion-exchange chromatography. Applications of other techniques are discussed such as NMR spectroscopy, osmometry, light scattering, visiometry, centrifugation thermal field flow fractionation electrophoresis and various mass spectrometric techniques including field desorption mass spectrometry, time of flight secondary ion mass (TOFSIMS) spectrometry and matrix assisted laser desorption Fourier transform mass spectrometry (MALDI).

Studies of the thermal and chemical stability of polymers are of paramount importance and instrumentation used in these studies discussed in Chapter 9 include thermogravimetric analysis, differential thermal analysis, differential scanning calorimetry, thermal volatilisation analysis and evolved gas analysis. Monitoring of resin cure is another important parameter in polymer processing in which dynamic mechanical analysis, dielectric thermal analysis and differential scanning calorimetry is used (Chapter 10).

Instrumentation used for polymer oxidative stability (Chapter 11) includes, thermogravimetric analysis, differential scanning calorimetry, evolved gas analysis, infrared spectroscopy and ESR spectroscopy, matrix assisted laser desorption/ionisation mass spectrometry and imaging chemiluminescence is included.

The application of differential photocalorimetry and other methods to the examination of photo-polymers is discussed in Chapter 12.

Differential scanning calorimetry, thermomechanical analysis, dynamic mechanical analysis, differential thermal analysis, dielectric thermal analysis, infrared and NMR spectroscopy, are some of the instrumental techniques that have been applied to the determination of glass transition and other transition temperatures in polymers (Chapter 13).

Instrumentation used in the measurement of polymer crystallinity includes light scattering, positron annihilation lifetime spectroscopy, differential scanning calorimetry, differential thermal analysis, infrared spectroscopy, NMR spectroscopy and wide and small angle x-ray diffraction (Chapter 14).

Measurement of viscoelastic and rheological properties of polymers are of increasing importance and recently developed instrumentation for measuring these properties discussed in Chapter 15 includes dynamic mechanical analysis, thermo-mechanical analysis and dielectric thermal analysis.

Various thermal material properties (as opposed to thermal stability, Chapter 9) are discussed in Chapter 16. These include coefficient of expansion, melting temperature, Vicat softening point, heat deflection/distortion temperature by thermomechanical analysis, also brittleness temperature, minimum filming temperature, delamination temperature, meltflow index, heat of volatilisation, thermal conductivity, specific heat and ageing in air.

Instrumentation for the measurement of the flammability characteristics of polymers intended for use in mining, electrical, transport, furniture, construction materials is discussed in Chapter 17, including methods of identifying combustion products and the measurement of oxygen consumption by oxygen cone calorimetry. New methods based on pyrolysis - gas chromatography - mass spectrometry and laser pyrolysis time of flight mass spectrometry are also discussed.

In Chapter 18 is discussed the measurement of mechanical, electrical and optical properties of polymers. Mechanical measurements include measurement of load bearing characteristics of polymers including stress/strain curves, stress temperature curves, recovery and rupture. Also measurement of impact strength characteristics by Izod and falling weight methods and many other polymer characteristics for polymer sheet, pipe, film, powders and rubbers and elastomers.

Electrical properties discussed include the measurement of volume and surface resistivity, dielectric strength and surface arc and tracking resistance. Optical properties and light stability includes a discussion of stress optical analysis and the effects of light and other influences on stabilised and unstabilised polymers.

Chapter 19 is dedicated to the measurement of the particle size distribution of polymer powders by a variety of methods.

The book gives an up-to-date and thorough exposition of the present state of the art of the theory and availability of instrumentation needed to effect chemical and physical analysis of polymers. Over 1,800 references up to late 2003 are included. The book should be of great interest to all those engaged in the subject in industry, university research establishments and general education. The book is intended for all staff who are concerned with instrumentation in the polymer laboratory, including laboratory designers, work planners, chemists, engineers, chemical engineers and those concerned with the implementation of specifications and process control.

T.R. Crompton

March 2006

1 Determination of Metals

Different techniques have evolved for trace metal analysis of polymers. Generally speaking, the techniques come under two broad headings:

- *Destructive techniques*: these are techniques in which the sample is decomposed by a reagent and then the concentration of the element in the aqueous extract determined by a physical technique such as atomic absorption spectrometry (AAS; Section 1.1.1), graphite furnace atomic absorption spectrometry (GFAAS; Section 1.1.2), cold vapour atomic absorption spectrometry (CVAAS; Section 1.1.4), Zeeman atomic absorption spectrometry (ZAAS; Section 1.1.5), inductively coupled plasma atomic emission spectrometry (ICP-AES; Sections 1.1.6 and 1.1.8), visible spectrometry (Section 1.1.13), and polarographic or anodic scanning voltammetric techniques (Section 1.114).

- *Non-destructive techniques*: these include techniques such as X-ray fluorescence (XRF; Section 1.2.1) and neutron activation analysis (NAA; Section 1.2.2.), in which the sample is not destroyed during analysis.

1.1 Destructive Techniques

1.1.1 Atomic Absorption Spectrometry

1.1.1.1 Theory

Since shortly after its inception in 1955, AAS has been the standard tool employed by analysts for the determination of trace levels of metals. In this technique a fine spray of the analyte is passed into a suitable flame, frequently oxygen–acetylene or nitrous oxide–acetylene, which converts the elements to an atomic vapour. Through this vapour is passed radiation at the right wavelength to excite the ground state atoms to the first excited electronic level. The amount of radiation absorbed can then be measured and directly related to the atom concentration: a hollow cathode lamp is used to emit light with the characteristic narrow line spectrum of the analyte element. The detection system consists of a monochromator (to reject other lines produced by the lamp and background flame radiation) and a photomultiplier. Another key feature of the technique involves modulation of the source radiation so that it can be detected against the strong flame and sample emission radiation.

This technique can determine a particular element with little interference from other elements. It does, however, have two major limitations. One of these is that the technique does not have the highest sensitivity. The other is that only one element at a time can be determined. This has reduced the extent to which it is currently used.

1.1.1.2 Instrumentation

Increasingly, due to their superior intrinsic sensitivity, the AAS currently available are capable of implementing the graphite furnace techniques. Available suppliers of this equipment are listed in Appendix 1.

Figures 1.1(a) and (b) show the optics of a single-beam flame spectrometer (Perkin Elmer 2280) and a double-beam instrument (Perkin Elmer 2380).

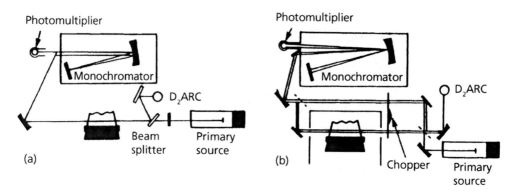

Figure 1.1 (a) Optics Perkin Elmer Model 2280 single beam atomic absorption spectrometer; (b) Optics Perkin Elmer 2380 double beam atomic absorption spectrometer. (*Source: Author's own files*)

1.1.2 Graphite Furnace Atomic Absorption Spectrometry

1.1.2.1 Theory

The GFAAS technique, first developed in 1961 by L'vov, was an attempt to improve the detection limits achievable. In this technique, instead of being sprayed as a fine mist into the flame, a measured portion of the sample is injected into an electrically heated graphite boat or tube, allowing a larger volume of sample to be handled. Furthermore, by placing the sample on a small platform inside the furnace tube, atomisation is delayed until the surrounding gas within the tube has heated sufficiently to minimise vapour phase interferences, which would otherwise occur in a cooler gas atmosphere.

The sample is heated to a temperature slightly above 100 °C to remove free water, then to a temperature of several hundred degrees centigrade to remove water of fusion and other volatiles. Finally, the sample is heated to a temperature near to 1000 °C to atomise it and the signals produced are measured by the instrument.

The problem of background absorption in this technique is solved by using a broad-band source, usually a deuterium arc or a hollow cathode lamp, to measure the background independently and subsequently to subtract it from the combined atomic and background signal produced by the analyte hollow cathode lamp. By interspersing the modulation of the hollow cathode lamp and 'background corrector' sources, the measurements are performed apparently simultaneously.

Graphite furnace techniques are about one order of magnitude more sensitive than direct injection techniques. Thus, lead can be determined down to 50 µg/l by direct atomic absorption spectrometry and down to 5 µg/l using the graphite furnace modification of the technique.

1.1.2.2 Instrumentation

The instrumentation is dealt with in Appendix 1.

1.1.3 Atom Trapping Technique

The sensitivity difference between direct flame and furnace atomisation has been bridged via the general method of atom trapping as proposed by Watling [1]. A silica tube is suspended in the air–acetylene flame. This increases the residence time of the atoms within the tube and therefore within the measurement system. Further devices such as water-cooled systems that trap the atom population on cool surfaces and then subsequently release them by temporarily halting the coolant flow are sometimes employed. The application of atom-trapping atomic absorption spectrometry to the determination of lead and cadmium has been discussed by Hallam and Thompson [2].

1.1.4 Vapour Generation Atomic Absorption Spectrometry

1.1.4.1 Theory

In the past certain elements, e.g., antimony, arsenic, bismuth, germanium, lead, mercury, selenium, tellurium, and tin, were difficult to measure by direct AAS [3–9].

A novel technique of atomisation, known as vapour generation via generation of the metal hydride, has been evolved, which has increased the sensitivity and specificity enormously for these elements [5–7, 9]. In these methods the hydride generator is linked to an atomic absorption spectrometer (flame graphite furnace) or inductively coupled plasma optical emission spectrometer (ICP-OES) or an inductively coupled plasma mass spectrometer (IPC-MS). Typical detection limits achievable by these techniques range from 3 µg/l (arsenic) to 0.09 µg/l (selenium).

This technique makes use of the property that these elements exhibit, i.e., the formation of covalent, gaseous hydrides that are not stable at high temperatures. Antimony, arsenic, bismuth, selenium, tellurium, and tin (and to a lesser degree germanium and lead) are volatilised by the addition of a reducing agent like sodium tetrahydroborate(III) to an acidified solution. Mercury is reduced by stannous chloride to the atomic form in a similar manner.

1.1.4.2 Instrumentation

Automating the sodium tetrahydroborate system based on continuous flow principles represents the most reliable approach in the design of commercial instrumentation. Thompson and co-workers [10] described a simple system for multi-element analysis using an ICP spectrometer, based on the sodium tetrahydroborate approach. PS Analytical Ltd developed a reliable and robust commercial analytical hydride generator system, along similar lines, but using different pumping principles from those discussed by Thompson and co-workers [10].

A further major advantage of this range of instruments is that different chemical procedures can be operated in the instrument with little, if any, modification. Thus, in addition to using sodium tetrahydroborate as a reductant, stannous chloride can be used for the determination of mercury at very low levels.

The main advantage of hydride generation atomic absorption spectrometry for the determination of antimony, arsenic, selenium, and so on, is its superior sensitivity.

More recently, PS Analytical have introduced the PSA 10.003 and the Merlin Plus continuous flow vapour generation atomic absorption and atomic fluorescence spectrometers [11–13]. These facilitate the determination of very low concentrations (ppt) of mercury, arsenic, and selenium in solution, enabling amounts down to 10–20 ppm of these elements to be determined in polymer digests.

1.1.5 Zeeman Atomic Absorption Spectrometry

1.1.5.1 Theory

The Zeeman technique, though difficult to establish, has an intrinsic sensitivity perhaps five times greater than that of the graphite furnace technique, e.g., 1 µg/l detection limit for lead.

The Zeeman effect is exhibited when the intensity of an atomic spectral line, emission or absorption, is reduced when the atoms responsible are subjected to a magnetic field, nearby lines arising instead (**Figure 1.2**). This makes a powerful tool for the correction of background attenuation caused by molecules or particles that do not normally show such an effect. The technique is to subtract from a 'field-off' measurement the average of 'field-on' measurements made just beforehand and just afterwards. The simultaneous, highly resolved graphic display of the analyte and the background signals on a video screen provides a means of reliable monitoring of the determination and simplifies method development.

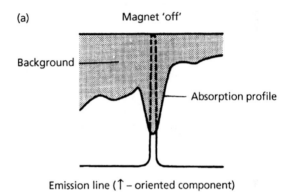

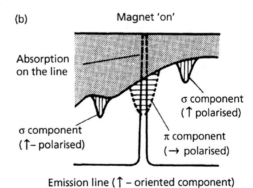

Figure 1.2 Zeeman patterns:
(a) analyte signal plus background;
(b) background only.
(*Source: Author's own files*)

The stabilised temperature platform furnace eliminates chemical interferences to such an extent that in most cases personnel- and cost-intensive sample preparation steps, such as solvent extractions, as well as the time-consuming method of additions are no longer required.

The advantages of Zeeman background correction are:

- Correction over the complete wavelength range.

- Correction for structural background.

- Correction for spectral interferences.

- Correction for high background absorptions.

- Single-element light source with no possibility of misalignment.

1.1.5.2 Instrumentation

The instrumentation for ZAAS is given in **Table 1.1** (see also Appendix 1).

Figure 1.2 illustrates the operating principle of the Zeeman 5000 system. For Zeeman operation, the source lamps are pulsed at 100 Hz (120 Hz) while the current to the magnet is modulated at 50 Hz (60 Hz). When the field is off, both analyte and background absorptions are measured at the unshifted resonance line. This measurement directly compares with a 'conventional' atom and absorption measurement without background correction. However, when the field is on, only the background is measured since the σ absorption line profiles are shifted away from the emission line while the static polariser, constructed from synthetic crystalline quartz, rejects the signal from the π components. Background correction is achieved by subtraction of the field-on signal from the field-off signal. With this principle of operation, background absorption of up to 2 absorbance units can be corrected most accurately even when the background shows a fine structure.

In assessing overall performance with a Zeeman effect instrument, the subject of analytical range must also be considered. For most normal class transitions, σ components will be completely separated at sufficiently high magnetic fields. Consequently, the analytical curves will generally be similar to those obtained by standard AAS. However, for certain anomalous transitions some overlap may occur. In these cases, curvature will be greater and may be so severe as to produce double-valued analytical curves. **Figure 1.3**, which shows calibration curves for copper, illustrates the reason for this behaviour. The Zeeman pattern for copper (324.8 nm) is particularly complex due to the presence of hyperfine structure. The dashed lines represent the separate field-off and field-on absorbance measurements. As sample concentration increases, field-off absorbance begins to saturate as in standard

Table 1.1 Available Zeeman atomic absorption spectrometers

Supplier	Model	Microprocessor	Type	Hydride and mercury attachment	Autosampler
Perkin-Elmer	Zeeman 3030	Yes (method storage on floppy disk)	Integral flame/ graphite furnace	-	Yes
	Zeeman 5000	Yes, with programmer	Fully automated integral flame/ graphite furnace double-beam operation roll-over protection	Yes	Yes
Varian	SpectrA A30/40	Yes, method storage on floppy disk	Automated analysis of up to 12 elements; roll-over protection	-	Yes
	SpectrA A300/400	Yes, total system control and colour graphics; 90 methods stored on floppy disk	Automated analysis of up to 12 elements; roll-over protection	-	Yes

Source: Author's own files

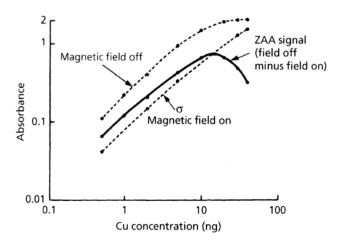

Figure 1.3 Copper calibration curves (324.8 nm) measured with the Zeeman 5000. (*Source: Author's own files*)

atomic absorption spectrometry. The σ absorbance measured with the field-on saturates at higher concentrations because of the greater separation from the emission line. When the increase in σ absorbance exceeds the incremental change in the field-off absorbance, the analytical curve, shown as the solid line, rolls over back towards the concentration axis. This behaviour can be observed with all Zeeman designs regardless of how the magnet is positioned or operated. The existence of roll-over does introduce the possibility of ambiguous results, particularly when peak area is being measured.

1.1.6 Inductively Coupled Plasma Atomic Emission Spectrometry

1.1.6.1 Theory

An inductively coupled plasma is formed by coupling the energy from a radio frequency (1–3 kW or 27–50 MHz) magnetic field to free electrons in a suitable gas. The magnetic field is produced by a two- or three-turn water-cooled coil and the electrons are accelerated in circular paths around the magnetic field lines that run axially through the coil. The initial electron 'seeding' is produced by a spark discharge but once the electrons reach the ionisation potential of the support gas further ionisation occurs and a stable plasma is formed.

The neutral particles are heated indirectly by collisions with the charged particles upon which the field acts. Macroscopically the process is equivalent to heating a conductor by a radio-frequency field, the resistance to eddy current flow producing joule heating.

The field does not penetrate the conductor uniformly and therefore the largest current flow is at the periphery of the plasma. This is the so-called 'skin' effect and coupled with a suitable gas-flow geometry it produces an annular or doughnut-shaped plasma. Electrically, the coil and plasma form a transformer with the plasma acting as a one-turn coil of finite resistance.

If mass spectrometric determination of the analyte is to be incorporated, then the source must also be an efficient producer of ions.

Greenfield and co-workers [14] were the first to recognise the analytical potential of the annular ICP.

Wendt and Fassel [15], reported early experiments with a 'tear-drop'-shaped inductively coupled plasma but later described the medium power, 1–3 kW, 18 mm annular plasma now favoured in modern analytical instruments [16].

The current generation of ICP emission spectrometers provides limits of detection in the range 0.1–500 µg/l of metal in solution, a substantial degree of freedom from interferences, and a capability for simultaneous multi-element determination facilitated by a directly proportional response between the signal and the concentration of the analyte over a range of about five orders of magnitude.

The most common method of introducing liquid samples into the ICP is by using pneumatic nebulisation [17], in which the liquid is dispensed into a fine aerosol by the action of a high-velocity gas stream. The fine gas jets and liquid capillaries used in ICP nebulisers may cause inconsistent operation and even blockage when solutions containing high levels of dissolved solids, or particular matter, are used. Such problems have led to the development of new types of nebuliser, the most successful being based on a principle originally described by Babington. In these, the liquid is pumped from a wide-bore tube and then to the nebulising orifice by a V-shaped groove [18] or by the divergent wall of an over-expanded nozzle [19]. Such devices handle most liquids and even slurries without difficulty.

Two basic approaches are used for introducing samples into the plasma. The first involves indirect vaporisation of the sample in an electrothermal vaporiser, e.g., a carbon rod or tube furnace or heated metal filament as commonly used in AAS [20–22]. The second involves inserting the sample into the base of the ICP on a carbon rod or metal filament support [23, 24].

Table 1.2 compares the detection limits claimed for AAS and the graphite furnace and ICP variants.

Table 1.2 Guide to analytical values for IL spectrometers (IL 157/357/457/451/551/951/Video 11/12/22/S11/S12 Atomic Absorption Spectrophotometers® and IL Plasma-100/-200/-300 ICP Emission Spectrometers)

Element	Wavelength (nm)		AA Lamp current (mA)	Flame AA		Furnace AA (IL755 CTF Atomiser)			ICP
	AA	ICP		Sensitivity[2] (µg/l)	Detection limit (µg/l)	Sensitivity[2]	(µg/l)	Detection limit (µg/l)	Detection limit (µg/l)
Aluminium (Al)[1]	309.3	396.15	8	400	25	4.0	0.04	0.01	10
Antimony (Sb)	217.6	206.83	10	200	40	8.0	0.08	0.08	40
Arsenic (As)	193.7	193.70	8	400	1403	12	0.12	0.08	30
Barium (Ba)[1]	553.5	455.40	10	150	12	4.0	0.04	0.04	0.5
Beryllium (Be)[1]	234.9	313.04	8	10	1	1.0	0.01	0.003	0.1
Bismuth (Bi)	223.1	223.06	6	200	30	4.0	0.04	0.01	35
Boron (B)[1]	249.7	249.77	15	9,000	700	-	-	-	3
Cadmium (Cd)	228.8	214.44	3	10	1	0.2	0.002	0.0002	1.5
Calcium (Ca)	422.7	393.37	7	50	2	1.0	0.01	0.01	0.2
Calcium[1]	422.7	-	7	10	1	-	-	-	-
Carbon (C)	-	193.09	-	-	-	-	-	-	40
Cerium (Ce)	-	413.77	-	-	-	-	-	-	40
Caesium (Cs)	852.1	455.53	10	150	20	-	-	-	-
Chromium (Cr)	357.9	205.55	6	40	3	4.0	0.04	0.004	3
Cobalt (Co)	240.7	238.89	8	50	4	8.0	0.08	0.008	3
Copper (Cu)	324.7	324.75	5	30	1.8	4.0	0.04	0.005	1
Dysprosium (Dy)[1]	421.2	353.17	8	600	60	-	-	-	4

Table 1.2 Continued

| Element | Wavelength (nm) | | AA Lamp current (mA) | Flame AA | | Furnace AA (IL755 CTF Atomiser) | | | ICP |
	AA	ICP		Sensitivity[2] (µg/l)	Detection limit (µg/l)	Sensitivity[2]	(µg/l)	Detection limit (µg/l)	Detection limit (µg/l)
Erbium (Er)1	400.8	337.27	8	400	40	50	0.5	0.3	3
Europium (Eu)	-	381.97	-	-	-	-	-	-	2
Gadolinium (Gd)[1]	368.4	342.25	9	13,000	2,000	1,600	16	8	4
Gallium (Ga)	287.4	294.36	5	400	50	5.2	0.05	0.01	15
Germanium (Ge)[1]	265.1	209.43	5	800	50	40	0.4	0.1	20
Gold (Au)	242.8	242.80	5	100	6	5.0	0.05	0.01	10
Holmium (Hf)[1]	307.3	339.98	10	14,000	2,000	-	-	-	5
Halmium (Ho)[1]	410.4	345.60	12	660	60	90	0.9	0.7	1
Indium (In)	303.9	325.61	5	180	30	11	0.11	0.02	15
Iridium (Ir)[1]	208.8	224.27	15	1,500	500	170	1.7	0.5	8
Iron (Fe)	248.3	238.20	8	40	5	3.0	0.03	0.01	2
Lanthanum (La)[1]	550.1	333.75	10	22,000	2,000	58	0.58	0.5	2
Lead (Pb)	217.0	220.35	5	100	9	4.0	0.04	0.007	2.5
Lithium (Li)	670.8	670.78	8	16	1	4.0	0.04	0.01	2.5
Lutetium (Lu)	-	261.54	-	-	-	-	-	-	0.2
Magnesium (Mg)	285.2	279.55	3	3	0.3	0.07	0.0007	0.0002	0.1
Manganese (Mn)	279.5	257.61	5	20	1.8	1.0	0.01	0.0005	1
Mercury (Hg)	253.7	253.65	3	2,500	140	40	0.4	0.2	12

17

Table 1.2 Continued

Element	Wavelength (nm)		AA Lamp current (mA)	Flame AA		Furnace AA (IL755 CTF Atomiser)			ICP
	AA	ICP		Sensitivity[2] (µg/l)	Detection limit (µg/l)	Sensitivity[2]	(µg/l)	Detection limit (µg/l)	Detection limit (µg/l)
Molybdenum (Mo)[1]	313.3	202.03	6	200	25	12	0.12	0.03	4
Neodymium (Nd)[1]	492.5	401.23	10	5,000	700	-	-	-	8
Nickel (Ni)	232.0	221.65	10	60	5	20	0.2	0.05	4
Niobium (Nb)[1]	334.9	309.42	15	12,000	2,000	-	-	-	6
Osmium (Os)[1]	290.9	225.59	15	1,000	90	270	2.7	2	0.6
Palladium (Pd)	247.6	340.46	5	140	20	13	0.13	0.05	13
Phosphorus (P)	213.6	213.62	8	125,000	30,000	4,900	49	20	16
Platinum (Pt)	265.9	214.42	10	1,000	50	80	0.8	0.2	16
Potassium (K)	766.5	766.49	7	10	1	0.4	0.004	0.004	305
Praseodymium (Pr)[1]	495.1	390.84	15	20,000	2,000	-	-	-	20
Rhenium (Re)	346.1	221.43	15	8,000	800	1,000	10	10	6
Rhodium (Rh)	343.5	343.49	5	200	2	20	0.20	0.1	8
Rubidium (Rb)	780.0	-	10	30	2	-	-	-	-
Ruthenium (Ru)	349.9	240.27	10	800	400	-	-	-	8
Samarium (Sm)[1]	429.7	359.26	10	3,000	500	-	-	-	8
Scandium (Sc)[1]	391.2	361.38	10	100	20	-	-	-	0.5
Selenium (Se)	196.0	196.03	12	300	803	8.0	0.08	0.05	30
Silicon (Si)[1]	251.6	251.61	12	800	60	60	0.60	0.6	6

Table 1.2 Continued

Element	Wavelength (nm)		AA Lamp current (mA)	Flame AA		Furnace AA (IL755 CTF Atomiser)			ICP
	AA	ICP		Sensitivity[2] (µg/l)	Detection limit (µg/l)	Sensitivity[2]	(µg/l)	Detection limit (µg/l)	Detection limit (µg/l)
Silver (Ag)	328.1	328.07	3	30	1.2	0.5	0.005	0.001	3
Sodium (Na)	589.0	589.59	8	3	0.4	0.4	0.004	0.004	7
Strontium (Sr)	460.7	407.77	12	80	6	1.8	0.018	0.01	0.2
Tantalum (Ta)[1]	271.5	240.06	15	10,000	800	-	-	-	13
Tellurium (Te)	214.3	214.28	7	200	30	7.0	0.07	0.03	20
Terbium (Tb)[1]	432.7	350.92	8	3,300	1,000	-	-	-	3
Thallium (Tl)	276.8	276.79	8	100	30	4.0	0.04	0.01	27
Thorium (Th)	-	283.73	-	-	-	-	-	-	8
Thulium (Tm)	-	313.13	-	-	-	-	-	-	0.9
Tin (Sn)1	235.5	189.99	6	1,200	90	7.0	0.07	0.03	30
Titanium (Ti)[1]	364.3	334.94	8	900	60	50	0.50	0.3	1
Tungsten (W)[1]	255.1	207.91	15	5,000	500	-	-	-	14
Uranium (U)[1]	358.5	263.55	15	100,000	7,000	3,100	31	30	70
Vanadium (V)[1]	318.5	309.31	8	600	25	40	0.40	0.1	3
Ytterbium (Yb)	398.8	328.94	5	80	-	1.3	0.01	0.01	1
Yttrium (Y)[1]	410.2	371.03	6	1,800	200	1,300	13	10	0.7
Zinc (Zn)	213.9	213.86	3	8	1.23	0.3	0.003	0.001	2
Zirconium (Zr)[1]	360.1	343.82	10	10,000	2,000	-	-	-	2

[1] Nitrous oxide/acetylene flame (AA)
[2] Sensitivity is concentration (or mass) yielding 1% absorption (0.0044 absorbance units)
[3] With background correction
©1985 Allied Analytical Sytems 7/85 5K

1.1.6.2 Instrumentation

Further details are given in Appendix 1.

There are two main types of ICP spectrometer systems. The first is the monochromator system for sequential scanning, which consists of a high-speed, high-resolution scanning monochromator viewing one element wavelength at a time. Typical layouts are shown in **Figure 1.4**. **Figure 1.4(a)** shows a one-channel air path double monochromator design with a pre-monochromator for order sorting and stray light rejection and a main monochromator to provide resolution of up to 0.02 nm. The air path design is capable of measuring wavelengths in the range 190–900 nm. The wide wavelength range enables measurements to be performed in the ultraviolet (UV), visible, and near-infrared regions of the spectrum (allowing determinations of elements from arsenic at 193.70 nm to caesium at 852.1 nm).

A second design (**Figure 1.4(b)**) is a vacuum monochromator design allowing measurements in the 160–500 nm wavelength range. The exceptionally low wavelength range gives the capability of determining trace levels of non-metals such as bromine at 163.34 nm as well as metals at low UV wavelengths, such as the extremely sensitive aluminium emission line at 167.08 nm. Elements such as boron, phosphorus or sulfur can be routinely determined using interference-free emission lines.

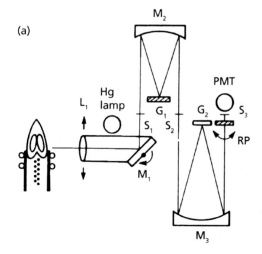

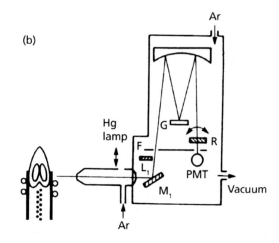

Figure 1.4 (a) A double monochromator consisting of an air-path monochromator with a pre-monochromator for order sorting and stray light rejection to determine elements in the 190-900 nm range; (b) the vacuum UV monochromator - an evacuated and argon-purged monochromator to routinely determine elements in the 160 to 500 μm range.

(*Source: Author's own files*)

The sequential instrument, equipped with either or both monochromators facilitates the sequential determination of up to 63 elements in turn, at a speed as fast as 18 elements per minute in a single sample. Having completed the analysis of the first sample, usually in less than a minute, it proceeds to the second sample, and so on.

The second main type of system is the polychromator system for simultaneous scanning. The polychromator systems scan many wavelengths simultaneously, i.e., several elements are determined simultaneously at higher speeds than are possible with monochromator systems. It then moves on to the next sample. A typical system is shown in **Figure 1.5**.

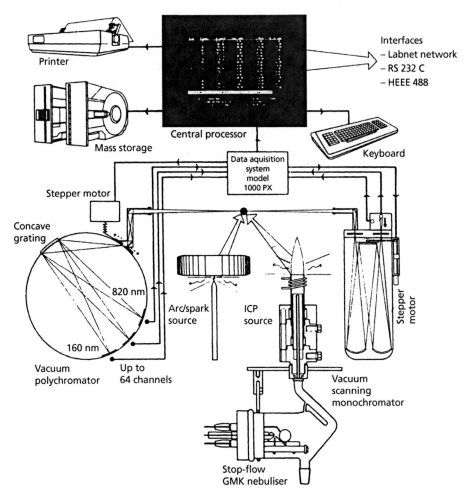

Figure 1.5 Polychromator system for inductively coupled plasma atomic emission spectrometer. (*Source: Author's own files*)

21

It is possible to obtain instruments that are equipped for both sequential and simultaneous scanning, e.g., the Labtam 8410.

1.1.6.3 Applications

Briseno and co-workers [25] quantified inorganic dopants in polypyrrole films by a combination of electrochemistry and ICP-AES.

1.1.7 Hybrid Inductively Coupled Plasma Techniques

1.1.7.1 Chromatography–Inductively Coupled Plasma

Direct introduction of a sample into an ICP produces information only on the total element content. It is now recognised that information on the form of the element present, or trace element speciation, is important in a variety of applications. One way of obtaining quantitative measurement of trace element speciation is to couple the separation power of chromatography to the ICP as a detector. Since the majority of interesting trace metal speciation problems concern either nonvolatile or thermally unstable species, high-performance liquid chromatography (HPLC) becomes the separation method of choice. The use of HPLC as the separation technique requires the introduction of a liquid sample into the ICP with the attendant sample introduction problem.

1.1.7.2 Flow Injection with Inductively Coupled Plasma

In conventional ICP-OES, a steady-state signal is obtained when a solution of an element is nebulised into the plasma. In flow injection [26] a carrier stream of solvent is fed continuously through a 1 mm id tube to the nebuliser using a peristaltic pump, and into this stream is injected, by means of a sampling valve, a discrete volume of a solution of the element of interest. When the sample volume injected is suitably small, a transient signal is obtained (as opposed to a steady-state signal which is obtained with larger sample volumes) and it is this transient signal that is measured. Very little sample dispersion occurs under these conditions, the procedure is very reproducible, and sample rates of 180 samples per hour are feasible.

1.1.7.3 Inductively Coupled Plasma with Atomic Fluorescence Spectrometry

Atomic fluorescence is the process of radiation activation followed by radiation deactivation, unlike atomic emission which depends on the collisional excitation of the spectral transition. For this, the ICP is used to produce a population of atoms in the ground state and a light source is required to provide excitation of the spectral transitions. Whereas a multitude of spectral lines from all the accompanying elements are emitted by the atomic emission process, the fluorescence spectrum is relatively simple, being confined principally to the resonance lines of the element used in the excitation source.

The ICP is a highly effective line source with a low background continuum. It is optically thin – it obeys Beer's law – and therefore exhibits little self-absorption. It is also a very good atomiser and the long tail flame issuing from the plasma has such a range of temperatures that conditions favourable to the production of atoms in the ground state for most elements are attainable. It is therefore possible to use two plasmas in one system: a source plasma to supply the radiation to activate the ground state atoms and another to activate the atomiser. This atomic fluorescence (AFS) mode of detection is relatively free from spectral interference, the main drawback of ICP-OES.

Good results have been obtained using a high-power (6 kW) ICP as a source and a low-power (<1 kW) plasma as an atomiser.

1.1.8 Inductively Coupled Plasma Optical Emission Spectrometry– Mass Spectrometry

1.1.8.1 Theory

ICP-MS combines MS with the established ICP to break the sample into a stream of positively charged ions which are subsequently analysed on the basis of their mass. ICP-MS does not depend on indirect measurements of the physical properties of the sample. The elemental concentrations are measured directly – individual atoms are counted giving the key attribute of high sensitivity. The technique has the additional benefit of unambiguous spectra and the ability to directly measure different isotopes of the same element.

The sample under investigation is introduced, most typically in solution, into the inductively coupled plasma at atmospheric pressure and a temperature of approximately 6,000 K. The sample components are rapidly dissociated and ionised and the resulting atomic ions are introduced via a carefully designed interface into a high-performance quadrupole mass spectrometer at high vacuum.

A horizontally mounted ICP torch forms the basis of the ion source. Sample introduction is via a conventional nebuliser optimised for general-purpose solution analysis and suitable for use with both aqueous and organic solvents.

Nebulised samples enter the central channel of the plasma as a finely dispersed mist which is rapidly vaporised; dissociation is virtually complete during passage through the plasma core with most elements fully ionised.

Ions are extracted from the plasma through a water-cooled sampling aperture and a molecular beam is formed in the first vacuum stage and passes into the high-vacuum stage of the quadrupole mass analyser.

In an ICP-MS system a compact quadrupole mass analyser selects ions on the basis of their mass-to-charge ratio, *m/e*. The quadrupole is a simple compact form of mass analyser which relies on a time-dependent electric field to filter the ions according to their mass-to-charge ratio. Ions are transmitted sequentially in order of their *m/e* with constant resolution, across the entire mass range.

1.1.8.2 Instrumentation

Several manufacturers, including VG Isotopes and Perkin Elmer, supply equipment for ICP-MS (see Appendix 1).

The Perkin Elmer Elan 500 instrument is designed for routine and rapid multi-element quantitative determinations of trace and ultra-trace elements and isotopes. The Elan 500 can determine nearly all of the elements in the periodic table with exceptional sensitivity.

The entire Elan 500 Plasmalok system is designed for simplicity of operation. A typical daily start-up sequence from the standby mode includes turning on the plasma and changing to the operating mode. After a brief warm-up period for the plasma, routine sample analysis can begin.

VG Isotopes Ltd., is another leading manufacturer of ICP-MS. The special features of their VG Plasmaquad PQ2 includes a multi-channel analyser which ensures rapid data acquisition over the whole mass range. The multi-channel analyser facilities include 4096 channels, 300 m facility for spectral analysis, user-definable number of measurements per peak in peak jumping mode, and the ability to monitor data as they are acquired. A multi-channel analyser is imperative for acquiring short-lived signals from accessories such as flow injection, electrothermal vaporisation, laser ablation, and so on, or for fast multi-element survey scans (typically 1 minute).

A variant of the Plasmaquad PQ2 is the Plasmaquad PQ2 plus instrument. This latter instrument has improved detector technology which incorporates a multimode system that can measure higher concentrations of elements without compromising the inherent

Table 1.3 Dynamic ranges of various techniques	
Graphite furnace AAS	0.1 µg/l to 1 µg/l
Plasmaquad PQ2	0.1 µg/l to 10 µg/l
Inductively coupled plasma - atomic absorption spectrometry	10.0 µg/l to 1000 µg/l
Plasmaquad PQ2 plus	0.1 µg/l to 1000 µg/l
Source: Author's own files	

sensitivity of the instrument. This extended dynamic range system (**Table 1.3**) produces an improvement in the effective linear dynamic range to eight orders of magnitude. Hence, traces at the microgram per litre level can be measured in the same analytical sequence as major constituents.

This technique has been applied to the analysis of aqueous digests of polymers containing up to 2% solids.

1.1.8.3 Applications

Dobney and co-workers [27] have used laser ablation ICP-MS to determine various metals in polyolefins.

1.1.9 Pre-concentration Atomic Absorption Spectrometry Techniques

Detection limits can be improved still further in the case of all three techniques mentioned previously by the use of a pre-concentration technique [28]. One such technique that has found great favour involves converting the metals to an organic chelate by reaction of a larger volume of sample with a relatively small volume of an organic solvent solution, commonly of diethyldithiocarbamates or ammonium pyrrolidone diethyldithiocarbamates. The chelate dissolves in the organic phase and is then back-extracted into a small volume of aqueous acid for analysis by either of the techniques mentioned previously. If 0.5–1 litre of sample is originally taken and 20 ml of acid extract finally produced then concentration factors of 25–50 are thereby achieved with consequent lowering of detection limits. Needless to say, this additional step in the analysis considerably increases analysis time and necessitates extremely careful control of experimental conditions.

Microscale solvent extractions involving the extraction of 2.5 ml sample with 0.5 ml of an organic solvent solution of a chelate give detection limits for lead and cadmium by the Zeeman graphite furnace atomic absorption spectrometry method of 0.6 and 0.02 µg/l, respectively. This is equivalent to determining 1.2 ppm cadmium in polymers (assuming the digest of 10 mg of polymer is made up to 20 ml).

1.1.10 Microprocessors

In recent years the dominating influence on the design and performance of AAS is that of the microprocessor. Even the cheapest instruments are expected to provide autosampling systems for both flame and furnace use and therefore a means of recording the data produced.

1.11 Autosamplers

Gilson and PS Analytical supply autosamplers suitable for automation of AAS and ICP. The Gilson autosampler can house up to 300 samples and is capable of operation 24 hours per day. PS Analytical supply 20- and 80-position autosamplers.

For many applications such as hydride analysis, conventional multi-element analysis, and repetitive analysis for major element quantification, conventional autosamplers are not sufficiently sophisticated. The PS Analytical 20.020 twenty-position autosampler has been specifically developed to fill this void. It is easily interfaced to computer systems via a transistor-transistor logic (TTL) interface.

1.1.12 Applications: Atomic Absorption Spectrometric Determination of Metals

1.1.12.1 Catalyst Remnants and Other Impurities

Two types of catalysts used in polymer manufacture are metallic compounds such as aluminium alkyls and titanium halides used in low-pressure polyolefin manufacture. As the presence of residual catalysts can have important effects on polymer properties it is important to be able to determine trace elements which reflect the presence of these substances.

1.1.12.2 Elemental Analysis of Polymers

Elements occurring in polymers and copolymers can be divided into three categories:

1. Elements that are a constituent part of the monomers used in polymer manufacture, such as nitrogen in acrylonitrile used in the manufacture of, for example, acrylonitrile–butadiene–styrene terpolymers.

2. Elements that occur in substances deliberately included in polymer formulations, such as, for example, zinc stearate.

3. Elements that occur as adventitious impurities in polymers. For example, during the manufacture of polyethylene (PE) by the low-pressure process, polymerisation catalysts such as titanium halides and organo-aluminium compounds are used, and the final polymer would be expected to, and indeed does, contain traces of aluminium, titanium, and chlorine residues.

The classic destructive techniques are generally based on one of three possible approaches to the analysis: (i) dry ashing of the polymer with or without an ashing aid, followed by acid digestion of the residue, alternatively acid digestion of the polymer without prior ashing, (ii) fusion of the polymer with an inorganic compound to effect solution of the elements, and (iii) bomb or oxygen flask digestion techniques.

Another method for avoiding losses of metals during ashing is the low-temperature controlled decomposition technique using active oxygen. This method has been studied in connection with the determination of trace metals in polyvinyl chloride (PVC), polypropylene (PP), and polyethylene terephthalate [29].

1.1.12.3 Trace Metals in Polymers

Sources of traces of metals in polymers are neutralising chemicals added to the final stages of manufacture to eliminate the effects of acidic catalyst remnants on polymer processing properties (e.g., hygroscopicity due to residual chloride ion). A case in point is high-density polyethylene (HDPE) and PP produced by the aluminium alkyl–titanium halide route which is treated with sodium hydroxide in the final stages of manufacture.

A technique that involves combustion of the polymer under controlled conditions in a platinum crucible, followed by dissolution of the residual ash in a suitable aqueous reagent prior to final analysis by spectrophotometry is of limited value. A quite complicated and lengthy ashing programme is necessary in this technique to avoid losses of alkali metal during ignition: 0–1 hour from start: heat to 200 °C; 1–2 hours from start: hold at 200 °C; 3–5 hours from start: heat to 450 °C; 5–8 hours from start: hold at 450 °C.

After ignition the residue is dissolved in warm nitric acid and made up to a standard volume prior to evaluation by flame photometry or AAS. Alternatively, the polymer is ashed overnight at 500 °C with sulfur and a magnesium salt of a long-chain fatty acid (Magnesium AC dope, Shell Chemical Co. Ltd), and the ash mixed with twice its weight of carbon powder containing 0.1% palladium prior to emission spectrographic evaluation of the sodium/palladium 330.3/276.31 line pair.

The results in **Table 1.4** show clearly that flame photometry following dope ashing at 500 °C gives a quantitative recovery of sodium relative to results obtained by a non-destructive method of analysis, i.e., NAA. Direct ashing without the magnesium ashing aid at 500 °C causes losses of 10% or more of the sodium while direct ashing at 800 °C causes even greater losses.

Table 1.4 The effects of modification of ashing procedure on the flame photometric determination of sodium in polyethylenes					
Sodium ppm by flame photometry					
Sample	By neutron activation	By emission spectrography	Original (ashed between 650 °C and 800 °C)	Dope ash at 500 °C	Direct ash at 500 °C
1	99, 96, 99	95	60, 76, 55	100	75
2	256, 247, 259	258, 259	160, 178, 271	225	208
3	343, 321, 339	339, 287	250, 312	282	265
4	213, 210, 212	218, 212	140, 196	210	191
5	194, 189, 192	209, 198	80, 158, 229	196	169
6	186, 191, 198	191, 191	96, 173	193	173
Source: Author's own files					

Dry ashing in platinum has been found to give reasonably good results for the determination of low concentrations of vanadium in an ethylene–propylene copolymer. An amount of 10 g of polymer is ashed in platinum by charring on a hot plate followed by heating over a Meker burner. Dilute nitric acid is added to the residue and any residue in the crucible dissolved by fusion with potassium persulfate. The vanadium is determined spectrophotometrically by the 3,3'-diaminobenzene method [30]. **Table 1.5** compares the results obtained by this method with those obtained by NAA, which in this case can be considered to be an accurate reference method. Good agreement is obtained between the two methods for samples containing vanadium.

Table 1.5 Vanadium (ppm)		
Sample	Dry ashing	Neutron activation
A	10.2	9.9 ± 0.2
B	14.0	14.1 ± 0.1
C	14.6	15.6 ± 0.3
D	0.5	0.14 ± 0.01
E	13.0	14.8 ± 0.2
F	0.9	0.27 ± 0.01
G	15.2	18.8 ± 0.3
H	18.2	17.9 ± 0.3
Source: Author's own files		

It has been shown [31, 32] in studies using a radioactive copper isotope that, when organic materials containing copper are ashed, losses of up to 10% of the copper will occur due to retention in the silica crucible; this could not be removed by acid washing. Virtually no retention of copper in the silica crucible occurred, however, when copper was ashed under the same conditions in the absence of added organic matter. This was attributed to reduction of copper to the metal by organic matter present, followed by partial diffusion of the copper metal into the crucible wall. Distinctly higher copper determinations are obtained for polyolefins by the procedure involving the use of a magnesium oxide ashing aid than are obtained without an ashing aid, or by the use of a molten potassium bisulfate fusion technique to take up the polymer ash.

Henn [33] has reported a flameless atomic absorption technique with solid sampling for determining trace amounts of chromium, copper and iron in polymers such as polyacrylamide with a detection limit of approximately 0.01 ppm.

AAS is a useful technique for the determination of traces of metals in polymers. Generally, the polymer is ashed at a maximum temperature of 450 °C: 0.1 hour from start: heat to 200 °C; 1–3 hours from start: hold at 200 °C; 3–5 hours from start: heat to 450 °C; 5–8 hours from start: hold at 450 °C. The ash is digested with warm nitric acid prior to spectrometric analysis. The detection limits for metals in polymers achievable by this procedure are given in **Table 1.6**.

Certain elements (such as arsenic, antimony, mercury, selenium, and tin) can, after producing the soluble digest of the polymer, be converted to gaseous metallic

Table 1.6 Analytical conditions. Metals in polymers					
Element	Wavelength (nm)	Band pass	Operating range (in polymer) ppm	Detection limit (in polymer) ppm	Concentration of standard solution ug/l
Iron	248.3	0.3	5	0.57	500
Manganese	279.5	0.5	1.25	0.03	250
Chromium	357.9	0.5	2.5	0.03	500
Cadmium	228.8	1.0	0.5	0.015	50
Copper	324.7	1.0	1.25	0.045	250
Lead	217.0	1.0	5	0.15	1250
Nickel	232.0	0.15	2.5	0.07	500
Zinc	213.9	1.0	0.5	0.015	125
Reprinted from L. Henn, Analytica Chimica Acta, 1974, 73, 273, with permission from Elsevier.					

hydrides by reaction of the digest with reagents such as stannous chloride or sodium borohydide:

$As_2O_3 + 3SnCl_2 + 6HCl = 2AsH_3 + 3H_2O + 3SnCl_4$

$NaBH_4 + 2H_2O = NaBO_2 + 4H_2$

$6H_2 + As_2O_3 = 2AsH_3 + 3H_2O$

These hydrides can be determined by AAS. To illustrate, let us consider a method developed for the determination of trace amounts of arsenic in acrylic fibres containing antimony oxide fire-retardant additive [34]. The arsenic occurs as an impurity in the antimony oxide additive and, as such, its concentration must be controlled at a low level.

In this method a weighed amount of sample is digested with concentrated nitric and perchloric acids and digested until the sample is completely dissolved. Pentavalent arsenic in the sample is then reduced to trivalent arsenic by the addition of titanium trichloride dissolved in concentrated hydrochloric acid:

$As^{5+} + 2Ti^{3+} = As^{3+} + 2Ti^{4+}$

The trivalent arsenic is then separated from antimony by extraction with benzene, leaving antimony in the acid layer. The trivalent arsenic is then extracted with water from the benzene phase. This solution is then extracted with a mixture of hydrochloric acid, potassium iodide, and stannous chloride to convert trivalent arsenic to arsine (AsH_3), which is swept into the AAS. Arsenic is then determined at the 193.7 nm absorption line. Recoveries of between 96 and 104% are obtained by this procedure in the 0.5–1.0 µg arsenic range, with a detection limit of 0.04 ppm.

The results for the arsenic obtained with various acrylic fibre samples containing antimony oxide are given in **Table 1.7**. Antimony, present as the trioxide, has been accurately determined in a concentrated hydrochloric acid extract of PP powder [35].

Table 1.7 Results for the determination of arsenic in acrylic fibres containing antimony oxide				
Sample no.	Supplier	Content of Sb_2O_5*%	No. of determinations	Arsenic content (ppm)
1	A	5.1	4	50 ± 1
2	A	5.1	3	84 ± 4
3	A	4.5	2	94 ± 7
4	A	4.6	5	10.3 ± 0.5
5	A	5.0	5	4.1 ± 0.2
6	B	3.0 (Sb_2O_3)	3	45 ± 0
7	C	-	2	180 ± 11
8	C	-	4	103 ± 6
9	D	2.4 (Sb_2O_3)	2	8.5 ± 0.5
10	E	1.0 (Sb_2O_3)	4	3.2 ± 0.1
11	-	Sb_2O_5 50 mg + acrylic fibre (no Sb) 1 g	2	0.47 ± 0.03
12	-	Sb_2O_5 50 mg + acrylic fibre (no Sb) 1 g	1	0.08
* Even when antimony (III) was present in acrylic fibres (samples 6, 9 and 10), antimony (III) was easily and completely oxidised to antimony (V) by the wet digestion with a mixture of nitric, perchloric and sulfuric acids. Hence, arsenic in acrylic fibres containing antimony (III) oxide could be determined as well as that in acrylic fibre samples containing antimony (V) oxide by this method. Reprinted from T. Korenaga, Analyst, 1981, 106, 40, with permission from the Royal Society of Chemistry [34]				

1.1.12.4 Pressure Dissolution

Pressure dissolution and digestion bombs have been used to dissolve polymers for which wet digestion is unsuitable. In this technique the sample is placed in a pressure dissolution vessel with a suitable mixture of acids and the combination of temperature and pressure effects dissolution of the sample. This technique is particularly useful for the analysis of volatile elements that may be lost in an open digestion.

1.1.12.5 Microwave Dissolution

More recently, microwave ovens have been used for polymer dissolution. The sample is sealed in a Teflon bottle or a specially designed microwave digestion vessel with a mixture of suitable acids. The high-frequency microwave temperature (~100–250 °C) and increased pressure have a role to play in the success of this technique. An added advantage is the significant reduction in sample dissolution time [36, 37].

1.1.12.6 Equipment for Sample Digestions

1.1.12.6.1 Pressure Dissolution Acid Digestion Bombs

Inorganic and organic materials can be dissolved rapidly in Parr acid digestion bombs with Teflon liners and using strong mineral acids, usually nitric and/or *aqua regia* and, occasionally, hydrofluoric acid. Perchloric acid must not be used in these bombs due to the risk of explosion.

For nitric acid, 200 °C (80 °C over the atmospheric boiling point) and 0.7 MPa can be achieved in 12 minutes and for hydrochloric acid 153 °C (43 °C over the atmospheric boiling point) and 0.7 MPa can be obtained in 5 minutes. The aggressive digestion action produced at the higher temperatures and pressures generated in these bombs results in remarkably short digestion times, with many materials requiring less than one minute to obtain a complete dissolution, i.e., considerably quicker than open-tube wet ashing or acid digestion procedures. Several manufacturers supply microwave ovens and digestion bombs (Parr Instruments, CEM Corporation, Prolabo).

1.1.12.6.2 Oxygen Combustion Bombs

Combustion with oxygen in a sealed Parr bomb has been accepted for many years as a standard method for converting solid combustible samples into soluble forms for chemical analysis. It is a reliable method whose effectiveness stems from its ability to treat samples quickly and conveniently within a closed system without losing any of the sample or its combustion products. Sulfur-containing polymers are converted to soluble forms and absorbed in a small amount of water placed in the bomb. Halogen-containing polymers are converted to hydrochloric acid or chlorides. Any mineral constituents remain as ash but other elements such as arsenic, boron, mercury, nitrogen and phosphorus, and all of the halogens are recovered with the bomb washings. In recent years the list of applications has been expanded to include metals such as beryllium, cadmium, chromium, copper, iron, lead, manganese, nickel, vanadium and zinc by using a quartz liner to eliminate interference from trace amounts of heavy metals leached from the bomb walls and electrodes [38-40].

Once the sample is in solution in the acid and the digest made up to a standard volume the determination of metals is completed by standard procedures such as AAS, ICP-OES, or any of the techniques listed in Sections 1.1.1–1.1.8.

1.1.12.7 Techniques for Sample Digestion

Table 1.8 shows results obtained for the digestion (in closed vessels) of 1 g samples digested (a) in 20 ml of 1:1 nitric acid:water and (b) in 5 ml of concentrated nitric acid and 3 ml of 30% hydrogen peroxide. In the former, at a power input of 450 W, the temperature and pressure

Table 1.8 Solid sample microwave digested in 1:1 HNO3:H2O			
Element	(a) in 1:1 $HNO_3:H_2O$	(b) in 5:3 $HNO_3:H_2O_2$	Certified value (%)
	Amount recovered (%)	Amount recovered (%)	
As	0.0060, 0.0060	0.0075, 0.0070	0.0066
Cd	0.0012, 0.0012	0.0011, 0.0012	0.0012 ± 0.00015
Cr	3.00, 2.98	3.04, 2.96	2.96 ± 0.28
Cu	0.0122, 0.0113	0.0118, 0.0119	0.0109 ± 0.0019
Mg	0.72, 0.72	0.70, 0.70	0.74 ± 0.02
Mn	0.0790, 0.0780	0.0720, 0.0725	0.0785 ± 0.0097
Ni	0.0050, 0.0050	0.0044, 0.0044	0.00458 ± 0.00029
Pb	0.0736, 0.0737	0.0736, 0.0733	0.0714 ± 0.0028
Se	0.0001, 0.0001	0.0001, 0.0001	(0.00015)
Zn	0.170, 0.168	0.160, 0.160	0.172 ± 0.017
Source: Author's own files			

rose to 180 °C and 0.7 MPa. At that point, microwave power was reduced to maintain the temperature and pressure at those values for an additional 50 minutes. In the latter case, 1 g samples were open-vessel digested in 1:1 nitric acid:water for 10 minutes at 180 W. After cooling to room temperature, 5 ml of concentrated nitric acid and 3 ml of 30% hydrogen peroxide were added to each. The vessels were then sealed and power was applied for 15 minutes at 180 W followed by 15 minutes at 300 W. The temperature rose to 115 °C after the first 15 minutes and to 152 °C at 0.3 MPa after the final 15 minutes of heating. With both reagent systems element recoveries are in good agreement with the values obtained using a hot plate total sample digestion technique, which typically requires 4–6 hours.

Flame and GFAAS techniques have adequate sensitivity for the determination of metals in polymer samples. In this technique up to 1 g of dry sample is digested in a microwave oven for a few minutes with 5 ml of *aqua regia* in a small polytetrafluoroethylene-lined bomb, and then the bomb washings are transferred to a 50 ml volumetric flask prior to analysis by flame AAS. Detection limits (mg/kg) achieved by this technique were: 0.25 (cadmium, zinc); 0.5 (chromium, manganese); 1 (copper, nickel, iron); and 2.5 (lead). Application of this technique gave recoveries ranging between 85% (cadmium) and 101% (lead, nickel, iron) with an overall recovery of 95%.

1.1.13 Visible and UV Spectroscopy

The theory of visible and UV spectroscopy is discussed in an HMSO publication [41]. The reader is also referred to Appendix 1.

Visible spectrophotometers are commonly used for the estimation of colour in a sample or for the estimation of coloured products produced by reacting a colourless component of the sample with a reagent.

Visible spectrophotometry is still used extensively in the determination of some anions such as chloride, phosphate and sulfate formed by the decomposition of chlorine, phosphorus and sulfur in polymers. An extensive modern application of visible spectrophotometry is in the determination of organic substances, including non-ionic detergents, in polymer extracts.

Some commercially available instruments, in addition to visible spectrophotometry, can also perform measurements in the UV and near-infrared regions of the spectrum.

1.1.14 Polarography and Voltammetry

A large proportion of trace metal analysis carried out in polymer laboratories is based on the techniques of AAS and ICP-AES. Both of these methods give estimates of the total concentration of metal present and do not distinguish between different valency states of the same metal. For example, they would not distinguish between arsenic and antimony in the tri- or pentavalent states present in water extracts of polymers. Polarographic techniques can make such distinctions and, as discussed in Chapter 7, can be used to determine electroreducible organic materials such as antioxidants and monomers in polymers.

1.1.14.1 Instrumentation

Three basic techniques of polarography are of interest. The basic principles of these are outlined below.

Universal: differential pulse (DPN, DPI, DPR). In this technique a voltage pulse is superimposed on the voltage ramp during the last 40 ms of controlled drop growth with a standard dropping mercury electrode - the drop surface is then constant. The pulse amplitude is increased then the peak current raised. The current is measured by integration over a 20 ms period immediately before the start of the pulse and again for 20 ms as the pulse nears completion. The difference between the two current integrals (l2 − l1) is recorded and this gives a peak-shaped curve. If the pulse amplitude is increased, but the peak current value is raised and the peak is broadened at the same time.

Classic: direct current (DCT). In this direct current method integration is performed over the last 20 ms of controlled drop growth (Tast procedure). During this time, the drop surface is constant in the case of a dropping mercury electrode. The resulting polarogram is step-shaped. Compared with classic DC polarography according to Heyrovsky, i.e., with a free-dropping mercury electrode, the DCT method offers great advantages: considerably shorter analysis times, no disturbance due to current oscillations, simpler evaluation, and larger diffusion-controlled limiting current.

Rapid: square wave (SQW). Five square-wave oscillations of frequency around 125 Hz are superimposed on the voltage ramp during the last 40 ms of controlled drop growth − with a dropping mercury electrode the drop surface is then constant. The oscillation amplitude can be pre-selected. Measurements are performed in the second, third, and fourth square-wave oscillation; the current is integrated over 2 ms at the end of the first and the end of the second half of each oscillation. The three differences of the six integrals (l1 − l2, l3 − l4, l5 − l6) are averaged arithmetically and recorded as one current value. The resulting polarogram is peak-shaped. Various suppliers of polarography equipment are summarised in Appendix 1.

1.1.14.2 Applications

Polarography is an excellent method for trace and ultra-trace analysis of inorganic and organic substances and compounds. The basic process of electron transfer at an electrode is a fundamental electrochemical principle, and for this very reason polarography can be used over a wide range of applications.

After previous enrichment at a ranging mercury drop electrode, metals can be determined using differential pulse-stripping voltammetry. Detection limits are of the order of 0.05 µg/l.

Mal'kova and co-workers [42] described an AC polarographic method for the determination of cadmium, zinc, and barium stearates or laurates in PVC. The samples are prepared for analysis by being ashed in a muffle furnace at 500 °C, a solution of the ash in hydrochloric acid being made molar in lithium chloride and adjusted to pH 4.0 ± 0.2. The solution obtained is de-aerated by the passage of argon and the polarogram is recorded. Cadmium, zinc, and barium give sharp peaks at –0.65, –1.01, and –1.90 V, respectively, against the mercury-pool anode.

1.1.15 Ion Chromatography [43]

When it is necessary to determine several metals in a polymer then application of ion chromatography has several advantages over AAS, ICP-AES, and polarography. These include specificity, freedom from interference, speed of analysis, and sensitivity. It is, of course, necessary to digest the polymer using suitable reagent systems to produce an aqueous solution of the ions to be determined. Ion chromatography can complement atomic absorption and plasma methods as a back-up technique.

At the heart of the ion chromatography system is an analytical column containing an ion exchange column on which various anions and/or cations are separated before being detected and quantified by various detection techniques such as spectrophotometry, AAS (metals), or conductivity (anions).

Ion chromatography is not restricted to the separate analysis of only anions or only cations. With the proper selection of the eluant and separator columns the technique can be used for the simultaneous analysis of both anions and cations.

The principles of ion chromatography are discussed in an HMSO publication [44].

1.1.15.1 Instrumentation

The reader is referred to Appendix 1 for further details.

Numerous manufacturers now supply instrumentation for ion chromatography. However,

Dionex are still the leaders in the field; they have been responsible for many of the innovations introduced into this technique and are continuing to make such developments.

Some of the features of the Dionex series 4000i ion chromatograph instruments are discussed next.

1.1.15.1.1 Chromatography module

- Up to six automated valves made of chemically inert, metal-free material eliminate corrosion and metal contamination.

- Liquid flow path is completely compatible with all HPLC solvents.

- Electronic valve switching, multi-dimensional, coupled chromatography, or multi-mode operation.

- Automated sample clean-up or pre-concentration.

- Environmentally isolates up to four separator columns and two suppressers for optimal results.

- Manual or remote control with Dionex Autoion 300 or Autoion 100 automation accessories.

- Individual column temperature control from ambient to 100 °C (optional).

1.1.15.1.2 Dionex Ion-Pac columns

- Polymer ion exchange columns are packed with new pellicular resins for anion or cation exchange applications.

- New 4 μm polymer ion exchange columns have maximum efficiency and minimum operating pressure for high-performance ion and liquid chromatography applications.

- New ion exclusion columns with bifunctional cation exchange sites offer more selectivity for organic acid separations.

- Neutral polymer resins have high surface area for reversed phase ion pair and ion suppression applications without pH restriction.

- 5 and 10 μm silica columns are optimised for ion pair, ion suppression and reversed phase applications.

1.1.15.1.3 Micromembrane suppressor

The micromembrane suppressor makes the detection of non-UV-absorbing compounds such as inorganic anions and cations, surfactants, fatty acids, and amines in ion exchange and ion pair chromatography possible.

Two variants of this exist: the anionic (AMMS) and the cationic (CMMS) suppressor. The micromembrane suppressor consists of a low dead volume eluent flow path through alternating layers of high-capacity ion exchange screens and ultra-thin ion exchange membranes. Ion exchange sites in each screen provide a site-to-site pathway for eluent ions to transfer to the membrane for maximum chemical suppression.

Dionex anion and cation micromembrane suppressors transform eluent ions into less conducting species without affecting sample ions under analysis. This improves conductivity detection, sensitivity, specificity, and baseline stability. It also dramatically increases the dynamic range of the system for inorganic and organic ion chromatography. The high ion exchange capacity of the micromembrane suppressor permits changes in eluent composition by orders of magnitude making gradient ion chromatography possible.

In addition, because of the increased detection specificity, preparation is dramatically reduced, making it possible to analyse most samples after simple filtering and dilution.

1.1.15.1.4 Conductivity detector

- High-sensitivity detection of inorganic anions, amines, surfactants, organic acids, group I and II metals, oxy-metal ions, and metal cyanide complexes (used in combination with micromembrane suppressor).

- Bipolar-pulsed excitation eliminates the non-linear response with concentration found in analogue detectors.

- Microcomputer-controlled temperature compensation minimises the baseline drift with changes in room temperature.

1.1.15.1.5 UV/visible detector

- High-sensitivity detection of metals, silica, and other UV-absorbing compounds using either post-column reagent addition or direct detection.

- Non-metallic cell design eliminates corrosion problems.

- Filter-based detection with selectable filters from 214 to 800 nm.

- Proprietary dual wavelength detection for ninhydrin-detectable amino acids and 2-pyridyl resorcinol-detectable transition metals.

1.1.15.1.6 Optional detectors

Dionex also offers visible, fluorescence, and pulsed amperometric detectors for use with the series 4000i. Dionex also supply a wide range of alternative instruments, e.g., single channel (2010i) and dual channel (2020i). The latter can be upgraded to an automated system by adding Autoion 100 or Autoion 300 controllers to control two independent ion chromatograph systems. Dionex also supply 2000i series equipped with conductivity pulsed amperometric, UV/visible, visible, and fluorescence detectors.

1.1.15.2 Applications

A typical system for the determination of metals is shown in **Figure 1.6**. A liquid sample is introduced at the top of the ion exchange analytical column (the separator column, **Figure 1.6**). An eluent (containing a complexing agent in the case of metal determination) is pumped through the system. This causes the ionic species (metal ions) to move through the column at rates determined by their affinity for the column resin. The differential migration of the ions allows them to separate into discrete bands.

As these bands move through the column they are delivered, one at a time, into the detection system. For metals, this comprises a post-column reactor that combines a colouring reagent (pyridyl azoresorcinol; PAR) with the metal bands. The coloured bands can then be detected by the appropriate detection mode. In the case of metal–PAR complex detection, visible wavelength absorbance is employed.

The detector is set to measure the complexed metal band at a pre-selected wavelength. The results appear in the form of a chromatogram, essentially a plot of the time the band was retained on the column *versus* the signal it produces in the detector. Each metal in the sample can be identified and quantified by comparing the chromatogram against that of a standard solution.

Because only the metal ions of interest are detected, ion chromatography is less subject to interferences compared with other methods. Since individual metals and metal compounds form distinct ions with differing retention times, it is possible to analyse several of them in a single run – typically less than 20 minutes.

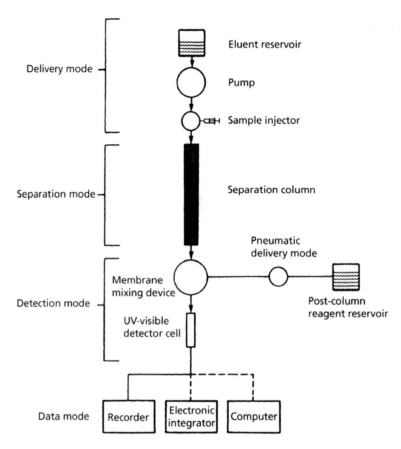

Figure 1.6 Ion chromatography with post-column reaction configuration for metals analysis. (*Source: Author's own files*)

By selecting the appropriate column for separating the ions of interest in a sample, it is possible to separate and analyse the oxidation state of many metals, and determine group I and II metals, metal complexes, and a complete range of inorganic and organic ions in a sample with excellent speed sensitivity.

Table 1.9 shows a comparison of the detection limits of ion chromatography *versus* flame AAS for ideal, single components in deionised water. On small-volume injections (50 µl) ion chromatography compares well with AAS. With sample pre-concentration techniques, the detection limits for ion chromatography can surpass those of GFAAS. While high concentrations of acids or bases can limit the applicability of AAS, ion chromatography allows direct injection of up to 10% concentrated acids or bases. This is extremely convenient in the direct analysis of acid-digested samples such as digests of

Table 1.9 Metal detection limits by ion chromatography				
Metal species detected by ion chromatography		Detection limit (mg/l)		
		Direct	Preconcentrated	Flame AA
Aluminium	Al^{3+}	56	0.5	20
Barium	Ba^{2+}	100	0.1	20
Cadmium	Cd^{2+}	10	0.1	1
Calcium	Ca^{2+}	50	0.5	3
Caesium	Cs^+	100	0.1	20
Chromium	Cr(III) as CrEDTA	1000	10	3
Chromium	Cr(VI) as CrO_4	50	1	3
Cobalt	Co^{2+}	3	0.03	5
Copper	Cu^{2+}	5	0.05	2
Dysprosium	Dy^{3+}	100	1	60
Erbium	Er^{3+}	100	1	60
Europium	Eu^{3+}	100	1	-
Gadolinium	Gd^{3+}	100	1	2000
Gold	Au(I) as $Au(CN)^-$	100	10	10
Gold	Au(III) as $Au(CN)_4^-$	100	10	10
Holmium	Ho^{3+}	100	1	60
Iron	Fe(II)	10	0.1	5
Iron	Fe(III)	3	0.03	5
Lead	Pb^{2+}	100	1	1
Lithium	Li^+	50	0.5	2
Lutetium	Lu^{3+}	100	1	-
Magnesium	Mg^{2+}	50	0.5	0.2
Molybdenum	as MoO_4	50	1	10
Nickel	Ni^{2+}	25	0.3	8
Palladium	as $PdCl_4^{2-}$	10	1	20
Platinum	as $PtCl_6^{2-}$	10	1	50
Potassium	K^+	50	0.5	1

Table 1.9 Continued				
Metal species detected by ion chromatography		Detection limit (mg/l)		
		Direct	Preconcentrated	Flame AA
Rubidium	Rb^+	100	1	2
Samarium	Sm^{3+}	100	1	700
Silver	as $Ag(CN)_2^-$	100	10	2
Sodium	Na^+	50	0.5	0.4
Strontium	Sr^{2+}	100	1	6
Terbium	Tb^{3+}	100	1	2000
Thulium	Tm^{3+}	100	1	-
Tin	$Sn(II)$	100	1	80
Tin	$Sn(IV)$	100	1	80
Tungsten	as WO_4^{2-}	50	1	100
Uranium	as UO_2^{2+}	5	0.05	7000
Ytterbium	Yb^{3+}	100	1	-
Zinc	Zn^{2+}	10	0.1	0.6
Source: Author's own files				

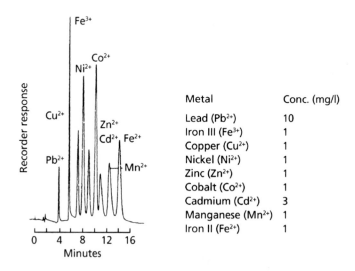

Metal	Conc. (mg/l)
Lead (Pb^{2+})	10
Iron III (Fe^{3+})	1
Copper (Cu^{2+})	1
Nickel (Ni^{2+})	1
Zinc (Zn^{2+})	1
Cobalt (Co^{2+})	1
Cadmium (Cd^{2+})	3
Manganese (Mn^{2+})	1
Iron II (Fe^{2+})	1

Figure 1.7 Ion chromatography: determination of nine transition metal ions. (*Source: Author's own files*)

polymers (**Figure 1.7**). Utilising ion exchange pre-concentration methods, extremely low concentrations of metals in polymer digests can be measured with ion chromatography. These detection limits are typically in the sub-picogram range.

1.2 Non-destructive Methods

1.2.1 X-ray Fluorescence Spectrometry

1.2.1.1 Theory

The XRF technique has a true multi-element analysis capability and requires no foreknowledge of the elements present in the sample. As such it is very useful for the examination of many of the types of samples encountered in the plastics laboratory.

This technique is very useful for solid samples especially if the main constituents (matrix) are made of low atomic weight elements and the sought impurities or constituents are of relatively high atomic weight.

Samples are irradiated with high-energy radiation, usually X-rays, to produce secondary X-rays which are characteristic of the individual elements present. The X-ray intensity due to a particular element is proportional to the concentration of that element in the sample. There are two types of instrument in production: those in which the emitted radiation is separated by wavelength using crystals as gratings, i.e., total reflection XRF [wavelength dispersive XRF (WDXRF) or total reflection XRF (TRXRF)], and those in which the radiation is not separated but identified by energy dispersive electronic techniques using solid-state detectors and multi-channel analysers, i.e., energy dispersive XRF (EDXRF).

Energy dispersive instruments rely on solid-state energy detectors coupled to energy discriminating circuitry to distinguish the radiation by its energy level and measure the amount at each level. Most X-ray detectors now in use are solid-state devices which emit electrons when X-rays are absorbed, the energy of the electrons being proportional to that of the incident X-rays and the quantity proportional to the intensity.

Typically, instruments will determine from a few percent down to parts per million in a solid sample.

WDXRF tends to be most accurate and precise method for trace element determinations. EDXRF instruments tend to lose precision for traces of light elements in heavy element matrixes unless longer counting times are used. With short counting times, for example, the coefficient of variation for a minor constituent element determination by an energy dispersive instrument should be better than 10%, but for a light trace element it may only be 50%. The advantage of the energy dispersive instrument is that it can be made so that almost all the radiation emitted hits the detector. Qualitative analysis is made by comparison with standard samples of known composition using total line energy. This is given by the total detector output of the line, or line peak area depending on the method of read-out used.

Due to the simple spectra and the extensive element range (sodium upwards in the periodic table) that can be covered using an Si(Li) detector and a 50 kV X-ray tube, EDXRF spectrometry is perhaps unparalleled for its qualitative element analysis power.

Qualitative analysis is greatly simplified by the presence of a few peaks that occur in predictable positions and by the use of tabulated element/line markers which are routinely available from computer-based analysers.

To date, the most successful method of combined background correction and peak deconvolution has been the method of digital filtering and least squares (FLS) fitting of reference peaks to the unknown spectrum [45]. This method is robust, simple to automate, and is applicable to any sample type.

The major disadvantage of conventional EDXRF has been poor elemental sensitivity, a consequence of high background noise levels resulting mainly from instrumental geometries and sample matrix effects. TRXRF is a relatively new multi-element technique with the potential to be an impressive analytical tool for trace element determinations for a variety of sample types. The fundamental advantage of TRXRF is its ability to detect elements in the picogram range in comparison to the nanogram levels typically achieved by traditional EDXRF spectrometry.

The principles of TRXRF were first reported by Yoneda and Horiuchi [46] and further developed by Aiginger and Wadbrauschek [47] and others [48–51]. In TRXRF the exciting primary X-ray beam impinges upon the specimen prepared as a thin film on an optically flat support of synthetic quartz or Perspex at angles of incidence in the region of 2 to 5 minutes of arc below the critical angle. In practice the primary radiation does not (effectively) enter the surface of the support but skims the surface, irradiating any sample placed on the support surface. The scattered radiation from the sample support is virtually eliminated, thereby drastically reducing the background noise. A further advantage of the TRXRF system, resulting from the geometry used, is that the solid-state energy dispersive detector can be accommodated very close to the sample (0.3 mm), which allows a large solid angle of fluorescent X-ray collection, thus enhancing a signal sensitivity and enabling the analysis to be carried out in air at atmospheric pressure.

Various suppliers of instruments are listed in Appendix 1.

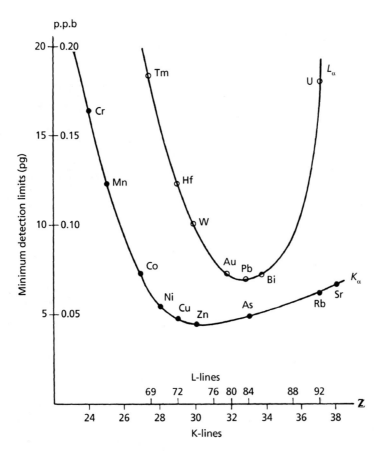

Figure 1.8 Minimum detection limits of TXRF spectrometer (Mo-tube 13 mA/60 kV; counting time 1000) (*Source: Phillips Electronic Instruments, Mahwah, NJ, USA*)

Instruments include the Philips PW 1404 spectrometer which is a powerful, versatile sequential X-ray spectrometer system developed from the PW 1400 series and incorporating many additional hardware and software features that further extend its performance. All system functions are controlled by powerful microprocessor electronics, which make routine analysis a simple, push-button exercise and provide extensive safeguards against operator error. The microprocessor also contains sufficient analytical software to permit stand-alone emergency operation, plus a range of self-diagnostic service-testing routines. The layout of the Philips PW 1404 instrument is shown in **Figure 1.8**.

An example of the detection limits achieved by the Seifert EXTRA III (3σ above background, counting time 1000 s) is shown in **Figure 1.9**, for a molybdenum anode

45

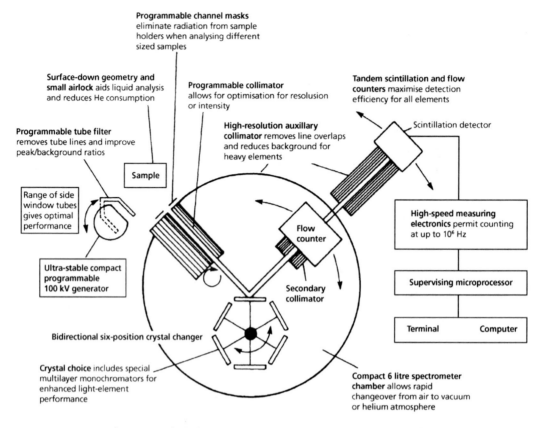

Programmable channel masks eliminate radiation from sample holders when analysing different sized samples

Surface-down geometry and small airlock aids liquid analysis and reduces He consumption

Programmable collimator allows for optimisation for resolusion or intensity

Tandem scintillation and flow counters maximise detection efficiency for all elements

Scintillation detector

Programmable tube filter removes tube lines and improve peak/background ratios

High-resolution auxillary collimator removes line overlaps and reduces background for heavy elements

Sample

Range of side window tubes gives optimal performance

Flow counter

High-speed measuring electronics permit counting at up to 10⁶ Hz

Ultra-stable compact programmable 100 kV generator

Secondary collimator

Supervising microprocessor

Bidirectional six-position crystal changer

Terminal   Computer

Crystal choice includes special multilayer monochromators for enhanced light-element performance

Compact 6 litre spectrometer chamber allows rapid changeover from air to vacuum or helium atmosphere

Figure 1.9 Layout of Philips PW 1404 energy-dispersive x-ray fluorescence spectrometer. (*Source: Author's own files*)

X-ray tube and for excitation with the filtered Bremsstrahlung spectrum from a tungsten X-ray tube. The data shown were obtained from diluted aqueous solutions which can be considered to be virtually free from any matrix effects. A detection limit of 10 pg for a 10 µl sample corresponds to a concentration of 1 µg/l. A linear dynamic range of four orders of magnitude is obtained for most elements; for example, lead at concentrations of 2–20,000 µg/l using cobalt as an internal standard at 2000 µg/l.

Seifert manufactures a TRXRF spectrometer [48–51]. Detection limits obtained for 60 elements by this technique are listed in **Table 1.10**.

Table 1.10 Detection limits using the Seifert Extra II x-ray spectrometer	
	Detection limit (pg)
Atomic numbers 18 - 38 (argon to strontium) and 53 - 57 (iodine - lanthanum) and 78 - 83 (osmium to bismuth)	<5
Chlorine atomic numbers 39 - 43 (yttrium to technetium), 47 - 52 (silver to tellurium), 65 - 71 (terbium to lutetium) and 90 - 92 (thorium to uranium)	5 - 10
Phosphorus (15)	10 - 30
Sulfur (16)	
Ruthenium (44)	
Rhodium (45)	
Palladium (46)	
Neptunium (93)	
Plutonium (94)	
Aluminium (13)	30 - 100
Silicon (14)	
Sodium (11)	>100
Magnesium (12)	
Source: Author's own files	

1.2.1.2 Applications

This technique can be used to conduct destructive or non-destructive analysis of polymers. XRF spectrometry has been used extensively for the determination of traces of metals and non-metals in polyolefins and other polymers. The technique has also been used in the determination of major metallic constituents in polymers, such as cadmium selenide pigment in polyolefins.

Specimen preparation is simple, involving compressing a disc of the polymer sample for insertion in the instrument, measurement time is usually less than for other methods, and X-rays interact with elements as such, i.e., the intensity measurement of a constituent element is independent of its state of chemical combination. However, the technique does have some drawbacks, and these are evident in the measurement of cadmium and selenium. For example, absorption effects of other elements present, e.g., the carbon and hydrogen of a polyethylene (PE) matrix, and excitation of one element by X-rays from another, e.g., cadmium and selenium affect one another. The technique has been applied to the determination of metals in polybutadiene, polyisoprene, and polyester resins [52]. The metals determined were cobalt, copper, iron, nickel and zinc. The samples were ashed and the ash dissolved in nitric acid prior to X-ray analysis. Concentrations as low as 10 ppm can be determined without inter-element interference.

Many investigators have found much higher recoveries using various ashing aids such as sulfuric acid [53], elemental sulfur [54, 55], magnesium nitrate [32, 56], and benzene and xylene sulfonic acids [57].

Leyden and co-workers [58] used XRF spectrometry to determine metals in acid digests of polymers. The aqueous solutions were applied to filter paper discs. They found that recoveries of metals by the X-ray technique were 101–110% compared to 89–94% by chemical methods of analysis.

XRF spectrometry has been applied very successfully, industrially, to the routine determination in hot-pressed discs of PE and PP down to a few parts per million of the following elements: aluminium, bromine, calcium, chlorine, magnesium, potassium, sodium, titanium and vanadium.

Ellis and Leyden (private communication) used dithiocarbamate precipitation methods to determine between 2 mg/l and 2 µg/l of five elements. **Table 1.11** shows the excellent

Table 1.11 The analysis of polymer digests by dibenzyldithiocarbamate precipitation - energy-dispersive x-ray spectrometry and electrothermal atomisation atomic absorption spectrometry										
Sample	Mn		Fe		Ni		Cu		Zn	
	XRF	AA	XRF	AA	XRF	AA	XRF	AA	XRF	AA
1	96	104	22	22	4	37	10	4	76	17
2	195	158	75	38	18	23	5	6	44	30
3	114	182	59	56	0.9	2.3	nd	nd	7	nd
4	450	450	46	42	0.9	2.2	1.8	5	61	51
5	2400	2700	15	22	20	25	5	14	1059	342
6	360	335	76	62	12	2.4	6	6	104	92
nd: not determined *Source: Author's own files*										

agreement generally found between XRF results obtained using a Link XR 200/300 instrument and AAS techniques in the analysis of pre-concentrates. Agreement does not extend over the whole concentration range examined for manganese. Some disparity also occurs in zinc determinations and it is believed that here the error is in the graphite furnace results.

Wolska [59] has reviewed recent advances in the application of XRF spectroscopy to the determination of antimony, bromine, copper, iron, phosphorus, titanium, and zinc, in various plastics. The new ED2000 high-performance EDXRF spectrometer manufactured by Oxford Instruments can determine up to 80 elements qualitatively and up to 50 elements quantitatively between sodium and uranium in various materials, including polymers [60].

1.2.2 Neutron Activation Analysis

This is a very sensitive technique. Due to the complexity and cost of the technique most laboratories do not have facilities for carrying out NAA. Instead, samples are sent to one of the organisations that possess the facilities.

An advantage of the technique is that a foreknowledge of the elements present is not essential. It can be used to indicate the presence and concentration of entirely unexpected elements, even when present at very low concentrations.

In NAA, the sample in a suitable container, often a pure PE tube, is bombarded with slow neutrons for a fixed time together with standards. Transmutations convert analyte elements into radioactive elements, which are either different elements or isotopes of the original analyte. After removal from the reactor the product is subjected to various counting techniques and various forms of spectrometry to identify the elements present and their concentrations.

1.2.2.1 Applications

This technique is capable of determining a wide range of elements, e.g., chlorine in polyolefins, metals in polymethylmethacrylate [61], total oxygen in polyethylene–ethylacrylate and polyethylene–vinylacetate copolymers [62], and total oxygen in polyolefins.

In many cases the results obtained by NAA can be considered as reference values and these data are of great value when these samples are analysed by alternative methods in the originating laboratory.

To illustrate this, some work is discussed on the determination of parts per million of sodium in polyolefins. It was found that replicate sodium contents determined on the same sample by a flame photometric procedure were frequently widely divergent. NAA offers an independent non-destructive method of checking the sodium contents which does not involve ashing.

In the flame photometric procedure the sample is dry ashed at 650–800 °C in a nickel crucible and the residue dissolved in hot water before determining sodium by evaluating the intensity of the line emission occurring at 589 nm.

NAA (flux of 1012 neutrons/cm/s) for sodium was carried out on PE and PP moulded discs containing up to about 550 ppm sodium which had been previously analysed by the flame photometric method. The results obtained in these experiments (**Table 1.12**) show that significantly higher sodium contents are usually obtained by NAA, and this suggests that sodium is being lost during the ashing stage of the flame photometric method. Sodium can also be determined by a further independent method, namely emission spectrographic analysis, which involves ashing the sample at 500 °C in the presence of an ashing aid consisting of sulfur and the magnesium salt of a long-chain fatty acid [55, 63–65]. **Table 1.13** shows the results by NAA and emission spectrography agree well with each other. The losses of sodium in the flame photometric ashing procedure were probably caused by the maximum ashing temperature used exceeding that used in the emission spectrographic method by some 150–300 °C.

Table 1.12 Interlaboratory variation of flame photometric sodium determinations in polyolefins (sodium content, ppm)		
Neutron activation analysis		Flame photometry
Powder	Moulded discs	Moulded discs
Polyethylene		
207	211, 204	35, 165, 140
177	175, 172	100, 140, 148
266	267, 263	85, 210, 221
203	187, 191	70, 160, 150
Polypropylene		
165	151, 161	50, 130, 133
198	186, 191	95, 173
322	333, 350	95, 138
Source: Author's own files		

Table 1.13 Comparison of sodium determination in polyolefins by neturon activation analysis, emission spectrography and flame photometry (sodium ppm)			
Sample description	By neutron activation analysis	By emission spectrography	By flame photometry
Polyethylene	99, 96, 99	95	60, 76, 55
Polyethylene	256, 247, 256	258, 259	160, 178, 271
Polyethylene	343, 321, 339	339, 287	250, 312
Polyethylene	213, 210, 212	218, 212	140, 196
Polypropylene	194, 189, 192	209, 198	80, 158, 229
Polypropylene	186, 191, 198	191, 191	95, 173
Source: Author's own files			

The results in **Table 1.14** show clearly that flame photometry following dope ashing at 500 °C gives a quantitative recovery of sodium. Direct ashing without an ashing aid at 500 °C causes losses of 10% or more of sodium, while direct ashing at 800 °C causes even greater losses.

Table 1.14 The effects of modification of ashing procedure on the flame photometric determination of sodium (sodium ppm)				
By neutron activation	By emission spectrography	By flame photometry		
		Original (ashed between 650 and 800 °C)	Dope ash at 500 °C	Direct ash at 500 °C
99, 96, 99	95	60, 75, 55	200	75
256, 247, 259	258, 259	160, 178, 271	225	208
343, 321, 339	339, 287	250, 312	282	265
213, 210, 212	218, 212	140, 196	210	191
194, 189, 192	209, 198	80, 158, 229	196	169
186, 191, 198	191, 191	95, 95, 173	193	173
Source: Author's own files				

Commonly, nowadays, the active catalyst (based on chromium, titanium, or vanadium) used in HDPE manufacture is adsorbed onto a highly porous silica support. Determination of the silica catalyst support content of the final polymer gives an assessment of the economic productivity of the reactor, i.e., its output of PE per gram of catalyst, and also enables the very low concentration of active catalyst metal in the polymer to be calculated.

Battiste and co-workers [66] have described three methods based on NAA, infrared spectroscopy, and ashing for the determination of silica catalyst supports in PE. In the NAA method approximately 2 g of PE powder is irradiated with neutrons obtained with a 500 keV neutron activator according to the following reaction:

3H(D,n) 4He

Silicon is activated by the reaction:

28Si(N,p) 28Al

and the concentration of silicon is then measured by the 1.78 MeV γ-ray emission from the decay of 28Al. A Conostan 5000 pm Si standard is used for the instrument calibration. Two 3-inch sodium iodide detectors are used to measure the 1.78 MeV γ-rays. Silica can be estimated by direct measurement of the silica absorbance at the 21.27 μm absorbance band of silica. This is the region of the infrared spectrum that is relatively free from polyethylene absorbance bands. The absorbance of the 21.27 μm band is calculated for each standard by use of the peak height determined by means of the baseline technique between minima near 28.57 and 17.24 μm.

Table 1.15 Catalyst productivity of some test samples				
Sample	IR at 21.27 micron	NAA	Ashing at 650 °C	Weight
1[a]	1893(3.3)d	1833	1923(4.9)	1923(4.9)
2[a]	4165(0)	4165	4545(9.1)	3571(14.3)
3[b]	1759(2.9)	1709	2222(30.0)	2007(17.4)
4[c]	1727(5.4)	1825	2000(9.6)	2000(9.6)
5[c]	2207(3.5)	2288	2778(21.4)	2941(28.5)
[a] *Catalyst A;* [b] *Catalyst B;* [c] *Cabosil S.17* *Values in parenthesis are percentage deviation from NAA* *Reprinted from D.R. Battiste, J.P. Butler, J.B. Cross and M. P. McDaniel, Analytical Chemistry, 1981, 53, 2232, with permission from the American Chemical Society [66]*				

Results of analysis of PE samples containing residual silica supports of three different catalysts are shown in **Table 1.15**. Infrared results differ from NAA results by 0–5% while ashing and weighing techniques differ from neutron activation by 5–21% and 5–28%, respectively.

References

1. R.J. Watling, *Analytica Chimica Acta*, 1977, **94**, 181.

2. C. Hallam and K.G. Thompson, *Determination of Lead and Cadmium in Potable Waters by Atom Trapping Atoms Absorbance Spectroscopy*, Divisional Laboratory, Yorkshire Water Authority, Sheffield, UK, 1986.

3. R.G. Godden and D.R. Thomenson, *Analyst*, 1980, **105**, 1137.

4. B.W.J. Rence, *Automatic Chemistry*, 1982, **4**, 61.

5. P.D. Goulden and P. Brooksbank, *Analytical Chemistry*, 1974, **46**, 1431.

6. P.B. Stockwell in *Topics in Automatic Chemical Analysis*, Ellis Horwood, Chichester, UK, 1979.

7. A.L. Dennis and D.G. Porter, *Automatic Chemistry*, 1981, **2**, 134.

8. B. Pahlavanpour, M. Thompson and L. Thorne, *Analyst*, 1981, **106**, 467.

9. B.W.J. Rence, *Automatic Chemistry*, 1980, **105**, 1137.

10. M. Thompson, B. Pahlavanpaur and L. Thorne, *Analyst*, 1981, **106**, 468.

11. W.T. Corus, L. Ebdon and S.J. Hill, *Analyst*, 1992, **117**, 717.

12. W.T. Corus, P.B. Stockwell, L. Ebdon and S.J. Hill, *Journal of Analytical Atomic Spectroscopy*, 1993, **8**, 71.

13. P.B. Stockwell and W.T. Corus, *Spectroscopy World*, 1992, **411**, 14.

14. S. Greenfield, I.L. Jones and C.T. Berry, *Analyst*, 1964, **89**, 713.

15. R.H. Wendt and V.A. Fassel, *Analytical Chemistry*, 1965, **37**, 920.

16. R.H. Scott, *Analytical Chemistry*, 1974, **46**, 75.

17. M. Thompson and J.N. Walsh in *A Handbook of Inductively Coupled Plasma Spectrometry*, Blackie, London, UK, 1983, p. 55.

18. R.F. Suddendorf and K.W. Boyer, *Analytical Chemistry*, 1978, **50**, 1769.

19. B.L. Sharp, inventor; B.L. Sharp, assignee; GB 8,432.338, 1984.

20. A.M. Gunn, D.L. Millard and G.F. Kirkbright, *Analyst*, 1978, **103**, 1066.

21. H. Matuslavicz and R.M. Barnes, *Applied Spectroscopy*, 1984, **38**, 745.

22. M.W. Tikkanen and K.M. Niemczyk, *Analytical Chemistry*, 1984, **56**, 1997.

23. E.D. Salin and G. Horlick, *Analytical Chemistry*, 1979, **51**, 2284.

24. E.D. Salin and R.L.A. Szung, *Analytical Chemistry*, 1984, **56**, 2596.

25. A.L. Briseno, A. Baca, Q. Zhou, R. Lai and F. Zhou, *Analytica Chimica Acta*, 2001, **441**, 123.

26. S. Ruzicka and E.H. Hansen, *Analytica Chimica Acta*, 1978, **99**, 37.

27. A.M. Dobney, A.J.G. Monk, K.H. Grobecker, P. Conneely and C.G. de Koster, *Analytica Chimica Acta*, 2000, **423**, 9.

28. A.M. Riquet and A. Feigenbaum, *Food Additives and Contaminants*, 1997, **14**, 53.

29. H. Narasaki and K. Umezawa, *Kobunski Kogaku*, 1972, **29**, 438.

30. A.J. Smith, *Analytical Chemistry*, 1964, **36**, 944.

31. T. Gorsuch, *Analyst*, 1962, **87**, 112.

32. T. Gorsuch, *Analyst*, 1959, **84**, 135.

33. E.L. Henn, *Analytica Chimica Acta*, 1974, **73**, 273.

34. T. Korenaga, *Analyst*, 1981, **106**, 40.

35. H. Ogure, *Bunseki Kogaku*, 1975, **24**, 197.

36. R. Reverz and E. Hasty, *Recovery Study Using an Elevated Pressure Temperature Microwave Dissolution Technique*, Pittsburgh Conference and Exposition of Analytical Chemistry and Applied Spectroscopy, 1987.

37. R.A. Nadkarni, *Analytical Chemistry*, 1984, 56, 2233.

38. H.M. Kingston and L.B. Jassie, *Analytical Chemistry*, 1986, 58, 2534.

39. *Parr Manual for 207M*, Parr Instruments Co., Moline, IL, USA.

40. R.A. Nadkarni, *American Laboratory*, 1981, **13**, 22.

41. *Ultra Violet and Visible Solution Spectrometry and Colorimetry 1980*, An Essay Review, Her Majesty's Stationery Office, London, 1981.

42. L.M. Mal'kova, A.I. Kalanin and E.M. Derepletchikove, *Zhurnal Analiticheskoi Kimii*, 1972, **27**, 56.

43. H. Small, T.S. Stevens and W.C. Bauman, *Analytical Chemistry*, 1975, **47**, 1801.

44. *High Performance Liquid Chromatography, Ion Chromatography, Thin-Layer and Column Chromatography of Water Samples*, Her Majesty's Stationery Office, London, UK, 1983.

45. P.J. Statham, *Analytical Chemistry*, 1977, **49**, 2149.

46. Y. Yoneda and T. Horiuchi, *Review Scientific Instruments*, 1971, **42**, 1069.

47. H. Aiginger and P. Wadbrauschek, *Nuclear Instruments and Methods*, 1974, **114**, 157.

48. J. Knoth and H. Schwenke, *Fresenius' Zeitschrift fürAnalytische Chemie*, 1978, **291**, 200.

49. J. Knoth and H. Schwenke, *Fresenius' Zeitschrift für Analytische Chemie*, 1980, **201**, 7.

50. H. Schwenke and J. Knoth, *Nuclear Methods*, 1982, **193**, 239.

51. D.A. Pella and R.C. Dobbyn, *Analytical Chemistry*, 1988, **60**, 684.

52. W.S. Cook, C.O. Jones and A.G. Altenau, *Canadian Spectroscopy*, 1968, **13**, 64.

53. A.A. Vasilieva, Yu V. Vodzinskii and I.A. Korschunov, *Zavodskaia Laboratoriia*, 1968, **34**, 1304.

54. J.S. Bergmann, C.H. Ehrhart, L. Grantelli and L.J. Janik in *Proceedings of the 153rd ACS Meeting*, Miami Beach, FL, USA, 1967.

55. W.A. Rowe and K.P. Yates, *Analytical Chemistry*, 1963, **35**, 368.

56. L.W. Gamble and W.H. Jones, *Analytical Chemistry*, 1955, **27**, 1456.

57. J.E. Shott Jr., T.J. Garland and R.O. Clarke, *Analytical Chemistry*, 1961, **33**, 506.

58. D.E. Leyden, J.C. Lennox and C.U. Pittman, *Analytica Chimica Acta*, 1973, **64**, 143.

59. J. Wolska, *Plastics Additives and Compounding*, 2003, **5**, 90.

60. *Rubber World*, 1999, **219**, 4, 79.

61. O.J. Kabayashi, *Polymer Science A-1*, 1979, **17**, 293.

62. D. Hull and J. Gilmore in the *Proceedings of the 141st ACS Division of Fuel Chemistry Meeting*, Washington, DC, USA, 1962.

63. J.S. Bergmann, C.H. Ekhart, L. Grantelli and J.L. Janik, Private communication, 1967.

64. E. Barendrecht, *Analytica Chimica Acta*, 1961, **24**, 498.

65. J.E. Shott Jr., T.J. Garland and R.O. Clarke, *Analytical Chemistry*, 1961, **33**, 507.

66. D.R. Battiste, J.P. Butler, J.B. Cross and M.P. McDaniel, *Analytical Chemistry*, 1981, **53**, 2232.

2 Non-metallic Elements

Non-metallic elements such as boron, the halogens, nitrogen, oxygen, phosphorus and sulfur, can occur in polymers either as major constituents present as impurities or as components of low-percentage additions of additives containing the element, e.g., the addition of 0.5% dilauryl thiodipropionate antioxidant to a polymer during processing will introduce parts per million concentrations of sulfur into the final polymer. Another source of non-metallic elements in polymers is catalyst residues and processing chemicals.

It is advisable when commencing the analysis of a polymer to determine the content of various non-metallic and metallic elements first. Initially, these tests could be qualitative, simply to indicate the presence or absence of the element. All that is required here is that the test is of sufficient sensitivity so that elements of importance are not missed. If in these tests an element is found it may then be necessary to determine it quantitatively as discussed next. The analytical methods used to determine elements should be sufficiently sensitive to determine about 10 ppm of an element in the polymer, i.e., should be able to detect in a polymer a substance present at 0.01% and containing down to 10% of the element in question.

This requirement is met for almost all the important elements by the use of optical emission spectroscopy and X-ray fluorescence spectrometry (XRFS). XRFS is applicable to all elements with an atomic number greater than 12. Using these two techniques, all metals and non-metals down to an atomic number of 15 (phosphorus) can be determined in polymers at the required concentrations [1-4].

Nitrogen is determinable by micro Kjeldahl digestion techniques. A cautionary note is that, in addition to the polymer itself, the polymer additive system may contain elements other than carbon, hydrogen, and oxygen. The detection of an element such as boron, halogens, nitrogen, phosphorus, silicon, or sulfur, in a polymer is indicative that the element originates in the polymer and not the additive system if the element is present at relatively high concentrations such as several percent. This is highlighted by the example of a high-density polyethylene which might contain 0.2–1% chlorine originating from polymerisation residues and polyvinyl chloride (PVC) homopolymer which contains more than 50% chlorine.

Commercial instrumentation available for the determination of the following total elements is the subject of this chapter: halogens; sulfur; halogens and sulfur; nitrogen; nitrogen, carbon, and sulfur; carbon, hydrogen, and nitrogen; and total organic carbon (TOC).

2.1 Instrumentation: Furnace Combustion Methods

The reader is also referred to Appendix 1.

2.1.1 Halogens

The Dohrmann DX 20B system is based on combustion of a sample to produce a hydrogen halide, which is then swept into a microcoulometric cell and estimated. It is applicable at total halide concentrations up to 1000 µg/l with a precision of ±2% at the 10 µg/l level. The detection limit is about 0.5 µg/l. Analysis can be performed in five minutes. A sample boat is available for carrying out analysis of solid samples.

Mitsubishi also supplies a microprocessor-controlled automatic total halogen analyser (model TOX-10) which is very similar in operating principles to the Dohrmann system discussed previously, i.e., combustion at 800–900 °C followed by coulometric estimation of the hydrogen halide produced.

2.1.2 Sulfur

The Mitsubishi trace sulfur analyser models TS-02 and TN-02(S) again involve a microcombustion procedure in which sulfur is oxidised to sulfur dioxide, which is then titrated coulometrically with triiodide ions generated from iodide ions:

$$SO_2 + I_3^- + H_2O \rightarrow SO_3 + 3I^- + 2H^+$$

$$3I^- \rightarrow I_3^- + 3e^-$$

2.1.3 Total Sulfur/Total Halogen

The Mitsubishi TSX-10 halogen–sulfur analyser expands the technology of the TOX-10 to include total chlorine and total sulfur measurement. The model TSX-10, which consists of the TOX-10 analyser module and a sulfur detection cell, measures total sulfur and total chlorine in liquid and solid samples over a sensitivity range of milligrams per litre to a percentage.

Dohrmann also produces an automated sulfur and chlorine analyser (models MCTS 130/120). This instrument is based on combustion microcoulometric technology.

2.1.4 Total Bound Nitrogen

Mitsubishi supply two total nitrogen analysers: the model TN-10 and the model TN-05 microprocessor-controlled chemiluminescence total nitrogen analysers. These measure down to micrograms per litre amounts of nitrogen in solid and liquid samples.

The sample is introduced into the combustion tube packing containing oxidative catalyst under oxygen carrier gas. High-temperature oxidation (800–900 °C) occurs and all chemically bound nitrogen is converted to nitric oxide (NO): $R–N \rightarrow CO_2 + NO$. Nitric oxide then passes through a drier to remove water formed during combustion and moves to the chemiluminescence detector, where it is mixed with ozone to form excited nitrogen dioxide (NO_2^*):

$$NO + O_3 \rightarrow NO_2^* + O_2 \rightarrow + O_2 + h\nu$$

Rapid decay of the NO_2^* produces radiation in the 590–2900 nm range. It is detected and amplified by a photomultiplier tube. The result is calculated from the signal produced and is given in milligrams per litre or as a percentage.

Dohrmann also supplies an automated nitrogen analyser with video display and data processing (model DN-1000) based on similar principles which is applicable to the determination of nitrogen in solid and liquid samples down to 0.1 mg/l.

The Dohrmann DN-1000 can be converted to the determination of sulfur and chlorine by adding the MCTS 130/120 microcoulometer detector modules. The control module, furnace module, and all the automated sample inlet modules are common to both detectors. The system automatically recognises what detector and sample inlet is present and sets the correct operating parameters for fast, simple conversion between nitrogen, sulfur, and chlorine detection.

Equipment for automated Kjeldahl determinations of organic nitrogen in water and solid samples is supplied by Tecator Ltd. Its Kjeltec system 1 streamlines the Kjeldahl procedure resulting in higher speed and accuracy compared to classic Kjeldahl measurements.

2.1.5 Nitrogen, Carbon, and Sulfur

The NA 1500 analyser supplied by Carlo Erba is capable of determining these elements in 3–9 minutes in amounts down to 10 mg/l with a reproducibility of ±0.1%. A 196-position autosampler is available.

'Flash combustion' of the sample in the reactor is a key feature of the NA 1500. This results when the sample is dropped into the combustion reactor which has been enriched with pure oxygen. The normal temperature in the combustion tube is 1020 °C and reaches 1700–1800 °C during the flash combustion.

In the chromatographic column the combustion gases are separated so that they can be detected in sequence by the thermal conductivity detector (TCD). The output signal is proportional to the concentration of the elements. A data processor plots the chromatogram, automatically integrates the peak areas, and gives retention times, percentage areas, baseline drift, and attenuation for each run. It also computes blank values, constant factors, and relative average elemental contents.

2.1.6 Carbon, Hydrogen, and Nitrogen

Perkin Elmer supplies an analyser (model 2400 CHN or PE 2400 series II CHNS/O analysers) suitable for determining these elements in polymers. In this instrument the sample is first oxidised in a pure oxygen environment. The resulting combustion gases are then controlled to exact conditions of pressure, temperature, and volume. Finally the product gases are separated under steady-state conditions and swept by helium or argon into a gas chromatograph for analysis of the components. The equipment is supplied with a 60-position autosampler and microprocessor controller covering all system functions, calculation of results, and on-board diagnostics. Analysis time is 6 minutes for the CHN mode, 8 minutes for the CHNS mode, and 4 minutes for the oxygen mode [5-10].

Table 2.1 gives theoretical *versus* determined carbon, nitrogen, and hydrogen values obtained by this instrument for three polymers.

Table 2.1 Automated determination of carbon, hydrogen and nitrogen in polymers						
Compound	Theory			Found, %		
	C	H	N	C	H	N
Nylon 6	63.68	9.80	12.38	63.58	9.85	12.35
				63.55	9.91	12.32
Styrene/25% acrylonitrile	86.10	7.24	6.60	86.00	7.28	6.62
				86.05	7.20	6.65
Teflon	24.00	–	–	23.97	–	–
				24.10	–	–
Reproduced by kind permission of Perkin Elmer Ltd., Beaconsfield, UK.						

2.1.7 Total Organic Carbon

Dohrmann supplies a TOC analyser. Persulfate reagent is continuously pumped at a low flow rate through the injection port (and the valve of the autosampler) and then into the ultraviolet reactor. A sample is acidified, sparged, and injected directly into the reagent stream. The mixture flows through the reactor where organics are oxidised by the photon-activated reagent. The light-source envelope is in direct contact with the flowing liquid. Oxidation proceeds rapidly, the resultant carbon dioxide is stripped from the reactor liquid and carried to the carbon dioxide-specific, non-dispersive infrared detector.

The Shimadzu TOC-500 total organic carbon analyser is a fully automated system capable of determining between 1 and 3000 µg/l TOC.

OIC Analytical Instruments produce the fully computerised model 700 TOC analyser. This is applicable to solids. Persulfate oxidation at 90–100 °C followed by non-dispersive infrared spectroscopy is the principle of this instrument.

2.2 Oxygen Flask Combustion Methods

2.2.1 Total Halogens

Oxygen flask combustion methods have been used to determine traces of chlorine in PVC [11] and in polyolefins and chlorobutyl rubber [12].

Traces of chlorine have been determined in polyolefins [11] at levels between 0 and 500 ppm. The Schoniger oxygen flask combustion technique requires a 0.1 g sample and the use of a 1 litre conical flask. Chlorine-free polyethylene (PE) foil is employed to wrap the sample, which is then supported in a platinum wire attached to the flask stopper. Water is used as the absorbent. Combustion takes place at atmospheric pressure in oxygen. The chloride formed is potentiometrically titrated in nitric acid/acetone medium with 0.01 M silver nitrate solution.

In the method of determining chlorine in chlorobutyl and other chlorine-containing polymers [12] the sample is combusted in a 1–2 litre oxygen-filled combustion flask containing 0.01 M nitric acid. After the combustion the flask is allowed to cool and 0.01 M silver nitrate added. The combustion solution containing silver chloride is evaluated turbidometrically at 420 mm using a grating spectrophotometer. Alternatively, to determine bromine, chlorine, iodine, or mixtures thereof, the combustion solution can be titrated with dilute standard silver nitrate solution or can be evaluated by ion chromatography (see Section 2.2.3).

A method for the determination of fluorine in fluorinated polymers such as polytetrafluoroethylene (PTFE) is based on decomposition of the sample by oxygen flask combustion followed by spectrophotometric determination of the fluoride produced by a procedure involving the reaction of the cerium(III) complex of alizarin complexan (1,2-dihydroxy-anthraquinone 3-ylmethylamine *N,N*-diacetic acid). The blue colour of the fluoride-containing complex (maximum absorption, 565 nm) is completely distinguishable from either the yellow of the free dye (maximum absorption, 423 nm) or the red of its cerium(III) chelate (maximum absorption, 495 nm).

A method has been described [13] for the determination of chlorine in polymers containing chlorine, fluorine, phosphorus and sulfur, which involves oxygen flask combustion over water, addition of ethanol, and titration to the diphenylcarbazide indicator end point with 0.005 M mercuric nitrate:

$$2HCl + Hg(NO_3)_2 = HgCl_2 + 2HNO_3$$

Using this method Johnson and Leonard [13] obtained from PTFE 75.8% of fluorine using a silica or boron-free glass combustion flask against a theoretical value of 76%. Using a borosilicate glass combustion flask they obtained a low fluorine recovery of 72.1%.

2.2.2 Sulfur

To determine sulfur in amounts down to 500 ppm in polyolefins, the sample is wrapped in filter paper and burnt in a closed conical flask filled with oxygen at atmospheric pressure. The sulfur dioxide produced in the reaction reacts with dilute hydrogen peroxide solution contained in the reaction flask to produce an equivalent amount of sulfuric acid:

$$H_2SO_3 + H_2O_2 = H_2SO_4 + H_2O$$

The sulfuric acid is estimated by visual titration with M/500 or M/50 barium perchlorate using Thorin indicator. The repeatability of this method is ±40% of the sulfur content determined at the 500 ppm sulfur level, improving to ±2% at the 1% level. Chlorine and nitrogen concentrations in the sample may exceed the sulfur concentration several times over without causing interference. Fluorine does not interfere unless present in concentrations exceeding 30% of the sulfur content. Phosphorus and metallic constituents interfere when present in moderate amounts.

2.2.3 Oxygen Flask Combustion: Ion Chromatography

Combustion of polymers in an oxygen-filled flask over aqueous solutions of appropriate reagents converts elements such as halogens, phosphorus and sulfur into inorganic anions.

For example:

- chlorine, bromine, iodine → chloride, bromide, iodide
- sulfur → sulfate
- phosphorus → phosphate

Subsequent analysis of these solutions by ion chromatography [14] enables the concentrations of mixtures of these anions (i.e., the original elements) to be determined rapidly, accurately, and with great sensitivity.

2.2.4 Instrumentation

Instrumentation for ion chromatography is discussed in Section 1.1.15. See also Appendix 1.

2.2.5 Applications

Figure 2.1(a) shows a separation of halides, nitrate, phosphate and sulfate, obtained in six minutes by ion chromatography using a Dionex A54A anion exchange separator.

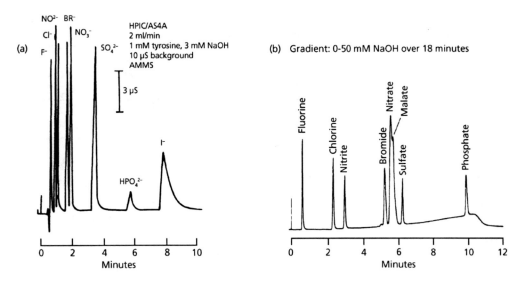

Figure 2.1 Ion chromatograms obtained with Dionex instrument using (anodic) AMMS and CMMS micromembrane suppression. (a) anions with micromembrane suppressor; (b) multi-component analysis by ion chromatography. (*Source: Author's own files*)

A further development is the Dionex HPLC AS5A-SU analytical anion exchange column. Quantitation of all the anions in **Figure 2.1(b)** would require at least three sample injections under different eluent conditions.

2.3 Acid and Solid Digestions of Polymers

2.3.1 Chlorine

Fusion with sodium carbonate is a very useful method for the fusion of polymers that, upon ignition, release acidic vapours – e.g., PE containing traces of chlorine or PVC, both of which, upon ignition, release anhydrous hydrogen chloride. To determine chlorine accurately in the polymer in amounts down to 5 ppm the hydrogen chloride must be trapped in a solid alkaline reagent such as sodium carbonate. In this method PE is mixed with pure sodium carbonate and ashed in a muffle furnace at 500 °C. The residual ash is dissolved in aqueous nitric acid, and then diluted with acetone. This solution is titrated potentiometrically with standard silver nitrate. **Table 2.2** compares results for chlorine determinations in PE obtained by this method with those obtained by XRFS. The averages of results obtained by the two methods agree satisfactorily to within ±15% of each other.

Table 2.2 Comparison of chlorine contents by X-ray method and chemical method (averages in parenthesis)		
X-ray on discs	Chemical methods on same discs as used for X-ray analysis	Chemical method on powder*
865, 841 (840)	700	786, 761 (773)
535, 570 (522)	606	636, 651
785, 675 (730)	598	650, 654 (652)
625, 675 (650)	600	637, 684 (660)
895, 870 (882)	733	828, 816 (822)
*Analysis carried out on samples which had been treated with alcoholic potash to avoid losses of chlorine when preparing discs.		

Sodium peroxide is another useful reagent for the fusion of polymer samples preparatory to analysis for metals such as zinc and non-metals such as chlorine [15, 16] and bromine. In this method the polymer is intimately mixed either with sodium peroxide in an open crucible or with a mixture of sodium peroxide and sucrose in a micro-Parr bomb. After acidification with nitric acid, chlorine can be determined [16]. In a method for the determination of traces of bromine in polystyrene in amounts down to 100 ppm bromine, a known weight of polymer is mixed intimately with pure sodium peroxide and sucrose in a micro-Parr bomb which is then ignited. The sodium bromate produced is converted to sodium bromide by the addition of hydrazine as the sulfate:

$$2NaBrO_3 + 3NH_2NH_2 = 2NaBr + 6H_2O + 3N_2$$

The combustion mixture is dissolved in water and acidified with nitric acid. The bromine content of this solution is determined by potentiometric titration with standard silver nitrate solution.

2.3.2 Nitrogen

Apart from the chemical Kjeldahl digestion procedure for the determination of organic nitrogen, acid digestion of polymers has found little application. One of the problems is connected with the form in which the polymer sample occurs. If it is in the form of a fine powder, or a very thin film, then digestion with acid might be adequate to enable the relevant substance to be quantitatively extracted from the polymer. However, low nitrogen results would be expected for polymers in larger granular form, and for the analysis of such samples classic microcombustion techniques are recommended.

Hernandez [17] has described an alternative procedure based on pyro-chemiluminescence which he applied to the determination of 250–1500 ppm nitrogen in PE. In this technique the nitrogen in the sample is subjected to oxidative pyrolysis to produce nitric oxide. This when contacted with ozone produces a metastable nitrogen dioxide molecule, which as it relaxes to a stable state emits a photon of light. This emission is measured quantitatively at 700–900 nm.

2.3.3 Phosphorus

Phosphorus has been determined [18, 19] in thermally stable polymers by mineralisation with a nitric–perchloric acid mixture and subsequent titration with lanthanum nitrate or by photometric determination of the phosphomolybdic blue complex.

2.3.4 Silica

Silica has been determined in PE films by a method based on near-infrared spectroscopy. For peak height measurements a single baseline point at the minimum near 525 cm^{-1} was found to be best. An additional baseline point below 430 cm^{-1} gave poorer results because of the increased noise at longer wavelengths due to atmospheric absorption. For the same reason peak area measurements were confined to the range 525–469 cm^{-1} [20]. Both height and area measurements gave an error index close to 1%, but derivative methods were considerably poorer. Derivative spectra generally show increased noise levels so that they are unlikely to be useful except when they are overlapping bands. The results obtained with the ratio program also showed a higher error index. The band index ratio method avoids uncertainty associated with measuring the film thickness, but in this case the error resulting from using a rather weak reference band appears greater.

2.4 X-ray Fluorescence Spectroscopy

The X-ray fluorescence (XRF) technique, already discussed in Chapter 1, has been applied extensively to the determination of macro- and micro-amounts of non-metallic elements in polymers.

An interesting phenomenon has been observed in applying the XRF method to the determination of parts per million of chlorine in hot-pressed discs of low-pressure polyolefins. In these polymers the chlorine is present in two forms, organically bound and inorganic, with titanium chloride compounds resulting as residues from the polymerisation

Table 2.3 Determination of chlorine by X-ray procedure		
A: Polymer not treated with alcoholic potassium hydroxide before analysis, ppm chlorine, X-ray fluorescence on polymer discs. Average of 2 discs (A)	B: Polymer treated with alcoholic potassium hydroxide before analysis, ppm chlorine, X-ray fluorescence on polymer discs. Average of 2 discs (B)	Difference between average chlorine contents obtained on potassium hydroxide-treated and untreated samples (B) – (A)
510	840	330
422	552	130
440	730	290
497	650	153
460	882	422

catalyst. The organic part of the chlorine is determined by XRF without complications. However, during hot processing of the discs there is a danger that some inorganic chlorine will be lost. This can be completely avoided by intimately mixing the powder with alcoholic potassium hydroxide, then drying at 105 °C before hot pressing into discs. The results illustrate this effect (**Table 2.3**). Considerably higher total chlorine contents are obtained for the alkali-treated polymers.

A further example of the application of XRF spectroscopy is the determination of tris(2,3-dibromopropyl) phosphate on the surface of flame retardant polyester fabrics [21]. The technique used involved extraction of the fabric with an organic solvent followed by analysis of the solvent by XRF for surface bromine and by high-pressure liquid chromatography for molecular tris(2,3-dibromopropyl) phosphate. The technique has been applied to the determination of hydroxy groups in polyesters [22, 23]:

with $n = 1-100$ and $x = 2$ (polyethylene terephthalate) or 4 (polybutylene terephthalate) and ester-interchange elastomers of 4-polybutylene terephthalate and polypropylene glycol. The hydroxyl groups in these products are determined by acetylation with an excess of dichloroacetic anhydride in dichloroacetic acid and measurement of the amount of acetylation by a chloride determination carried out on the derivative.

The XRF method of Wolska [24] discussed in Section 1.2.1 has been applied to the determination of bromine and phosphorus in polymers. Various other workers have applied this technique to the determination of chlorine and sulfur [25] and various other elements [26, 27].

2.5 Antec 9000 Nitrogen/Sulfur Analyser

This instrument [28] can provide analysis of nitrogen- and sulfur-containing additives in polyolefins.

References

1. W.S. Cook, C.O. Jones and A.G. Altenau, *Canadian Spectroscopy*, 1968, **13**, 64.

2. H.W. Hank and L. Silverman, *Analytical Chemistry*, 1959, **31**, 1069.

3. B.J. Mitchell and H.J. O'Hear, *Analytical Chemistry*, 1962, **34**, 1621.

4. J.S. Bergmann, C.H. Ekhart, L. Grantelli and J.L. Janik in *Proceedings of the 153rd ACS Meeting*, Miami Beach, FL, USA, 1967.

5. *The Elemental Analysis of Various Classes of Chemical Compounds Used with Perkin Elmer PE2400 CHN Elemental Analyser*, Perkin Elmer Elemental Newsletter EAN-5, Perkin Elmer, Norwalk, CT, USA.

6. *Principles of Operation – the Perkin Elmer PE2400 CHN Elemental Analyser*, Perkin Elmer Elemental Newsletter EAN-2, Perkin Elmer, Norwalk, CT, USA.

7. *Principles of Operation – Oxygen Analysis Accessory for the PE2400 CHN Elemental Analyser*, Perkin Elmer Elemental Newsletter EAN-4, Perkin Elmer, Norwalk, CT, USA.

8. *Calculation of Weight Percentages for Organic Elemental Analysis*, Perkin Elmer Elemental Newsletter EAN-6, Perkin Elmer, Norwalk, CT, USA.

9. *Analysis of a Copolymer or Polymer Blend when One Component Contains a Heteroelement*, Perkin Elmer Elemental Analysis Newsletter EAN-13, Perkin Elmer, Norwalk, CT, USA.

10. *Application of the PE2400 CHN Elemental Analyser for the Analysis of Plasticizers*, Perkin Elmer Elemental Analysis Newsletter EAN-23 Perkin Elmer, Norwalk, CT, USA.

11. K. Tanaka and T. Morikawat, Kagaku *To Kogyo (Osaka)*, 1974, **48**, 387.

12. J.Z. Falcon, J.L. Love, L.T. Yacta and A.G. Altenau, *Analytical Chemistry*, 1975, **47**, 141.

13. C.A. Johnson and H. Leonard, *Analyst*, 1961, **86**, 101.

14. H. Small, T.S. Stevens and W.C. Bauman, *Analytical Chemistry*, 1975, **47**, 1801.

15. H. Mitterberger and H. Gross, *Kunststoff-Technic*, 1973, **12**, 76.

16. M.L. Bakroni, N.K. Chakavarty and S. Chopra, *Indian Journal of Technology*, 1975, **13**, 576.

17. H.A. Hernandez, *International Laboratory*, 1981, September, 84.

18. H. Narasaki, V. Hijaji and A. Unno, *Bunseki Kagaku*, 1973, **22**, 541.

19. L.S. Kalinina, N.I. Nikitina, M.A. Matorina and I.V. Sedova, *Plasticheskie Massy*, 1976, **5**, 66.

20. F. Mocker, *Kautschuk und Gummi*, 1964, **11**, 1161.

21. T.L. Smith and B.N. Whelihan, *Textile Chemist and Colorist*, 1978, 10, 35.

22. G.D.B. Van Houwelingen, *Analyst*, 1981, **106**, 1057.

23. G.D.B. Van Houwelingen, N.W.M.G. Peters and W.G.B. Huysmens, *Fresenius' Zeitscrift für Analytical Chemistry*, 1978, **293**, 396.

24. J. Wolska, *Plastics Additives and Compounding*, 2003, **5**, 90.

25. F. Blockhuys, M. Claes, R. Van Grieken and H.J. Geise, *Analytical Chemistry*, 2000, **72**, 3366.

26. L.M. Sherman, *Plastics Technology*, 1998, **44**, 56.

27. J. M. Bruna and Izasa SA, *Revista de Plásticos Modernos*, 1995, **65**, 550.

28. R.A. Morton and A.L. Stubbs, *Analyst*, 1946, **71**, 348.

3 Functional Groups and Polymer Structure

A variety of instrumental techniques have been used to determine functional groups in polymers and to elucidate the detail of polymer structure. These include infrared spectroscopy, near-infrared spectroscopy including Raman spectroscopy, Fourier transform infrared spectroscopy (FT-IR), nuclear magnetic resonance (NMR) and proton magnetic resonance (PMR) spectroscopy, chemical reaction gas chromatography, pyrolysis gas chromatography (PGC), pyrolysis gas chromatography–mass spectrometry, pyrolysis–NMR spectroscopy, X-ray fluorescence spectroscopy (XRFS), and also newer techniques such as time-of-flight secondary ion mass spectrometry (ToF-SIMS), X-ray photoelectron spectroscopy (XPS), tandem mass spectrometry (MS-MS), matrix-assisted laser desorption/ionisation mass spectrometry (MALDI-MS), microthermal analysis, atomic force microscopy (AFM), and various X-ray methods including scanning electron microscopy (SEM)–energy dispersive analysis using X-rays (EDAX).

Functional groups that may have to be determined include: hydroxy, carbonyl, carboxyl, alkyl, aryl, alkoxy, oxyalkylene, nitrile, ester, amino, nitro, amide, amido, imino, and epoxy groups. Comonomer ratios, isomers, and short-chain branching are all structural features of polymers that may have to be elucidated in order to obtain a complete picture of polymer structure.

3.1 Infrared and Near-Infrared Spectroscopy

3.1.1 Instrumentation

Suppliers of infrared spectrometers are listed in Appendix 1.

3.1.2 Applications

Some applications of infrared spectroscopy to the determination of functional groups in polymers are discussed next.

3.1.2.1 Determination of Bound Ethylene in Ethylene–Propylene Copolymers

Lomonte and Tirpak [1] have developed a method for the determination of the percentage of ethylene incorporated in ethylene–propylene block copolymers. Standardisation is done from mixtures of the homopolymers. Both standards and samples are scanned at 180 °C in a spring-loaded demountable cell. The standardisation is confirmed by the analysis of copolymers of known ethylene content prepared with ^{14}C-labelled ethylene. By comparison of the infrared results from the analyses performed at 180 °C and also at room temperature, the presence of the ethylene homopolymer can be detected. These workers derived an equation for the quantitative estimation of the percentage of ethylene present as copolymer blocks.

The method distinguishes between true copolymers and physical mixtures of copolymers. The method makes use of a characteristic infrared rocking vibration due to sequences of consecutive methylene groups. Such sequences are found in polyethylene (PE) and in the segments of ethylene blocks in ethylene–propylene copolymers. This makes it possible to detect them at 13.70 and 13.89 µm. There are bands at both these locations in the infrared spectrum of the crystalline phase but only at 13.89 µm in the amorphous phase. The ratio of these two bands in the infrared spectrum of a polymer film at room temperature is a rough measure of crystallinity. As seen by this ratio, the infrared spectra of the copolymers show varying degrees of PE-type crystallinity, dependent on the ethylene concentration and method of incorporation. It is this varying degree of crystallinity that allows the qualitative detection of ethylene homopolymer in these materials. A calibration curve of absorbance at 13.89 µm *versus* ethylene is made from known mixtures for both hot and cold runs. Both plots result in straight lines from which the following equations are calculated:

% Ethylene at 180 °C = A/(0.55b)

% Ethylene at room temperature = A/(3.0b)

where A is the absorbance measured at 13.89 µm and b the thickness of the specimen in centimetres.

A series of ethylene–propylene block copolymers prepared with ^{14}C-labelled ethylene was analysed for percentage of ethylene incorporation by radiochemical methods. These samples when scanned at 180 °C gave results that agreed with the radiochemical assay quite well. However, when the cooled samples were scanned, the results from the cold calibration were low in comparison with the known ethylene content. These data are shown in **Table 3.1**.

Table 3.1 ^{14}C Labelled ethylene-propylene copolymers ethylene, %			
Sample	Radiochemistry	Hot infrared scan	Cold infrared scan
3401	2.4	2.9	0.9
3402	4.0	3.65	1.3
3403A	22.4	20.7	14.7
3403B	24.5	22.2	15.3
3404	12.4	13.0	7.1
3405	14.0	14.1	7.5
Reprinted from J.N. Lomonte and G.A. Tirpak, Journal of Polymer Science, 1964, A2, 705, with permission from Wiley Interscience.			

A pair of samples were prepared in which the active sites on the growing propylene polymer were eliminated by hydrogen before addition of ethylene. Practically identical values for percentage of ethylene incorporation were calculated for both the hot and cold scans. These data are shown in **Table 3.2**.

Table 3.2 Samples with hydrogen-reduced active sites		
Sample No.	Ethylene, %	
	Hot infrared scan	Cold infrared scan
1487	5.0	5.3
1553	7.1	6.9
Reprinted from J.N. Lomonte and G.A. Tirpak, Journal of Polymer Science [1], 1964, A2, 705, with permission from Wiley Interscience.		

Ciampelli and co-workers [2] have developed two methods based on infrared spectroscopy of carbon tetrachloride solutions of polymers at 7.25, 8.65, and 2.32 μm for the analysis of ethylene–propylene copolymers containing greater than 30% propylene. One method can be applied to copolymers soluble in solvents for infrared analysis, the other can be applied to solvent-insoluble polymer films. The absorption band at 7.25 μm due to methyl groups is used in the former case, whereas the ratio of the band at 8.6 μm to the band at 2.32 μm is used in the latter. Infrared spectra of polymers containing 55.5 and 85.5% ethylene are shown in **Figure 3.1**.

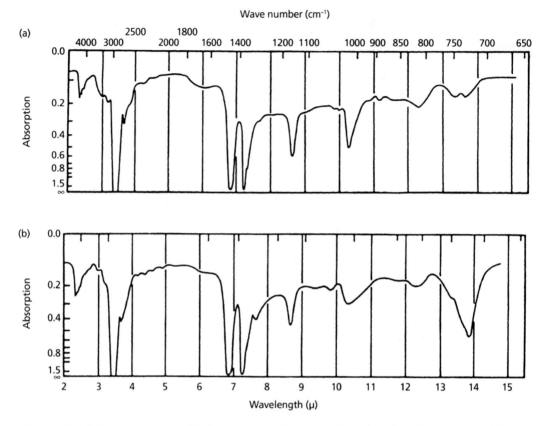

Figure 3.1 Infrared spectra of ethylene-propylene copolymers of various compositions:
(a) 85.5% polyethylene; (b) 55.5% polyethylene
(*Reprinted from F. Ciampelli and co-workers, La Chimica l'Industria, 1962, 44, 489 [2]*)

3.1.2.2 Determination of Bound Propylene in Ethylene–Propylene Copolymers

Tosi and Simonazzi [3] have described an infrared method for the evaluation of the propylene content of ethylene-rich ethylene–propylene copolymers. This is based on the ratio between the absorbance of the 7.25 μm band and the product of the absorbances by the half width of the 6.85 μm band obtained on diecast polymer film at 160 °C (Table 3.3).

The calibration curve, based on a series of standard copolymers prepared with either ^{14}C-labelled ethylene or propylene, is obtained by plotting the 7.25:6.85 absorbance ratio against the C_3 weight fraction. The basis for the calibration of many methods is

Table 3.3 Calibration points for the composition analysis of C_2-C_3 copolymers (spectra recorded at 160 °C)		
Sample	C_3 (wt-%)[a]	$^A7.25/^A6.85$
3137-25	40.5	0.652 ± 0.011
3629-48	32.9	0.561 ± 0.053
3629-44	32.0	0.548 ± 0.032
3137-38	30.7	0.606 ± 0.024[b]
3137-31	20.7	0.416 ± 0.018
3297-55	18.2	0.365 ± 0.037
3297-59	18.0	0.357 ± 0.015
3274-53	14.2	0.318 ± 0.015
Blend No. 1	11.0	0.315 ± 0.022
Blend No. 2	6.0	0.204 ± 0.013
Polyethylene	6.0	0.036 ± 0.007

[a] Radiochemical analysis

[b] The standard deviation has been multiplied by the student's $t_{95\%}$ coefficient to the number of replications (on the average 5) for each calibration point

Reprinted from C. Tosi and T. Simmonazi, Die Angewandte Makromolekulare Chemie, 1973, 32, 153 [3]

the work published by Natta and co-workers [4] which involves measuring the infrared absorption of polymer solutions at 7.25 µm presumably due to methyl vibrations which are related to the propylene concentration in the copolymer. In some cases the dissolution of copolymers with low propylene content or some particular structures is difficult [5, 6]. Moreover, Natta's solution method was calibrated against his radiochemical method [4], for which the precision of the method was not stated, a considerable amount of scatter is evident in the data presented. Typical methods that have used Natta's solution procedure [4] for calibration are described in publications by Wei [7] and Gössl [8]. These infrared methods avoid the problems found with the solution method by employing intensity measurements made on pressed films. The ratio of the absorption at 13.95 µm to that at 8.70 µm is related to the propylene content of the copolymer. Some objections [9] to the use of solid films have been raised because of the effect of crystallinity on the absorption spectra in copolymers with low propylene content. These film methods are reliable only over the range 30–50 mol% propylene.

3.1.2.3 Determination of Unsaturation in Ethylene–Propylene–Diene Terpolymers and Styrene–Butadiene Copolymers

Infrared spectroscopy has been used for the determination of unsaturation in ethylene–propylene–diene terpolymers [10]. Determination of extinction coefficients for the various terpolymers is required for quantitative work.

Fraga [11] has also described an infrared thin-film area method for the analysis of styrene–butadiene copolymers. The integrated absorption area between 6.6 and 7.2 µm has been found to be essentially proportional to total bound butadiene, and is independent of the isomeric type of butadiene structure present. This method can be calibrated for bound styrene contents ranging from 25 to 100%.

3.1.2.4 Determination of Hydroxy Groups in Dinitropropyl Acrylate Prepolymer

The dinitropropyl acrylate polymer has the following structure:

$$HO\!-\!CH_2\text{\Large\sim}(CH_2\!-\!CH_n\text{\Large\sim}CH_2OH)$$

$$
\begin{array}{c}
C=O \\
| \\
O \\
\diagdown \\
CH_2\!-\!\underset{\displaystyle NO_2}{\overset{\displaystyle NO_2}{C}}\!-\!CH_3
\end{array}
$$

The method [12-15] utilises the strong infrared absorption band at 2.90 µm. The hydroxyl concentration of approximately 30 mequiv/l is low enough that the hydroxyl groups are completely associated with the tetrahydrofuran spectroscopic solvents, and there are no apparent free self-associated hydroxy peaks. It is essential in this method that the sample is dry, as water absorbs strongly in the 2.9 µm region of the spectrum. A further limitation of the method is that any other functional group in the sample, such as phenols, amides, amines, and sulfonic acid groups, that absorb in the 2.94–2.86 µm region are likely to interfere in the determination of hydroxyl groups. Kim and co-workers [15] compared hydroxy equivalent weights obtained by this method and chemical methods of analysis for a range of prepolymers. **Table 3.4** shows that good agreement was obtained.

Prepolymers	Vendor's eq. wt.	IR method	Chemical methods	
			PA/PY 614*	AA/NMIM 615 616*
Hydroxy terminated polybutadiene	1300	1280	1350	1440
Hydroxy terminated butadiene acrylonitrile copolymer	1820	1770		
Hydroxy terminated polycaprolactone	980	970	970	1000
Polyethylene glycol	1660	1630a	1670	1740
Hydroxy terminated polytetrahydrofuran	500	480	500	850

Table 3.4 Hydroxy equivalent weights of prepolymers: comparison of IR method with other methods

a Alkoxyethanols used for calibration

* Reaction with excess acid anhydride in presence of base catalyst PA - phthalic anhydride, AA - acetic anhydride. Excess anhydride hydrolysed at end of reaction and carboxylate groups titrated with standard potassium hydroxide

Reprinted from C.S.Y. Kim and co-workers, Analytical Chemistry, 1982, 54, 232, with permission from the American Chemical Society [15]

3.1.2.5 Determination of Bound Ester Groups in Polyacrylates and Polymethacrylates

Anderson and co-workers [16] and Puchalsky [17] have described an infrared procedure for distinguishing between copolymerised acrylic and methacrylic acids in acrylic polymers containing more than 10% of the acid. This method is based on measurement of the wavelength of the carboxylic acid absorption maximum at about 5.9 µm.

3.1.2.6 Determination of Bound Nitrile Groups in Styrene–Acrylonitrile Copolymers

An infrared method has been described [18] for the compositional analysis of styrene–acrylonitrile copolymers. In this method the relative absorbance between a nitrile $\nu(CN)$ mode at 4.4 µm and a phenyl $\nu(CC)$ mode at 6. 2 µm is used.

3.1.2.7 Determination of Free and Combined Vinyl Acetate Groups in Vinyl Chloride–Vinyl Acetate Copolymers

Infrared spectroscopy has been applied to the determination of free and combined vinyl acetate in vinyl chloride–vinyl acetate copolymers [19]. This method is based upon the quantitative measurement of the intensity of absorption bands in the near-infrared spectral region arising from vinyl acetate. A band at 1.63 µm due to vinyl groups enables the free vinyl acetate content of the sample to be determined. A band at 2.15 µm is characteristic for the acetate group and arises from both free and combined vinyl acetate. Thus, the free vinyl acetate content may be determined by difference at 2.15 µm. Polymerised vinyl chloride does not influence either measurement.

3.1.2.8 Determination of Bound Vinyl Acetate in Ethylene– Vinyl Acetate Copolymers

There are two methods for this determination, depending on the concentration of vinyl acetate. At levels below 10% a band at 2.89 µm is used. This band is not suitable for higher concentrations because the necessary film thickness is less than 0.1 mm. For higher concentrations the carbonyl overtone band at 16.39 µm can be used, since much thicker films are needed to give suitable absorbance levels. Here the carbonyl overtone band was used for a series of standards with vinyl acetate concentrations up to 35%, the nominal thickness being about 0.5 mm.

A combination of CDS and QUANT software on a Perkin Elmer model 683 infrared spectrometer was used to establish the calibration for this analysis. Once the calibration has been carried out, the simplest way to measure an unknown sample would be with an OBEY routine in the CDS II software. This would incorporate the calibration data so that the single routine would measure the spectrum and calculate the vinyl acetate concentration with an error of approximately 5%.

3.1.2.9 Determination of Low Concentrations of Methyl Groups in Polyethylene

This analysis is based on the 7.24 µm methyl absorption which appears on the side of the strong methylene bands at 7.29 and 7.58 µm.

A wedge of polymethylene, or high-density polyethylene (HDPE) with very low methyl group content, is placed in the reference beam and moved until the methylene absorptions are cancelled. A computerised equivalent of the wedge method was used, in which the

PECDS program was employed to subtract the methylene absorption bands before applying the QUANT program. Subtraction of the methylene absorption left an uneven baseline and visual inspection of this baseline proved the most satisfactory method of choosing the optimum scaling factor.

A set of six standards was prepared with methyl group concentrations ranging from 0.8 to 28.9 CH_3 groups per thousand carbon atoms. The film thickness was approximately 0.12 mm.

When the QUANT program is used directly the best results are clearly given by the first derivative maximum (D1MAX). The first derivative spectrum shows that the maximum at 7.22 μm is relatively free from overlap by the CH_2 bands, and is much larger than the minimum at 7.26 μm. The relative success of the D1MAX measurement is therefore not surprising. Despite considerable band overlap, the simple peak height measurement gives an error index of only 2.6%. This implies that the background due to the CH_2 bands is fairly constant or varies with CH_2 concentration in a roughly linear fashion.

3.1.2.10 Determination of Polyisobutene in Polyethylene

In this case the analytical band is at 10.52 μm, and for band ratio the 4.95 μm band is again used as the reference. Five standard blends with polyisobutene contents from 0 to 20% were prepared, with a nominal film thickness of 0.40 mm. When the measured film thickness was used the results were similar for height, area, and first derivative, with error indices around 2.5%. The smallest errors were found for area measurements using a 20 cm^{-1} slice around the peak maximum.

3.1.2.11 Determination of Composition and Microstructure in Polybutadiene and Styrene–Butadiene Copolymers

Near-infrared spectroscopy has been used to determine as 1,4, *trans*-1,4, and 1,2 butadiene groups in these polymers [20].

Miller and co-workers [21] used near-infrared spectroscopy to determine the microstructure and composition of polybutadiene and styrene–butadiene copolymers. The procedure was capable of distinguishing between *cis*-1,4, *trans*-1,4, and 1,2 butadiene groups. Geyer [22] has given details of a Bruker Spectrospin P/ID. 28 used for the identification of plastics using mid-infrared spectroscopy.

Pandey and co-workers [23] describe a novel approach to the non-destructive evaluation of various physicomechanical properties of propylene copolymer. A single-step measurement of the infrared spectrum followed by chemometric operation of the predetermined characteristic region of the spectrum enables one to obtain various properties/parameters, e.g.,, melt flow index (MFI), Izod impact, and flexural modulus, required for quality evaluation of the copolymer.

3.2 Fourier Transform Near-Infrared Raman Spectroscopy

3.2.1 Theory

Raman spectroscopy is an emission technique which involves irradiating a sample with a laser and collecting the scattered radiation. Most of the scattered radiation has the same wavelength as the laser. A very small fraction (approximately one millionth) of the scattered radiation is displaced from the laser wavelength by values corresponding to the vibrational frequencies of the sample. This is the Raman signal.

The major features of Raman spectroscopy are:

1. The information on molecular structure is complementary to that obtained from infrared spectroscopy.

2. Functional groups that give weak absorptions in the infrared, such as S–S, C=C, and N=N, give strong Raman signals.

3. It is a non-contact, non-destructive technique.

4. No sample preparation is required.

6. Glass is an ideal optical window material.

6. Dangerous or delicate samples may be examined in sealed containers.

Conventional Raman spectrometers use visible lasers to irradiate the sample with wavelengths between 400 and 800 nm. At these wavelengths 90% of 'real-world' samples fluoresce and no useful Raman data can be collected. When a near-infrared laser (i.e., near-infrared Raman spectroscopy) is used to irradiate a sample, fluorescence is unlikely to occur and at least 80% of samples will give useful Raman spectra. Thermal degradation of coloured and delicate samples is also reduced. The intensity of Raman scattering is dependent on the wavelength of the excitation source, therefore a near-infrared laser produces a much weaker Raman signal than a visible laser. However, Fourier transform technology provides the signal-to-noise advantage necessary to overcome this low signal level. In addition, circular apertures in a Fourier transform spectrometer make sample alignment much less critical than with a dispersive system.

The advantages of Fourier transform near-infrared Raman spectroscopy are:

1. Fluorescence is minimised.

2. Thermal degradation of coloured or delicate samples is reduced.

3. 80% of 'real-world' samples give useful Raman spectra.

4. Multiplex advantage: complete spectra, fast.

5. Real-time display of complete spectra.

6. Signal-to-noise optimisation in real time.

7. Abscissa accuracy: Raman shift calibration uses HeNe laser referencing.

Figure 3.2 shows a Raman spectrum of nitrile rubber demonstrating the sensitivity of the C=C stretching vibration to its environment in the polymer molecule.

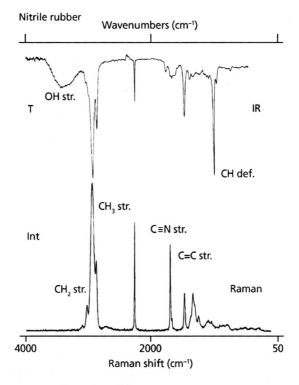

Figure 3.2 Raman spectra of nitrile rubber. In Raman spectroscopy the C=C stretching vibration is particularly sensitive to its environment in a similar way to C=O in the IR. (*Source: Author's own files*)

3.2.2 Instrumentation

Suppliers of this equipment are listed in Appendix 1.

3.2.3 Applications

Aghenyega and co-workers [24] have investigated the application of Fourier transform near-infrared Raman spectroscopy in the synthetic polymer field. Their investigations covered the following areas:

1. Crystallinity in polymers, e.g., atactic and crystalline isotactic polystyrene (**Figure 3.3**), polypropylene (PP), PE, polyurethane sulfide, polytetrafluoroethylene (PTFE).

2. Raman spectra of different polyamides showing that the spectra are sufficiently different to be of use for characterisation purposes and evaluation of crystallinity.

3. Raman spectra of poly(aryl ether ether ketone) and poly(aryl ether ketones).

4. Raman spectra of poly(aryl ether ether sulfone) and poly(aryl ether sulfones).

5. Curing of epoxy resins, study of chemical reactions occurring during cure, and effect of temperature.

6. Determination of monomer ratios in copolymers, e.g., ethylene oxide–vinyl chloride (**Figure 3.4**), tetrafluoroethylene–hexafluoropropylene.

7. Raman spectra of polythioethers.

8. Study of the structure of polyvinyl chloride (PVC) gels.

These workers concluded that this technique has a great future in polymer analysis problems.

Fourier transform near-infrared spectroscopy had been used to determine traces of hydroxy and carboxy functional groups and water in polyesters. Bowden and co-workers [25] monitored the degradation of PVC using Raman microline focus spectrometry. They demonstrated that PVC decomposition is accompanied by the formation of modal polyene chains containing 11–12 or 13–19 double bonds. Bloor [26] has discussed the Fourier transform Raman spectroscopy of polydiacetylenes. Koenig [27] discusses results obtained by the application of infrared and Raman spectroscopy to polymers.

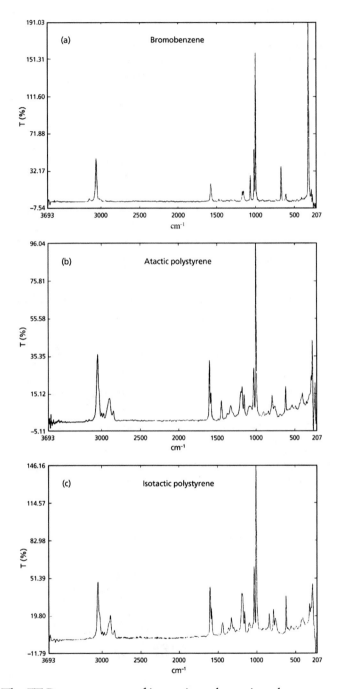

Figure 3.3 The FT Raman spectra of isotactic and atactic polystyrene compared to bromobenzene. (*Reprinted from J.K. Aghenyega and co-workers, Spectrochimica Acta, 1990, 46A, 197, with permission from Elsevier*)

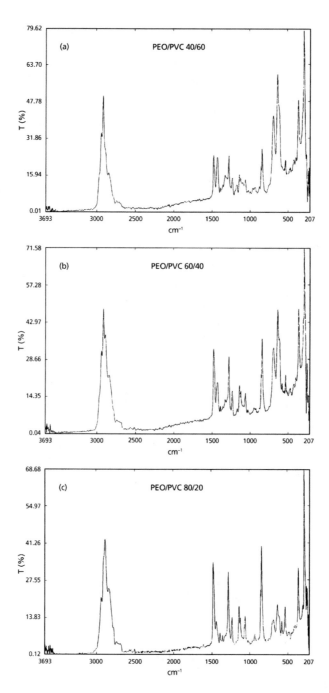

Figure 3.4 The Raman spectra of ethylene oxide/vinyl chloride containing 80, 60 and 40% oxide. (*Reprinted from J.K. Aghenyega and co-workers, Spectrochimica Acta, 1990, 46A, 197, with permission from Elsevier*)

3.3 Fourier Transform Infrared Spectroscopy

3.3.1 Instrumentation

Fourier transform infrared spectroscopy, a versatile and widely used analytical technique, relies on the creation of interference in a beam of light. A source light beam is split into two parts and a continually varying phase difference is introduced into one of the two resultant beams. The two beams are recombined and the interference signal is measured and recorded as an interferogram. A Fourier transform of the interferogram provides the spectrum of the detected light.

A Fourier transform infrared spectrometer consists of an infrared source, an interference modulator (usually a scanning Michelson interferometer), a sample chamber, and an infrared detector. Interference signals measured at the detector are usually amplified and then digitised. A digital computer initially records and then processes the interferogram and also allows the spectral data that result to be manipulated.

The principal reasons for choosing Fourier transform infrared spectroscopy are that (i) the instruments record all wavelengths simultaneously and thus operate with maximum efficiency and (ii) Fourier transform infrared spectrometers have a more conventional optical geometry than do dispersive infrared instruments. These two factors lead to the following advantages:

1. Much higher signal-to-noise ratios can be achieved in comparable scanning times.

2. Wide spectral ranges can be covered with a single scan in a relatively short scan time, thereby permitting the possibility of kinetic time-resolved measurements.

3. Higher resolution is possible without undue sacrifice in energy throughput or signal-to-noise ratios.

4. None of the stray light problems usually associated with dispersive spectrometers are encountered.

5. A more convenient focus geometry – circular rather than slit-shaped – is provided at the sample focus.

Perkin Elmer supplies a range of Fourier transform infrared spectrometers. The equipment is reviewed in Appendix 1.

3.3.2 Applications

Figure 3.5 shows the Fourier transform infrared spectrum of polymethylmethacrylate (PMMA). **Figure 3.6** (top spectrum) shows a spectrum of an acrylonitrile–butadiene copolymer of unknown composition. The lower spectrum is a match obtained from one of a range of standard copolymers of known composition demonstrating that the unknown copolymer contains between 30 and 32% acrylonitrile and between 70 and 68% butadiene.

A further example of the application of Fourier transform infrared spectroscopy is the determination of total silanol (SiOH), silane hydrogen (SiH), and tetrapropoxysilane and diphenylmethylsilanol crosslinking agents in room temperature vulcanised silicone foams [28]. Total SiOH and SiH are determined by Fourier transform infrared spectroscopy. The SiOH peak at 2.71 µm and the SiH peak at 4.61 µm are used for quantitation. The tetrapropoxysilane content is determined by gas chromatography using a solid capillary open tubular (SCOT) column and linear programmed temperature control. The diphenylmethylsilanol content is determined by gel permeation chromatography using a tetrahydrofuran solvent.

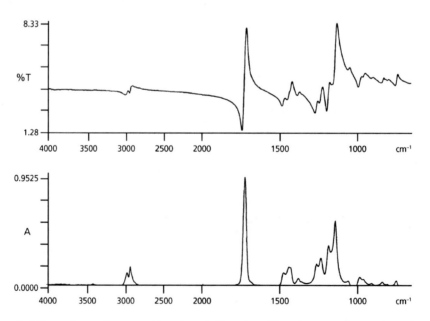

Figure 3.5 Specular reflectance (top) and Kramers-Kronig transformation (bottom) spectra of polymethylmethacrylate. (*Source: Author's own files*)

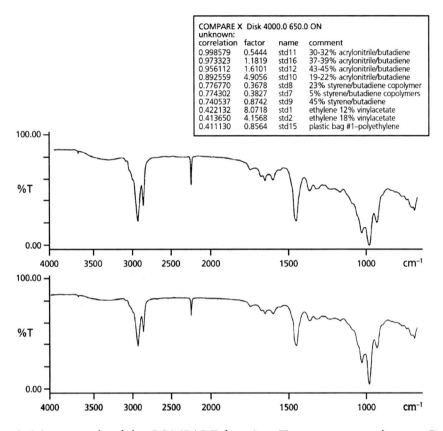

Figure 3.6 An example of the COMPARE function. Top spectrum: unknown. Bottom spectrum: Match from hits list shown above. (Sample is 30-32% acrylonitrile/butadiene copolymer. (*Source: Author's own files*)

A new technique for the identification of components of polymer laminates is the diamond anvil cell technique supplied by Perkin Elmer [29]. A laminate is separated by cutting a small portion of the sample with a razor blade. The layers are then separated by sectioning each piece with the blade. All sample preparation is performed under a stereo microscope. The separated layers are then individually placed in the diamond cell and a spectrum is obtained. This work was performed using a Perkin Elmer model 1650 instrument (a DTGS detector) at 8 cm⁻¹ resolution and 16 scans. **Figure 3.7** shows spectra identifying the components of a three-layer polymer laminate.

Garside and Wyeth [30] have used Fourier transform infrared spectroscopy to characterise cellulose fibres such as jute, sisal, and cotton. The technique has also been used to determine

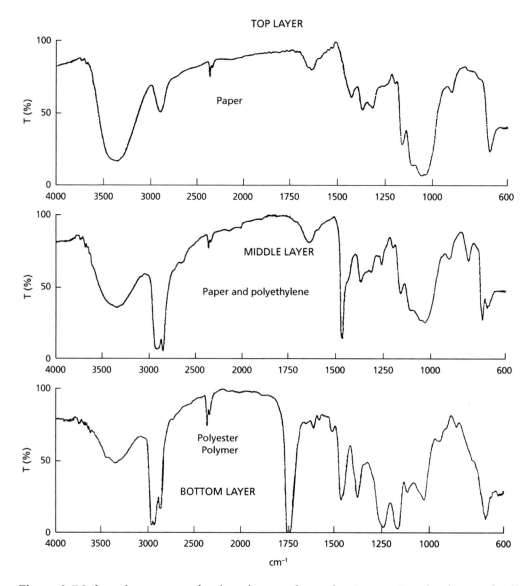

Figure 3.7 Infrared spectrum of a three layer polymer laminate using the diamond cell technique. (*Source: Author's own files*)

low levels of polyvinyl pyrrolidinone in polysulfone [31]. Weiss and co-workers [32] used Fourier transform infrared microspectroscopy in the study of organic and inorganic phases of an injectable hydroxypropylmethylcellulose–calcium phosphate composite for bone and dental surgery.

3.4 Nuclear Magnetic Resonance (NMR) Spectroscopy

3.4.1 Instrumentation

This technique is at its most useful in organic structure identification and has numerous applications in polymer analysis. Instrument suppliers are listed in Appendix 1.

One instrument on the market is the Varian Gemini superconducting NMR spectrometer. This instrument combines excellent computer and radio frequency technology. It uses a cryogenic superconducting magnet system providing high stability and low operating cost, excellent field homogeneity, and stable drift characteristics. The Gemini can be used with a wide variety of probes, including the standard computer switchable $^1H/^{13}C$ or broadband probes. The broadband probe provides complete coverage of the most common NMR nuclei, including fluorine. Zero susceptibility probe design provides improved lineshape. Advanced coil design means highly efficient power transfer to the sample, resulting in complete spin excitation and decoupling of much lower power levels than is typically required. An automatic sample manager unit uses robotics to ensure reliable operation with a high degree of flexibility. The removable 50- or 100-sample tray can be added or removed at any time, permitting loading of one tray whilst another is in use.

3.4.2 Applications

Scientists studying polymers use NMR as a major method of analysis. The repeating structure of polymer chains, for example, provides very simple NMR spectra that clearly identify important structural characteristics. Tacticity, chain branching, composition, and residual monomer content are but a few of the many parameters that can be positively determined from a single NMR spectrum.

Deuterium and proton wideline are important techniques for studying polymer dynamics to provide insights into bulk properties.

3.4.2.1 Determination of Unsaturation

NMR spectroscopy is capable of distinguishing between the different types of NMR unsaturation that can occur in a polymer. NMR spectroscopy has been used to determine unsaturation in acrylonitrile–butadiene–styrene terpolymers (ABS) [33, 34], 1,2-polybutadiene [34, 35], ethylene–propylene terpolymers [35], and vinyl chloride–vinylidene chloride copolymers [10, 36, 37].

Regarding ABS terpolymers [29], NMR is capable of determining ungrafted butadiene rubber in solvent extracts of these polymers. No aromatic protons of styrene or acrylonitrile protons are seen in the NMR spectra. The vinyl content of the polybutadiene is about 20%.

Sewell and Skidmore [38] used time averaged NMR spectroscopy at 60 MHz to identify low concentrations of non-conjugated dienes such as cyclopentadiene, 1,4-hexadiene, or ethylidene norbornene introduced into ethylene–propylene copolymers to permit vulcanisation. Although infrared spectroscopy [39] and iodine monochloride unsaturation methods [40] have been used to determine or detect such dienes, these two methods can present difficulties. The spectra obtained by time averaged NMR are usually sufficiently characteristic to allow identification of the particular third monomer incorporated in the terpolymer. Moreover, as the third monomer initially contains two double bonds, differing in structure and reactivity, the one used up in copolymerisation may be distinguished from the one remaining for subsequent use in vulcanisation. Therefore, information concerning the structure of the remaining unsaturated entity may be obtained. **Table 3.5** shows the chemical shifts of olefinic protons of a number of different third monomers in the copolymers.

Table 3.5 Determination of non-conjugated dienes in copolymers	
Third monomer	**Chemical shift ppm**
1,5-Cyclooctadiene	4.55
Dicyclopentadiene	4.55
1,4 Hexadiene	4.7
Methylene norbornene	5.25 and 5.5
Ethylidene norbornene	4.8 and 4.9
Reprinted from P.R. Sewell and D.W. Skidmore, Journal of Polymer Science, A1, 1968, 6, 8, 2425 [38], with permission from Wiley Interscience	

Altenau and co-workers [41] applied time averaging NMR to the determination of low percentages of termonomer such as 1,4 hexadiene, dicyclopentadiene, and ethylidene norbornene in ethylene–propylene termonomers. They compared results obtained by NMR and the iodine monochloride procedure of Lee and co-workers [42]. The chemical shifts and splitting pattern of the olefinic response were used to identify the termonomer. **Table 3.6** compares the amount of termonomer found by the NMR method of Altenau and co-workers [41] and by iodine monochloride procedures. The termonomers were identified by NMR and IR spectroscopy.

NMR	Lee, Koltoff and Johnson [42]	Kemp and Peters [43]	Termonomer
7.3	3.0		dicyclopentadiene
1.1	1.6		1,4-hexadiene
5.7	5.9	9.0	ethylidene norbornene
2.8	3.6	4.5	ethylidene norbornene
1.7	2.3	4.8	ethylidene norbornene
4.6	5.4	6.0	ethylidene norbornene

Table 3.6 Determination of termonomer in ethylene-propylene diene terpolymers - comparison of methods (data shown as weight % terpolymer)

Reprinted from H.G. Altenau and co-workers, Analytical Chemistry, 1970, 42, 1280, with permission of the American Chemical Society [41]

Table **3.6** shows that the data obtained by the NMR method agree more closely with the Lee [42] iodine monochloride method than with the iodine monochloride method of Kemp and Peters [43]. The difference between the latter two methods is best explained on the basis of side reactions occurring between the iodine monochloride and the polymer because of branching near the double bond [42]. The reason for the difference between the NMR method and the method of Lee and co-workers [42] is not clear. The reproducibility of the NMR method was ±10–15%.

3.4.2.2 Determination of Hydroxy Groups

Most of the chemical methods for determining hydroxy groups in polymers do not distinguish between primary, secondary, or tertiary hydroxy groups.

A method for distinguishing between different types of hydroxy groups in polymers is the NMR method described by Ho [14]. This procedure is based on the reaction of the hydroxy compound with hexafluoracetone to form an adduct which is amenable to ^{19}F NMR spectroscopy:

$$(CF_3)_2CO + ROH \rightleftharpoons CH_3-\underset{\underset{OR}{|}}{\overset{\overset{OH}{|}}{C}}-CF_3$$

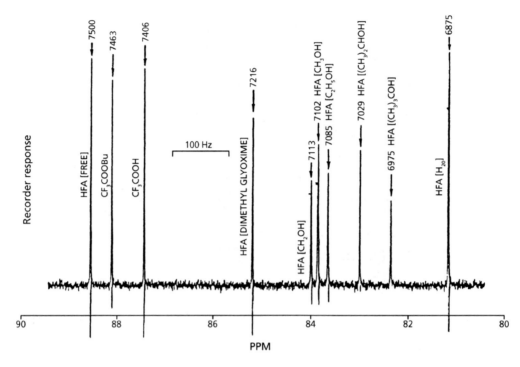

Figure 3.8 [19]Fluorine NMR spectra of several HFA addicts. The numbers on top are "3 downfield from C_6F_6 lock. (*Source: Author's own files*)

A fluorine resonance spectrum of a mixture of several hexafluoroacetone adducts is shown in **Figure 3.8**. This spectrum illustrates clearly the high resolution and the information on structural aspects of the molecules one can obtain by this method. The hydroxyl adduct from each alcohol gives a sharp resonance, with the tertiary adducts at high field, followed by secondary and then primary. The chemical shift of the hexafluoroacetone alcohol adduct is determined by the structural environment of the hydroxyl. For example, the gradual upfield shift observed in the series methanol, ethanol, isopropanol, and *tert*-butanol results from the increase in shielding caused by replacing a hydrogen atom with a methyl group. The high degree of resolution of **Figure 3.8** was also observed for polymeric materials. Therefore, not only can the total hydroxyl concentration be determined but also the type or types of hydroxyl present in the polymer.

The quantitative aspect of hydroxyl determination is illustrated in **Table 3.7** for the analysis of hydroxyl in some polymeric materials. In most cases, the fluorine resonance from *n*-butyl trifluoroacetate (at 7463 Hz in **Figure 3.8**) was used as internal standard to calibrate the spectral integral. From the data shown in **Table 3.7**, it seems that the adduct

Table 3.7 Quantitative determination of hydroxyls in several commercial polymers		
Polymers	**Hydroxyls**	
	Found	Expected
Aliphatic polyethers		
Polyethylene glycol		
Carbowax 200	17.40	16.2-17.9[a]
Carbowax 1000	3.54	3.24-3.58[b]
Polypropylene glycol		
Voranol P1010	3.46	3.37[d]
Voranol P4000	0.85	0.85[e]
Polyepichlorohydrin[f]	0.84	0.85
Aromatic polyethers		
Bakelite phenoxy resin PKHC	5.8	6.0[g]
Monsonto R, J-100 resin	5.4	5.4-6.0
Polyesters		
Atlac 382E	0.92	0.93
Aromatic ester resin No. 1[h]	8.20	8.30
Aromatic ester resin No. 2[h]	12.20	12.35

[a] *Based on literature molecular weight of 190 to 210*

[b] *Based on literature molecular weight of 950 to 1050*

[c] *Determined after addition of 4.7 molar equivalent of trifluroacetic acid and 7.6 molar equivalent of water*

[d] *Based on an average molecular weight of 1010*

[e] *Based on an average molecular weight of 4000*

[f] *Polyepichlorihydrin 2000 from Shell Chemical Co.*

[g] *Calculated from the idealised molecular structure with a unit weight of 284*

[h] *Experimental resin from Hercules Incorporated*

Reprinted from F.F. Ho, Analytical Chemistry, 1973, 45, 603, with permission from the American Chemical Society [14]

formation is quantitative for primary and secondary hydroxyls. Tertiary hydroxyl was found to react only partially, and the reaction for *tert*-butanol required about 24 hours to reach equilibrium. The method can be used for the determination of hydroxyl groups in polymers of unknown structure.

3.4.2.3 Determination of Ester Groups

NMR has been used to determine ethyl acrylate in ethyl acrylate–ethylene and vinyl acetate–ethylene copolymers [44]. Measurements were made on 10% solutions in diphenyl ester at an elevated temperature. Resolution improved with increasing temperature and lower polymer concentration in the solvent. NMR spectra for an ethylene–ethyl acrylate copolymer indicate clearly both copolymer identification and monomer ratio. A distinct ethyl group pattern (quartet, triplet), with methylene quartet shifted downfield by the adjacent oxygen, is observed. The oxygen effect carries over to the methyl triplet which merges with the aliphatic methylene peak. No other end-group would give this characteristic pattern. The area of the quartet is a direct and quantitative measure of the ester content. All features are consistent with an identification of an ethylene-rich copolymer with ethyl acrylate. Ethyl acrylate contents obtained by NMR (6.0%) agreed well with those obtained by PMR (6.2%) and neutron activation analysis for oxygen (6.1%).

NMR spectroscopy has been applied to the determination of ester groups in polyethylene terephthalate (PET) [45]. A blend of a protic solvent (dimethyl sulfoxide), sodium hydroxide, and methanol hydrolyses ester groups in this polymer much more rapidly than do hydrolysis reagents previously used.

3.4.2.4 Repeat Units in Hexafluoropropylene–Vinylidene Fluoride Copolymers

When it is desired to measure the composition of a single component in a mixture, it is necessary to relate the component resonance to that of another compound, an internal standard, which is of known chemical composition and has been added in known weight to a known weight of unknown. Brame and Yeager [46] used dichlorobenzotrifluoride as an internal standard in the continuous wave method for determining the composition of both repeat units in hexafluoropropylene–vinylidene fluoride copolymers. This work demonstrated the utility of the Fourier transform NMR method for the quantitative analysis of the copolymer in relation to results obtained by continuous wave ^{19}F NMR and ^{1}H NMR.

In the Fourier transform ^{19}F NMR spectrum of polyhexafluorpropylene–vinylidene fluoride copolymer, the lines observed are attributed to the following: CF_3 group ($\delta = -70$ to -75), CF_2 group ($\delta = -90$ to -120), and CF group ($\delta = -180$ to -185). The results of the determination for a copolymer sample are given in **Table 3.8**. The values obtained are in excellent agreement with those obtained by mass balance.

Table 3.8 HFP/VF$_2$ copolymer compositions			
Sample	Mass balance, wt.% VF$_2$	FT-NMR	
		wt.% VF$_2$	wt.% HFP
A	48 ± 2	46.6 ± 1.1[a]	53.4 ± 1.2[a]

[a] *Precision at 95% confidence level*

Adapted and reprinted from E.G. Brame and F.W. Yeager, Analytical Chemistry, 1976, 48, 709, with permission from the American Chemical Society [46]

3.4.2.5 Characterisation of Perfluoro-tetradecahydrophenanthroline Oligomers

Tuminello and co-workers [47] characterised complex mixtures of these oligomers using NMR spectroscopy, electron spin resonance spectroscopy, and ToF-SIMS. Infrared and ultraviolet visible spectroscopy were also used. The following distributions were established: olefin: 8%; CF$_3$: 8%; CF$_2$: 51%; and CF: 33%.

3.4.2.6 Sequence Distribution in Ethylene–Propylene Copolymers

A structure of $C_{14}F_{23}$ $(C_{14}F_{22})_n$ $C_{14}H_{23}$ is proposed, where n = 0, 1, or 2 for the dimer, trimer, or tetramer, respectively. Ethylene–propylene copolymers can contain up to four types of sequence distribution of monomeric units. These are propylene to propylene (head-to-tail and head-to-head), ethylene to propylene, and ethylene to ethylene. These four types of sequence and the average sequence lengths of both monomer units, i.e., the value of n in the structures opposite, can be measured by the Tanaka and Hatada [48] method.

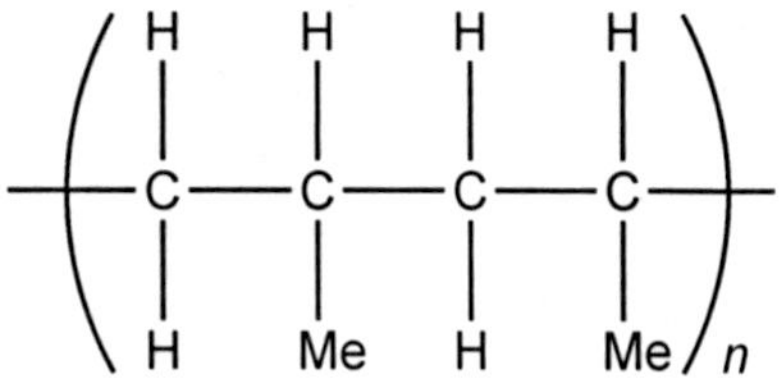

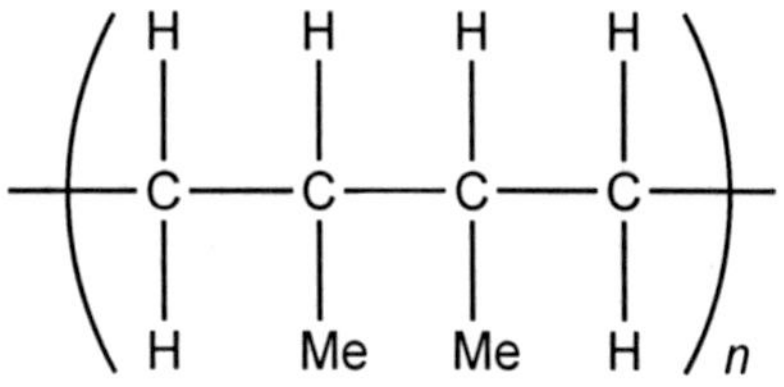

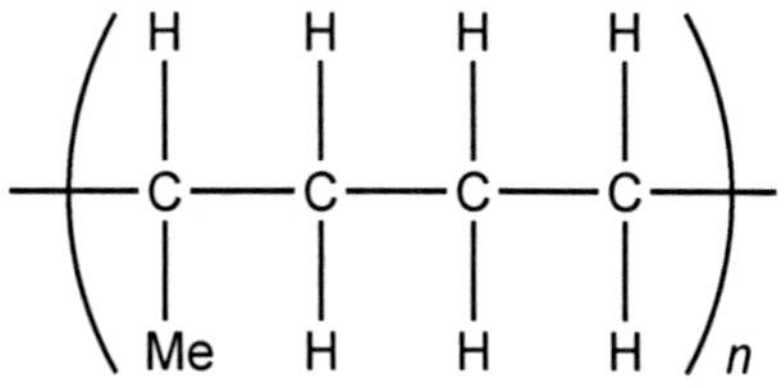

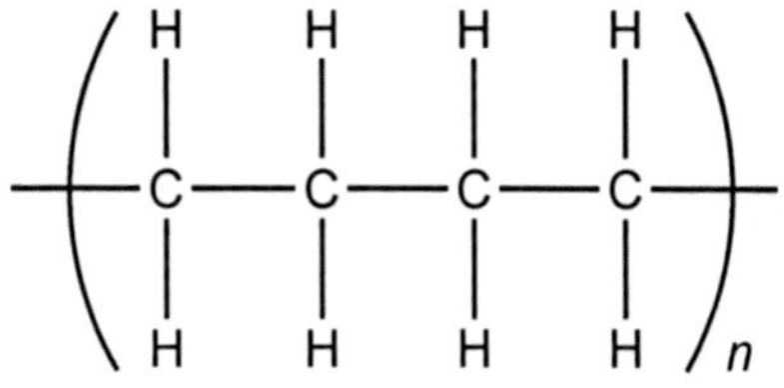

95

Measurements were made at 15.1 MHz. Assignments of the signals were carried out using the method of Grant and Paul [49] and also by comparing the spectra with those of squalane, hydrogenated natural rubber (isoprene), PE, and atactic polypropylene (aPP). The accuracy and precision of intensity measurements were acceptable at a maximum of 12%.

The investigation of the chain structure of copolymers by ^{13}C NMR has several advantages because of the large chemical shift involved. Crain and co-workers [50] and Cannon and Wilkes [51] have demonstrated this for the determination of the sequence distribution in ethylene–propylene copolymers. In these studies the assignments of the signals were deduced using model compounds and the empirical equation derived by Grant and Paul for low molecular weight linear and branched alkanes.

Randall [52] has developed a quantitative ^{13}C NMR method for measuring ethylene–propylene mole fractions and methylene number average sequence lengths in ethylene–propylene copolymers. He gives methylene sequence distributions from one to six and larger consecutive methylene carbons for a range of ethylene–propylene copolymers and uses this to distinguish copolymers that have either random, blocked, or alternating comonomer sequences.

3.4.2.7 Comonomer Distribution in Ethylene–α-Olefin Copolymers

Hadfield and co-workers [53] used ^{13}C magic angle spinning (MAS) NMR spectroscopy to determine comonomer types and composition in ethylene–α-olefin copolymers. They used melt state ^{13}C NMR with magic angle spinning and dipolar coupling to perform this analysis. The advantages of the technique include reduced analysis time and the ability to analyse samples not amenable to solution-state NMR spectroscopy. Examination of a range of copolymers including insoluble crosslinked copolymers showed good agreement with the results obtained by solution NMR spectroscopy. A recent example of the application of NMR spectroscopy is a study of the structural distribution in trimethoylpropane/o-phthalic anhydride condensates [54].

3.5 Proton Magnetic Resonance (PMR) Spectroscopy

3.5.1 Instrumentation

The instrumentation is detailed in Appendix 1.

3.5.2 Applications

Randall [55, 56] carried out a detailed study of the PP methyl group in triad and pentad configuration environments. **Figure 3.9** shows the PMR spectrum of a hot *n*-heptane solution of aPP containing up to 40% syndiotactic placement, and which by Natta's definition may be stereoblock. The spectrum is inherently complex, as a first-order theoretical calculation, and this indicated the possibility of at least 15 peaks with considerable overlap between peaks, because differences in chemical shifts are about the same magnitude as the splitting due to spin–spin coupling.

The largest peak, at high field in **Figure 3.9(a)**, represents pendant methyl moieties in propylene units. It is characteristically split by the tertiary hydrogen. By peak area integration, about 20% of the nominal methyl proton peak is due to overlap of absorption from chain methylenes. This overlap is consistent with a reported syndiotactic triplet, two peaks of which are close to A and B in **Figure 3.9(a)** and a third peak that falls with the low-field branch of the methyl split. The absence of a strong singlet peak in the methylene range indicates the virtual absences of 'amorphous' polymer in the aPP shown in **Figure 3.9(a)**, which could possibly be due to the head-to-head and tail-to-tail units. The low-field peak represents the partial resolution of tertiary protons that are opposite the methyls on the hydrocarbon chain.

Figure 3.9(a) also shows a spectrum for an isotactic polypropylene (iPP) (>95% isotactic by solubility, <5% soluble in boiling heptane). This spectrum has the same general character as the aPP. The important difference is a marked decrease of peak intensity in the chain methylene region. This decrease is caused by extensive splitting, and the difference in chemical shifts for the non-equivalent methylene hydrogens in isotactic environments. This is in accord with the study of Stehling [57] on deuterated PP, which indicated that much of isotactic methylene absorption is buried beneath the methyl and tertiary hydrogen peaks. The fractional area in the nominal methylene region of the spectrum is thus sensitive to the number of isotactic and syndiotactic diads, and therefore may be used as a measure of PP and linear PE. The low-field absorption in **Figure 3.9(b)**, characteristic of aromatic hydrogens, is due to the polymer solvent, diphenyl ether, which was used throughout. Polymer concentrations in solution can be readily calculated from the ratios of peak areas adjusted to the same sensitivity. The superpositions of spectra of the homopolymers, that were obtained separately, show the same pattern as spectra of physical mixtures but with different intensities. The PE absorption falls on the peak marked A of the chain methylene complex in PP. Peak B, also due to chain methylenes in PP, is resolved in both spectra in **Figure 3.9(b)**, which also gives the spectrum of an ethylene–propylene block copolymer. The ethylene contribution again falls on peak A.

Various workers have developed analyses for physical mixtures and block copolymers based on the ratio of the incremental methylene area to the total polymer proton

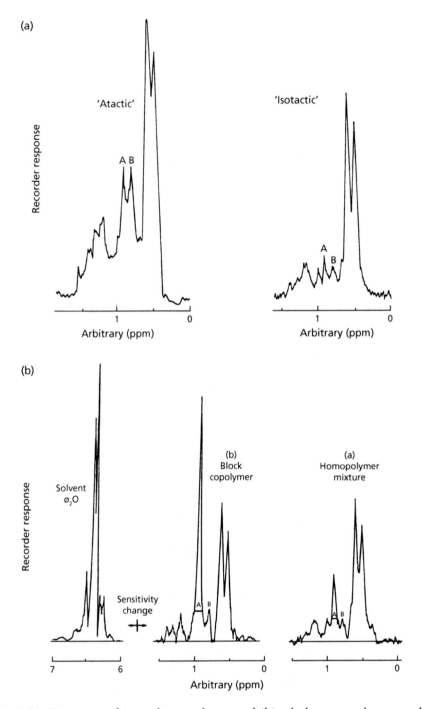

Figure 3.9 PMR spectra of (a) polypropylenes and (b) ethylenepropylene copolymers. (*Source: Author's own files*)

absorption. This concept has been tested by Barrall and co-workers [58, 59] using PMR analyses of a series of physical mixtures and block copolymers synthesised with ^{14}C-labelled propylene and others with ^{14}C-labelled ethylene. A most important feature of this analysis is that the methylene peaks A and B have virtually the same relative heights in PP with a variety of tacticities (note **Figure 3.9(a)**). This is also true for PMR spectra given by Satoh and others for atactic series of PP [60]. This suggests that PMR analyses for ethylene are independent of tacticity, since the area increment of peak A above peak B has been used for analysis.

Qualitative PP tacticities can be estimated by PMR not only for homopolymers (**Figure 3.9(a)**) but also in the presence of PE and ethylene copolymer blocks. The relative heights of the peaks for secondary and for tertiary hydrogen in **Figure 3.9(b)** indicate that the PP in the copolymer and the physical mixture is dominantly isotactic.

3.6 Reaction Gas Chromatography

Polymers being with few exceptions, solid substances, cannot be directly analysed using gas chromatography. However, it is possible by the application of well-controlled chemical reactions to decompose polymers to simpler volatile substances that are amenable to gas chromatography and thereby one can obtain information concerning the original polymer.

A further special case of reaction gas chromatography involves pyrolysis (or photolysis) of the polymer in the absence of oxygen (Section 3.7) and examination of the volatiles produced by gas chromatography to provide information on the structure of the original polymer.

Further complementary techniques can be applied to obtain even more information. Thus PGC can be coupled with mass spectrometry or NMR spectroscopy.

3.6.1 Instrumentation

Information on the availability of gas chromatographs is supplied in Appendix 1.

3.6.2 Applications

3.6.2.1 Saponification Methods

Ester groups occur in a wide range of polymers, e.g., PET. The classic chemical method for the determination of ester groups, namely saponification, can be applied to some types of polymer. For example, copolymers of vinyl esters and esters of vinyl esters and esters of acrylic acid can be saponified in a sealed tube with 2 M sodium hydroxide. The free acids from the vinyl esters can then be determined by gas chromatography. The alcohols formed by the hydrolysis of the acrylate esters are determined by gas chromatography. Polymethyl acrylate can be hydrolysed rapidly and completely under alkaline conditions; however, the monomer units in PMMA prepared and treated similarly are resistant to hydrolysis although their benzoate end-groups react readily. Thus, saponification techniques should be applied with caution to polymeric materials.

3.6.2.2 Zeisel Procedures

Hydrolysis using hydriodic acid has been used for the determination of the methyl ethyl, propyl, and butyl esters of acrylates, methacrylates, or maleates [61] and the determination of polyethyl esters in methyl methacrylate (MMA) copolymers [62, 63]. First, the total alcohol content is determined using a modified Zeisel hydriodic acid hydrolysis [64]. Second, the various alcohols, after being converted to the corresponding alkyl iodides, are collected in a cold trap and then separated by gas chromatography. Owing to the low volatility of the higher alkyl iodides the hydriodic acid hydrolysis technique is not suitable for the determination of alcohol groups higher than butyl alcohol. This technique has also been applied to the determination of alkoxy groups in acrylate esters [61].

Anderson and co-workers [16] used combined Zeisel reaction and gas chromatography to analyse acrylic copolymers. Acrylic esters were cleaved with hydriodic acid and gas chromatography was used for analysing the alkyl iodides so formed:

$R = H, CH_3, C_2H_5, C_4H_9, R^1 = H, CH_3$

Using this procedure, the recovery of alkyl iodides is greater than 95% for polymers containing between 10 and 90% of the methyl, ethyl, and butyl esters of acrylic and methacrylic acid. In addition, the use of isopropylbenzene as the trapping solvent allows the determination of all C_1 to C_4 alkyl iodides.

Polymer	Methyl acrylate %	Methyl methacrylate %	Ethyl acrylate %	Ethyl methacrylate %	Butyl acrylate %	Butyl methacrylate %
Table 3.9 Recovery of alkyl iodides from the Zeisel cleavage of acrylic polymers[a]						
1	33.7 (33.3)		33.0 (33.3)		32.7 (33.3)	
2		33.6 (33.3)		33.4 (33.3)		31.9 (33.3)
3			19.5 (20.0)			
4		50.1 (50.0)	30.5 (30.0)			
5		59.9 (60.0)			39.8 (40.0)	
6		29.7 (30.0)			29.8 (30.0)	
7		59.8 (60.0)				9.8 (10.0)
8	19.9 (20.0)		19.8 (20.0)			19.8 (20.0)
9		30.1 (30.0)		30.0 (30.0)	29.8 (30.0)	
10		29.9 (30.0)	29.6 (30.0)		39.7 (40.0)	
11		60.1 (60.0)	30.1 (30.0)			9.5 (10.0)

[a] *All values are the average of at least three determinations and are reported as:*
% monomer found (% monomer in polymer)
Reprinted from D.G. Anderson and co-workers, Analytical Chemistry, 1971, 43, 894, with permission from the American Chemical Society [16]

Table 3.9 shows some results obtained by applying this method to a range of acrylic polymers. The calculated recoveries are greater than 95% for polymers containing between 10 and 100% acrylic monomer. The method has 99% confidence interval of 0.8. The presence of comonomers such as styrene, acrylonitrile, vinyl acetate, acrylamide, or acrylic acid does not change the recovery of acrylate or methacrylate esters. Non-quantitative results are obtained, however, for polymers containing hydroxypropyl methacrylate.

The Ziesel reaction has been used for the determination of alkoxyl groups in cellulosic materials [65-67] and the determination of ether groups in cellulose and polyvinyl ethers [68]. However, hydriodic acid also cleaves any ester linkages on the polymer backbone, giving positive interference.

3.6.2.3 Determination of Alkyl Groups

The technique of potassium hydroxide fusion–reaction gas chromatography has been applied to the determination of alkyl and aryl groups in polysiloxanes. The method involves the quantitative cleavage of all organic substituents bonded to silicon, producing the corresponding hydrocarbons:

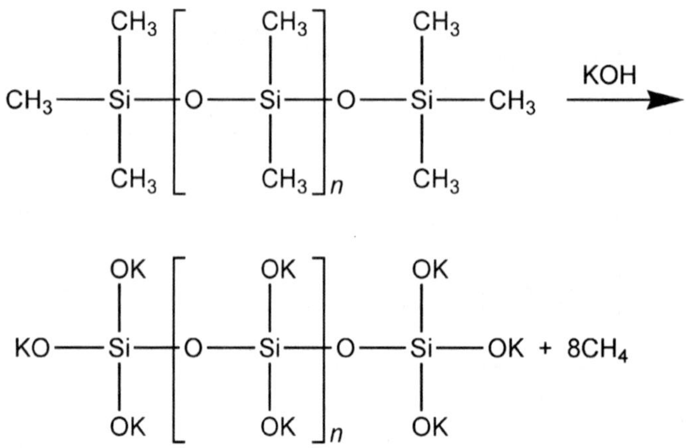

After concentration of the volatile products, they are separated and determined by gas chromatography. The percent relative standard deviation of the method is 1.00%; the average deviation between experimental and theoretical results is 0.5% absolute.

Various other reagents have been used for the determination of organic substituents bonded to silicon in organosilicon polymers (**Table 3.10**).

Table 3.10 Reaction methods for the quantitative determintion of organic substituents bonded to silicon					
Group determined	Reagent	Reaction conditions	Product	Analysis	Ref.
Phenyl	60% aqueous KOH in DMSO	2 h at 120 °C	Benzene	GC	[146]
Phenyl	Bromine in glacial acetic acid	Boiling solution	Bromobenzene	Titration of excess bromine	[147]
Ethyl and phenyl	Phosphorus pentoxide and water	30-580 °C over 45 min	Ethane and benzene	GC-FID	[148]
Methyl and ethyl	Powdered potassium hydroxide	2 h at 250-270 °C	Methane and ethane	Gas burette	[149]
Methyl	Sulfuric acid	20 min at 280-300 °C	Methane	Gas burette	[150, 151]
Phenyl	Ethylbromide in the presence of aluminium chloride	–	Hexaethyl benzene	Gravimetric	[152]
Vinyl	Phosphorus pentoxide and water	80-600 °C over 40 min	Ethylene	GC-FID	[153]
Vinyl	Phosphorus pentoxide	Ambient to 500 °C	Ethylene	GC-FID	[154]
Vinyl	90% sulfuric acid	75-250 °C at 10 °C/min and 1 h at 250 °C	Ethylene	GC-TC	[155]
Vinyl	Sodium hydroxide pellets	300 °C for 15 min	Ethylene	Colorimetric	[156]
Vinyl	Potassium hydroxide pellets	Heat with Meker burner	Ethylene	GC-FID	[157]
Reprinted from D.D. Schlueter and S. Siggia, Analytical Chemistry, 1977, 49, 2343, with permission from the American Chemical Society [69]					

3.6.2.4 Determination of Amide and Imide Groups in Polyamides, Polyimides, and Polyamides/Imides

Schleuter and Siggia [69-71] and Frankoski and Siggia [72] used the technique of alkali fusion reaction gas chromatography for the analysis of imide monomers and aromatic polyimides, polyamides, and poly(amide-imides). Samples are hydrolysed with a molten potassium hydroxide reagent at elevated temperatures in a flowing inert atmosphere:

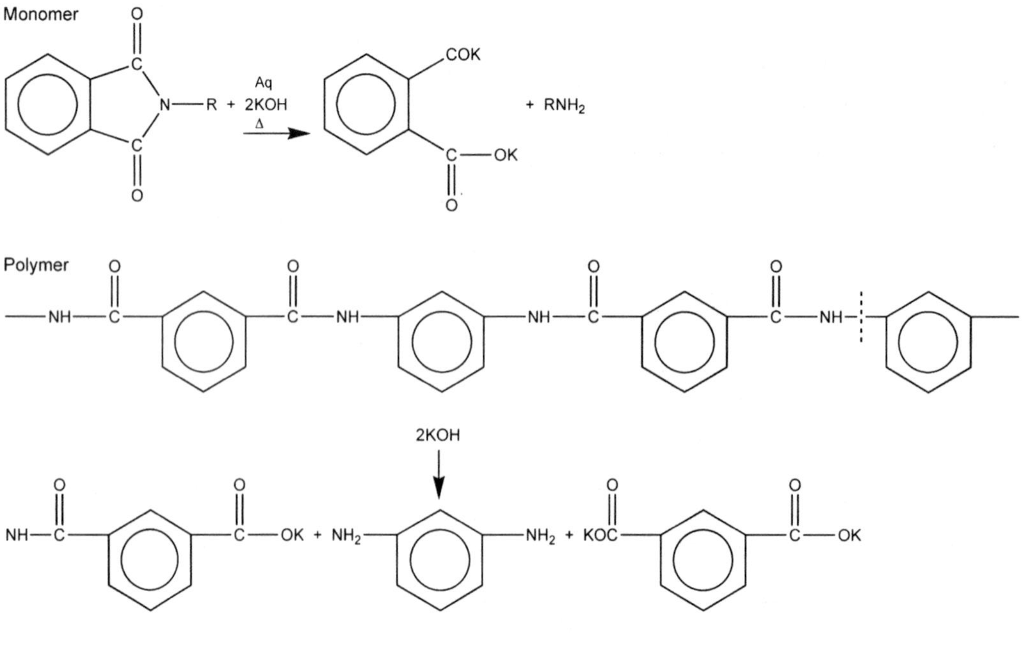

Volatile reaction products are concentrated in a cold trap before separation by gas chromatography. The identification of the amine and/or diamine products aids in the characterisation of the monomer or polymer; the amount of each compound generated is used as the basis for quantitative analysis. The average relative standard deviation of the method is ±1.0%.

Table 3.11 summarises the chemical structures and sample designations of the polymers studied. **Table 3.12** summarises the diamine recoveries obtained. These data represent the mole percentage of theoretical diamine based on the dry weight of sample and the idealised linear polymer repeat units depicted in **Table 3.11**. Recoveries were, in all cases, between 90 and 101% of theoretical values.

	Structure of repeat unit	Wt.% water	Decomposition temperature, °C
PI-1		6.5	385
PI-2		2.2	410
PI-3		3.4	410
PI-4	4,4'-Methylenedianiline	0.6	310
PA-1		5.8	315

Table 3.11 Structure, water content and decomposition temperature of the polymers studied

Table 3.11 *Continued*			
	Structure of repeat unit	Wt.% water	Decomposition temperature, °C
PA-2		9.7	330
PAI-1		6.2	340
PAI-2		12.6	395
PAI-3		8.9	340

Reprinted from D.D. Schlueter and S. Siggia, Analytical Chemistry, 1977, 49, 2349, with permission from the American Chemical Society [70]

Sample	Diamine produced	Mol% of theoretical[a] ± RSDb			
		5 min at 380 °C		30 min from 100 to 390 °C	
PI-1	4,4′-Methylenedianiline	98.3 ± 0.8		97.5 ± 0.9	
PI-2	*m*-Phenylenediamine	89.1 ± 0.7 91.2 ± 0.9 91.6 ± 1.3		91.0 ± 0.8 91.0 ± 0.7	
PI-3	2,4-Toluenediamine	73.1 ± 0.2		73.2 ± 1.0	
			97.8 ± 0.5		99.1 ± 1.2
	4,4′-Methylenedianiline	24.8 ± 0.3		25.9 ± 0.2	
PA-1	*m*-Phenylenediamine	100.7 ± 0.6		97.0 ± 1.7 97.7 ± 1.7 97.9 ± 1.0 99.6 ± 2.6	
PA-2	*m*-Phenylenediamine	95.9 ± 0.6 96.4 ± 1.5		93.1 ± 0.7 93.3 ± 0.5	
PAI-1	*m*-Phenylenediamine	93.5 ± 0.9		93.1 ± 0.7 95. 3 ± 0.5	
PAI-2	*m*-Phenylenediamine	93.5 ± 1.1 94.1 ± 0.6 95.1 ± 1.2		97.4 ± 0.9	
PAI-3	4,4′-Methylenedianiline	98.0 ± 1.1		98.8 ± 1.1	

Table 3.12 Analysis of polyimides, polyamides and poly(amide-imides) by alkali fusion reaction gas chromatography

[a] These recovery values are based on the structures shown in Table 3.11, which assume that one mole of diamine is produced for each mole of repeat unit

[b] The relative standard deviation is based on five or more determinations

Reprinted from D.D. Schlueter and S. Siggia, *Analytical Chemistry*, 1977, 49, 2349, with permission from the American Chemical Society [70]

3.7 Pyrolysis Gas Chromatography

3.7.1 Theory

Pyrolysis is simply the breaking of large, complex molecules into smaller fragments by the application of heat. When the heat energy applied to a molecule is greater than the energy of specific bonds in that molecule, those bonds will dissociate in a predictable, reproducible way. The smaller molecules generated in this bond-breaking process can be identified by a number of analytical tools, including gas chromatography and mass spectrometry. Once identified, they help in understanding the structure of the original macromolecule. Thus, a copolymer of isoprene and styrene is readily identified by a pyrogram, whose dominant peaks are isoprene, styrene, and dipentene, a dimer of isoprene.

A good pyrolysis instrument must be able to heat a sample reproducibly, to a preset temperature at a known rate for a specific amount of time. Inability to control any of these variables will result in a pyrogram that cannot be reproduced. If required, the separated pyrolysis products can each be fed into a mass spectrometer to obtain detailed information on pyrolysis product identity (PGC–MS (Section 3.8)) or into an NMR spectrometer (PGC–NMR spectroscopy) or into a FT-IR spectrometer (Section 3.9).

3.7.2 Instrumentation

Gas chromatography instrumentation is discussed in Chapter 5; see also Appendix 1. As examples, the following pyrolyser units, produced by CDS Instruments, are available.

Pyroprobe 1000. This provides complete choice of thermal processing parameters. Pulse pyrolysis is permitted at rates up to 20 °C per second to temperatures as high as 1400 °C. In addition to accuracy and reproducibility, the temperature versatility allows for uninterrupted sequential runs on the same sample under different thermal conditions without removing the sample probe. Two probe designs are available: coil element for solid polymers and ribbon element for polymer film and solvent deposits on polymers.

Pyroprobe 2000. This instrument has independently programmable interface and probe temperatures, and is capable of performing all the functions of the Pyroprobe 1000 and more. It has the ability to program at rates in degrees centigrade per millisecond, second, or minute, as low as 0.01 °C per minute. It can use methods employing up to five independent steps. This flexibility allows complete temperature control of the sample including absorption, volatilisations, and pyrolysis, for pulse or time-resolved experiments. Slow heating rates are ideal for thermal evolution studies and for programmed pyrolysis with continuous analysis by MS or FT-IR spectroscopy.

3.7.3 Applications

During pyrolysis, polymeric materials may degrade via a number of mechanisms that are generally grouped into three classes: random scission, depolymerisation, and side group elimination.

Random scission results from the production of free radicals along the backbone of the polymer which causes the macromolecule to be fragmented into smaller molecules of varying chain lengths. On chromatographic analysis these fragments reveal a repeating series of oligomers frequently differing in chain length by the number of carbons in the original monomer:

Polyolefins generally degrade through a random scission mechanism, and PE is a good example of this behaviour. When a free radical is formed along the chain of PE, chain scission occurs, producing a molecule with an unsaturated end and another with a terminal free radical. This free radical may abstract a hydrogen from a neighbouring carbon, producing a saturated end and a new radical, or combine with another free radical to form an alkane. Multiple cleavages produce molecules small enough to be volatile, with double bonds at both ends, one end, or neither end. Since the scission is random, molecules are made with a wide variety of chain lengths. These appear in the pyrogram as a series of triplet peaks. Each triplet consists of an alkane, an alkene, and a diene of a specific chain

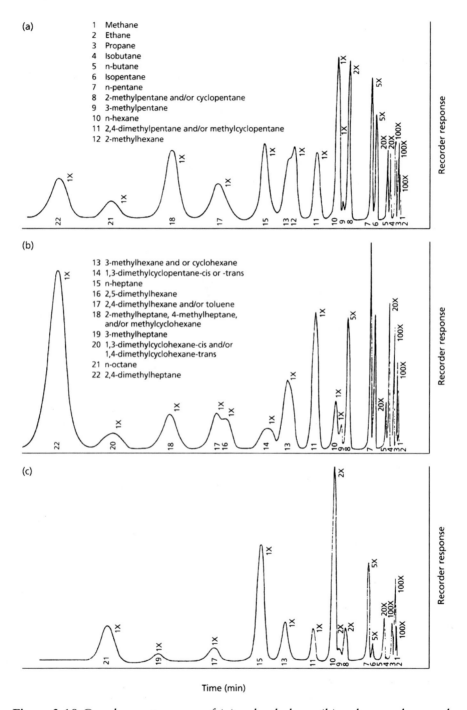

Figure 3.10 Gas chromatograms of (a) polyethylene, (b) polypropylene and (c) ethylene-propylene copolymer. (*Source: Author's own files*)

length. The hydrocarbons in each triplet have one more carbon than the molecules in the triplet that eluted just prior to it.

The chromatogram resulting from the pyrolysis of PE at 750 °C shows oligomers containing up to 30 carbons.

An example of the results obtainable by PGC of polyolefins is shown in **Figure 3.10**, which compares the pyrograms of PE, PP, and an ethylene–propylene copolymer. To obtain these results the sample (20 mg), in a platinum dish, was submitted to controlled pyrolysis in a stream of hydrogen as carrier gas. The pyrolysis products were then hydrogenated at 200 °C by passing through a small hydrogenation section containing 0.75% platinum on 30/50 mesh aluminium oxide. The hydrogenated pyrolysis products were then separated on squalane on a fireback column, and the separated compounds detected by a katharometer. Under the experimental conditions used in this work only alkanes up to C_9 could be detected.

It can be seen that major differences occur between the pyrograms of these three similar polymers. PE produces major amounts of normal C_2 to C_8 alkanes and minor amounts of 2-methyl and 3-methyl compounds such as isopentane and 3-methylpentane, indicative of short-chain branching on the polymer backbone. For PP, branched alkanes predominate, these peaks occurring in regular patterns, e.g., 2-methyl, 3-ethyl, and 2,4-dimethlpentane and 2,4-dimethylheptane, which are almost absent in the PE pyrolysate. Minor components obtained from PP are normal paraffins present in decreasing amounts up to *n*-hexane. This is to be contrasted with the pyrogram of PE, where *n*-alkanes predominate. The ethylene–propylene copolymer, as might be expected, produces both normal and branched alkanes. The concentrations of 2,4-dimethylpentane and 2,4-dimethylheptane are lower than those that occur in PP.

Depolymerisation is a free radical mechanism in which the polymer essentially reverts to a monomer or monomers. Unlike random scission, which produces fragments of a variety of chain lengths, depolymerisation generates a simple chromatogram consisting of large peaks for the monomers from which the polymer or copolymer was produced (see overleaf).

Several polymers degrade primarily by a free radical depolymerisation, including polystyrene (PS) and polymethacrylates. When a free radical is produced in the backbone of polyethyl methacrylate, for example, the molecule undergoes scission to produce an unsaturated small molecule (ethyl methacrylate) and another terminal free radical. This radical will also cleave to form ethyl methacrylate and propagate the free radical. The net effect is often referred to as 'unzipping' the polymer.

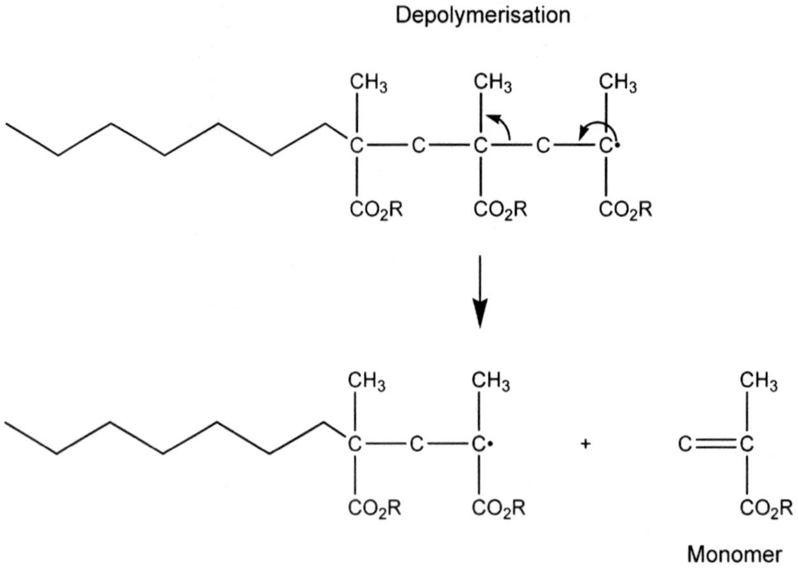

Depolymerisation

Monomer

Polyethyl methacrylate unzips extensively when heated to 600 °C for ten seconds. Copolymers of two or more methacrylate monomers will undergo the same degradation mechanism, producing a peak for each of the monomers used in the original polymerisation.

Side group elimination is usually a two-stage process in which the polymer chain is first stripped of atoms or molecules attached to the backbone of the polymer, leaving an unsaturated chain. This polyene then undergoes further reactions, including scission, aromatisation, and char formation:

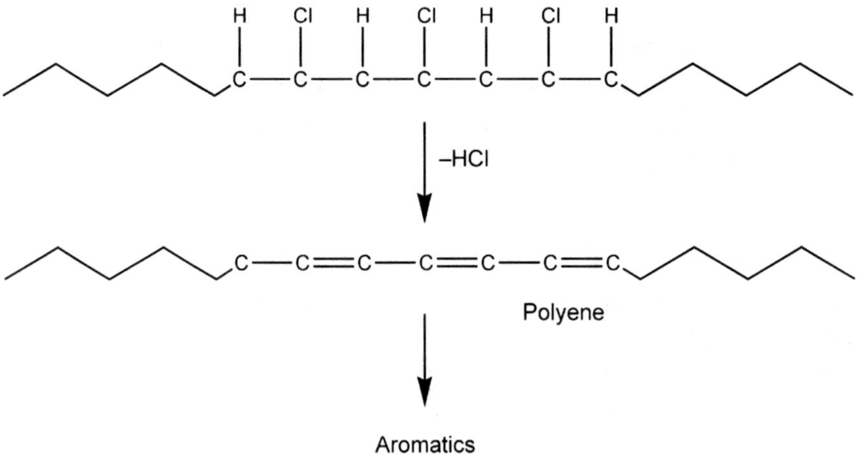

Polyene

Aromatics

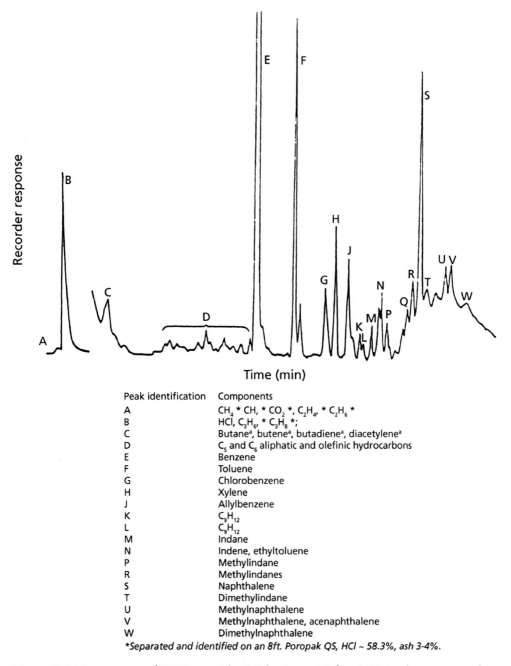

Peak identification	Components
A	CH_4 * CH, * CO_2 *, C_2H_4, * C_2H_6 *
B	HCl, C_3H_6, * C_3H_8 *;
C	Butane[a], butene[a], butadiene[a], diacetylene[a]
D	C_5 and C_6 aliphatic and olefinic hydrocarbons
E	Benzene
F	Toluene
G	Chlorobenzene
H	Xylene
J	Allylbenzene
K	C_9H_{12}
L	C_9H_{12}
M	Indane
N	Indene, ethyltoluene
P	Methylindane
R	Methylindanes
S	Naphthalene
T	Dimethylindane
U	Methylnaphthalene
V	Methylnaphthalene, acenaphthalene
W	Dimethylnaphthalene

Separated and identified on an 8ft. Poropak QS, HCl ~ 58.3%, ash 3-4%.

Figure 3.11 Pyrograms of PVC resin obtained using a 20 ft × 0.19 in chromatographic column containing 10% SE32 on 80/100 CRW. Lettered peaks refer to identification below. (*Reprinted from M.M. O'Mara, Journal of Applied Polymer Science, Part A1: Polymer Chemistry, 1970, 8, 1887, with permission from John Wiley and Sons*)

113

A good example of a material that pyrolyses in this way is PVC. PVC first undergoes a loss of hydrogen chloride to form a conjugated polyene backbone. This unsaturated chain is further degraded, mostly to form aromatics, as well as some smaller, unsaturated hydrocarbon fragments. The principal pyrolysis products produced from PVC (in addition to hydrogen chloride) are benzene, toluene, and naphthalene. Small amounts of chlorinated aromatics may also be produced, which indicate that some chlorine is still attached to the polymer chain during aromatisation. This results from defects in the PVC which place two chlorine atoms either on the same carbon or on neighbouring carbons, so that one remains after hydrogen chloride is eliminated from the original polymer.

O'Mara [85] carried out pyrolysis of PVC (Geon 1-3, CI = 57.5%) by two general techniques. The first method involved heating the resin in the heated (325 °C) inlet of a mass spectrometer in order to obtain a mass spectrum of the total pyrolysate. The second, more detailed, method consisted of degrading the resin in a PGC interfaced with a MS through a molecule enricher. Samples of PVC and plastisols were pyrolysed at 600 °C in a helium carrier gas flow. Since a stoichiometric amount of hydrogen chloride is released (58.3%) from PVC when heated at 600 °C, over half of the degradation products, by weight, is hydrogen chloride. A typical pyrogram of a PVC resin obtained by this method using an SE32 column is shown in **Figure 3.11**. The major components resulting from the pyrolysis of PVC are benzene, hydrogen chloride, naphthalene and toluene. In addition to these major products, a homologous series of aliphatic and olefinic hydrocarbons ranging from C_1 to C_4 are formed. O'Mara [85] obtained a linear correlation between weight of PVC pyrolysed and weight of hydrogen chloride obtained by GC.

3.7.3.1 Determination of Unsaturation in Ethylene–Propylene– Diene Terpolymers

PGC has been used to determine the overall composition of ethylene–propylene–diene terpolymers [10]. In attempting to determine the third component in these materials, difficulties might be anticipated, since this component is normally present in amounts around 5 wt%. However, dicyclopentadiene was identified in ethylene–propylene–diene terpolymers even when the amount incorporated was very low.

Van Schooten and Evenhuis [86, 87] applied their pyrolysis (500 °C)–hydrogenation-gas chromatography technique to unsaturated ethylene–propylene copolymers, i.e., ethylene–propylene–dicyclopentadiene and ethylene–propylene–norbornene terpolymers. The pyrograms show that very large cyclic peaks are obtained from unsaturated rings: methylcyclopentane is found when methylnorbornadiene is incorporated; cyclopentane when dicyclopentadiene is incorporated; methylcyclohexane and 1,2-dimethylcyclohexane when the addition compounds of norbornadiene with isoprene and dimethylbutadiene, respectively, are incorporated; and methylcyclopentane when the dimer of methylcyclopentadiene is incorporated. The saturated cyclopentane rings present in the same ring system in equal concentrations, however, give rise to peaks that are an order of magnitude smaller. Obviously, therefore, the peaks that form the termonomer could be used to determine its content if a suitable calibration procedure could be found.

3.7.3.2 Stereoregularity of Polypropylenes

Sugimura and co-workers [88] used pyrolysis–hydrogenation–glass capillary gas chromatography to obtain high-resolution pyrograms of isotactic, syndiotactic, and atatic polypropylene. They interpreted assigned characteristic peaks in the pyrograms in terms of the stereoregularity and degree of chemical inversion of the monomer units along the polymer chains. This method can be used for the routine characterisation of PP.

3.7.3.3 Tacticity of Polypropylenes

Audisto and Bajo [89] pyrolysed stereoregular isotactic and syndotactic polypropylenes and identified some components of the tetramer (C_{11}–C_{13}) fraction and resolved these into two diastereoisomers, the relative intensities of which varied according to the tacticity of the original PP.

3.7.3.4 Analysis of Copolymers

PGC has been used to determine the composition (i.e., monomer ratios) of a wide range of copolymers including ethylene–propylene [90, 91], natural rubber and styrene butadiene rubbers [92, 93], styrene–divinyl benzene [94], polyhexafluoropropylene–vinylidene fluoride [95], acrylic and methacrylic acid [96], PE–ethyl acrylate and PE–vinyl acetate [97], and MMA–ethyl acrylate copolymers [98].

3.7.3.5 Analysis of Comonomer Units in Polyhexafluoropropylene–Vinylidene Chloride Copolymers

Curie point pyrolysis has been used to carry out quantitative analysis of monomer units in polyhexafluorpropylene–vinylidene fluoride [95]. The polymer composition is calculated from the relative amounts of monomer regenerated and the trifluoromethane produced during pyrolysis. A calibration curve is obtained using samples whose compositions are measured by ^{19}F NMR as standards and a least squares fit calculated. The reproducibility of the pyrolysis step achieved by the Curie point pyrolyser permitted the monomer composition to be determined with a reproducibility of ±1%.

3.7.3.6 Analysis of Polyethylene Acrylate and Polyethylene–Vinyl Acetate Copolymers

Barrall and co-workers [97] have described a PGC procedure for the analysis of PE–ethyl acrylate and PE–vinyl acetate copolymers and physical mixtures thereof. The pyrolysis chromatogram of polyethylene–vinyl acetate contains two principal peaks. The first is methane and the second acetic acid:

$$\text{——OCH}_2\text{——CH}_2\text{——CH}_2\text{——CH} \quad \xrightarrow[\text{300–480 °C}]{\text{Pyrolysis}} \quad CH_3COOH + CH_4$$
$$|$$
$$OOCCH_3$$

The pyrolysis chromatogram of polyethylene–ethyl acrylate at 475 °C shows one principal peak due to ethanol. No variation in peak areas is noted in the temperature range 300–480 °C. **Table 3.13** shows the analysis of 0.05 g samples of polyethylene–ethyl acrylate and polyethylene–vinyl acetate obtained at a pyrolysis temperature of 475 °C.

Table 3.13 Pyrolysis results on physical mixtures of polyethylene-ethyl acrylate and polyethylene-vinyl acetate						
Mixture	Acetic acid found	wt.% calcd	Ethylene found	wt.% calcd[a]	Oxygen found	wt.% calcd
50% PEEA-1 and 50% PEVA-2	9.10	9.05	2.65	2.62	7.88	8.25
33.3% PEEA-2 and 66.6% PEEA-3	12.15	12.33	0.75	0.70	7.33	7.49
[a] *Calculated from results from acetic acid and ethylene content for individual samples on weight per cent basis* *Source: Author's own files*						

3.7.3.7 Structure Determination of Styrene–n-Butyl Acrylate Copolymers

Wang and co-workers [99] carried out structure determinations of styrene–*n*-butyl acrylate copolymers. They established the 'degree of structure', i.e., the number average sequence length for styrene and *n*-butyl acrylate repeat units, and compared the results with those obtained from a homogenous (i.e., non-structured) random copolymer. Number average sequence lengths were calculated using formulae that incorporated pure trimer peak intensities and hybrid trimer peak intensities.

3.7.3.8 Microstructure of Styrene–Methyl Methacrylate Copolymers

Wang and Smith [100] applied PGC to determine the composition and microstructure of styrene–MMA copolymers. The composition of these copolymers was quantified by monomer peak intensities obtained from pyrolysis. Because of the poor stability of MMA oligomers, neither MMA dimers nor trimers were detected under normal pyrolysis conditions. The number average sequence length for styrene was determined from pure and hybrid styrene trimer peak intensities. The number average sequence length for MMA was determined using formulae that incorporate the composition and number average sequence length of styrene.

3.7.3.9 Composition of Vinyl Chloride–Vinylidene Chloride Copolymers

Wang and Smith [101] used PGC to carry out compositional and structural studies of these copolymers. Composition and number average sequence lengths (which reflects monomer arrangements) in the copolymers were calculated using formulae that incorporate pure trimer and hybrid trimer peak intensities.

Because of the differences in reactivity between vinyl chloride and vinylidene chloride monomers, the structure of the copolymer was further investigated on the basis of the percentage of grouped monomers (i.e., the number average sequence length for vinyl chloride and vinylidene chloride repeat units). The results obtained were in excellent agreement with those obtained by PMR spectroscopy.

For this method 2.5 mg of polymer was placed in a quartz tube, which had been equilibrated for 5 minutes at 180 °C. The sample was then pyrolysed for 20 seconds at 700 °C using a pyroprobe CDS 190 with a platinum coil. Gas chromatography was conducted using a flame ionisation or a GC–MS system.

3.7.3.10 Determination of Short-Chain Branches in PVC

Ahlstrom and co-workers [102] applied the techniques of PGC and pyrolysis–hydrogenation–gas chromatography to the determination of short-chain branches in PVC and reduced PVC. Their attempts to determine the short-chain branches in PVC by PGC were complicated by an inability to separate all of the parameters affecting the degradation of the polymer. Not only does degree of branching change the pyrolysis pattern, but so do tacticity and crosslinking [103].

For the pyrolysis of PVC, a ribbon probe was used. On-line hydrogenation of the pyrolysis products was accomplished using hydrogen as the carrier gas with 1% palladium on Chromasorb-P catalyst inserted in the injection port liner. Maximum triplet formation occurred at C_{14} for low-density polyethylene (LDPE) and for reduced PVC and at C_{15} for HDPE. The occurrence of the peak maxima at C_{14} for reduced PVC indicates that the total branch content is higher than that of HDPE. However, aside from the C_{14}/C_{15} peak maxima difference, the pyrolysis pattern for even the most highly branched PVC resembles HDPE more than LDPE. These data indicate that the type of short-chain branch in PVC is qualitatively more like that of HDPE, but that the sequence length between branch sites is shorter in LDPE and PVC. Since LDPE contains a large amount of ethyl and butyl branches, and PVC and HDPE contain mainly methyl and some ethyl branches, this qualitative resemblance would be expected. A relative measure of the total amount of short-chain branches for these polymers can be obtained by calculating the percentage of branched products formed (**Table 3.14**).

More information about the specific type of short-chain branch in PVC can be found from an examination of the $C_1–C_{11}$ hydrocarbons (**Figure 3.12**). Here quantitative differences between the reduced PVC become more apparent. The most obvious differences occur in the amounts of iso-C_7 and iso-C_8 products formed, which indicate differences in the

Table 3.14 Relative total branch content of high-density polyethylene, LAH-reduced PVC and low-density polyethylene	
Sample	**Branched products (%)**
High-density polyethylene	12.0
Reduced PVC*	19.0
Low-density polyethylene	26.0
Average value Reprinted from D.H. Ahlstrom, S.A. Liebman and K.B. Abbas, *Journal of Polymer Science, Polymer Chemistry Edition*, 1976, **14**, 2479, with permission from Wiley Interscience [102]	

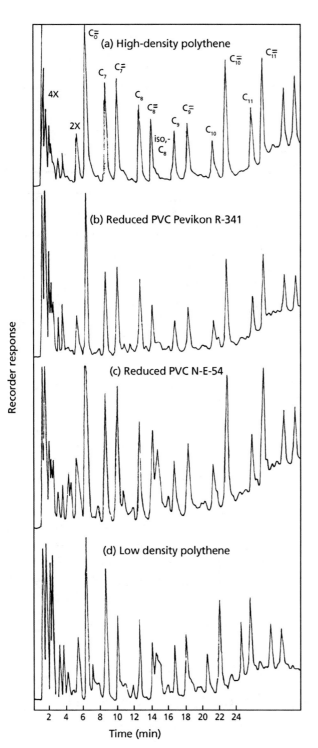

Figure 3.12 Pyrolysis of polyethylenes and reduced PVC; (a) high density polyethylene; (b) reduced PVC Pevikron R-341; (c) reduced PVC N-E-54; (d) low density polyethylene Durapak column fragments C_1–C_{11}.

Reprinted from D.H. Ahlstrom, S.A. Liebman and K.B. Abbas, J. Polymer Science, Polymer Chemistry Edition, 1976, 14, 2479, with permission from Wiley Interscience [102]

119

Table 3.15 Short chain branch content of lithium aluminium hydride reduced PVC		
Sample	*iso* C_8^- / n-C_8^-	C_{10}^- %
High density polyethylene	0.1	8.6
Pevikon R341	0.52	7.7
Nordforsk E-80	0.71	7.2
Nordforsk S-80	1.20	7.0
Nordforsk S-54	1.59	6.8
Nordforsk E-54	1.63	6.5
Ravinil R100/650	1.63	6.5
Shinttsu TK1000	1.87	6.3
Low density polyethylene	1.83*	5.2

* This ratio does not include the C4 branch content
Reprinted from D.H. Ahlstrom, S.A. Liebman and K.B. Abbas, J. Polymer Science,
Polymer Chemistry Edition, 1976, 14, 2479, with permission from Wiley Interscience [102]

total branch content. As the amount of short-chain branching (C_{10}) in the reduced PVC increases, there is a decrease in the amount of *iso*-alkanes formed (**Table 3.15**). The data in **Table 3.15** show small but distinguishable differences in the short-chain branch content of the reduced PVC.

3.7.3.11 Determination of Degree of Cure of Rubber

PGC can be used to determine the degree of cure of natural rubber. **Figures 3.13(a)–(c)** are pyrograms of three samples of natural rubber cured for successively longer intervals. Most of the chromatographic peaks produced by pyrolysis of rubber are insensitive to the degree of cure. The peaks labelled A and B in **Figures 3.13(a–c)** are sensitive to the degree of cure and are affected oppositely by pyrolysis temperature: peak A increases with increasing temperature and peak B decreases with increasing temperature. The ratio of the areas of peak B to peak A is a measure of the degree of cure.

3.7.3.12 Determination of Styrene Copolymer Sizing Agents in Paper

PGC has been applied to the determination of styrene copolymer sizing agents [104]. PGC using a multi-detector on the gas chromatograph has been used to determine end-group functionality in PMMA.

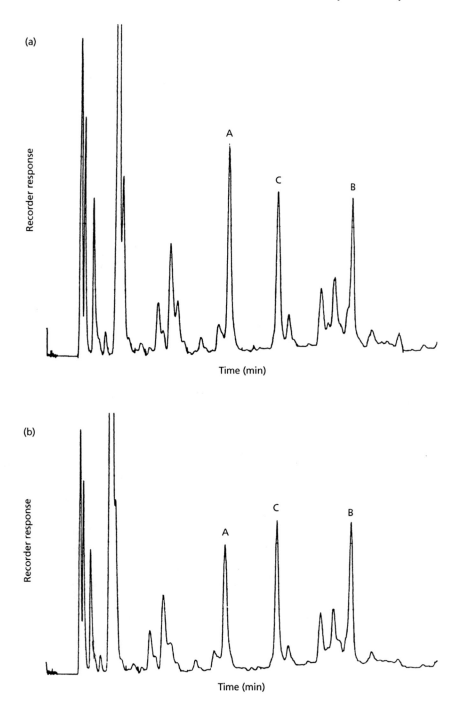

Figure 3.13 Pyrogram of (a) under cured rubber 30 µg; (b) optimally cured rubber 20 µg; (c) over cured rubber 20 µg – Continued overleaf. (*Source: Author's own files*)

121

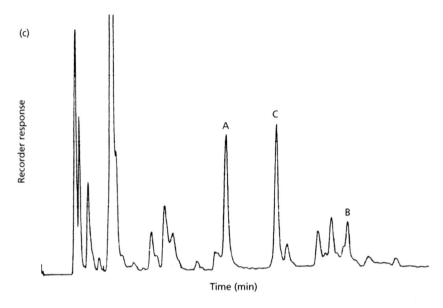

Figure 3.13 *Continued from previous page*

3.8 Pyrolysis Gas Chromatography–Mass Spectrometry

The time has long since passed when one could rely on gas chromatographic data alone to identify unknown compounds in gas chromatograms. The sheer number of compounds that could be present would invalidate reliance on the use of such techniques. The practice nowadays is to link a mass spectrometer (or NMR or FT-IR spectrometers, see following sections) to the outlet of the GC so that a mass spectrum is obtained for each chromatographic peak as it emerges from the separation column. If the peak contains a simple substance then computerised library searching facilities attached to the MS will rapidly identify the substance. If the emerging peak contains several substances, then the mass spectrum will provide information on the substances present. Alternatively, different gas chromatographic columns can be tried that will resolve the mixture of substances in question.

The use of GC-MS grew rapidly during the early 1970s as discussed by Shackleford and McGuire [105].

3.8.1 Instrumentation

There are now several suppliers of equipment for carrying out this technique (see Appendix 1). One supplier is Finnigan MAT, which supplies single-stage quadruple mass spectrometers.

3.8.1.1 SSQ70 Series Single-Stage Quadrupole Mass Spectrometer

This offers premium single-stage performance, with the option of being upgraded to a triple-stage quadrupole system (i.e., the TSQ70). The SSQ70 features a network of distributed microprocessors with more than 1.5 megabytes of memory linked to a powerful DEC 11/73 processor with 2.0 megabytes of memory for data-processing operations. Instrument control tasks can be displayed in up to eight windows on a colour display terminal. The hyperbolic quadrupole analyser gives the SSQ70 a mass range of up to 4000 μm; system performance is specified to 200 m/z. The cradle vacuum system with three large inlet points at the ion source accommodates a variety of sample inlets such as capillary gas chromatography, thermospray, liquid chromatography, MS, supercritical fluid chromatography, and solids probe. Standard features of the instrument also include a high-performance electron impact/chemical ionisation (EI/CI) ion source with exchangeable ion volumes, a PPI NICI with high-voltage conversion dynode multiplier for positive and negative ion detection and fast ion bombardment.

The Varian 3400 GC gas chromatograph supplied with this instrument incorporates a high-performance capillary column with multi-linear temperature programming in up to eight sequences, a data-control and recording system for temperatures in the GC oven and for interface temperatures, and also for controlling and recording valve timing, a data system control of optional GC accessories, and a split/splitless capillary injector. The Micro VIP computer data system comprises a Dec 11/73 processor with video colour display, dot matrix printer, and a data system for control of instrument parameters and user-initiated diagnostics.

3.8.1.2 Mass Spectrometry–Mass Spectrometry

In high-performance MS-MS (as opposed to GC-MS) the separator as well as the analysis is performed by the mass spectrometer. One advantage of this technique over combined chromatography–mass spectrometry is that separation is a spatial process rather than being dependent on time. This can lead to improved analysis times and/or greater specificity. MS–MS also opens up other areas such as the study of complete structures.

3.8.1.3 H-SQ30 Hybrid Mass Spectrometer–Mass Spectrometer

This instrument combines a reverse-geometry (BE) magnetic section instrument with a quadrupole (QQ) analyser. This hybrid combination provides MS–MS operation with a high-resolution first stage (BE) and a unit resolution second stage (QQ). The four available collision regions allow experiments of low (2–100 eV) and high (3 keV) collision energy, as well as consecutive collision induced dissociation (CID) experiments using two separate collision regions. The H-SQ30 is an ideal instrument for structural elucidation studies and ion physics.

3.8.1.4 MAT-90 High Mass–High-Resolution Mass Spectrometer

This is a very high-performance instrument in which control resides in a multiprocessor system manager leaving only the analytically important parameters to be defined by the operator. It utilises a new concept of ion optics for double focusing equipment for MS–MS, equipment for EI, CI, and fast ion bombardment studies. An ion trap detector is also supplied. The ion trap detector detects any compound that can be chromatographed - it is a universal detector that can replace several conventional gas chromatography detectors such as the type used in the Varian model 3400 gas chromatogram included in the Finnigan MAT SSA-70 and TSQ-70 instruments. Electron capture, flame ionisation, element specific, detectors used in the latter instruments are not universal in this sense and will not respond to all types of organic compounds, i.e., some compounds will be missed. The ion trap detector obviates this difficulty by responding to all types of organic compounds. In the ion trap technique one does not have to rely on retention data for identification. The mass spectrum indicates the identity with certainty. Various aspects of ion trap detectors have been discussed by workers at Finnigan MAT and elsewhere [106–128].

During development of the ion trap detector it was found that the low voltage previously used for storage encouraged the production and storage of the H_2O^+ and H_3O^+ ions which occasionally led to an increase of the M + 1 molecular ions. This problem has been eliminated by adjusting the storage voltage so that the H_2O^+ and H_3O^+ ions are no longer stored.

This scanning method produces standard EI spectra which can be rapidly searched through the standard NBS library (42,222 spectra). In each case the number of ions stored in the trap would be the optimum required to produce a conventional EI spectrum. In order that the procedure does not affect quantitative results, ion intensities are stored after application of an adjustment factor which is always related to the true size of the peak as measured by the original fast scan. Scaling is controlled by a computer, and the net result is a system with a dynamic range between 104 and 105.

The ion trap detector may be operated as a universal detector (when full scans are stored) or, with the application of multiple ion monitoring, as a specific detector. Because approximately 50% of the ions formed in the trap are analysed, the sensitivity of the instrument in full-scan mode can be much higher than conventional mass spectrometers, in which only 0.1–0.2% of the ions formed may be detected. Thus the instrument can be used to detect 2–5 pg (in full scan) of compounds eluting from the column, a performance that compares extremely favourably with those of the most sensitive specific detectors (e.g., the electron capture detector) and easily outstrips that of the flame ionisation detector (FID). This sensitivity is not achieved at the expense of dynamic range, as the instrument can produce linear calibration graphs for quantities within the range 5–10 pg to 1000 ng on the column. This again compares favourably with the performance of the FID.

The Finnigan MAT Chem Master Workstation is a gas chromatography and gas chromatography–mass spectrometry data-processing system that speeds the flow of data through the laboratory and provides essential quality assurance and quality control review. It is a PC-based integrated hardware/software system that converts gas chromatographic and GC-MS data into reliable analytical reports.

3.8.2 Applications

In the mass spectra of products of the nickel chloride-catalysed PGC of cellulose, compounds from benzofuran (peak 1) to trimethylxanthone (peak 33) were identified.

Qian and co-workers [129] used in-source direct pyrolysis mass spectrometry to identify 150 different polymers and copolymers. Library searching facilities were included to enable positive polymer identifications to be carried out.

In this procedure the unknown polymer was pyrolysed within the ion source of a mass spectrometer by a coiled filament designed for desorption chemical ionisation/desorption electron ionisation applications. Pyrolysis products ionised by 70 eV electron impact yielded highly reproducible mass spectra which were characteristic of the polymer.

Some specific recent applications of the chromatography–mass spectrometry technique to various types of polymers include the following: PE [130, 131], poly(1-octene), poly(1-decene), poly(1-dodecene) and 1-octene–1-decene–1-dodecene terpolymer [132], chlorinated polyethylene [133], polyolefins [134, 135], acrylic acid, methacrylic acid copolymers [136, 137], polyacrylate [138], styrene–butadiene and other rubbers [139-141], nitrile rubber [142], natural rubbers [143, 144], chlorinated natural rubber [145, 146], polychloroprene [147], PVC [148-150], silicones [151, 152], polycarbonates (PC) [153], styrene–isoprene copolymers [154], substituted PS [155], polypropylene carbonate [156], ethylene–vinyl acetate copolymer [157], Nylon 6,6 [158], polyisopropenyl cyclohexane–α-methylstyrene copolymers [195], cresol-novolac epoxy resins [160], polymeric flame retardants [161], poly(4-N-alkylstyrenes) [162], polyvinyl pyrrolidone [31, 163], vinyl pyrrolidone–methacryloxysilicone copolymers [164], polybutylcyanoacrylate [165], polysulfide copolymers [1669], poly(diethyl-2-methacryloxy) ethyl phosphate [167, 168], ethane–carbon monoxide copolymers [169], polyetherimide [170], and bisphenol-A [171].

3.9 Pyrolysis Gas Chromatography–Fourier Transform NMR Spectroscopy

This is another complementary technique that provides more detailed information on the structure of polymers than does PGC alone. As an example of the application of this technique Leibman and co-workers [172] applied it to a study of short-chain branching in PE and PVC. The nature and relative quantities of the short branches along the polymer chains were determined, providing detailed microstructural information. Down to 0.1 methyl, ethyl, or *n*-butyl branches per 1000 methylene groups can be determined by this procedure. Other structural defects can be determined, providing significant information on polymer microstructure that is not otherwise readily obtained. Instrumentation is discussed in Appendix 1.

3.10 High-Performance Liquid Chromatography

Kawai and co-workers [173] determined the composition of butyl acrylate–ethyl acrylate copolymers with a narrow chemical composition distribution by ^{1}H NMR spectroscopy and the components of the copolymers separated by normal and reversed phase high-performance liquid chromatography (HPLC) using crosslinked acrylamide and styrene beads. Samples containing higher butyl acrylate content eluted faster with normal phase HPLC while the opposite occurred with reversed phase HPLC, indicating that butyl acrylate is less polar than ethyl acrylate.

Kase and co-workers [174] investigated an isocyanate prepolymer composition using HPLC and field desorption MS.

Sandra and co-workers [175] present an overview of the use of MS as a detection system for liquid chromatography. Enjalbal and co-workers [176] characterised soluble polymer-supported organic compounds by liquid chromatography/electroscopy ionisation mass spectrometry. Julian and Domingo [177] have pointed out that the coupling of spectroscopic instruments to liquid chromatography instruments has not been as rapidly developed or successful as techniques such as GC-MS and GC-IR spectroscopy. A new liquid chromatography–infrared spectroscopy interface is described which promises to be a successful marriage of these techniques. Although it does not provide a real-time analysis, it has the advantages of total solvent removal before infrared analysis, analysis of the total liquid chromatography eluate as opposed to incremental fractions, and the ability to produce infrared spectra at high signal-to-noise ratios.

3.11 Mass Spectrometric Techniques

3.11.1 Time-of-Flight Secondary Ion Mass Spectrometry (ToF-SIMS)

Instrumentation for carrying out this mass spectrometric technique is discussed in Appendix 1.

SIMS is the most commonly used of the surface mass spectrometry techniques. SIMS analyses the secondary ions ejected from a sample following bombardment with a primary ion beam, usually argon ions. The impact of the primary ion causes an atomic-scale collision cascade within the surface layers of the sample and, at points remote from impact, secondary ions are ejected from the surface. These ions are then determined by MS.

ToF-SIMS uses a primary ion beam, usually argon ions, which is pulsed and focused onto the sample. The secondary ions emitted from the sample are accelerated by a field to the same kinetic energy, and mass separation is achieved by measuring the flight time necessary for the passage of the ions along a path of known length. The secondary ions reach a detector in order of increasing mass, the detector consisting of a multi-channel plate in combination with a scintillation counter and a photomultiplier. The relationship between kinetic energy and flight time may be used to determine the mass of the secondary ion and therefore identify it. The sensitivity of the instrument requires great care to avoid sample contamination.

There are two modes of operating SIMS: static and dynamic.

Static SIMS (SSIMS). This operates under non-damaging conditions. It gives good information on molecular content and, like Auger spectroscopy, can be used for surface micro-analysis at high spatial resolution.

The most widely used method of static SIMS is ToF-SIMS. This uses ToF mass spectrometry which provides optimum sensitivity and mass resolution for secondary ions.

With the latest ToF-SIMS instruments mass resolutions of 10^{-3} to 10^{-4} amu are achievable with sampling depths of greater than 1 nm. This has two main advantages. Firstly, the sensitivity is increased, allowing peaks to be resolved where previously there had been overlap (i.e., two peaks at nominally the same mass). This now enables ToF-SIMS to be sensitive to ppm/ppb levels of surface species. Secondly, high mass resolution enables very accurate mass measurement so that empirical formulae can be calculated from the measured accurate mass, making unknown peak identification more certain.

Dynamic SIMS (DSIMS). This operates under conditions designed to remove surface layers sequentially during the analysis. This is achieved by rastering a primary ion beam over the area of analytical interest and collecting the emitted secondary elemental or cluster ions in the MS system. The erosion rates can be controlled from a few nanometres per hour to several tens of micrometres. The technique provides high sensitivity, quantitative elemental information in the form of mass spectra, depth profiles, and two- and three-dimensional images. All elements in the periodic table can be detected with sensitivities in the parts per million to parts per billion range. High mass resolution analysis can be used to separate elemental and molecular species with the same nominal mass values.

This technique has been used in the examination of structural details, and in some cases molecular weight determination of various types of polymers, and is extremely useful in polymer surface studies (see Chapter 4). The technique has also been used to provide new information and establish relations between species in monolayers, e.g., in the bonding of plastic films to metal or glass.

The ToF-SIMS technique is used either on its own or, as is the growing trend, in conjunction with XPS (Section 3.11.2). Often the combination of these two techniques provides better solutions to problems than either technique alone.

3.11.1.1 Applications of ToF-SIMS Alone to Problem Solving (i.e., XPS Not Used)

A study has been carried out of hydrogen–deuterium exchange in PS using ToF-SIMS [178].

Cox and co-workers [179] analysed PS/polyvinyl methyl ether blends by coincidence counting ToF mass spectrometry. This technique gave information on the chemical and spatial relationships between secondary ions. Thompson [180] carried out a quantitative surface analysis of organic polymer blends (e.g., miscible polycarbonate/PS blends) using ToF-SIMS. Lin and co-workers [181] used supersonic beam/multiphoton ionisation/ToF mass spectrometry to analyse photoablation products resulting from styrene-containing polymers such as styrene–butadiene, ABS, and PS foams. Photoablation products were examined by supersonic beam spectrometry and the results were compared with those obtained by thermal decomposition.

The high selectivity provided by supersonic beam spectrometry allowed the detection of minor species such as styrene, resulting from ablation of poly-α-methylstyrene produced by cleavage of a methyl group and by proton rearrangement.

Because ablation is carried out at high temperatures it is possible to examine thermally static polymers such as poly-*p*-methylstyrene. Other recent applications of ToF-SIMS include the examination of PS [182-184], PE [185], carbon fibre-reinforced epoxy resins [186], polyalkylacrylates [187], alkylketene dimers [188], perfluorinated polymers [189], siloxanes [190, 191], perfluorinated ethers [192], polyethylene glycol (PEG) oligomers [193-197], rubbers, ethylene–tetrafluorethylene copolymers [198], Nylon 6 [199], PC [200], polydimethyl siloxane [201], polypyrrole-coated PS [202], poly-*p*-phenylene vinylene [203], butyl rubber [204], poly(4-vinyl phenol)/poly(4-vinyl pyridine) blends [205], polypyrrole–silica gel composites [206], γ-glycidoxypropyltrimethyoxy silane [207], triblock copolymer poly(ethylene glycol)-β-poly(phenylene ethynylene)-β-poly(ethylene glycol) [208], and ethylene terephthalate–hydroxybenzoate copolymer [209].

3.11.1.2 Applications of Both ToF-SIMS and XPS [210]

These techniques have been applied to PTFE [211, 212], polybutadiene [213], rubbers [214, 215], acrylics [216, 217], PE [218–220], polyurethane [221], PS [222], polyvinyl carbazole [223], polymalic acid [224], poly-β-hydroxybutyrate and poly-β-hydroxyvalerate [225], γ-glycidoxy propyltrimethoxysilane [226], polypyrrole [227], and acrylonitrile–butadiene rubber [228].

3.11.1.3 Further Applications of ToF-SIMS

Adhesion studies

ToF-SIMS, and also in some cases XPS, has been applied to a range of polymer problems such as adhesion studies of elastomers based on brominated poly(isobutylene-*co*-4-methylstyrene) and diene elastomers [229], PVC [230], epoxy resin aluminium [231, 232], glass rubber [233], squalene brass [234], and sealants [235].

- Chain diffusion studies

Lin and co-workers [233] studied the chain diffusion behaviour and microstructure at an interface between the rubbery polymer PS and the glassy polymer polyphenylene oxide by depth-resolved SIMS. The interface region demonstrated a very sharp symmetric profile before annealing, with a thickness of 25–30 nm. As chain diffusion took place the interfacial region widened asymmetrically, moving into the polyphenylene oxide region, and the chain diffused across the interface in a manner that simultaneously showed both Fickian and Case II non-Fickian behaviour. The interfacial region could be divided into a rubbery and glassy region using the local glass transition temperature, which could be calculated by the Flory–Fox equation. The Fickian behaviour predominated at the rubbery side, and the non-Fickian behaviour at the glassy side. The effect of temperature on the interfacial diffusion agreed with the WLF equation. The average velocity of the interface decreased with the molecular weight according to a negative power law. The precise exponents of the power law were different above and below the critical molecular weight of PS, which was 38,000. The mutual diffusion coefficients were calculated from the MS data, and showed a good agreement with the predictions of the 'slow theory', which deals with the case where the diffusion coefficient is dominated by the effect of the slower-moving component.

3.11.2 XPS

Electron spectroscopy encompasses two main techniques, XPS (or ESCA) and Auger electron spectroscopy (AES). Both techniques identify and quantify elements present and can indicate the chemical state or functionality of elements at the surface. For best results AES relies upon the sample being electrically conducting and consequently it is very rarely used in polymer analysis. Hence, it is not discussed further here [236].

XPS uses an X-ray beam to cause the emission of electrons from the surface of the sample. The electrons analysed do not have enough energy to escape from a depth of more than ~3–5 nm, so the XPS technique is inherently surface sensitive. The binding energy of the emitted electrons is measured and used to identify the elements present.

XPS is best performed using a monochromatic X-ray source. Monochromatic XPS inflicts the least damage to delicate materials and optimises chemical state sensitivity. A monochromatic X-ray source is an essential requirement to extract the maximum information content from most polymer systems.

The thickness of surface layers and the depth distributions in the extreme surface region can be measured using angle-dependent XPS. By simply changing the angle of the sample to the detector, the effective depth of analysis can be varied in the range 1–10 nm.

XPS is a good technique for elemental analysis of solids and low-vapour-pressure liquids, with a detection limit to most elements of ~1000 ppm. It can be directly quantified, showing not only what is present, but also how much. This technique can be used to analyse polymers to a surface depth of 1–10 nm.

Swift [237] has discussed the fundamental principles of XPS (and SIMS) and demonstrates the use of these surface analytical techniques in problem solving. XPS can supply information on factors such as surface coating integrity (e.g., in medical polymer applications or plastic coating of metals or wire), contamination, molecular diffusion and adsorption phenomena [238], and characterisation of polymer surfaces. Often XPS is used in conjunction with SIMS to solve problems, as discussed in Section 3.11.1.

3.11.2.1 Applications in Which Only XPS is Used

Structural studies of polymer surfaces. Materials that have been studied include PMMA [239], PMMA–polypyrrole composites [240], poly(chloromethyl styrene) bound 1,4,8,11-tetrazacyclotetradecane, poly(chloromethyl styrene) bound thenoyl trifluoroacetone [241], poly(dimethyl siloxane)–polyamide copolymers [242], PS [243], ion-implanted PE [244], monoazido-terminated polyethylene oxide [245], polyurethanes [246], polyaniline [247], fluorinated polymer films [248], poly(o-toluidine) [249], polyetherimide and polybenzimidazole [250], polyfullerene palladium [251], imidazole-containing imidazolylethyl maleamic acid–octadecyl vinyl ether copolymer [252], polyphenylene vinylene ether [253], thiophene oligomers [254], fluorinated styrene–isoprene derivative of a methyl methacrylate–hydroxyethyl methacrylate copolymer [255], polythiophene [256], dibromoalkane–hexafluorisopropylidene diphenol and bisphenol A [257], and geopolymers [258].

Bulk polymer structural studies. These include studies on reactions between polyacrylic acid and polymethacrylic acid and polyacrylic acid with poly(4-vinyl pyridine) and poly(2-vinyl pyridene) [259], polybithiophene [260], polyvinyl carbazole sulfonation products [223], β-hydroxybutyrate and β-hydroxyvalerate containing Biopol [225].

Adhesion studies: Rattana and co-workers [231] and Watts and co-workers [232] have studied the adhesion of aluminium and epoxy resins using amine and organosilane primers. Leadly and co-workers [261] give details of coating delamination in a hot, humid environment for a cationic radiation-cured coating of cycloaliphatic epoxy resin. XPS and ToF mass spectroscopy of the delaminated surface showed that the phosphorus hexafluoride anion of the photoinitiator segregates to the interface. The durability of the coating was improved by reformulation with a reduced concentration of photoinitiator.

Elemental composition: Niino and Yabe [262] used XRF to determine the chlorine content of products obtained in the photoirradiation of polyvinylidene chloride film.

Carbon black studies: A determination has been carried out of Buckminsterfullerene in carbon black [263].

Particle identification: Stickle [264, 265] used XPS, ToF-SIMS, and Auger electron microscopy to carry out particle analysis. It is concluded that XPS has limited utility in this application, but that it can be used as a complementary characterisation tool to ToF-SIMS and AES for particle identification.

Pyrolysis studies: XPS has been applied to the study of pyrolysis products of poly(1,4-diphenyl-1-buten-3-yne) [266] and poly(phenyl silsesquioxane) [267, 268].

3.11.2.2 Applications in Which Both XPS and ToF-SIMS are Used

Some examples of these applications include studies on *p*-hydroxybenzoic acid-6-hydroxy-2-naphthoic acid copolyester-based adhesives [269] and further miscellaneous studies on adhesion [270, 271]. West [272] has characterised medical polymers using XPS and ToF-SIMS. These two techniques have also been used to characterise carbon black surfaces [273] and carbon fibres [274]. Other workers have reviewed various aspects of the application of ToF-SIMS to polymer surface studies [237, 275–277] (See also Section 3.11.1).

3.11.3 Tandem Mass Spectrometry (MS/MS)

Jackson and co-workers [278] have used tandem mass spectrometry in an AutoSpec 5000-orthogonal acceleration-time of flight (oa-ToF) instrument to generate end group and structural information from a variety of polymers. These systems include PMMA, polyethyl methacrylate, and polybutyl methacrylate. Matrix-assisted laser desorption/ionisation–collision-induced dissociation (MALDI-CID) data have been obtained from polymeric precursor ions with mass-to-charge ratios of up to approximately 5000. It is shown that end group and structural information may also be obtained from copolymers using MALDI-CID. Sequence information is generated for a MMA/butyl methacrylate block copolymer synthesised by group transfer polymerisation.

Bouajila and co-workers [279] used a technique based on liquid chromatography–ultraviolet spectroscopy–mass spectrometry–mass spectrometry (LC-UV-MS-MS) to elucidate the structure of phenol oligomers in resols by fragmenting the monomers. The progression of resin crosslinking was determined by solid-state ^{13}C NMR (constant potential (CP)/MAS), and the residual percentage of monomers and oligomers was determined in leachates and characterised by LC-UV-MS-MS. Results for crosslinking advancement were then correlated with the various synthesis parameters.

3.11.4 Fourier Transform Ion Cyclotron Mass Spectrometry

Van der Hage and co-workers [280] combined MALDI and Fourier transform ion cyclotron mass spectrometry (FT-ICR-MS) for the characterisation of polyoxyalkyleneamines. MALDI FT-ICR-MS was used to resolve intact, sodium ion cationised oligomer ions in the mass range from *m/z* 500 to 3500. NMR was used to measure the average end-group distribution to provide insight into conformational differences. In this respect, FT-ICR-MS and NMR data were complementary. Combined results yielded detailed information about chemical composition distributions of polyalkyleneamines that hitherto it was not possible to obtain with either technique separately. Merits and limitations of the data produced with MALDI FT-ICR-MS are discussed and compared with those of ^{1}H and ^{13}C NMR data [280].

3.11.5 MALDI-MS

MALDI-MS represents a unique tool for the simultaneous determination of molecular weight distribution and chemical structure of oligomers and polar and non-polar low molecular weight homopolymers. This includes end-group analysis fragmentation studies, molar mass analysis, analysis of copolymers, and fundamental studies. However, MALDI analysis of multifunctional polymers and block copolymers has some drawbacks, which are caused by the different ionisation behaviour of structurally different polymers. To obtain both chemical and structural information a coupling of chromatography with MALDI-MS can be applied. Special attention is paid to the principle of chromatography used for polymer separation. Besides size exclusion chromatography (SEC), which can be used for the separation of polymers according to their hydrodynamic volume, some other modes of chromatography can be applied.

The structural and molecular analysis of copolymers is a difficult problem that often cannot be easily dealt with by analysing the intact copolymer, and therefore much effort has been directed to the task of converting SEC traces to molar masses of copolymers. Multiple detectors and SEC/light scattering techniques may provide a solution to the problem, but difficulties remain. Being able to discriminate among different masses and possessing a remarkably high sensitivity, the MALDI-MS detector does not suffer these limitations and, coupled with analytical SEC, MALDI-MS is found to give mass spectra with average molar mass values in excellent agreement with those obtained by conventional techniques.

Advances in methods used in the structural characterisation of redox-active polymers will aid in developing structure/property relationships and in guiding synthetic efforts. MALDI-MS analysis has proved to be a powerful structural characterisation tool for biopolymers. The application of MALDI-MS to synthetic polymers has been primarily limited to polar or polarisable polymers that can be protonated or can form salt/metal adducts. Electron transfer matrices for MALDI-MS of small, non-polar, redox-active analytes have been evaluated and it was found that anthracene and terthiophene are effective MALDI matrices for these analytes and assist in producing analyte molecular ions (not protonated molecules). Thus, non-polar, redox-active polymers might also be similarly ionised.

Combined MALDI-MS and ion exclusion chromatographic techniques. In most of these techniques described in the literature the MALDI is not directly coupled off-line with the ion exclusion column. An exception is that of the work of Esser and co-workers [281], in which the two units are interfaced via a robotic interface. This technique was applied to studies of PS, PMMA, and butyl(methacrylate–methylmethacrylate) copolymers. Mehl and co-workers [282] combined ion exclusion with MALDI-MS to provide accurate molecular weight determinations on polyether and polyester polyurethane soft blocks.

133

Other applications in which the two techniques are not connected in line mainly include the determination of molecular weights of copolymers of MMA, butyl acrylate styrene, and maleic anhydride [283, 284], cyclic PS [285-288], thiophene–phenylene copolymers [289], methacryloxypropyltrimethoxy silane [290], various copolymers [291], PEG [287], polyetherimide photooxidation products [292], polyester–polyurethane [293], biodegradable polymers [294], and polyethylene–propylene oxide–ethylene oxide triblock polymers [295].

In most of the published work the ion exclusion chromatographic technique was not used, the work being limited to the application of ToF MALDI MS as reviewed here to silicone elastomers [283, 288, 296], PMMA [297, 298], acyclic dienes [299], polyesters [300, 301], polycaprolactone [302, 303], polyhexylthiophene [304], polyether and polyester–polyurethane soft blocks [305], polisobutylene [306], rubbers [307], PS [308], PE [309], polyalkylene oxides [310], polypropylene glycol [311], polyisobutyl vinyl ether [312], polypropylene oxide [313], hindered amine light stabilisers [314], poly-1,2:3,4-di-o-isopropylidene-6-O-(2-vinyloxyethyl)-D-galactopyranose [315], cyclic carbonates [316], vinyl polyperoxide end groups [317], poly(amidoamine) dendrimers [318], ethylene di-oxy substituted bithiophene alt-thiophene-S-S-dioxide copolymers [319], blue-light-emitting polymers [320], polylactides [321], furanylene vinyl or cyclopentadienyl vinyl polymers [322], poly-3-methylenehydroxy-bile acid derivatives [323], poly(dimethyl 5-(2-hydroxyethoxy)isophthalate) [324], polyacrylic acid [325], and rubbers [326].

Theoretical studies using ToF MALDI MS have been reported by various workers [326-330].

3.11.6 Radio Frequency Glow Discharge Mass Spectrometry

Shick and co-workers [331] used this technique for the characterisation of bulk polymers such as PTFE–methyl vinyl ether and PTFE-hexafluoropropylene–polyvinylidene fluoride.

A radio frequency (RF) powered glow discharge atomisation/ionisation source was used to determine the applicability of the technique for direct polymer analysis. The technique provided fingerprint mass spectra for a series of PTFE-based polymers, the mass spectrum being obtained by observing the different base peaks and relative peak intensities. Discharge stabilisation and internal stability were very reproducible (with a standard deviation of less than 5%). On-depth profiles could be obtained by observing the polymers following deposition of a layer of copper on the polymer surface.

3.12 Microthermal Analysis

This is a technique that combines thermal analysis and atomic force microscopy. This new technique is discussed in a series of papers published by ASTM in 1999 [332] and other workers [333-338].

Workers at TA Instruments UK, manufacturer of thermal micro-analysis equipment, have given details of the model TA 2990 system. They give examples of its application to thermal and morphological problems [339].

Murray discusses the use of microthermal analysis using a heated probe for thermal conductivity imaging. The same probe performs local calorimetric measurements through microthermal mechanical analysis and micro-differential thermal analysis. Application examples include copolymer identification in food packaging materials, effects of embedded carbon fibres on matrix crosslinking, and transparency films. Alternatively, a sample heating stage can be used to track temperature-dependent morphology, stiffness, and adhesion changes using pulsed force mode.

Microthermal analysis combines the visualisation power of AFM with the characterisation ability of thermal analysis [333]. The AFM head is fitted with an ultra-miniature thermal probe, which provides not only topographic and thermal contrast information, but also information similar to traditional thermal analysis on a sub-micrometre scale.

Microthermal analysis is being developed as a tool for carrying out studies in a number of areas including morphology, topography, glass transitions, depth probing, and phase separation studies.

Morphology: Morphological studies have been conducted on multi-block copolymers [340] and Nylon 6,6/polytetrafluorethylene/silicone blends [341].

Topography: Abad and co-workers [342] used the TA Instruments TA2990 microthermal analyser to carry out topography studies including fabrication failure on biaxially oriented PP film.

Glass transition: Glass transition studies have been reported on amorphous PS [343], araldite 2011, adhesive–metal bonds [344], thin PS films [345], and PEG entrapped into polylactic acid [346].

Depth profiling studies: Grossette and co-workers [347] used microthermal analysis and FT-IR spectroscopy to study the physical heterogeneity induced by the photooxidation of PP at a sub-micrometric level.

Phase separation studies: Song and co-workers [348] used microthermal analysis to study the phase separation process in a 50:50 (by weight) PS/polyvinylmethylether blend and nitrile rubber. Microthermal analysis will image the composition in the near-surface region or surface region of multi-component materials if the resolution is high enough.

3.13 Atomic Force Microscopy

Scanning probe microscopy initially provided three-dimensional visualisation of surfaces down to the atomic scale using scanning tunnelling and AFM. Today, a range of imaging modes and spectroscopic techniques can be used to determine additional information on physical, chemical, and thermal properties of polymeric materials. Examples of the uses of lateral force microscopy to determine surface friction and force modulation to elucidate surface stiffness have been presented. Pulsed force mode now enables both these properties to be displayed simultaneously. Intermittent contact atomic force microscopy (tapping) combined with phase imaging provides fast imaging of soft polymers combined with simultaneous materials contrast based on surface viscoelastic properties. Hence, the spatial distribution of multi-component polymers can be determined.

The past decade has witnessed an explosion of techniques used to pattern polymers on the nano- and sub-micrometre scale, driven by the extensive versatility of polymers for diverse applications such as molecular electronics, data storage, and all forms of sensors. Lyuksyutov and co-workers [349] demonstrate a novel lithography technique – electrostatic nanolithography using AFM – that generates features by mass transport of polymer within an initially uniform, planar film without chemical crosslinking, substantial polymer degradation or ablation.

3.13.1 Applications

3.13.1.1 Polymer Characterisation Studies and Polymer Structure

The application of AFM and other techniques has been discussed in general terms by several workers [350–353]. Other complementary techniques covered in these papers include FT-IR spectroscopy, Raman spectroscopy, NMR spectroscopy, surface analysis by spectroscopy, GC-MS, scanning tunnelling microscopy, electron crystallography, X-ray studies using synchrotron radiation, neutron scattering techniques, mixed crystal infrared spectroscopy, SIMS, and XPS. Applications of atomic force spectroscopy to the characterisation of the following polymers have been reported: polythiophene [354], nitrile rubbers [355], perfluoro copolymers of cyclic polyisocyanurates of hexamethylene diisocyanate and isophorone diisocyanate [356], perfluorosulfonate [357], vinyl polymers

[353], polyhydroxybutyrate [358], polyacrylic acid nanogels [359], acrylic copolymers [360, 361], polyurethanes [362, 363], ethylene–methacrylate copolymer [365], polyamides [365], polyvinyl pyrrolidone [366], and polyethyl methacrylate dispersions [367].

3.13.1.2 Morphology

Merrett and co-workers [368] give a brief review of various techniques for analysing surface properties of polymeric biomaterials. Mention is made of scanning electron microscopy, electron microscopy, scanning tunnelling microscopy, atomic force microscopy, confocal scanning microscopy, Auger electron spectroscopy, XPS, SIMS, FT-IR, MALDI ToF spectroscopy, contact angle, and ellipsometry.

The application of atomic force microscopy to surface morphological studies has been covered in terms of the following polymers: polyesters, PE, PS [369], polycarbonate, polyimide, PTFE [370], polyurethane [371], rubbers [372], PEG [373], PS and poly(N-butyl methacrylate) [371], PS [375], PP [376], polyethers [377], polyorthoesters [378], poly(*p*-phenylene-vinylene) [379], bisphenol A–1,8-dibromooctane copolymer [380], and polycatechol [381].

3.13.1.3 Surface Defects

Brogly and co-workers [382] propose a method using AFM for the determination of the persistence length of molecular orientation of thin films adsorbed onto highly reflecting metals. The approach was based on the fact that molecular orientation persisted only over a given distance from the geometrical interface, this distance being called the 'persistence length of molecular orientation'. It was supposed that the nanofilm adsorbed was stratified and consisted of an oriented layer (in the near-interface region) plus an isotropic one. The correlation between infrared reflection absorption band intensities and simulated band intensities permitted determination of accurate molecular orientation and persistence length of orientation of a considered functional group. This was accomplished using various infrared reflection angles and polarisation state of the incident infrared wave. Film thickness and complex refractive index spectra were only needed to deduce calculated specular reflectance intensities.

3.13.1.4 Adhesion Studies

AFM has been used in studies of adhesion in the cases of PS/PMMA [383], PET/PP [384], and polyurethane- and epoxy-based adhesives [385]. General discussions of the use of this technique in adhesion studies have been reported [355, 386].

3.13.1.5 Polydispersivity

Zhang and co-workers [387] in their discussion of the polydispersivity of ethylene sequence length in metallocene ethylene–α-olefin copolymers characterised by the thermal fractionation technique used AFM to study crystal morphology.

3.13.1.6 Sub-surface Particle Studies

Feng and co-workers [388] used tapping mode AFM in the imaging of sub-surface nanoparticles in poly(N-vinyl-2-pyrrolidone) thin films.

3.13.1.7 Size of Nanostructures

The visualisation and size determination of polyacrylate nanostructures has been determined by AFM [389].

3.13.1.8 Visualisation of Molecular Chains

Individual chains of poly(2-ethynyl-9-substituted carbazoles) have been visualised by AFM [390].

3.13.1.9 Compositional Mapping

AFM and electric force microscopy are commonly used for compositional mapping of elastomers and multi-component rubber materials. Several aspects of the optimisation of atomic force microscopic experiments on polymers have been discussed by Yerina and Magonov [372]. Images reveal changes of ethylene–propylene–diene (EPDM) morphology caused by crosslinking and by loading with fillers (carbon black and silicon particles) and oil. It is shown that the morphology of iPP/EPDM vulcanisates, studied by AFM and electric force microscopy, depends on the ratio of components, degree of cure, and processing conditions. Diffusion of oil from the rubber component to the matrix is evident in the AFM images. Selective distribution of carbon black in the iPP matrix is responsible for the electrical conductivity of the thermoplastic vulcanisate material.

3.13.1.10 Surface Roughness

Sukhadia and co-workers [391] have used angle light scattering and wide-angle X-ray scattering to investigate the origins of surface roughness and haze in PE blown films.

3.13.1.11 Microthermal Analysis

Variable temperature pulsed force mode atomic force microscopy has been found to be a practical technique [392] for carrying out microthermal analyses of materials. Studies on a PS/PMMA blend showed that the pull-off force of PMMA was relatively insensitive to changes in temperature, whereas PS showed a large step increase above its glass transition temperature. The phase separated morphology could be characterised. However, before the technique can be applied more generally as a characterising tool, further investigation of the force affecting the pull-off force and its relationship with temperature is needed. At room temperature, the technique produced high-contrast images, showing the phase separated morphology of three segmented polyurethane elastomers. The results indicated that these materials had a complex structure with phases an order of magnitude larger than the domain size usually obtained using techniques such as small-angle X-ray scattering.

3.14 Scanning Electron Microscopy and Energy Dispersive Analysis using X-rays

The earliest paper identified on this technique was published in 1996, in which EDAX was used for elemental analysis in order to map the non-crystalline regions in semicrystalline PET [393].

Dilsiz and co-workers used SEM and EDAX to examine silver coatings of spindle- and filament-type particles for conductive adhesive properties [394]. The same technique was used by Lambert and co-workers [395] to examine the silicate structure deep in PET comonomer/silicate hybrid materials.

Start and Mauritz [396] used environmental SEM-EDAX, and also AFM and transmission electron microscopy, to study the formation of organic–inorganic nanocomposites within surlyn(PE-*co*-methacrylate-cation forms) random copolymers. SEM-EDAX has also been used to study of thin films of Prussian blue and *N*-substituted polypyrroles [397], epoxy resins [399], and the cause of failure in acetal plumbing fittings caused by exposure to chlorine [400].

References

1.   J.N. Lomonte and G.A. Tirpak, *Journal of Polymer Science*, 1964, **A2**, 705.

2.   F. Ciampelli, G. Bucci, A. Simonazzi and A. Santambroglio, *La Chimica e l'Industria*, 1962, **44**, 489.

3.   C. Tosi and T. Simonazzi, *Die Angewandte Makromolekulare Chemie*, 1973, **32**, 153.

4.   G. Natta, G. Mazzanti, A. Valvassori and A. Pajaro, *Chimica e L'Industria (Milan)*, 1957, **29**, 773.

5.   H.V. Drushel and F.A. Iddings, *Analytical Chemistry*, 1963, **35**, 28.

6.   H.V. Drushel and F.A. Iddings in *Proceedings of the 142nd ASC Meeting*, Atlantic City, NJ, USA, Fall 1962, Paper No.20.

7.   P.W. Wei, *Analytical Chemistry*, 1961, **33**, 215.

8.   T. Gössl, *Die Makromolekulare Chemie*, 1961, **42**, 1.

9.   P.J. Corish, *Analytical Chemistry*, 1961, 33, 1798.

10. R. Hank, *Rubber Chemistry and Technology*, 1967, **40**, 936.

11. D.W. Fraga, Shell Chemical Co. Ltd., Emerville Research Laboratory, Emerville, CA, USA, private communication.

12. I.I. Kaduji and J.H. Rees, *Analyst*, 1974, **99**, 435.

13. R.M. Smith and M. Dauson, *Analyst*, 1980, **105**, 85.

14. F. F-L. Ho, *Analytical Chemistry*, 1973, **45**, 603.

15. C.S.Y. Kim, A.L. Dodge, S-F. Lau and A. Kawsaki, *Analytical Chemistry*, 1982, **54**, 232.

16. D.G. Anderson, K.E. Isakson, D.L. Snow, D.J. Tessari and J.T. Vandeberg, *Analytical Chemistry*, 1971, **43**, 894.

17. C.B. Puchalsky, *Analytical Chemistry*, 1979, **51**, 1343.

18. A. Krishen, *Analytical Chemistry*, 1972, **44**, 494.

19. J. Helmuth, *Poly Vehromarium Plast*, 1973, **3**, 7.

20. E. Klesper, A. Corwin and D. Turner, *Journal of Organic Chemistry*, 1962, **27**, 700.

21. C.E. Miller, B.E. Eichinger, T.W. Gurley and J.G. Hermiller, *Analytical Chemistry*, 1990, **62**, 1778.

22. I. Geyer, *Revue Générale des Caoutchoucs et Plastiques*, 1996, **751**, 34.

23. G.C. Pandey, A. Kumar and R.K. Gary, *European Polymer Journal*, 2002, **38**, 745.

24. J.K. Aghenyega, G. Ellis, P.J. Hendra, W.F. Maddams, C. Parsingha, H.A. Wills and J. Chalmers, *Spectrochimica Acta, Part A*, 1990, **46A**, 197.

25. M. Bowden, P. Donaldson, D.J. Gardiner, J. Birnie and D.L. Gerrard, *Analytical Chemistry*, 1991, **63**, 2915.

26. D. Bloor, *Polymer*, 1999, **40**, 3901.

27. J.L. Koenig, *Infrared and Raman Spectroscopy of Polymers*, Rapra Review Report No.134, Rapra Technology Ltd., Shrewsbury, UK, 2001, **12**, 2.

28. S.V. Dubiel, G.W. Griffiths, L. Long, G.M. Baker and R.E. Smith, *Analytical Chemistry*, 1983, **55**, 1533.

29. S.C. Palticini and T.J. Porro, *Identification of Polymer Laminates Using the Diamond Cell Technique*, Perkin Elmer Infrared Bulletin, No.IRP121, Perkin Elmer, Boston, MA, USA.

30. P. Garside and P. Wyeth, *Polymer Preprints*, 2000, 41, 1792.

31. W. K. Way and C. Gloeckner in *Proceedings of 60th SPE Annual Technical Conference, ANTEC 2002*, San Francisco, CA, USA, 2002, Session W9, Paper No. 759.

32. P. Weiss, S. Bohic, M. Lapkowski and G. Daculsi, *Journal of Biomedical Materials Research*, 1998, **41**, 167.

33. R.R. Turner, S.W. Carlson and A.G. Altenau (unpublished work).

34. H.J. Sloane and R. Branstone-Cooke, *Applied Spectroscopy*, 1973, **27**, 217.

35. C. Shibota, M. Yamazaki and M. Takenchi, *Bulletin of the Chemical Society of Japan*, 1977, **50**, 311.

36. S. Chujo, S. Satoh, T. Ozeka and E. Nagai, *Journal of Polymer Science*, 1962, **61**, 171, S12.

37. S. Chujo, S. Satoh and E. Nagai, *Journal of Polymer Science: Part A General Papers*, 1964, A2, 895.

38. P.R. Sewell and D.W. Skidmore, *Journal of Polymer Science*, A1, 1968, **6**, 8, 2425.

39. W. Cooper, D.E. Eaves, M.E. Tunnicliffe and G. Vaughan, *European Polymer Journal*, 1965, **1**, 121.

40. M.E. Tunnicliffe, D.A. MacKillop and R. Hank, *European Polymer Journal*, 1965, **1**, 259.

41. A.G. Altenau, L.M. Headley, C.O. Jones and H.C. Ransaw, *Analytical Chemistry*, 1970, **42**, 1280.

42. T.C. Lee, I.M. Kolthoff and E. Johnson, *Analytical Chemistry*, 1950, **22**, 995.

43. A.R. Kemp and H. Peters, *Industrial and Engineering Chemistry Analytical Edition*, 1943, **15**, 52.

44. R.S. Porter, S.W. Nickosic and J.F. Johnson, *Analytical Chemistry*, 1963, **35**, 1948.

45. G.W. Tindall, R.L. Perry and J.L. Little, *Analytical Chemistry*, 1991, **63**, 1251.

46. E.G. Brame, Jr., and F.W. Yeager, *Analytical Chemistry*, 1976, **48**, 709.

47. W.H. Tuminello, J.V. Bletsos, T. Davidson, F.J. Wengert, I.N. Simpson and D. Slinn, *Analytical Chemistry*, 1995, **67**, 1955.

48. T. Tanaka and K. Hatada, *Journal of Polymer Science*, 1973, **11**, 2057.

49. D.M. Grant and E.C. Paul, *Journal of the American Chemical Society*, 1964, **86**, 2984.

50. W.O. Crain, A. Zambelli and J.P. Roberts, *Makromolecules*, 1971, **4**, 330.

51. C.J. Cannon and C.E. Wilkes, *Rubber Chemistry and Technology*, 1971, **44**, 781.

52. J.C. Randall, *Makromolecules*, 1978, **11**, 33.

53. G.R. Hadfield, W.E. Killinger and R.C. Zeigler, *Analytical Chemistry*, 1995, **67**, 3082.

54. M.J. Callego Cudero, M.M.C. López-Gonzáles and J.M. Barrales-Rienda, *Polymer International*, 1997, **44**, 61.

55. J.C. Randall in *Carbon 13 NMR in Polymer Science*, Ed., W.M. Pasika, ACS Symposium Series No. 103, American Chemical Society, Washington, DC, USA, 1979.

56. J.C. Randall, *Journal of Polymer Science: Polymer Physics Edition*, 1976, **14**, 2083.

57. F.C. Stehling, *Journal of Polymer Science A-1*, 1966, **4**, 189.

58. E.M. Barrall, R.S. Porter and J.F. Johnson in *Polymer Preprints*, 1964, **5**, 2, 816.

59. E.M. Barrall, R.S. Porter and J.F. Johnson *Journal of Applied Polymer Science*, 1965, **9**, 3061.

60. S.R. Satoh, T. Chujo and E. Nagai, *Journal of Polymer Science*, 1962, **62**, 510.

61. D.L. Miller, E.P. Samsel and J.G. Cobler, *Analytical Chemistry*, 1961, **33**, 677.

62. J. Haslam, J.B. Hamilton and A.R. Jeffs, *Analyst*, 1958, **83**, 66.

63. J. Haslam and A.R. Jeffs, *Analytical Chemistry*, 1957, **7**, 24.

64. E.P. Samsel and J.A. McHard, *Industrial and Engineering Chemistry Analytical Edition*, 1942, **14**, 750.

65. R. Fritz, *Fresenius' Zeitscrift für Analytical Chemistry*, 1960, **176**, 421.

66. A. Steyermark, *Journal of the Association of Official Agricultural Chemists*, 1955, **38**, 367.

67. S. Ehrlich-Rogozinsky and A. Patchornik, *Analytical Chemistry*, 1964, **36**, 840.

68. F. Viebock and C. Brechner, *Berich Deutscher Chemie*, 1930, **63**, 3207.

69. D.D. Schlueter and S. Siggia, *Analytical Chemistry*, 1977, **49**, 2343.

70. D.D. Schlueter and S. Siggia, *Analytical Chemistry*, 1977, **49**, 2349.

71. D.D. Schlueter, University of Massachusetts, 1976 [Ph.D. Thesis].

72. S.P. Frankoski and S. Siggia, *Analytical Chemistry*, 1972, **44**, 507.

73. R.D. Parker, Dow Corning Corporation, Barry, UK, unpublished procedure.

74. G. Gritz and H. Hurcht, *Zeitschrift fur Anorganische und Allgemeine Chemie*, 1962, **317**, 35.

75. V.M. Krasikova, A.N. Kaganova and V.D. Lobtov, *Journal of Analytical Chemistry USSR*, 1971, **28**, 1458. (English translation)

76. M.G. Voronlov and V.T. Shemyatenkova, *Bulletin of the USSR Academy of Sciences, Division of Chemical Science*, 1961, 178. *Chemical Abstracts*, 1961, **55**, 16285b.

77. J. Franc and K. Placek, *Collection of Czechoslovak Chemical Communications*, 1973, **38**, 513.

78. J. Franc, *Chemical Abstracts*, 1975, **82**, 67923q.

79. A.P. Kreshkov, V.T. Shemyatenkova, S.V. Syavtsillo and N.A. Palarmarchuck, *Journal of Analytical Chemistry USSR*, 1960, **15**, 727. (English translation)

80. G.W. Heylmun, R.L. Bujalski and H.B. Bradley, *Journal of Gas Chromatography*, 1964, **2**, 300.

81. V.M. Krasikova and A.N. Kaganova, *Journal of Analytical Chemistry USSR*, 1970, **25**, 1212. (English translation)

82. E.R. Bissell and D.B. Fields, *Journal of Chromatographic Science*, 1972, **10**, 164.

83. J. Franc and K. Placek, *Mikrochimica Acta*, 1975, **1**, 31.

84. C.L. Hanson and R.C. Smith, *Analytical Chemistry*, 1972, **44**, 1571.

85. M.M. O'Mara, *Journal of Applied Polymer Science, Part A1: Polymer Chemistry*, 1970, **8**, 1887.

86. J. van Schooten and J.K. Evenhuis, *Polymer (London)*, 1965, **6**, 561.

87. J. van Schooten and J.K. Evenhuis, *Polymer (London)*, 1965, **6**, 343.

88. Y. Sugimura, T. Nagaya, S. Tsuge, T. Murata and T. Takeda, *Macromolecules*, 1980, **13**, 928.

89. G. Audisto and G. Bajo, *Die Angewandte Makromolekulare Chemie*, 1975, **176**, 991.

90. H. Hoer, *Analytica Chimica Acta*, 1951, **5**, 550.

91. J.E. Brown, M. Tyron and J. Manel, *Analytical Chemistry*, 1963, **35**, 2173.

92. N. Tyron, S. Horowicz and E. Mondel, *Journal of Research of the National Bureau of Standards*, 1955, **55**, 219.

93. C.G. Smith and R. Beaver, *TAPPI*, 1980, **63**.

94. N. Svob and F. Flajsnman, *Croatia Chemica Acta*, 1970, **42**, 417.

95. J.L. Blackwell, *Analytical Chemistry*, 1976, 48, 1883.

96. J.L. Sharp and G. Paterson, *Analyst*, 1974, **105**, 135.

97. E.M. Barrall, R.S. Porter and D.E. Johnson, *Analytical Chemistry*, 1963, **35**, 73.

98. S. Paul, *Journal of Coatings Technology*, 1980, **52**, 47.

99. F.C-Y. Wang, B.B. Gerhart and P.B. Smith, *Analytical Chemistry*, 1995, **67**, 3536.

100. F.C-Y. Wang and P.B. Smith, *Analytical Chemistry*, 1996, **68**, 3033.

101. F.C-Y. Wang and P.B. Smith, *Analytical Chemistry*, 1996, **68**, 425.

102. D.H. Ahlstrom, S.A. Liebman and K.B. Abbas, *Journal of Polymer Science, Polymer Chemistry Edition*, 1976, **14**, 2479.

103. M. Suzuki, S. Tsuge and T. Takeychi, *Journal of Polymer Science*, 1972, **A10**

104. A.C. Jones, *Analytical Chemistry*, 1994, **66**, 1444.

105. W.W. Shackleford and J.M. McGuire, *Spectroscopy*, 1986, **10**, 17.

106. P.E. Kelly, *Ion Trap Detector Literature Reference List*, Finnegan MAT IDT Publication 21.

107. P.E. Kelly, *New Advances in the Operation of the Ion Trap Mass Spectrometer*, Finnegan MAT IDT Publication 10.

108. C. Campbell, *The Ion Trap Detector for Gas Chromatography; Technology and Application*, Finnegan MAT IDT Publication 15.

109. G.C. Stafford, *Recent Improvements in Analytical Applications of Advanced Trap Technology,* Finnegan MAT IDT Publication 16.

110. G.C. Stafford, *Advanced Ion Trap Technology in an Economical Detector for GC*, Finnegan MAT IDT Publication 20.

111. G.C. Stafford, *The Finnigan MAT Ion Trap Mass Spectrometer (ITMS) – New Developments with Ion Technology,* Finnegan MAT IDT Publication 24.

112. B.F. Rordorf, *An Automated Flow Tube Kinetics Instrument with Integrated GC – ITD Analysis,* Finnegan MAT IDT Publication 13.

113. J.E.P. Syka, *Positive Ion Chemical Ionization with an Ion Trap Mass Spectrometer*, Finnegan MAT IDT Publication 19.

114. R.A. Yost, N. McClennan and H.L.C. Menzzelaan, *Enhanced Full Scale Sensitivity and Dynamic Range in Finnigan MAT Ion Trap Detector with New Automatic Gain Control Software*, Finnegan MAT IDT Publication 22.

115. C. Camp, *Ion Trap Advancements, Higher Sensitivity and Greater Dynamic Range with Automatic Gain Control Software*, Finnegan MAT IDT Publication 23.

116. J.M. Richards and M.C. Bradford, *Development of a Cure Point Pyrolyser Inlet for the Finnigan MAT Ion-Trap Detector*, Finnegan MAT IDT Publication 25.

117. P. Bishop, *The Ion-Trap Detector, Universal and Specific Detection in One Detector*, Finnegan MAT IDT Publication 28.

118. P. Bishop, *The Use of an IDT 50 GLC Ion-Trap Detector Combination*, Finnegan MAT IDT Publication 36.

119. P. Bishop, *Low Cost Mass Spectrometer for GC*, Finnegan MAT IDT Publication 42.

120. C. Campbell and S. Evans, *The Ion-Trap Detector – The Techniques and its Application*, Finnegan MAT IDT Publication 29.

121. E. Olsen, *Serially Interfaced Gas Chromatography Fourier Transform Infrared Spectrometer Ion-Trap Mass Spectrometer*, Finnegan MAT IDT Publication 35.

122. J. Allison, *The Hows and Whys of Ion Trapping*, Finnegan MAT IDT Publication 41.

123. J. Todd, C. Mylchreest, T. Berry and D. Graves, *Supercritical Chromatography Mass Spectroscopy with an Ion-Trap Detector*, Finnegan MAT IDT Publication 46.

124. J.W. Eichelberger and W.L. Budd, *Studies in Mass Spectroscopy with an Ion-Trap Detector*, Finnegan MAT IDT Publication 47.

125. J.W. Eichelberger and L.E. Slivan, *Existence of Self-Chemical Ionization in the Ion-Trap Detector*, Finnegan MAT IDT Publication 48.

126. E. Genin, *Le Detecteur à Plegeage D'Jous de Chromatographic en Phase Gazeuse Technologie et Applications*, Finnegan MAT IDT Publication 53.

127. M. Le Leir, *The Use of the IDT a Low Cost GC/MS System for the Identification of Trace Compounds*, Finnegan MAT IDT Publication 51.

128. J.M. Richards, W.H. McClennan, J.A. Burger and H.H.C. Menza, *Pyrolysis-Short Column GC/MS Using the IDT and ITMS*, Finnegan MAT IDT Publication 56.

129. K. Qian, W.E. Killinger, M. Casey and G.R. Nicol, *Analytical Chemistry*, 1996, **68**, 1019.

130. *Pyrolysis GC/MS/IR Analysis of Polyethylene*, Hewlett Packard Application Note 228-971989.

131. D.L. Xoller, F.J. Cox, M.V. Johnston and K. Qian, *Polymer Preprints*, 2000, **41**, 669.

132. R. Yang, L. Guoping and K. Wang, *Journal of Applied Polymer Science*, 2001, **81**, 359.

133. F.C-Y. Wang and P.B. Smith, *Analytical Chemistry*, 1997, **69**, 618.

134. C. Westphal, C. Perrot and S. Karlsson, *Polymer Degradation and Stability*, 2001, **73**, 281.

135. M. Predel and W. Kaminsky, *Polymer Degradation and Stability*, 2000, **70**, 373.

136. J.L. Sharpe and G. Paterson, *Analyst*, 1974, **105**, 135.

137. O. Chiantore, D. Scalarone and T. Learner, *International Journal of Polymer Analysis and Characterization*, 2003, **8**, 67.

138. H. Zhang, P.R. Westmoreland, R.J. Farris, E.B. Coughlin, A. Plichta and Z.K. Brzozowski, *Polymer*, 2002, **43**, 5463.

139. S.R. Shield, G.N. Ghebremeskel and C. Hendrix, *Proceedings of the 159th ACS Rubber Division Meeting*, Providence, RI, USA, Spring 2001, Paper No.19.

140. G.N. Ghebremeskel and C. Hendrix in *Proceedings of the 152nd ACS Rubber Division Meeting*, Cleveland, OH, USA, Fall 1997, Paper No.72.

141. *Analysis of Polybutadiene by GC/IR/MS and Pyrolysis*, Hewlett Packard Application Brief IRD87-7, 1987.

142. M.R.S. Fuh and G-Y. Wang, *Analytica Chimica Acta*, 1998, **371**, 89.

143. *Rubber Technology International*, 1997, p.129.

144. M. Phair and T. Wampler in *Proceedings of the 150th ACS Rubber Division Meeting*, Louisville, KY, USA, Fall 1996, Paper No.69.

145. D. Yang, S-D. Li and D-M. Jia, *China Synthetic Rubber Industry*, 2001, **24**, 375.

146. D. Yang, S-D. Li, W-W. Fu, J.P. Zhong and D-M. Jia, *Journal of Applied Polymer Science*, 2003, **87**, 199.

147. R.S. Lehrle, N. Dadvand, I.W. Parsons, M. Rollinson and A.R. Skinner, *Polymer Degradation and Stability*, 2000, **70**, 395.

148. R.P. Lattimer and W.J. Keoenke, *Journal of Applied Polymer Science*, 1980, **25**, 101.

149. N. Dadvand, R.S. Lehrle, M. Parsons and M. Rollinson, *Polymer Degradation and Stability*, 1999, **66**, 247.

150. D. Fabbri, D. Tartari and C. Tronbini, *Analytica Chimica Acta*, 2000, **413**, 3.

151. J.C. Kleinert and C.J. Weschler, *Analytical Chemistry*, 1980, **52**, 1245.

152. M. Ezrin and G. Lavigne in *Proceedings of the 60th SPE Annual Technology Conference ANTEC 2000*, San Francisco, CA, USA, 2002, Session W9, Paper No.577.

153. K. Oba, H. Ohtani and S. Tsuge, *Polymer Degradation and Stability*, 2001, **74**, 171.

154. *Pyrolysis GC/MS/IR Analysis of Kraton 1107*, Hewlett Packard Application Note 228-100, 1989.

155. V.V. Zuev, F. Bertini and G. Audisio, *Polymer Degradation and Stability*, 2001, **71**, 213.

156. X.H. Li, Y.Z. Meng, Q. Zhu and S.C. Tjong, *Polymer Degradation and Stability*, 2003, **81**, 157.

157. L. Haussler, G. Pompe, V. Albrecht and D. Voigt, *Journal of Thermal Analysis and Calorimetry*, 1998, **52**, 131.

158. *Pyrolysis GC/MS/IR Analysis of Nylon 6/6*, Hewlett Packard Application Note 228-106, 1989.

159. R.J. Gritter, E. Gipstein and G.E. Adams, *Journal of Polymer Science*, 1979, **17**, 3959.

160. J.B. Maynard, J.E. Twichell and J.Q. Walker, *Journal of Chromatographic Science*, 1979, **17**, 82.

161. F.C-Y. Wang, *Analytical Chemistry*, 1999, **71**, 2037.

162. V.V. Zuev, F. Bertini and G. Audisio, *Polymer Degradation and Stability*, 2000, **69**, 169.

163. T.M.H. Cheng and G.E. Malawar, *Analytical Chemistry*, 1999, **71**, 468.

164. J.C. Kim, M.E. Song, S.K. Park, E.J. Lee, M.J. Rang and H.J. Ahn, *Journal of Applied Polymer Science*, 2002, **85**, 2244.

165. A. Hickey, J.J. Leahy and C. Birkinshaw, *Macromolecular Rapid Communications*, 2001, **22**, 1158.

166. S. Sundarrajan, M. Surianarayanan, K.S.V. Srinivasan and K. Kishore, *Macromolecules*, 2002, **35**, 3331.

167. L-H. Perng, *Journal of Polymer Research*, 2000, **7**, 195.

168. L-H. Perng, C.J. Tsai, Y.C. Ling, S.D. Wang C.Y. Hsu and T. Hwa, *Journal of Applied Polymer Science*, 2002, **85**, 821.

169. J. Kiji, A. Yamada, F. Bertini and G. Audisio, *Macromolecular Rapid Communications*, 2001, **22**, 598.

170. S. Carroccio, C. Puglisi and G. Montaudo, *Polymer Preprints*, 2000, **41**, 1, 684.

171. Y. Haishima, Y. Hayashi, T. Yagami and A. Nakamura, *Journal of Biomedical Materials Research (Applied Biomaterials)*, 2001, **58**, 209.

172. C.A. Liebman, D.H. Ahlstrom, W.H. Starnes and F.C. Schilling, *Journal of Macromolecular Science - Chemistry*, 1982, **A17**, 935.

173. E. Kawai, H.C. Lee, Y. Takao, K. Ogino and H. Sato, *Nippon Gomu Kyokaishi*, 2002, **75**, 465.

174. M. Kase, K. Kurihara and Y. Tachikawa, *Kobunishi Ronbunshu*, 1999, **56**, 8.

175. P. Sandra, G. Vanhoenacker, F. Lynen, L. Li and M. Schelfaut, *LCGC Europe*, 2001, **14**, 81.

176. C. Enjalbal, F. Lamaty, P. Sanchez, E. Suberchicot, P. Ribiere, S. Varray, R. Lazao, N. Yadav-Bhatnager, J. Martinez and J.L. Aubagnac, *Analytical Chemistry*, 2003, **75**, 175.

177. J.M. Julian and R.C. Domingo in *Proceedings of RadTech '96*, Nashville, TN, USA, 1996, Volume 1, p.423.

178. T. Yoshini and M.J. Shimoniya, *Polymer Science*, 1965, **A3**, 2811.

179. B.D. Cox, M.A. Park, G. Kaecher and E.A. Schweibert, *Analytical Chemistry*, 1992, **64**, 843.

180. P.M. Thompson, *Analytical Chemistry*, 1991, **63**, 2447.

181. C.H. Lin, Y. Murata and T. Imasaka, *Analytical Chemistry*, 1996, **68**, 1153.

182. F. David, L. Vanderroost and P. Sandra in *Proceedings of the 13th International Symposium on Capillary Chromatography*, Riva del Garda, Italy, 1991, p.1539.

183. F. Pisciotti, J. Lausmaa, A. Boldizar and M. Rigdahl, *Polymer Engineering and Science*, 2003, **43**, 1289.

184. X. van den Eynde, P. Bertrand and R. Jerome, *Macromolecules*, 1997, **30**, 6407.

185. D.W. Abmayr Jr., in *Proceedings of 2002 PLACE Conference*, Boston, MA, USA, 2002, Session 30, Paper No.95.

186. A.C. Prickett, P.E. Vickers, P.A.Smith and J.F. Watts in *Proceedings of Adhesion '99 Conference*, Cambridge UK, 1999, p.369.

187. P.A. Zimmerman, D.M. Hercules and A. Benninghoven, *Analytical Chemistry*, 1993, **65**, 983.

188. P.A. Zimmerman, D.M. Hercules and W.O. Loos, *Analytical Chemistry*, 1995, **67**, 2901.

189. I.V. Bletsos, J.M. Hercules, D. Fowler, D. van Leyen and A. Benninghoven, *Analytical Chemistry*, 1990, **62**, 1275.

190. I.V. Bletsos and A. Benninghoven, *Analytical Chemistry*, 1991, **63**, 2466.

191. X. Dong, A. Proctor and D.M. Hercules, *Macromolecules*, 1997, **30**, 63.

192. D.E. Fowler, R.D. Johnson, D. van Leyen and A. Benninghoven, *Analytical Chemistry*, 1990, **62**, 2088.

193. I.V. Bletsos, D.M. Hercules, D. van Leyen, B. Hagenhoff, E. Niehuis and A. Benninghoven, *Analytical Chemistry*, 1991, **63**, 1953.

194. R.J. Ratway and C.M. Balik, *Journal of Polymer Science, Polymer Physics Edition*, 1997, **35**, 1651.

195. W.J. van Ooij, J.M. Kim, S. Luo and S. Borros in *Proceedings of the 156ᵗʰ ACS Rubber Division Meeting*, Orlando, FL, USA, 1999, Paper No.57.

196. P. Bertrand, L.T. Weng, W. Lauer and R. Zimmer, *Rubber Chemistry and Technology*, 2002, **75**, 627.

197. S. Borros, E. Vidal, N. Agullo and W.J. van Ooij, *Kautschuk und Gummi Kunststoffe*, 2000, **53**, 711.

198. L.T. Weng, T.L. Smith. J. Feng and C.M. Chan, *Macromolecules*, 1998, **31**, 928.

199. S.S. Reddy, X. Dong, R. Murgasova and A.I. Gusev, *Macromolecules*, 1999, **32**, 1367.

200. V.H. Perez-Luna, K.A. Hooper, J. Kohn and B.D. Ratner, *Journal of Applied Polymer Science*, 1997, **63**, 1467.

201. J.T. Cherian and D.G. Castner, *Journal of Advanced Materials*, 2000, **32**, 28.

202. S.F. Lascelles, S.P. Armes, P.A. Zhdan, S.J. Greaves, A.M. Brown, J.F. Watts, S.R. Leadley and S.Y. Luk, *Journal of Materials Chemistry*, 1997, **7**, 1349.

203. W. Bijnens, J. Manca, T-D. Wu, M. D'Olicslacger, D. Vanderzande, J. Gelan, W. De Ceuninck, L. De Schepper and L.M. Stals, *Synthetic Metals*, 1996, **83**, 261.

204. X. van den Eynde, P. Bertrand and R. Jerome, *Macromolecules*, 1998, **31**, 6409.

205. X.M. Zeng, C.M. Chan, L.T. Weng and L. Li, *Polymer*, 2000, **41**, 8321.

206. C. Perruchot, M.M. Chehimi, M. Delamar and M.A. Eccles, *Synthetic Metals*, 2000, **113**, 53.

207. M.L. Abel, I.W. Fletcher, R.P. Digby and J.F. Watts in *Proceedings of Adhesion '99 Conference*, Cambridge, UK, 1999, p.87.

208. W.Y. Huang, S. Matsuoka, F.K. Kwei, Y. Okamoto, X. Hu, M.H. Rafailovich and J. Sokolov, *Macromolecules*, 2001, **34**, 7809.

209. M.J. Stachowski and A.T. Dibenedetto, *Polymer Engineering and Science*, 1997, **37**, 252.

210. A. Dhanabalan, S.S. Talwar, A.Q. Contractor, N.P. Kumar, S.N. Narang, S.S. Major, K.P. Muthe and J.C. Vyas, *Journal of Materials Science Letters*, 1999, **18**, 603.

211. V. Boittiaux, F. Boucetta, C. Combellas, F. Konoufi, A. Thiebault, M. Delamar and P. Bertrand, *Polymer*, 1999, **40**, 2011.

212. J. Eccles, R. Williams, T. Markkula and S. Hird in *Proceedings of the Rapra Polymers for the Medical Industry – Innovations in Healthcare Delivery Conference*, Brussels, Belgium, 2001, Paper No.19.

213. D.V. Patwardhan, H. Zimmer and J.E. Mark, *Journal of Macromolecular Science*, 1998, **35**, 1941.

214. Y. Tsai, F.J. Boerio, W.J. van Ouij and D.K. Kim, *Journal of Adhesion*, 1997, **62**, 127.

215. J. Sidwell in *Proceedings of the Rapra, Rubber Bonding 2001 Conference*, Cologne, Germany, 2001, Paper No.10.

216. M.R. Alexander and T.M. Due, *Journal of Materials Chemistry*, 1998, 8, 937.

217. X. van den Eynde, P. Bertrand and J. Penelle, *Macromolecules*, 2000, **33**, 5674.

218. D. Pleul, S. Schneider, F. Simon and H.J. Jacobasch, *Journal of Adhesion Science and Technology*, 1998, **12**, 47.

219. K. Endo, N. Kobayashi, M. Aida and T. Hoshi, *Polymer Journal (Japan)*, 1996, **28**, 901.

220. H.H-K. Lo, C-M. Chan and S-H. Zhu, *Polymer Engineering and Science*, 1991, **39**, 721.

221. Y. Deslandes, G. Pleizlier, D. Alexander and P. Santerre, *Polymer*, 1998, **309**, 2361.

222. J.B. Lhoest, E. Detrait, P. van den Bosch de Aguilar and P. Bertrand, *Journal of Biomedical Materials Research*, 1998, **41**, 95.

223. L.T. Weng, D.C.L. Wong, K. Ho, S. Wang, Z. Zeng and S. Yang, *Analytical Chemistry*, 2000, **72**, 4908.

224. S.R. Leadley, M.C. Davies, M. Vert, C. Braud, A.J. Paul, A.G. Shard and J.F. Watts, *Macromolecules*, 1997, **30**, 6920.

225. E.R. Lang, D. Leonard, H.J. Mathieu, E.M. Moser and P. Bertrand, *Macromolecules*, 1998, **21**, 6177.

226. M.L. Abel, A. Rattana and J.F. Watts, *Journal of Adhesion*, 2000, **73**, 313.

227. C. Perruchot, M.M. Chehimi, M. Delamar, P.C. Lacaze, A.J. Eccles, T.A. Steele and C.D. Mair, *Synthetic Metals*, 1999, **102**, 1194.

228. H. Hirahara, S. Abe, K. Mori, Y. Oishi and I. Tanaka, *Kobunshi Ronbunshu*, 2000, **57**, 95.

229. M.F. Tse, K.O. McElrath, H-C. Wang and W. Hu in *Proceedings of the 158th ACS Rubber Division Meeting*, Cincinnati, OH, USA, Fall 2000, Paper No.94.

230. J.W. Burley in *Proceedings of the ANTEC 2001 Conference*, Dallas, TX, USA, 2001, Paper No.606.

231. A. Rattana, J.D. Hermes, M.L. Abel and J.F. Walls, *International Journal of Adhesion and Adhesives*, 2002, **22**, 205.

232. J.F. Watts, A. Rattana, M.L. Abel and J. Hermes in *Proceedings of the 39th Annual Conference on Adhesion and Adhesives*, Oxford, UK, 2001, Paper No.3.

233. H.C. Lin, I.F. Tsai, A.C.M. Yang, M.S. Hsu and Y.E. Ling, *Macromolecules*, 2003, **36**, 2464.

234. J.M. Kim and W.J. van Ooij, *Journal of Adhesion Science and Technology*, 2003, **17**, 165.

235. D.A. Cole, *Journal of the Adhesive and Sealant Council*, Conference Proceedings, San Francisco, CA, Nov. 1996, Vol. 1, 503.

236. A.J. Swift in *Proceedings of a Rapra Conference on Developments in Polymer Analysis and Characterisation*, Shawbury, UK, 1998, Paper No.3.

237. A.J. Swift in *Proceedings of a Rapra Conference on Developments in Polymer Analysis and Characterisation*, Shawbury UK, 1999, Paper No.3.

238. R. West *in Proceedings of a Rapra Conference on Polymers for the Medical Industry – Characterising Medical Polymer Supply Chain by Surface Analysis*, London, UK, Paper No. 18.

239. S. Gross, G. Trimmel, U. Schubert and V. Di Noto, *Polymers for Advanced Technologies*, 2002, **13**, 254.

240. M. Omastova, J. Pavlinec, F. Simon and S. Kosina, *Polymer*, 1998, **39**, 6559.

241. Y-P. Wong, Y. Luo, R-M. Wang and L. Yuan, *Journal of Applied Polymer Science*, 1997, **66**, 755.

242. K. Shenshu, T. Furuzona, N. Koshizaki, S. Yamashita, T. Matsumato, A. Kishida and M. Akashi, *Macromolecules*, 1997, **30**, 4221.

243. Z. Lei, X. Han, Y. Hu, R-M. Wang and Y-P. Wang, *Journal of Applied Polymer Science*, 2000, **75**, 1068.

244. O.N. Tretinnikov and Y. Ikada, *Journal of Polymer Science*, 1998, **36**, 715.

245. X.D. Huang, S.G. Goh and S.Y. Lee, *Macromolecular Chemistry and Physics*, 2000, **201**, 2660.

246. X. Duan, C.M. Griffith, M.A. Dube and H. Sheardown, *Journal of Biomaterials Science, Polymer Edition*, 2002, **13**, 667.

247. Z. Luping, K.G. Neoh and E.T. Kong, *Chemistry of Materials*, 2002, **14**, 1098.

248. V.N. Vasilets, I. Hirata, H. Iwata and Y. Ikada, *Journal of Polymer Science, Polymer Chemistry Edition*, 1998, **36**, 2215.

249. M. Wan and J. Li, *Polymers for Advanced Technologies*, 2003, **14**, 320.

250. T-S. Chung, Z-L. Xu and C-H.A. Huan, *Journal of Polymer Science, Polymer Physics Edition*, 1999, **37**, 1575.

251. K. Winkler, K. Noworyta, A. de Bettencourt-Dias, J.W. Sobczak, C.T. Wu, L.C. Chen, W. Kutner and A.L. Balch, *Journal of Materials Chemistry*, 2003, **13**, 518.

252. H. Jeoung, Y.S. Kwan, B.J. Lee and C.S. Ha, *Molecular Crystals and Liquid Crystals, Section A*, 1997, **294** and **295**, 443.

253. K. Alimi, P. Molinie, V. Blel, J.L. Fave, J.C. Bernede and M. Ghedira, *Synthetic Metals*, 2002, **126**, 19.

254. J.C. Bernede, Y. Tregouet, E. Gourmelon and F. Martinez, *Polymer Degradation and Stability*, 1997, **55**, 55.

255. A. Boker, T. Herweg and K. Reihs, *Macromolecules*, 2002, **35**, 4929.

256. L. Ugalde, J.C. Bernede, M.A, Del Valle, F.B. Diaz and P. Leray, *Journal of Applied Polymer Science*, 2002, **84**, 1799.

257. L. Li, C.M. Chan, S. Liu, K.M. Ng, L.T. Weng and K.C. Ho, *Macromolecules*, 2000, **33**, 8002.

258. T.H. Xu and J.S. Van Deventer, *Industrial and Engineering Chemistry Research*, 2003, **42**, 1698.

259. X. Zhon, S.H. Go, S.Y. Lee and K.L. Ton, *Polymer*, 1998, **39**, 3631.

260. F. Samir, M. Morsli, J.C. Bernede, A. Bonnet and S. Letrant, *Journal of Applied Polymer Science*, 1997, **66**, 1839.

261. S.R. Leadly, J.F. Watts, A. Rodríguez and C. Lowe, *International Journal of Adhesion and Adhesives*, 1998, **18**, 193.

262. H. Niino and A. Yabe, *Journal of Polymer Science, Polymer Chemistry Edition*, 1998, **36**, 2483.

263. P. Johnson, R.W. Locke, J.B. Donnet, T.K. Wang, C.C. Wang and P. Bertrand in *Proceedings of the 156th ACS Rubber Division Meeting*, Orlando, FL, USA, Fall 1999, Paper No.179.

264. W.F. Stickle, *Particles on Surfaces 7: Detection, Adhesion and Removal*, Ed., K. Mittal, VSP BV, Utrecht, The Netherlands, 2002, p.3-10.

265. *Applying Surface Analysis Techniques for Particle Identification*, Hewlett Packard Ltd.

266. S.C. Shim, M.C. Suh and D.S. Kim, *Journal of Polymer Science, Polymer Chemistry Edition*, 1996, **34**, 3131.

267. J. Ma, L.H, Shi, J.M. Zhang, B.Y. Li, D.Y. Shen and J. Xu, *Journal of Applied Polymer Science, Polymer Chemistry Edition*, 2002, **20**, 573.

268. J. Ma, L. Shi, Y. Shi, S. Luo and J. Xu, *Journal of Applied Polymer Science*, 2002, **85**, 1077.

269. D. Frich, A. Hall and J. Economy, *Macromolecular Chemistry and Physics*, 1998, **199**, 913.

270. M.C. Biesinger, P-Y. Paepegaey, N.S. McIntyre, R.R. Harbottle and N.O. Petersen, *Analytical Chemistry*, 2002, **74**, 5711.

271. J.F. Watts and J.E. Castle, *Advanced Composite Materials*, 1999, **19**, 435.

272. R. West in Proceedings *of the Polymers for the Medical Industry Conference*, London, UK, 1999, Paper No.18.

273. P. Berrand and L.T. Weng in *Proceedings of the 153rd ACS Rubber Division Meeting*, Indianapolis, IN, Spring 1998, Paper No.38.

274. J.F. Watts, P.E. Vickers, A.C. Prickett and P.A. Smith in *Proceedings of the ACS Polymeric Materials Science and Engineering Meeting*, New Orleans, LA, USA, 1999, Volume 81, p.380.

275. M. Brenda, R. Doring and U. Schernau, *Progress in Organic Coatings*, 1999, **35**, 183.

276. D.A. Cole, *Journal of the Adhesive and Sealant Council*, 1996, 1, 503.

277. R. Michel, R. Luginbuehl, D. Graham and B.D. Ratner, *Polymer Preprints*, 1999, **40**, 2, 591.

278. A.T. Jackson, J.H. Scrivens, W.J. Simonsick, M.R. Green and R.H. Bateman, *Polymer Preprints*, 2000, **41**, 1, 641.

279. J. Bouajila, G. Raffin, H. Waton, C. Sanglar, J.O. Paisse and M.F. Grenier-Loustalot, *Polymers and Polymer Composites*, 2003, **11**, 233.

280. E.R.E. van der Hage, M.C. Duursma, R.M.A. Heeren, J.J. Boon, M. Nielen, A.J.M. Weber, C.G. de Koster and N.K. de Vries, *Macromolecules*, 1997, **30**, 4302.

281. E. Esser, C. Keil, D. Braun, P. Montag and H. Pasch, *Polymer*, 2000, **41**, 4039.

282. J.T. Mehl, R. Murgasova, X. Dong, D.M. Hercules and H. Nefzger, *Analytical Chemistry*, 2000, **72**, 2490.

283. A.M. Hawkridge and J.A. Gardella, *Polymer Preprints*, 2000, **41**, 635.

284. M.S. Montaudo, *Polymer News*, 2002, **27**, 115.

285. C.M. Guttman, S.J. Welzel, W.R. Blair and B.M. Fanconi, *Analytical Chemistry*, 2001, **73**, 1252.

286. B. Lepoittevin, X. Perrot, M. Masure and P. Hemery, *Macromolecules*, 2001, **34**, 425.

287. T.J. Kemp, R. Berridge, M.D. Eason and D.M. Hoddleton, *Polymer Degradation and Stability*, 1999, **64**, 329.

288. S. Hunt, G. Cash, H. Liu, G. George and D. Birtwistle, *Journal of Macromolecular Science*, 2002, **A39**, 1007.

289. M. Jayakannan, J.L.J. van Dongen and R.A.J. Janssen, *Macromolecules*, 2001, **34**, 5386.

290. P. Eisenberg, R. Erra-Balsells, Y. Ishikawa, J.C. Lucas, H. Nonami and R.J.J. Williams, *Macromolecules*, 2002, **35**, 1160.

291. M.S. Montaudo, C. Puglisi, F. Samperi and G. Montaudo, *Polymer Preprints*, 2000, **41**, 1, 686.

292. S. Carroccio, C. Puglisi and G. Montaudo, *Polymer Degradation and Stability*, 2003, **80**, 459.

293. R. Murgasova, E.L. Brantley, D.M. Hercules and H. Nefzger, *Macromolecules*, 2002, **35**, 8338.

294. S. Karlsson, *Polymer News*, 2002, **27**, 305.

295. G. Gallet, S. Carroccio, P. Rizzarelli and S. Karlsson, *Polymer*, 2002, **43**, 1081.

296. H. Goetz, U. Maschke, T. Wagner, C. Rosenauer, K. Martin, S. Ritz and B. Ewen, *Macromolecular Chemistry and Physics*, 2000, **201**, 1311.

297. M.S. Montaudo and G. Montaudo, *Polymer Preprints*, 2000, **41**, 1, 661.

298. D. Cho, S. Park, T. Chang, K. Ute, I. Fukuda and T. Kitayama, *Analytical Chemistry*, 2002, **74**, 1928.

299. V.T. Petkovska, D. Powell and K.B. Wagener, *Polymer Preprints*, 2002, **43**, 1, 427.

300. H.R. Kricheldorf, M. Rabenstein, M. Kaskos and M. Schmidt, *Macromolecules*, 2001, **34**, 713.

301. H.R. Kricheldorf and O. Petermann, *Journal of Polymer Science, Polymer Chemistry*, Edition, 2002, **40**, 4537.

302. T. Hamaide, M. Pantiru, H. Fessi and P. Boullanger, *Macromolecular Rapid Communications*, 2001, **22**, 659.

303. T. Hamaide and E. Lavit, *Polymer Bulletin*, 2002, **48**, 173.

304. T.D. McCarley, C.J. Dubois, R.L. McCarley, C. Cardona and A.E. Kailer, *Polymer Preprints*, 2000, **41**, 1, 674.

305. D.M. Hercules, J.T. Mehl, R. Murgasova and X. Dong, *Polymer Preprints*, 2000, **41**, 1, 639.

306. H. Ji, N. Sato, Y. Nakamura, Y. Wan, A. Howell, Q.A. Thomas, R.F. Storey, W.K. Nonidez and J.W. Mays, *Macromolecules*, 2002, **35**, 1196.

307. P.B. Smith, P.J. Pasztor, M.L. McKelvey, D.M. Meunier, S.W. Froelicher and F.C-Y. Wang, *Analytical Chemistry*, 1999, **71**, 61R.

308. C.M. Guttman, S.J. Wetzel, W.E. Wallace, W.R. Blair, R.M. Goldschmidt, D.L. Vanderhart and B.M. Fanconi, *Polymer Preprints*, 2000, **41**, 1, 678.

309. S. Lin-Gibson, L. Brunner, D.L. Vanderhart, B.J. Bauer, B.M. Franconi, C.M. Guttman and W.E. Wallace, *Polymer Preprints*, 2002, **43**, 1, 1331.

310. R.P. Lattimer and B.F. Goodrich, *Polymer Preprints*, 2000, **41**, 1, 667.

311. S.M. Weidner, J. Falkenhagen, H. Much, R.P. Krueger and J.F. Freidrich, *Polymer Preprints*, 2000, **41**, 1, 655.

312. H. Katayama, M. Kamigaito and M. Sawamoto, *Journal of Polymer Science, Polymer Chemistry Edition*, 2001, **39**, 1249.

313. A. Sunder, H. Frey and R. Mulhaupt, *Macromolecular Symposia*, 2000, **153**, 187.

314. G.J. Sun, H.J. Jang, S. Kaang and K.H. Chae, *Polymer*, 2002, **43**, 5855.

315. F. D'Agosto, M-T. Charreyre, F. Delolme, G. Dessalces, H. Cramail and A. Defficux, *Macromolecules*, 2002, **35**, 7911.

316. G. Rokicki, T. Kowalczyk and M. Glinski, *Polymer Journal (Japan)*, 2000, **32**, 381.

317. A.K. Nanda, K. Ganesh, K. Kishore and M. Surinarayanani, *Polymer Preprints*, 2000, **41**, 26, 9063.

318. J. Peterson, V. Allikmaa, J. Subbi T. Pehk and M. Lopp, *European Polymer Journal*, 2003, **39**, 33.

319. A. Berlin, G. Zotti, S. Zecchin, G. Schiavon, M. Cocchi, D. Virgili and C. Sabatini, *Journal of Materials Chemistry*, 2003, **13**, 27.

320. H. Chen, M. He, J. Pei and B. Liu, *Analytical Chemistry*, 2002, **74**, 6252.

321. S. Sosnowski, S. Slomkowski, A. Lorene and H.R. Kricheldorf, *Colloid and Polymer Science*, 2002, **280**, 107.

322. H.S Bazzi and H.F. Sleiman, *Polymer Preprints*, 2002, **43**, 1, 694.

323. F. Zuluaga, M. Larrahondo and K.B. Wagener, *Polymer Preprints*, 2002, **43**, 1, 627.

324. W.J. Feast and D. Parker, *Proceedings of the ACS Polymeric Materials Science and Engineering Meeting*, San Diego, CA, USA, 2001, Volume 84, p.299.

325. C. Ladaviere, N. Dorr and J.P. Claberie, *Macromolecules*, 2001, **34**, 5370.

326. B.V. Rozynov, R.J. Liukkonen, D.O. Becklin, A.L. Noreen and S.D. Ponto, *Polymer Preprints*, 2000, **41**, 1, 692.

327. C.N. McEwen and P.M. Peacock, *Analytical Chemistry*, 2002, **74**, 2743.

328. S. Karlsson, *Polymer News*, 2002, **27**, 305.

329. A.E. Giannakopulos, A.R. Bottrill, K.S. Lee and P.J. Derrick, *Polymer Preprints*, 2000, **41**, 1, 643.

330. U.S. Schubert and C. Eschbaumer, *Polymer Preprints*, 2000, **41**, 1, 676.

331. C.R. Shick, P.A. de Palma and R.K. Marcus, *Analytical Chemistry*, 1996, **68**, 2113.

332. *Proceedings of the SPE Joint Regional Technical Conference on Thermal and Mechanical Analysis of Plastics in Industry and Research*, Newark, DE, USA, 1999.

333. *Advanced Materials and Processes*, 1998, **153**, 10.

334. I.R. Harrison, *Proceedings of the SPE Joint Regional Technical Conference on Thermal and Mechanical Analysis of Plastics in Industry and Research*, Newark, DE, USA, 1999, p.142.

335. Loughborough University, UK, *Advances in Polymer Technology*, 1999, **18**, 181.

336. H.M. Pollock and A. Hammiche, *Journal of Physics*, 2001, **34**, R23.

337. V.V. Gorbunov, N. Fuchigami and V.V. Tsukruk, *Polymer Preprints*, 2000, **41**, 2, 1495.

338. H.M. Pollock, A. Hammiche, L. Bozec, E. Dupas, D.M. Price and M. Reading, *Polymer Preprints*, 2000, **41**, 2, 1421.

339. *Industria della Gomma*, 1999, **43**, 45.

340. M. El Fray and V. Altstädt, *Designed Monomers and Polymers*, 2002, **5**, 353.

341. A. Gupper, P. Wilhelm, N. Schmeid, S.G. Kazarian, K.L.A. Chan and J. Reussner, *Applied Spectroscopy*, 2002, **56**, 1515.

342. M. Abad, A. Ares, L. Barral, J. Cano, F.J. Diez, J. Lopez and C. Ramirez, *Journal of Applied Polymer Science*, 2002, **85**, 1553.

343. M.S. Tillman, J. Takatoya, B.S. Hayes and J.C. Seferis, *Journal of Thermal Chemistry and Calorimetry*, 2000, **62**, 559.

344. R. Haessler and H. Kleinert, *Adhasion Kleben und Dichten*, 2000, **44**, 36

345. V.V. Gorbunov, N. Fuchigami and V.V. Tsukruk, *High Performance Polymers*, 2000, **12**, 603.

346. J. Zhang, C.J. Roberts, K.M. Shakesheff, M.C. Davies and S.J.B. Tendler, *Macromolecules*, 2003, **36**, 1215.

347. T. Grossette, L. Gonon and V. Verney, *Polymer Degradation and Stability*, 2002, **78**, 203.

348. M. Song, D.J. Hourston, D.B. Grandy and M. Reading, *Journal of Applied Polymer Science*, 2001, **81**, 2136.

349. S.F. Lyuksyutov, R.A. Vaia, P.B. Paramonov, S. Juhl, L. Waterhouse, R.M. Ralich, G. Sigalov and E. Sancaktor, *Nature Materials*, 2003, **2**, 468.

350. M.L. Cerrada, *Revista de Plásticos Modernos*, 2002, **83**, 501.

351. *Characterisation of Solid Polymers: New Technologies and Developments*, Ed., H.J. Spells, Chapman and Hall, London, UK, 1994.

352. M. Ezrin, *Plastics Engineering*, 2002, **58**, 40.

353. J.D. Isner and R.W. Eiden in *Proceedings of Vinyltec 2000 Conference*, Philadelphia, PA, USA, 2000, p.198.

354. R.C. Advencula, C. Xia, J. Locklin and X. Fan, *Polymer Preprints*, 2002, **43**, 2, 345.

355. T.C. Ward, D.S. Parker and R.E. Jensen in *Proceedings of Adhesion '99*, Cambridge, UK, 1999, p.81.

356. S. Turri, A. Sanguineti, S. Novelli and R. Lecchi, *Molecular Materials and Engineering*, 2002, **287**, 319.

357. K. Adachi, W. Hu, H. Matsumoto, K. Ito and A. Tanioka, *Polymer*, 1998, **39**, 2315.

358. M. Nitschke, G. Schmack, A. Janke, F. Simon, D. Pleul and C. Werner, *Journal of Biomedical Materials Research*, 2002, **59**, 632.

359. S. Kadlubowsky, J. Grubelny, W. Olejniczak, M. Cichomsky and P. Ulanski, *Macromolecules*, 2003, **36**, 2484.

360. R.R. Thomas, K.G. Lloyd, K.M. Stika, L.E. Stephans, G.S. Magallanes, V.L. Dinonie, E.D. Sudol and M.S. El-Aasser, *Macromolecules*, 2000, **33**, 8828.

361. P. Polanowski, R. Ulanski, R. Wojciechowski, A. Tracz, J.K. Jeszka, S. Matejcek, S. Dorman, B. Pongs and H.W. Helberg, *Synthetic Metals*, 1999, **102**, 1789.

362. X. Chen, L. Wu, S. Zhon and B. You, *Polymer International*, 2003, **52**, 993.

363. A. Aneja and G.L. Wilkes, *Polymer*, 2002, **43**, 5551.

364. D.A. Suizdak, R.P. Start and K.A. Mauritz, *Journal of Polymer Science, Polymer Physics Edition*, 2003, **41**, 1.

365. T. McNally, W.R. Murphy, C.Y. Lew and P.J. Turner, *Polymer*, 2003, **44**, 2761.

366. D.K. Hood, L. Senak, S.L. Kopolow, M.A. Tallon, Y.T. Kwak, D. Patel and J. McKittrick, *Journal of Applied Polymer Science*, 2003, **89**, 734.

367. M.J. Krupers, H.R. Fischer, A.G.A. Schuurman and F.F. Vercauteren, *Polymers for Advanced Technologies*, 2001, **12**, 561.

368. A. Merrett, P.M. Cornelius, W.G. McClung and L. Unsworth, *Journal of Biomaterials Science*, 2002, **13**, 593.

369. R. Ganapathy, S. Manolache, M. Sarmadi and W.J. Simonsick, *Journal of Applied Polymer Science*, 2000, **78**, 1783.

370. V. Zaporojtchenko, T. Strunskus, K. Behnke, C. von Bechtolsheim, K. Kiene and F. Faupel, *Journal of Adhesion Science and Technology*, 2000, **14**, 467.

371. N.D. Kaushiva, S.R. McCartnet, G.R. Rossmy and G.L. Wilkes, *Polymer*, 2000, **41**, 285.

372. M. Yerina and S. Magonov in *Proceedings of the 161st ACS Rubber Division Meeting*, Savannah, GA, USA, Spring 2002, Paper No.31.

373. Y.G. Ko, Y.H. Kim, K.D. Park, H.J. Lee, W.K. Lee, H.D. Park, G.S. Lee and D.J. Ahn, *Biomaterials*, 2001, **22**, 2115.

374. S. Aftrossman, R. Jerome, S.A. O'Neill, T. Schmitt and M. Stamm, *Colloid and Polymer Science*, 2000, **278**, 993.

375. V. Shapovalov, V.S. Zaitsev, Y. Strzhemechny, J. Choudhery and M.H. Rafailovich, *Polymer International*, 2000, **49**, 432.

376. J. Hautojarvi and A. Leijala, *Journal of Applied Polymer Science*, 1999, **74**, 1242.

377. C.M. Chan, L. Li, K. Ng, J. Li and L.T. Weng, *Macromolecular Symposia*, 2000, **159**, 113.

378. S.R. Leadly, K.M. Shakesheff, M.C. Davies, J. Heller, N.M. Franson, A.J. Paul, A.M. Brown and J.F. Watts, *Biomaterials*, 1998, **19**, 1353.

379. W. Wang, Q. He, J. Zhai, J. Yang and F. Bai, *Polymers for Advanced Technologies*, 2003, **14**, 341.

380. L. Li, K-M. Ng, C-M. Chan, J-Y. Feng, X-M. Zeng and L-T. Weng, *Macromolecules*, 2000, **33**, 5588.

381. Y. Kong, S.L. Mu and B.W. Mao, *Chinese Journal of Polymer Science*, 2002, **20**, 517.

382. M. Brogly, G. Bistac and J. Schulz, *Macromolecular Theory and Simulations*, 1998, 7, 65.

383. H. Kim, M.H. Rafailovich and J. Sokolov, *Journal of Adhesion*, 2001, 77, 81.

384. M.J. Walzak, J.M. Hill, C. Huctworth, M.L. Wagter and D.H. Hunter in *Proceedings of the 20th Annual Anniversary Meeting of the Adhesion Society*, Hilton Head Island, SC, USA, 1997, p.89.

385. W.C. Wang, Y. Zhang, E.T. Kang and K.G. Neoh, *Plasmas and Polymers*, 2002, 7, 207.

386. A. Hussain, *Adhesive Age*, 2001, **44**, 24.

387. F. Zhang, J. Liu, Q. Fu, H. Huang, Z. Hu, S. Yao, X. Cai and T. He, *Journal of Polymer Science, Polymer Physics Division*, 2002, 40, 813.

388. J. Feng, L.T. Weng, C.M. Chan, J. Xhie and L. Li, *Polymer*, 2001, **42**, 2259.

389. Q. Ma and K.L. Wooley, *Journal of Polymer Science, Polymer Chemistry Edition*, 2000, **38**, 4805.

390. V. Percec, M. Obata, J.G. Rudick, B.B. De, M. Glodde, T.K. Bera, S.N. Magonov, V.S.K. Balagurusamy and P.A. Heiney, *Journal of Polymer Science, Polymer Chemistry Edition*, 2002, **40**, 3509.

391. A.M. Sukhadia, D.C. Rohlfing, M.B. Johnson and G.L. Wilkes, *Journal of Applied Polymer Science*, 2002, **85**, 2396.

392. D.B. Grandy, D.J. Hourston, D.M. Price, M. Reading, G.G. Silva, M. Song and P.A. Sykes, *Macromolecules*, 2000, **33**, 9348.

393. M. Sundararajan, H.L. Lee, P. Schwartz and S.K. Obendorf in *Proceedings of the ACS Polymeric Materials Science and Engineering Meeting*, Orlando, FL, USA, Fall 1996, Volume 75, p.295.

394. N. Dilsiz, R. Partch, E. Matijevic and E. Sancaktar, *Journal of Adhesive Science and Technology*, 1997, **11**, 1105.

395. A.A. Lambert, K.A. Mauritz and D.A. Schiraldi, *Journal of Applied Polymer Science*, 2002, **84**, 1749.

396. P.R. Start and K.A. Mauritz, *Polymer Preprints*, 2001, **42**, 2, 75.

397. R. Koncki and O.S. Wolfbeis, *Analytical Chemistry*, 1998, **70**, 2544.

398. A.J. Shields, D.M. Hepburn, I.J. Kemp and J.M. Cooper, *Polymer Degradation and Stability*, 2000, **70**, 253.

399. P.R. Lewis in *Proceedings of the ANTEC 2000 Conference*, Orlando, FL, USA, Paper No.607.

4 Examination of Polymer Surfaces and Defects

4.1 Introduction

It is often necessary for the polymer chemist in addition to examining the bulk properties of a polymer to examine particular features of the polymer: inclusions, defects, fractures, polymer surface, and so on. The features being looked at might include metallic inclusions such as catalyst remnants or impurities, or organic materials, such as weathered surfaces of polymer mouldings, thin laminate films on the surface of polymer film or sheet, or migrated additives on the surface of polymer mouldings. Various instrumental techniques are available for examining inclusions, etc., including electron microprobe X-ray emission spectrometry, NMR micro-imaging and Fourier transform infrared (FT-IR) microscopy, external reflectance spectroscopy, photoacoustic spectroscopy, time-of-flight secondary ion mass spectrometry (ToFSIMS), LIPI with laser desorption, atomic force microscopy (AFM), electron microscopy, and microthermal analysis. The availability of instrumentation for carrying out these measurements is reviewed in Appendix 1. The techniques are discussed in further detail in the following sections.

4.2 Electron Microprobe X-ray Emission Spectrometry

This is very much a specialised method, but is very useful for the identification of small inclusions in solids. The basic instrument is an electron microscope, but all or part of the electron beam can be focused onto any desired part of the sample causing it to emit X-rays characteristic of that material. These are then collimated and analysed in a conventional X-ray fluorescence spectrograph, thus giving at least a partial ratio analysis of that part of the sample. These instruments are expensive and are not for routine analysis, but are very useful for the identification of small problem areas in solid materials.

Energy dispersive measuring devices are most commonly used. Interference effects do occur, which are chiefly caused by fluorescent emissions and matrix defects. Corrections can be made.

Auger electron, electron energy loss, and X-ray photoelectron spectroscopies are related methods suitable for the determination of elements with atomic numbers down as far

165

as lithium, especially in small particles and surface films. When atoms are excited by electron beams or X-rays, electrons are emitted with energies characteristic of the emitting element. Analysis of these electrons by energy and rate of emission allows identification and determination of the emitting element of the target sample. These techniques require special equipment and expertise. The Atomic Energy Research Establishment, Harwell, UK, and similar laboratories can be contacted for information and advice.

4.2.1 Applications

Electron probe microanalysis has been used for the identification of metallic inclusions in polymer film and sheet. The sample is bombarded with a very narrow beam of X-rays of known frequency and the backscattered electron radiation is examined. An image is produced of the distribution of elements of any particular atomic numbers.

The technique can be illustrated by its application to the elucidation of a phenomenon known as lensing in polypropylene (PP) films. When the film in question is drawn down to between 25 and 50 μm in thickness any imperfection becomes apparent (**Figure 4.1**). Electron probe microanalysis was used as a direct check on whether or not a higher concentration of chlorine is associated with a lens than is present in the surrounding lens-free polymer. The PP film was first coated with copper to keep the sample cool by conduction during electron bombardment. A number of imperfections were seen to have a speck at the centre. **Figure 4.2** shows a backscattered electron image of such a speck; part of the lens is also seen. The specks examined were shown to contain sodium chloride. The X-rays emitted by the sample in electron probe microanalysis can be detected and counted by a proportional counter. It was demonstrated that chlorine and sodium only occur in high concentrations in the speck occurring in a lens, and that these elements are not generally distributed throughout the polymer. The material that surrounds the central speck within the lens is not very different from pure sodium chloride.

Murray [1] showed that scanning probe microscopy provided three-dimensional visualisation of surfaces down to the atomic scale using scanning tunnelling and AFM. The range of managing modes and spectroscopic techniques now available can be used to determine additional information on physical, chemical, and thermal properties of polymeric materials. Examples of the uses of lateral force microscopy to determine surface friction and force modulation to elucidate surface stiffness have been presented. Pulsed force mode now enables both these properties to be displayed simultaneously. Intermittent contact AFM (tapping) combined with phase imaging provides fast imaging of soft polymers combined with simultaneous materials contrast based on surface viscoelastic properties. Hence, the spatial distribution of multi-component polymers can be determined. Typical examples include identification of filler/binder materials and calculation of percentage

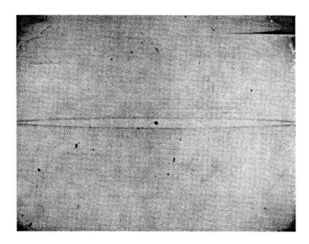

Figure 4.1 Lens in polypropylene film. Magnification × 18.
(*Source: Author's own files*)

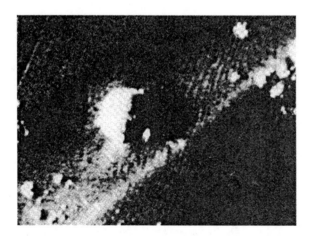

Figure 4.2 Back-scattered electron microanalysis of a speck within a lens in polypropylene. (*Source: Author's own files*)

distribution of copolymers. Microthermal analysis (see Section 4.16) uses a heated probe for thermal conductivity imaging. The same probe can be used for local calorimetric measurements through microthermal mechanical analysis and micro-differential thermal analysis. Alternatively, a sample heating stage can be used to track temperature-dependent morphology, stiffness, and adhesion changes using pulsed force mode.

4.3 NMR Micro-imaging

NMR micro-imaging is a powerful tool for examining the chemical composition of samples in a non-invasive way. It is one of the techniques available with the Varian UNITY spectrometer system. This is a high-performance research tool that is completely modular and fully integrated. UNITY NMR spectrometers enable spectroscopists to switch easily from micro-imaging liquids to solids and back, depending on the scope of their work.

The use of horizontal back imaging has extended this technique to small samples at high magnetic field strength. Because the field of micro-imaging is evolving so rapidly, Varian built modularity into its UNITY NMR spectrometers, giving the spectroscopist the ability to add new experimental capabilities as they are introduced.

Uses of this technique include the examination of microscopic changes of the physical state of polymeric materials, the distribution and binding of water in polymers, diffusion experiments of liquids in polymers, structural integrity investigations, and material homogeneity, porosity, and pore size.

Gussoni and co-workers [2] used proton NMR imaging techniques to characterise a silicone rubber. The effect of the polymerisation on the silicone properties and of the dispersion of a guest polymer inside the host matrix was investigated. Any source of inhomogeneities such as air bubbles or defects was localised and measured. The occurrence of an *in situ* polymerisation reaction was proved by IR spectroscopy.

4.4 Fourier Transform Infrared Spectroscopy

FT-IR spectroscopy is becoming a technique of choice for the analysis and characterisation of materials. It provides a unique fingerprint of the sample, which can lead to a positive identification within minutes. In the past, the problem was how to handle the sample and whether there was sufficient sample to obtain an adequate spectrum. As early as 1949 it was appreciated that one solution was to utilise the handling and viewing capabilities of a microscope, i.e., FT-IR microscopy.

The FT-IR microscope is a high-performance research-grade microscope offering all of the optical features and functions of an optical microscope. In addition to the high-quality visual image, the diagnostic powers of IR spectroscopy become available to the researcher. The ability to select accurately a specific region of a specimen and acquire spectroscopic data from a few hundred picograms of material makes FT-IR microscopy one of the most powerful analytical techniques available.

When studying samples by FT-IR microscopy, there are two major optical factors to be considered if the best results are to be obtained. The two primary sources of error are:

1. Diffraction arising from the measurement of a small object at or near the classic diffraction limit.

2. Optical aberrations introduced by the sample itself.

In order to obtain spectral data with the most accurately defined sampling area and with the highest energy throughput it is necessary to minimise both of these errors.

Redundant aperturing overcomes these problems and ensures high-quality data from free-standing polymer films, from small inclusions in polymer films, and from laminates composed of thin films.

Redundant aperturing involves placing remote apertures at image places located above and below the sample. The effect is to square the integral of the function describing the diffraction at the aperture, resulting in much improved spatial resolution. This can only be successfully implemented when an on-axis lens system is used.

4.4.1 Instrumentation

Perkin Elmer and Spectra-Tech are two major suppliers of FT-IR microscopes (see Appendix 1). Spectra-Tech supply four types of microscopes ranging from the most expensive research microscope, to an analytical microscope, to a laboratory microscope, to the relatively low cost spectroscope designed for routine transmission analysis of microsamples. Grazing angle microscopy of monomolecular films can also be carried out using the Spectra-Tech microscope.

The Perkin Elmer FT-IR microscope features compact Cassegrain optics, fixed stereo, zoom stereo, and video viewing points, a vernier calibrated sample stage, multiple illumination positions, high sensitivity, reproducibility, and ease of operation.

4.4.2 Applications

FT-IR microscopy has been applied to the examination of polymer films, polymer laminates, and polymeric printing inks or coatings on metal cans. It has been used to identify defects, imperfections, and inclusions in films and laminates. Inclusions not on the surface can be revealed for analysis by cutting across sections with a microtome. Polarised FT-IR microscopy studies have been performed, using a grid wire polariser, on single polymer films to establish the effects of molecular orientation on their physical performance. FT-IR microscopy has been combined with differential scanning calorimetry to follow the thermal behaviour of polymers [3].

The effects of film thickness on the drying of alkyd resin-based paints can be directly monitored by ratioing the intensity of the carbonyl band at 1735 cm^{-1} against the alkene stretch band at 1636 cm^{-1}. To avoid saturation of the strong carbonyl band, a 60° ZnSe crystal was used. The cured resin was readily cleaned from the trough with either methylene chloride or acetone. The unique out-of-compartment design permits the use of UV lamps or other radiation sources to evaluate their effects on rate of cure.

FT-IR microscopy has been applied to a study of the effect of weathering on the degradation of PP sheet after two years of outdoor exposure. Plots of 1712 cm^{-1}/1163 cm^{-1} absorption ratio (1712 cm^{-1} carbonyl absorption) *versus* depth from the exposed polymer surfaces showed that degradation extended deep within the polymer sheet showing an exponential decrease from the outer surface typical of a diffusion-controlled process.

4.5 Diffusion Reflectance FT-IR Spectroscopy (Spectra-Tech)

The high throughput and signal-to-noise ratios of FT-IR spectrometers, in combination with on-axis diffuse reflectance accessories, have made diffuse reflectance a powerful IR sampling technique. Diffuse reflectance permits the rapid analysis of many types of solid samples including powders, fibres, and rough surfaces. With the use of a Spectra-Tech Si-Card sampling kit (which provides finely divided polymer produced by abrasion with silicon carbide paper) diffuse reflectance can be extended to the analysis of many large intractable polymer samples. Diffuse reflectance spectroscopy (DRS) can also be used to study materials under a variety of environmental conditions such as elevated temperatures and pressures. An advantage of this technique is the speed of sample preparation and analysis (2 minutes) compared to the time taken for conventional sample preparation and methods involving scraping, grinding, pellet pressing, or solvent casting on film.

A brief overview [4] is given next of one theory of DRS for particles that absorb IR radiation. Incident radiation from the spectrometer is focused onto the surface of the sample and reflected energy collected. The reflected energy can either be classified as specular or diffuse reflectance. Specular reflectance arises from energy that is reflected by the particles but is not absorbed. The energy that penetrates one or more particles and is collected is called diffuse reflectance. With these experimental parameters optimised, a high-quality spectrum is obtained. For some samples, the specular component can be large and difficult to avoid. This can lead to problems in the measurement of the diffuse reflectance spectra. This is particularly true when using reflectance for any quantitative measurements. To eliminate the specular component in the diffuse reflectance measurement the Spectra-Tech collector employs the patented BLOCKER which ensures that radiation reflected without penetrating into the sample is not collected. Only the diffusely scattered radiation is collected.

4.6 Attenuated Total Infrared Internal Reflectance (ATR) Spectroscopy (Spectra-Tech)

ATR spectroscopy is a versatile and powerful technique for IR sampling. Minimal or no sample preparation is usually required for rapid analysis. ATR is ideal for those materials that are strong absorbers. In addition, ATR is a useful technique for providing information about the surface properties or condition of a material.

The phenomenon of internal reflection was first reported in IR spectroscopy in 1959. It was observed that if certain conditions are met, IR radiation entering a prism made of a high refractive index IR transmitting material (ATR crystal) will be totally internally reflected (**Figure 4.3**). This internal reflectance creates an evanescent wave which extends beyond the surface of the crystal into the sample held in intimate contact with the crystal. In regions of the IR spectrum where the sample absorbs energy, the evanescent wave will be attenuated.

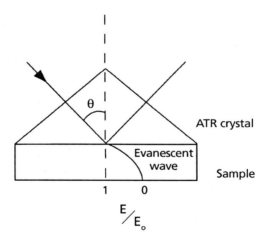

Figure 4.3 Theory of attenuated total internal reflectance spectroscopy.
(*Source: Author's own files*)

The condition that must exist for total internal reflectance is that the angle of the incident radiation, θ, must exceed the critical angle, θ_c. The critical angle is a function of the refractive indices of the sample and ATR crystal and is defined as:

$$\theta_c = \sin^{-1}(n_2/n_1)$$

where n_1 is the refractive index of the ATR crystal and n_2 is the refractive index of the sample. High-refractive-index materials are chosen for the ATR crystal to minimise the critical angle.

A property of the evanescent wave that makes ATR a powerful technique is that the intensity of the wave decays exponentially with distance from the surface of the ATR crystal. The distance, which is of the order of micrometres, makes ATR generally insensitive to sample thickness, allowing for the analysis of thick or strongly absorbing samples.

- *Horizontal ATR spectroscopy (Spectra-Tech)*. ATR spectroscopy is one of the most versatile and powerful sampling techniques in IR spectroscopy. Spectra-Tech's horizontal ATR accessories use a patented crystal designed to increase total sensitivity. ATR has been established as a method of choice for the analysis of samples that are either too thick or too strongly absorbing to analyse by transmission techniques without complicated sample preparation. A traditional limitation of the technique was in the 'vertical' mounting of the sampling crystal which limited the types of samples that could be analysed. Introduction of the horizontal ATR technique has extended the technique by increasing the possible sample types. Horizontal ATR has also greatly simplified the way in which samples are introduced and subsequently cleaned from the crystal surface. Because the evanescent wave decays very rapidly with distance from the surface, it is important to have the sample in intimate contact with the crystal. This is easily achieved with most liquids since they wet the surface of the ATR crystal. For solids, such as polymers, it is important to use a pressure device that presses the sample against the crystal.

4.7 External Reflectance Spectroscopy (Spectra-Tech)

Specular reflectance provides a non-destructive method for measuring surface coatings without sample preparation. These include surface-treated metals, resin and polymer coatings, paints, semiconductors, and many more. Specular reflection is a mirror-like reflection from the surface of the sample. The IR radiation is directed onto the surface of the sample at an angle of incidence, I. For specular reflection, the angle of reflection, R, is equal to the angle of incidence, θ. The amount of radiation reflected depends on the angle of incidence and the refractive index, surface roughness, and absorption properties of the sample.

172

The angle of incidence is selected depending on the thickness of the coating being studied. For very thin coatings in the nanometre thickness range, an 80° angle of incidence would be chosen. Reflectance measurements at this angle of incidence are often referred to as grazing angle measurements. For more routine samples having coatings in the micrometre range, a 30° angle of incidence is normally chosen. Many beverage and food containers are coated with a polymer coating on the inside of the container to preserve the flavour and freshness of the products. The spectrum is obtained by simply placing a piece of the inside of the sample on the sampling surface of the accessory and obtaining the sample spectrum. The sample spectrum is ratioed against a background spectrum obtained with the alignment mirror in place. The analysis time is less than two minutes.

Figure 4.4 shows some internal reflectance IR spectra of polymer films obtained using a Perkin Elmer 1760 FT-IR spectrometer. Using a room temperature detector, spectra measurements were made for two minutes at 4 cm^{-1} resolution. Two different reflection accessories were used. Pellets, powders, and fibres were measured with a Spectra-Tech Collector accessory. This is well suited to measuring surface reflection from small samples. Larger samples such as moulded products are measured with a Harrick specular reflection accessory with an horizontal sample stage.

The appearance of reflection spectra depends markedly on the physical form of the sample. For the materials considered here one can group the samples into three classes: continuous solids, powders and fibres, and foams.

For continuous solids with a shiny surface a specular reflection spectrum is obtained. In principle, the absorption spectrum can be derived from this by the Kramers–Kronig transformation. Examples of this are shown for both PP (**Figure 4.4(a)**) and polyester (**Figure 4.4(b)**) samples. Equally good results are obtained from both unfilled and carbon-filled samples. The presence of inorganic fillers does not alter the nature of the spectra and any contribution from the filler appears specular. The presence of glass as a filler has little effect on the spectra. The spectra obtained are adequate for qualitative identification, but there are limitations. Band shapes often appear unsymmetrical and baselines are uneven. When the surface is not shiny the spectra are weaker and may contain a diffuse component. When there are surface species the reflection spectrum may be unrepresentative of the bulk material. No spectra were obtainable from those carbon-filled samples that did not have shiny surfaces.

The spectra of fibres and powders are generally similar (**Figure 4.4(c)**). They are primarily diffuse reflection spectra with surface reflection significant in regions where individual particles are totally absorbing. These spectra are of limited use because their appearance is so dependent on particle size or fibre diameter.

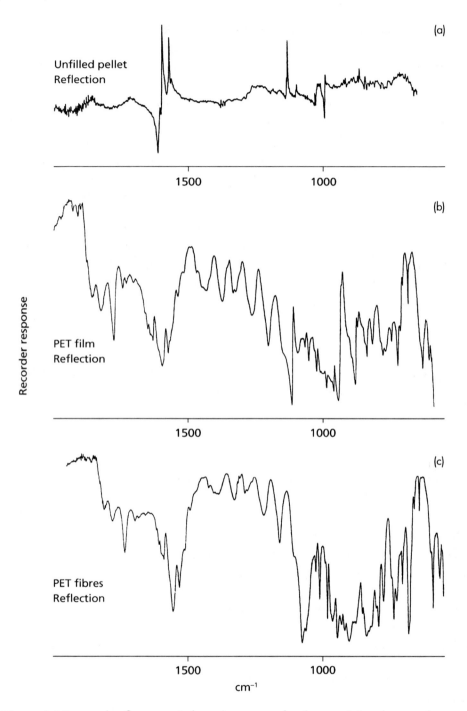

Figure 4.4 Internal reflectance infrared spectra of polymers (a) polypropylyene, (b) and (c) polyethylene terephthalate. (*Source: Author's own files*)

Foams generally give good diffuse reflection spectra. However, the surface of foam products is often a continuous skin which may introduce a significant specular component into the spectrum.

4.8 Photoacoustic Spectroscopy

The photoacoustic technique was discovered over a century ago when it was found that modulated light impinging on a solid in a sealed cell can produce a sound [5]. This is because energy absorbed by the sample is converted to heat which is transferred to the gas surrounding it. Since the gas is enclosed in a sealed volume the resulting pressure fluctuations can be detected as sound by a microphone. Note that only the energy absorbed by the sample contributes to the photoacoustic signal, unlike normal transmission measurements which recall the total attenuation of the source energy by the sample, including effects of scatter and reflection. As a result, unwanted saturation effects and Christiansen-type effects often observed in transmission and reflectance measurements can be overcome (or are much less prominent) in photoacoustic spectroscopy.

For many solid samples presented to the instrument as powders, photoacoustic spectra more closely resemble the true transmission spectra than do the corresponding diffuse reflectance spectra.

4.8.1 Instrumentation

The widespread acceptance of FT-IR spectroscopy has resulted in considerable advances in the design of IR sampling accessories. The high signal-to-noise (S/N) performance available with modern FT-IR spectrometers allows very rapid sample measurement; for many materials the measurement time is actually dominated by the time required for sample preparation and presentation rather than IR measurement. With the growing use of FT-IR as a routine measurement technique, analysts are increasingly looking towards sampling techniques that combine minimal sample preparation with convenient sample presentation. The use of photoacoustic detectors with FT-IR spectrometers has gained a reputation as a 'no-preparation' method. However, while the analyst is often spared lengthy sample preparation time, the total measurement time has often been comparable with conventional techniques because relatively low S/N performance offered by many FT-IR photoacoustic spectroscopy systems necessitates longer measurement times. For these reasons, analysts have tended to avoid the use of photoacoustic spectroscopy when 'difficult' samples have been encountered, i.e., it has been regarded as a research technique. Scan times tend to be of the order of minutes rather than seconds, which makes it imperative to use interferometers that can obtain the highest S/N ratio possible.

To overcome the problems discussed previously it is necessary to use an FT-IR instrument with very good S/N performance and good stability at lower optical path difference velocities. Perkin Elmer 1800 series or 1700 series (particularly models 1720/1760) FT-IR spectrometers coupled to an MTEC model 100 PA photoacoustic detector adequately meet these requirements.

4.8.2 Applications

Conventional IR measurements might require solvent casting and hot compression moulding of the sample into thin films prior to measurement. Photoacoustic spectroscopy requires no such sample preparation. **Figure 4.5** shows a photoacoustic spectrum of butyl maleate-stabilised polyvinyl chloride (PVC) granules. The spectrum shown is ratioed against a carbon black background spectrum. Total measurement time is one minute. Deeley and co-workers [6] compared Fourier transform near-IR Raman spectroscopy (see Chapter 3) with FT-IR photoacoustic and external reflection measurements for a range of solids including

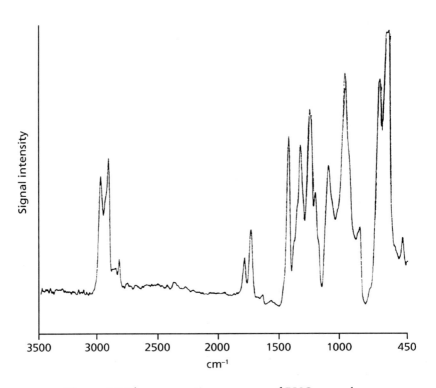

Figure 4.5 Photoacoustic spectrum of PVC granules.
(*Source: Author's own files*)

polyethylene terephthalate (PET) and PP. For Raman spectroscopy they used a Perkin Elmer 1720 Fourier transform spectrometer with a 1.064 μm Nd:YAG laser and an InGaAs detector at 77 K. For IR spectroscopy they used a Perkin Elmer 1760 FT-IR spectrometer using a DTGS detector for reflection measurements and a MTEC model 200 detector for photoacoustic measurements.

Raman spectra appear essentially independent of the physical form of the sample (**Figures 4.6(a)–(d)**). The only obvious difference between spectra of pellets, powders, films, fibres, and foams is the signal intensity. Powders give stronger spectra than solid chunks of polymer. Fibres and films give poor spectra presumably because of inefficient packing. The appearance of IR external spectra (**Figures 4.6(e)–(g)**) depends markedly on the physical form of the polymer. Solids with a shiny surface give specular reflection spectra from which the absorption spectra can in principle be derived by the Kramers–Kronig transformation. Good spectra were obtained even from carbon-filled polymers but unsymmetrical band shapes and uneven baselines were sometimes seen. Unless there are surface species the reflection spectrum may be unrepresentative of the bulk material. Solids with a non-shiny surface give weaker spectra which may contain a diffuse component. The spectra of fibres and powders are generally similar. They are primarily diffuse reflection spectra with surface reflection significant in regions where individual particles are totally absorbing. Foams generally give good diffuse reflection spectra unless the surface is a continuous skin which introduces a significant reflection contribution.

Good photoacoustic spectra were obtained from all forms of sample examined except for materials containing a high concentration of carbon black. The spectra showed relatively little variation with sample form or surface texture. This was the only technique used which showed clearly the presence of glass filler.

All three techniques (i.e., FT-IR, near-IR Raman, and photoacoustic spectroscopies) provided spectra suitable for qualitative identification quickly and with little sample preparation. Raman and external reflection spectra can be obtained directly from large objects but for photoacoustic measurements the samples must be small.

Raman spectra show the least dependence on physical form and so provide the most straightforward means of identification. However, they are not readily obtained from carbon-filled materials. For these samples photoacoustic spectroscopy was the most versatile alternative. Provided that the sample has a shiny surface, external reflection is a useful identification method. Its usefulness is restricted by the dependence of the spectra on the surface texture and form of the sample.

Urban [7] studied the quantitative aspects of alternated total reflectance and photoacoustic step-scan FT-IR and interfacial analysis of polymeric films. There are significant implications arising from the ability to stratify polymer surfaces and interfaces using attenuated total reflectance (ATR) and photoacoustic step-scan FT-IR spectroscopy. Theoretical concepts and

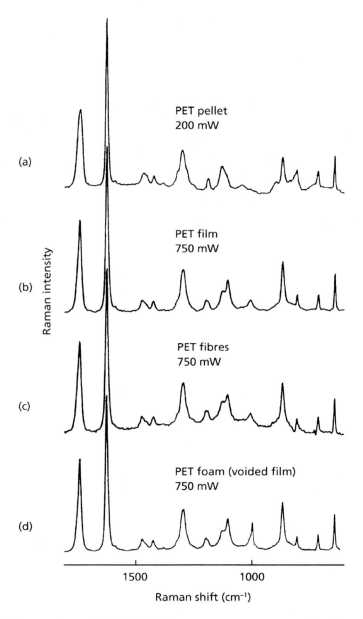

Figure 4.6 Comparison of FTIR NIR Raman spectroscopy with infrared external
spectra (a) – (d), Raman spectra, (e) – (h) infrared external spectra.
(*Source: Author's own files*)

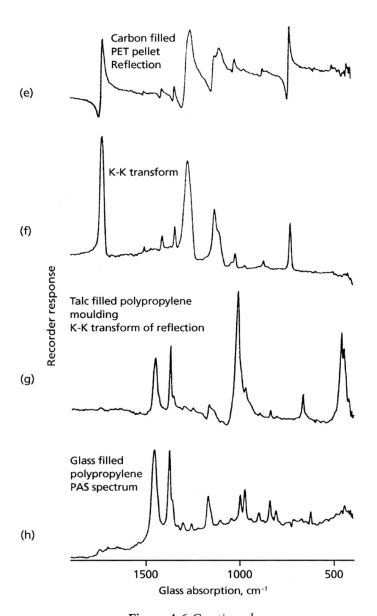

Figure 4.6 *Continued*

experimental methods involved in the stratification approach to polymer films are discussed. Quantitative analysis of interfacial regions in multi-layered polymers is given particular emphasis. The utilisation of ATR and step-scan photoacoustic spectroscopy, together with the stratification multi-layer concept, enables depth-profiling experiments to be carried out on multi-layered polymer films. It also provides molecular level data relating to interfacial interactions between the layers.

Dias and co-workers [8] carried out a photoacoustic study to monitor the crosslinking process in a grafted polymer based on ethylene and vinyl trimethoxysilane when crosslinked using saturated water vapour. Photoacoustic spectra at overtones of $-CH_2-$, $-CH_3$, and $-OH$ showed that the crosslinking processes were more efficient when the samples were prepared at 80 °C with the catalyst concentration in the range 5–7%. Using a different light modulation frequency procedure, it was observed that $-OH$ and $-CH_2-$ groups were more concentrated near the surface, showing a larger concentration gradient for $-CH_2-$ than $-OH$. The thermal diffusivity analysis did not exhibit a tendency that could be used for understanding the crosslinking process. The grafted samples appeared to exhibit a higher thermal conductivity.

4.9 X-ray Diffraction/Infrared Microscopy of Synthetic Fibres

In order to establish structure–property relationships for high-performance fibres, X-ray diffraction has been the analytical technique most often chosen. For most cases, X-ray diffraction was more than adequate in analysing structure in the fibres industry, however, with the discovery of the new generation of high-performance fibres with higher strength and orientation the sensitivity of X-ray diffraction to structure is being pushed to the limit. There is a definite need for a more sensitive technique capable of both orientation measurements and general fibre analysis.

High-performance fibres have historically been difficult to analyse by transmission FT-IR spectroscopy. Until a few years ago, the analysis of samples smaller than 25 µm was virtually impossible even with an IR microscope because the visible light portion did not have sufficient depth of field to focus the sample, and diffraction effects would result in a sloping baseline and stray light. Filament physical geometry, especially for highly orientated samples, causes optical aberrations resulting in band distortions that can lead to poor spectral quality.

Some of the new IR microscopes on the market are capable of minimising these problems. An IR-Plan IR microscope (obtained from Spectra-Tech, Stamford, CT) has been used to obtain IR spectra of Kevlar filaments and also to carry out orientation measurements. The microscope uses a 15× IR/visible reflecting lens with Cassegrain optics to focus the IR beam onto the sample (Refleachromat, Spectra-Tech). A 10× reflecting Cassegrain beam condenser is used to collect the transmitted beam after it passes through the sample. The beam condenser also has a sample compensator, which can be used to correct for optical aberrations caused by the fibre samples. The microscope uses its own narrow band liquid nitrogen-cooled Cassegrain detector to maximise sensitivity. The FT-IR system used was a Nicolet 5DXB bench model with a 2 cm⁻¹ maximum resolution capability (Nicolet Instrument Corporation, Madison, WI). In both the sample and the background files, 256 scans were run at 4 cm⁻¹ resolution for a spectral sweep from 2.5 to 15.3 μm. The background was acquired after the sample was scanned using the same aperture width. The microscope uses remote slits to mask above and below the sample, which speeds up sample preparation without sacrificing stray light rejection. The dichroic ratios were measured using a silver chloride wire-grid polariser placed before the 15× Cassegrain reflecting lens. The single filament should be mounted under some tension because any slack in the filament will tend to cause it to vibrate in and out of the path of the beam. There should also be no twist in the filament, especially if orientation measurements are to be taken.

Figure 4.7 shows an IR spectrum of a single 12 μm diameter filament of highly crystalline Kevlar. The modes that are most characteristic of Kevlar are the very intense amide vibrations. The intense band at 3.01 μm is the N–H stretching mode. The characteristic amide-I and amide-II bands are at 6.07 and 6.49 μm, respectively. The 6.07 μm band has been attributed to the C–O and C–N stretches from the carbonyl carbon of amide. The 6.49 μm band is a combination of vibration due to the N–H bending, the C–N stretching from the carbonyl carbon, and the aromatic C–C stretch. The peaks that confirm the presence of an aromatic amide are the aromatic ring vibrations at 6.20 and 6.61 μm. Para-substitution of the rings is confirmed by the 12.12 μm band. The intensity and sharpness of the bands are characteristic of the radial structure and the large crystallite

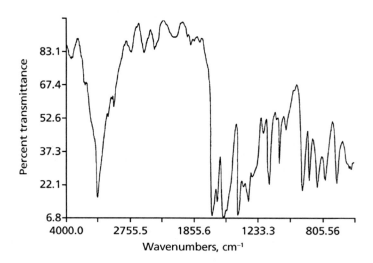

Figure 4.7 Infrared spectrum of single filament of Kevlar with the microscope optimised. (*Source: Author's own files*)

size of Kevlar. A final feature of Kevlar is the broad base of the 3.01 μm vibration. This has been associated with moisture and also with non-hydrogen-bonded N–H stretch. The IR spectra of Kevlar films are different from those of fibres in terms of band intensity and sharpness, due to crystallinity differences and the absence of radial sheets, but the same degree of orientation can be achieved in a Kevlar film cast from a spinning solution.

Molecular orientation measurements of polymers are an important way of determining how a product will perform in service. The overall bulk orientation of Kevlar has been directly related to its stiffness and indirectly related to its strength. As far as structure is concerned, the orientation is a measure of the degree of order and how the macromolecular chains are placed in the crystal lattice.

Orientation as measured by X-ray diffraction is well documented as being directly related to fibre stiffness (modulus). X-ray techniques yield bulk measurements that are sensitive only to the crystalline phase of the material and for most cases this is adequate. However, it has been observed that filaments of very different moduli sometimes cannot be distinguished using X-ray measurements. Clearly, this causes a dilemma as to whether orientation controls modulus or X-ray diffraction is not sensitive enough to detect the orientation difference.

A silver chloride wire-grid polariser placed before the 15× IR/visible reflecting lens of the set-up described previously was used to orient the beam. The filament orientation was very sensitive to tension, so each sample was measured in a set of callipers to keep the tension constant. This was found to be most critical for low-modulus samples because they seemed to be slightly easier to stretch, thereby changing the orientation. The standard deviation of the dichroic ratios (DR) measured ranged from 0.3 to 0.55 and is attributed more to small fibre structure differences than to the reproducibility of the technique. The DR were calculated using the following formula:

$$DR = \frac{A_1 + A_{11}}{A_1 - A_{11}}$$

where A_1 is the IR absorbance when the electric vector of the source is perpendicular to the fibre axis and A_{11} is the IR absorbance when the electric vector of the source is parallel to the fibre axis.

The samples were then broken on a recording stress–strain analyser using a 2.5 cm break to determine the modulus. The results, shown in **Figure 4.8**, are calculated for the 3.01 μm band of Kevlar. (The ideal orientation of Kevlar would be the polymer chains parallel to the fibre axis and the hydrogen bonding perpendicular to the fibre axis, as shown in **Figure 4.9**.) It can be seen from **Figure 4.8** that as the modulus increases, the hydrogen bonding of Kevlar preferentially becomes more highly oriented perpendicular to the fibre axis.

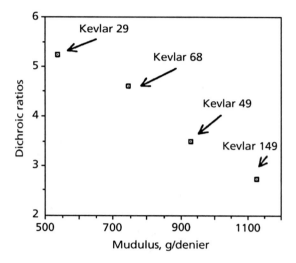

Figure 4.8 Plot of dichroic ratios of the 3320 cm⁻¹ band measured by infrared microscopy of the different stiffnesses of Kevlar products (mean of 10 measurements per sample). (*Source: Author's own files*)

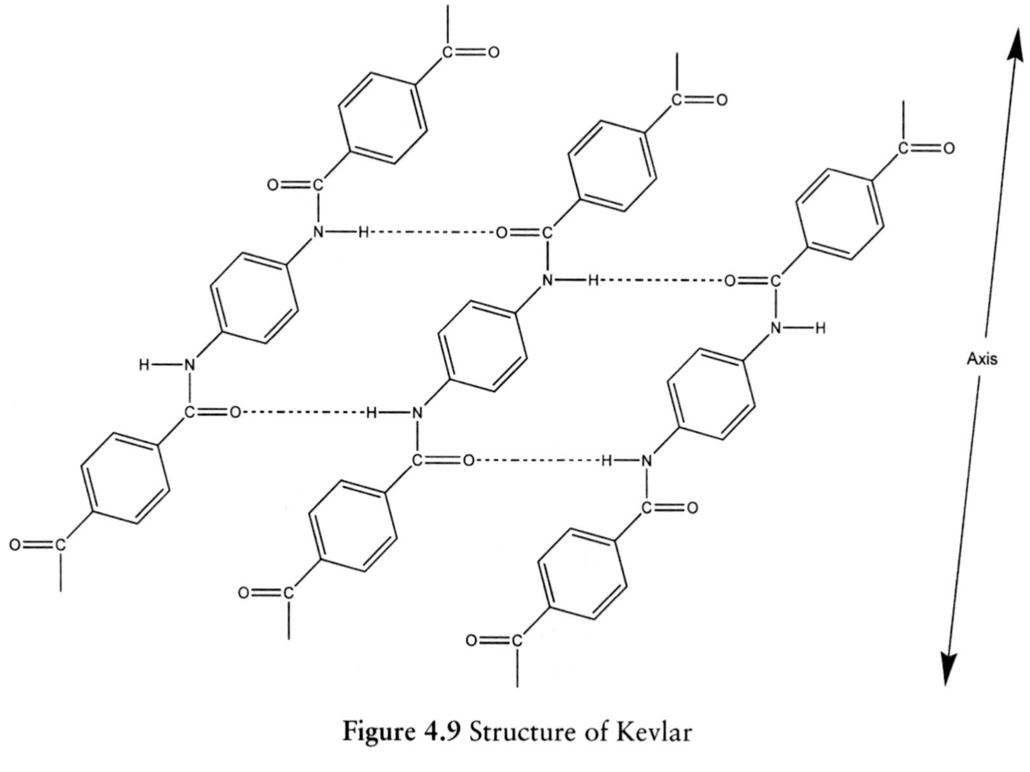

Figure 4.9 Structure of Kevlar

183

4.10 Scanning Electrochemical Microscopy (SECM)

This technique has been applied to the examination of polymer surfaces [9].

4.11 Scanning Electron Microscopy (SEM)

The surface chemical composition of polymer blends is often different from that of the bulk [10]. The general trend is such that the lower surface energy component tends to migrate toward the surface. Various experimental techniques probe different surface depths. Some techniques such as X-ray photoelectron spectroscopy (ESCA) and SIMS probe in the tens of angstroms range of the surface, ATR-IR probes usually in the micrometre range. SEM is a technique capable of probing in the hundreds to thousands of angstroms range of the surface.

Nagano and Nishimoto [11] used this technique to study the degradation due to weathering of styrene–butadiene vulcanisates. The degradation of these vulcanisates subjected to outdoor exposure in the Arizona desert for a year was studied. The dynamic viscoelasticity of sample sheets 2 mm thick and sliced films 0.2 mm thick was measured. SEM images of the surface and sheet cross-sections were also taken to observe microscopic changes in the samples.

SEM has also been used to study the structure of polyurethane foam–hydroxyapatite ceramics used in artificial bone materials [12] and for the surface analysis of rubber bonding [13].

4.12 Transmission Electron Microscopy (TEM)

Castellano and co-workers [14] investigated elastomer–silica interactions by image analysis-aided TEM.

4.12.1 Electron Microscopy and Inverse Gas Chromatography

At infinite dilution, inverse gas chromatography (IGC) was carried out using precipitated silica as the stationary phase to determine the surface properties and to calculate the adsorption free energy and enthalpy of low-molecular-weight analogues of the elastomers used as the eluents. Vulcanised samples based on Neoprene rubber, butadiene rubber, and styrene butadiene rubber were studied by image analysis-aided TEM to study the dispersion of the filler within the matrix, together with the aggregate size and shape distribution of silica, and the polymer–filler interactions. The thermodynamic predictions obtained via IG Cagreed well with the results of morphological studies.

Garbassi and co-workers [15] have the reviewed the subject of polymer surface analysis including surface characterisation techniques using a wide variety of spectroscopies and measurement of contact angles and surface force. Also, studies on surfaces and applications of surface science, including wettability, adhesion, barrier properties, biocompatibility, reduction of friction, and wear resistance, were carried out.

The volatile characteristics of all three amide-cured epoxy powders studied were an unidentified compound and melamine. The source of the melamine is not clear but it has been suggested that dicyandiamide decomposes to cyanide during curing. Cyanide can react with dicyanamide to form melamine. However, cyanamide was not observed among the volatilised products.

Gas chromatograms for a black paint with a resin made by the solvent method are shown in Figure 9.12. Table 9.18 lists the compounds identified. Even in this case, the most abundant compounds were isobutyl methyl ketone and xylenes. The origin of 2-phenyl-4-H-imidazole is not certain, but it might have been added as a catalyst or it may arise from the anhydride adduct during the curing process.

Most identified compounds gave observable molecular ions in electron impact mass spectrometry. Chemical ionisation mass spectrometry (CIMS) with isobutane was carried out to determine the relative molecular masses. In all instances the quasi-molecular ion was M + 1 with a relative abundance of 10% - no M + 57 was observed. Some compounds were identified on the basis of their mass spectra only because no reference compounds were obtainable.

High-resolution mass spectra were obtained for unidentified compounds a, b, c, d, and f. The molecular formula of compounds a, b, c, and d is $C_{12}H_{22}O$. Proposed structures are isomers of ketones, e.g., a and b might be 2- or 4-isomers of ethyl isopropylcyclohexyl ketone and c and d might be cyclohex-4-yl-hexan-2-one and cyclohex-4-yl-4-methylpentan-2-one, respectively. One unidentified compound (e) was found, also the quasi-molecular ion M + 1105. The only fragment was M – 18 in CIMS. The molecular formula of unidentified compound f is $C_{15}H_{14}O$. The proposed structure is 2-biphenyl-2,3-epoxypropane. The eight most intense ions of the unidentified compounds and their relative abundances, together with the M – 1 ions in their chemical ionisation spectra, are given in Tables 9.17 and 9.18 (see Chapter 9).

Other polymers that have been examined by evolved gas analysis include PVC [16–18], polystyrene (PS) [19-22], styrene–acrylonitrile copolymers [23–25], polyethylene and PP [26–32], polyacrylates and copolymers [33–39], PET and polyphenylenes and polyphenylene oxides and sulfides [36, 40–44]. Studies involving the use of chromatography include the thermal degradation of PVC [45], and vinyl plastics [46] and polysulfone [47].

4.12.2 Supersonic Jet Spectrometry

Franc and Placek [48] studied the chemical species of PS and polycarbonate (PC) resulting from supersonic jet spectrometry at temperatures up to 400 °C. The decomposition products from PS were styrene monomer, dimer, and trimer and toluene. The decomposition product from PC was *p*-cresol.

4.13 ToF SIMS

This technique has been applied to the analysis of the surface of polymer blends such as miscible PC/PS blends [49] (See also Section 3.11.1.).

Surface mass spectrometry (MS) techniques measure the masses of fragment ions that are ejected from the surface of a sample to identify the elements and molecules present. The techniques are complementary to electron spectroscopy since they provide extra absolute and surface sensitivity and give very specific molecular information. For unknown samples it is common to use a combination of electron spectroscopy and MS for surface characterisation. There are two methods of surface mass spectrometry used in polymer analysis: SIMS and laser ionisation mass analysis (LIMA). Of these SIMS is by far the most important.

LIMA is a similar technique to ToFSIMS except that the ionising primary beam is a laser. One of the main benefits of LIMA is that it has a small analysis area which allows the study of small particles on or in sample surfaces. The data are obtained quickly and from a deeper depth (~1 mm) than SIMS. However, because the laser/surface interaction can be damaging, only elemental and simple molecular information is available.

LIMA is a good technique for rapid micro-point analysis of any material and can be used to examine buried features. It is especially sensitive to low-mass elements which makes LIMA a highly complementary technique to SEM/energy dispersive analysis using X-rays.

4.14 Laser-Induced Photoelectron Ionisation with Laser Desorption

Schrimmer and Li [50] carried out surface analysis of bulk polymers using laser-induced photoelectron ionisation (LIPI) with laser desorption in a ToFS. In this method the laser beam is focused on the polymer surface to induce fast thermal dissociation and/or photodissolution of the polymer. The resulting neutral species are then ionised using photoelectron ionisation. This process involves a laser–metal interaction in which a low-power pulsed ultraviolet laser beam is directed to an appropriate metal surface placed exterior to the source region of a Wiley–McLaren ToF-MS. The photoelectrons generated are accelerated as a narrowly distributed beam by the fringing fields of the source region. The electron beam, travelling in a direction almost parallel to the extraction grid, is used to ionise the desorbed neutrals that are entrained into the ionisation region.

Signal intensity versus *m/z* plots all give readily interpretable MS suitable for structural analysis and chemical identification of the polymer being examined. End-group analysis can be performed. A high detection sensitivity is achievable. The following polymers were examined by Schrimmer and Li [50]: PS, polyvinyl pyridine, PS-2-vinyl pyridine, polyethylene glycol (PEG), and polymethyl methacrylate.

4.15 Atomic Force Microscopy

One of the applications of this technique is the examination of polymer surfaces. For further details, see Section 3.13.

4.16 Microthermal Analysis

Microthermal analysis has been used in surface and depth profiling studies of PP [51, 52], multi-block copolymers [53], polytetrafluoroethylene/silicone blends [14], PEG/polylactic acid blends [55], and PS/polyvinyl methyl ether blends [56]. See also Section 3.12.

References

1. A. Murray in *Proceedings of a Rapra Conference on Polymer Rheology 2001: A Practical Approach to Quality Control for the Rubbers and Plastics Industries*, Shawbury, UK, 2001, Paper No.9.

2. M. Gussoni, F. Greco, M. Mapelli, A. Vezzoli and E. Ranucci, *Macromolecules*, 2002, 35, 1714.

3. F.M. Mirabella in *Infrared Microspectroscopy Theory and Applications*, Eds., G. Messerschmidt and M.A. Harthcock, Practical Spectroscopy 6, Marcel Dekker Inc., New York, NY, USA, 1988.

4. *Applications of FT-IR Microscopy to Polymer Analysis*, Spectratech Application Note Volume No.10, Spectratech Lord, 1988.

5. J.F. McClelland, *Analytical Chemistry*, 1983, 55, 89A.

6. C. Deeley, J. Sellors and R. Spragg, *A Comparison of FT Near IR Raman Spectroscopy, with FT-IR Photoacoustic and Reflection Measurements of Polymers*, Perkin Elmer Infrared Spectroscopy Applications 14.3, Perkin Elmer Corporation, Norwalk, CT, USA, 1988.

7. M.W. Urban, *Proceedings of ACS Polymeric Materials Science and Engineering Meeting*, Orlando, FL, Fall 1996, Volume 75, p.11.

8. D.T. Dias, A.M. Medina, M.L. Baesso, A.C. Beuto, M.F. Porto and A.T. Rubiro, *Journal of Physics D*, 2002, 35, 3240.

9. C. Lee and A.J. Bard, *Analytical Chemistry*, 1990, **62**, 906.

10. S. Lee and C.S.P. Sung, *Polymer Preprints*, 1999, 40, 2, 649.

11. E. Nagano and K. Nishimoto, *Nippon Gomu Kyokaishi*, 2000, **73**, 399.

12. S. Padilla, J. Roman and M. Vallet-Regi, *Journal of Materials Science: Materials in Medicine*, 2002, **13**, 1193.

13. J. Sidwell in *Proceedings of a Rapra Conference on Rubber Bonding 2001*, Cologne, Germany, 2001, Paper No.10.

14. M. Castellano, L. Falqui, G. Costa, A. Turturro and R. Valenti, *Journal of Macromolecular Science B*, 2002, **B41**, 451.

15. F. Garbassi, M. Morra and E. Occhiello, *Polymer Surfaces: from Physics to Technology*, 1994, John Wiley & Sons, Chichester, UK.

16. W.H. Tuminello, I.V. Bletsos, F. Davidson, F.J. Weigert, N. Simpson and D. Slinn, *Analytical Chemistry*, 1995, 67, 1955.

17. T. Tanaka and K. Hatada, *Journal of Polymer Science*, 1973, **11**, 2057.

18. D.M. Grant and E.C. Paul, *Journal of the American Chemical Society*, 1964, 86, 2984.

19. W.O. Crain, A. Zambelli and J.P. Roberts, *Macromolecules*, 1971, **4**, 330.

20. C.J. Cannon and C.E. Wilkes, *Rubber Chemistry and Technology*, 1971, **44**, 781.

21. J.C. Randall, *Makromolecules*, 1978, **11**, 33.

22. G.R. Hatfield, W.E. Killinger and R.C. Zeigler, *Analytical Chemistry*, 1995, **67**, 3082.

23. J.C. Randall in *Carbon 13 NMR in Polymer Science*, Ed., W.M. Pasika, ACS Symposium Series No.103, ACS, Washington, DC, USA, 1979.

24. J.C. Randall, *Journal of Polymer Science: Polymer Physics Edition*, 1976, **14**, 2083.

25. F.C. Stehling, *Journal of Polymer Science A-1*, 1966, **4**, 189.

26. E.M. Barrall, R.S. Porter and J.F. Johnson in *Proccedings of the 148th National Meeting of the ACS Division of Polymer Chemistry*, Chicago, IL, USA, 1964, 5, 2, 816.

27. E.M. Barrall, R.S. Porter and J.F. Johnson, *Journal of Applied Polymer Science*, 1965, **9**, 3061.

28. S.R. Satoh, T. Chujo and E. Nagai, *Journal of Polymer Science*, 1962, **62**, 510.

29. D.L. Miller, E.P. Samsel and J.G. Cobler, *Analytical Chemistry*, 1961, **33**, 677.

30. E.P. Samsel and J.A. McHard, *Industrial Engineering Chemistry Analytical Edition*, 1942, **14**, 750.

31. J. Haslam, J.B. Hamilton and A.R. Jeffs, *Analyst*, 1958, **83**, 66.

32. J. Haslam and A.R. Jeffs, *Analytical Chemistry*, 1957, **7**, 24.

33. R. Fritz, *Fresenius' Zeitscrift für Analytical Chemistry*, 1960, **176**, 421.

34. A. Steyermark, *Journal of the Association of Official Agricultural Chemists*, 1955, **38**, 367.

35. S. Ehrlich-Rogozinsky and A. Patchornick, *Analytical Chemistry*, 1964, **36**, 840.

36. F. Viebock and C. Brechner, *Berich Deutscher Chemie*, 1930, **63**, 3207.

37. J. Franc and K. Placek, *Collection of Czechoslovak Chemical Communications*, 1973, **38**, 513.

38. J. Franc, *Chemical Abstracts*, 1975, **82**, 67923q.

39. A.P. Kreshkov, V.T. Shemyatenkova, S.V. Syavtsillo and N.A. Palarmarchuck, *Journal of Analytical Chemistry USSR*, 1960, **15**, 727. (English translation)

40. D.D. Schleuter and S. Siggia, *Analytical Chemistry*, 1977, **49**, 2343.

41. R.D. Parker, Dow Corning Corporation, Barry, Wales, UK, unpublished procedure.

42. G. Gritz and H. Hurcht, *Zeitschrift für Anorganische Allegmeine Chemie*, 1962, **317**, 35.

43. V.M. Krasikova, A.N. Kaganova and V.D. Lobtov, *Journal of Analytical Chemistry USSR*, 1971, **28**, 1458. (English translation)

44. M.G. Voronlov and V.T. Shemyatenkova, *Bulletin of the Academy of Sciences of the USSR Division of Chemical Sciences*, 1961, 178 (Chemical Abstracts, 1961, 55, 16285b).

45. G.W. Heylmun, R.L. Bujalski and H.B. Bradley, *Journal of Gas Chromatography*, 1964, **2**, 300.

46. V.M. Krasikova and A.N. Kaganova, *Journal of Analytical Chemistry USSR*, 1970, **25**, 1212. (English translation)

47. E.R. Bissell and D.B. Fields, *Journal of Chromatographic Science*, 1972, **10**, 164.

48. J. Franc and K. Placek, *Mikrochimica Acta 11*, 1975, **1**, 31.

49. P.M. Thompson, *Analytical Chemistry*, 1991, **63**, 2447.

50. D.C. Schrimmer and L. Li, *Analytical Chemistry*, 1996, **68**, 250.

51. T. Grossette, L. Gonon and V. Verney, *Polymer Degradation and Stability*, 2002, **78**, 203.

52. M. Abad, A. Ares, L. Barral, J. Cano, F.J. Diez, J. Lopez and C. Ramirez, *Journal of Applied Polymer Science*, 2002, **85**, 1553.

53. M. El Fray and V. Altstadt, *Designed Monomers and Polymers*, 2002, **5**, 353.

54. A. Gupper, P. Wilhelm, N. Schmeid, S.G. Kazarian, K.L.A. Chan and J. Reussner, *Applied Spectroscopy*, 2002, **56**, 1515.

55. J. Zhang, C.J. Roberts, K.M. Shakesheff, M.C. Davies and S.J.B. Tendler, *Macromolecules*, 2003, **36**, 1215.

56. M. Song, D.J. Hourston, D.B. Grandy and M. Reading, *Journal of Applied Polymer Science*, 2001, **81**, 2136.

5 Volatiles and Water

In this chapter the instrumentation used for the determination of volatile materials present in polymers is discussed. These include:

- Monomers, e.g., styrene, acrylonitrile, vinyl chloride.

- Volatile dimers, trimers of monomers, i.e., oligomers.

- Residual polymerisation solvents, e.g., aliphatic hydrocarbons in low-pressure polyethylene.

- Unreacted impurities from the original monomer, e.g., ethylbenzene in polstyrene (PS).

Methods for the determination of volatiles in polymers are based principally on direct gas chromatography (GC) of solutions or extracts of the polymer, high-performance liquid chromatography (HPLC), polarography of solutions or extracts of the polymer, headspace analysis, headspace–gas chromatography–mass spectrometry (MS), and purge and trap analysis. These various techniques are discussed in the following sections.

5.1 Gas Chromatography

5.1.1 Instrumentation

The sample is introduced to the column in an ideal state, i.e., uncontaminated by septum bleed or previous sample components, without modification due to distillation effects in the needle, and quantitatively, i.e., without hold-up or adsorption prior to the column. The instrument parameters that influence the chromatographic separation are precisely controlled. Sample components do not escape detection, i.e., highly sensitive and reproducible detection and subsequent data processing are essential.

There are two types of separation column used in GC: capillary and packed. Packed columns are still used extensively, especially in routine analysis. They are essential when sample components have high partition coefficients and/or high concentrations. Capillary columns provide a high number of theoretical plates, hence a very high resolution, but they cannot be used in all applications because there are not many types of chemically bonded capillary columns.

Recent advances in capillary column technology presume stringent performance levels for the other components of a gas chromatograph, as column performance is only as good as that of the rest of the system. One of the most important factors in capillary column GC is that a high repeatability of retention times be ensured even under adverse ambient conditions.

Another important factor for reliable capillary column GC is the sample injection method. Various types of sample injection ports are available. The split/splitless sample injection port unit series is designed so that the glass insert is easily replaced and the septum is continuously purged during operation. This type of sample injection unit is quite effective for the analysis of samples containing high boiling point compounds as the major components.

Shimadzu gas chromatographs are typical examples of high-performance gas chromatography systems (see **Table 5.1** for further details). Both capillary and packed columns can be used.

The inner chamber of the oven has curved walls for smooth circulation of air; the radiant heat from the sample injection port units and the detector oven is completely isolated. These factors combine to provide demonstrably uniform temperature distribution. (The temperature variance in a column coiled in a diameter of 20 cm is less than ±0.75 °C at a column temperature of 250 °C.)

When the column temperature is set to a near ambient temperature, external air is brought into the oven via a computer-controlled flap, providing rigid temperature control stability. (The lowest controllable column temperature is 24 °C when the ambient temperature is 18 °C and the injection port temperature is 250 °C. The temperature fluctuation is less than ±0.1 °C even when the column temperature is set at 50 °C.)

This instrument features five detectors (**Table 5.1**). In the flame ionisation detector, the high-speed electrometer, which ensures a very low noise level, is best suited to trace analysis and fast analysis using a capillary column. Samples are never decomposed in the jet, which is made of quartz.

The flows of carrier gas, hydrogen, air, and make-up gas are separately controlled. Flow rates are read from the pressure flow-rate curves.

In the satellite system, one or more satellite gas chromatographs (GC-14 series) are controlled by a core gas chromatograph (e.g., GC-16A series). Since the control is made externally, the satellite gas chromatographs are not required to have control functions (the keyboard unit is not necessary).

Table 5.1 Commercial gas chromatographs

Manufacturer	Model	Packed column	Capillary column	Detectors	Sample injection point system	Keyboard control	Link to computer	Visual display	Printer	Core instrument amenable to tap automation	Temp. programming/isothermal	Cryogenic unit (sub-ambient chromatography)
Shimadzu	GC-14A	Yes	Yes	FID ECD FTD FPD TCD (all supplied)	1. Split-splitters 2. Glass insert for single column 3. Glass insert to dual column 4. Cool on column system unit 5. Moving needle system 6. Rapidly ascending temperature vaporizer	Yes	Yes	No	No	No	Yes/Yes	No
Shimadzu	GC-15A	Yes	Yes	FID ECD FTD FPD TCD	1. Split-splitters 2. Direct sample injection (capillary column)	No	Yes	Yes	Yes	No	Yes/Yes	No
Shimadzu	GC-16A	Yes	Yes	FID FCD FTD FPD TCD (all supplied)	Split-splitters	Yes	Yes	Yes	Yes	No	Yes/Yes	No
Shimadzu	GC-8A	Yes	Yes	FID FCD FPD FTD TCD Single detector instruments (detector chosen on purchase)	1. Point for packed columns 2. Point for capillary columns 3. Split-splitters	Yes Not built in	Option	No	Option	No	Yes Temp. programming GC 8APT (TCD detector) GC 8APF (FID detector) GV 9APFD (FID detector) Isothermal: GC 8A1T (TCD detector) GC 8A1F (FID detector) GC A1E (ECD detector)	No
Varian	34009C	Yes	Yes	Up to 4 types	Temp. programme SPI system	Yes	Yes	No	Yes	No	Yes/Yes	No
Varian	3600	Yes	Yes	FID ECD TCD TSD FPD photoionisation Hall EC	–	Yes	Yes	No	Yes	No	Yes/Yes	Yes

Table 5.1 Continued

Manufac-turer	Model	Packed column	Capillary column	Detectors	Sample injection point system	Key-board control	Link to computer	Visual display	Printer	Core instrument amenable to tap automation	Temp. programming/isothermal	Cryogenic unit (sub-ambient chromato-graphy)
Perkin-Elmer	8410	Yes	Yes	Single detector instrument (detector chosen on purchase) FID ECD FTD FPD TCD	1. Flash vaporisation 2. Split-splitless injector 3. Manual or automatic as sampling valves 4. Manual or automatic liquid sampling valves	No	No	No	No	No	Yes/Yes	Yes, down to -80 °C
Perkin-Elmer	8420	No	Yes	Single detector instrument (chosen on purchase) FID ECD FTD FPD TCD	1. Programmable temperature vaporiser 2. Split-splitless injector 3. Direct on column injector	No	No	No	No	No	Yes/Yes	Yes, down to -80 °C
Perkin-Elmer	8400 and 8500	Yes	Yes	Dual detector instrument (detectors chosen from following) FID ECD FTD FPD TCD	Can be fitted with any combination of above injection systems	Yes	Yes	Yes	Yes (GP100 printer plotter)	Yes	Yes/Yes	Yes, down to -80 °C
Perkin-Elmer	8700	Yes	Yes	FID ECD FPD TCD Hall E.C. photoionisation dual detector instrument (Detectors chosen from above list)	1. Flash vaporisation 2. Split-splitless 3. Programmable temperature vaporiser 4. Gas sampling valve 5. Liquid sampling valve	Yes	Yes	Yes	Yes	–	Yes/Yes	Yes

Table 5.1 *Continued*

Manufac-turer	Model	Packed column	Capillary column	Detectors	Sample injection point system	Key-board control	Link to computer	Visual display	Printer	Core instrument amenable to tap automation	Temp. programming/ isothermal	Cryogenic unit (sub-ambient chromato-graphy)
Nordion	Micromat HRGC 412	No	Yes	Dual simultaneous detector combinations from the following: FID ECD FTD Photoinitiation Hall E.C.	1. Split-splitless 2. In-column injector	Yes	Yes	Yes	Yes	–	Yes/Yes	No
Siemens	SiChromat 1-4 (single oven) SiChromat 2-8 (dual oven) for multi-dimensional GC	Yes	Yes	FID ECD FTD EPD TCD Helium detector	1. Liquid-liquid packed columns 2. Split-splitless 3. Temperature programmable 4. On-column 5. Liquid injector valve on-line 6. Gas injection valve 7. Rotary as injection valve	No	No	Yes	Yes	–	Yes/Yes	No

Source: Author's own files

When a Shimadzu GC-16A series gas chromatograph is used as the core, various laboratory automation-orientated attachments such as barcode and magnetic card readers become compatible: a labour-saving system can be built, in which the best operational parameters are automatically set. Each satellite gas chromatograph (GC-14A series) operates as an independent instrument when a keyboard unit is connected.

The IC card-operated GC system consists of a GC-14A series gas chromatograph and a C-R5A Chromatopac data processor. All of the chromatographic and data processing parameters are automatically set simply by inserting a particular IC card. This system is very convenient when one GC system is used for the routine analysis of several different types of samples.

One of the popular trends in laboratory automation is to arrange for a PC to control the GC and to receive data from it to be processed as desired. Bilateral communication is made via an RS-232C interface built in a GC-14A series gas chromatograph. A system can be built to meet requirements.

A multi-dimensional GC system (multi-stage column system) is effective for analysis of difficult samples and can be built up by connecting several column ovens, i.e., tandem GC systems, each of which has independent control functions such as for temperature programming.

The Shimadzu GC-15A and GC-16A systems are designed not only as independent high-performance gas chromatographs but also as core instruments (see previously) for multi-GC systems or computerised laboratory automation systems. Other details of these instruments are given in **Table 5.1**. The Shimadzu GC-8A range of instruments do not have a range of built-in detectors but are ordered either as temperature-programmed instruments with thermal conductivity detection (TCD), flame ionisation detection (FID), or flame photometric detectors (FPD) detectors or as isothermal instruments with TCD, FID, or electron capture detectors (ECD) (**Table 5.1**).

There are many other suppliers of GC equipment, some of which are shown in **Table 5.1**.

5.1.2 Applications

The styrene monomer used in the manufacture of PS generally contains low concentrations of aromatic impurities. These fall into two categories: (a) saturated compounds such as benzene, toluene, ethylbenzene, xylenes, propylbenzene, butylbenzene, ethyltoluene, and diethylbenzenes, traces of which remain in the final polymer and (b) unsaturated impurities such as styrene and α-methylstyrenes.

Crompton and co-workers [1, 2] have applied the gas chromatographic technique to the determination in PS of styrene and a wide range of other aromatic volatiles in amounts down to the 10 ppm level. In this method a weighed portion of the sample is dissolved in propylene oxide containing a known concentration of pure *n*-undecane as an internal standard. After allowing any insolubles to settle an approximately measured volume of the solution is injected into the chromatographic column which contains 10% Chromasorb 15-20 M supported on 60-70 BS Celite.

Figure 5.1 shows a device that is connected to the GC in order to prevent the deposition of polymeric material in the injection port with consequent blockages. **Figure 5.2** shows a chromatogram obtained for a PS sample, indicating the presence of benzenes, toluene, ethylbenzene, xylene, cumene, propylbenzenes, ethyltoluenes, butylbenzenes, styrene, and α-methylstyrene.

Halmo and co-workers [3] used capillary GC to determine the *iso* paraffin distribution in unvulcanised rubbers.

Wide-bore fused silica column capillary GC has been used to determine impurities in vinyl chloride monomer used in the manufacture of polyvinyl chloride (PVC) [4]. The analysis was performed using a Perkin Elmer model 8500 GC using a 25 m × 0.53 mm fused silica column coated with 5 μm immobilised methyl silicone phase, the column being run at –10 °C for 4 minutes, programmed at 10 °C/min to 10 °C for 1 minute, and programmed at 30 °C/min to 170 °C for 2 minutes. A FID was employed. The three principal impurities present in the monomer, and, consequently, expected to occur in the PVC, were methyl chloride, monovinyl acetylene, and ethyl chloride.

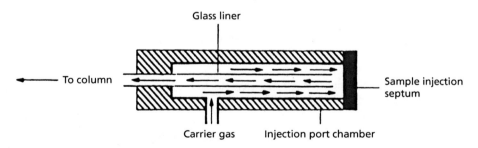

Figure 5.1 Injection port glass liner fitted to F & M Model 1609 gas chromatograph. The glass liner measures 60 mm × 40 mm od and 2 mm id and is very loosely packed with glass fibre. (*Source: Author's own files*)

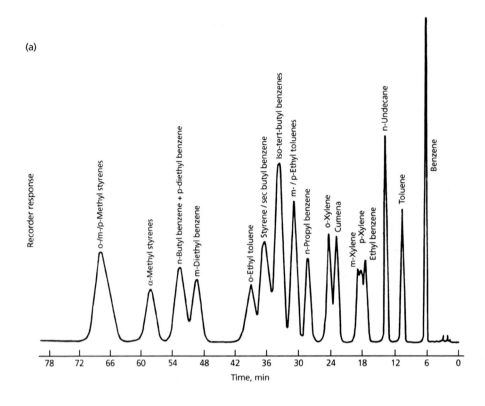

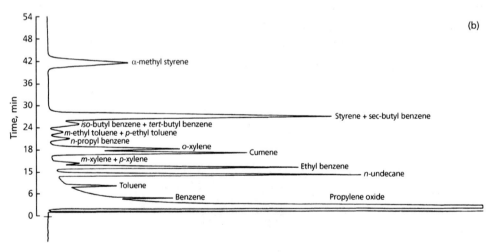

Figure 5.2 Gas chromatography of aromatics in polystyrene. (a) synthetic mixture; (b) propylene oxide solution of polymer in propylene oxide. Column Carbowax 15-20 M at 90 °C, glass liner on injection port; internal standard *n*-undecane concentrations as percentage polymer. (*Source: Author's own files*)

5.2 High-Performance Liquid Chromatography

5.2.1 Instrumentation

For instrumentation details see Section 7.7.2 and Appendix 1.

5.2.2 Applications

Brown [5] has described a HPLC method for the determination of acrylic acid monomer in polyacrylates in which a known mass of polymer is added to a methanol:distilled water mixture (1:1) and allowed to stand overnight to complete the extraction of the monomer from the polymer. An aliquot of the sample is injected into the liquid chromatograph's Rheodyne valve. The analysis was carried out under the following conditions: column, Whatman PXS 10/25 µm PAC (250 × 4 mm id); mobile phase – 0.01% (*v/v*) orthophosphoric acid in distilled water; flow rate – 4 ml/min; pressure – 13.MPa; detector wavelength – 195 nm; chart speed – 0.5 cm/min; and absorbance scale – 0.02. The concentration of acrylic acid is found by comparison with a previously prepared calibration graph of total absorbance *versus* original acrylic acid concentration.

HPLC using the reverse phase mode, has been used [6, 7] to determine the acrylamide monomer and related compounds, including methacrylonitrile, in polyacrylamide. By employing a low-wavelength UV detector, these compounds can be measured with high sensitivity. The relative precision of the 95% confidence level for acrylamide is ±7.5%. The retention times for acrylamide and related compounds are given in **Table 5.2**. No known impurities are observed at the retention time of acrylamide.

Table 5.2 HPLC retention times[a] for acrylamide and related compounds	
Compound	Minutes
Acrylic acid	1.4
β-hydroxypropanamide	2.1
Acetamide	3.0
Acrylamide	5.4
Propanamide	7.3
Acrylonitrile	11.8
Methacrylamide	18.0
Butanamide	20.8
Methacrylonitrile	46.0
[a] Partisil-10 ODS-2, water: 2.0 ml/min, 208 nm, 0.04 aufs. Source: Author's own files.	

Harrison and co-workers [8] used field desorption MS and liquid chromatography (LC)–MS to determine individual cyclics of polyethylene terephthalate (trimer, tetramer, pentamer, hexamer heptamer, and octamer) obtained by a polymer-supported reaction. Further structural information on the fragmentation of cyclic polyethylene terephthalate trimer, tetramer, pentamer, and hexamer was obtained using LC–tandem MS. The results showed that the cyclic trimer was particularly stable.

5.3 Polarography

5.3.1 Instrumentation

Instrumentation for this technique is discussed in Chapter 1 and Appendix 1.

5.3.2 Applications

Residual amounts of styrene and acrylonitrile monomers usually remain in manufactured batches of styrene–acrylonitrile copolymers and acrylonitrile–butadiene–styrene terpolymers (ABS). As these copolymers have a potential use in the food packaging field, it is necessary to ensure that the content of both of these monomers in the finished copolymers is below a stipulated level. In a polarographic procedure [9, 10] for determining acrylonitrile (down to 2 ppm) and styrene (down to 20 ppm) monomers in styrene–acrylonitrile copolymer, the sample is dissolved in 0.2 M tetramethylammonium iodide in dimethyl formamide base electrolyte and polarographed at start potentials of –1.7 V and –2.0 V, respectively, for the two monomers. Excellent results are obtained by this procedure. **Table 5.3** shows the results obtained for determinations of acrylonitrile monomer in some copolymers by the polarographic procedure.

Betso and McLean [11] have described a differential pulse polarographic method for the determination of acrylamide and acrylic acid in polyacrylamide. A measurement of the acrylamide electrochemical reduction peak current is used to quantify the acrylamide concentration. The differential pulse polarographic technique also yields a well-defined acrylamide reduction peak at ~2.0 V *versus* SCA (reduction potential), suitable for qualitatively detecting the presence of acrylamide. The procedure involves extraction of the acrylamide monomer from the polyacrylamide, treatment of the extracted solution on mixed resin to remove interfering cationic and anionic species, and polarographic reduction in an 80/20 (*v/v*) methanol/water solvent with tetra-*n*-butylammonium hydroxide as the supporting electrolyte. The detection limit of acrylamide monomer by this technique is less than 1 ppm.

Table 5.3 Determination of acrylonitrile: comparison of polarographic and chemical methods	
Acrylonitrile content, percent *w/w*, determined by:	
Polarograph	Titration with dodecyl mercaptan
0.06 0.07	0.09 0.09
0.08	0.12
0.11	0.15
0.11	0.19
0.12	0.15
0.12	0.17
0.14	0.18
0.21	0.31 0.29
Reprinted from T.R. Crompton and D. Buckley, Analyst, 1965, 90, 76, with permission from the Royal Society of Chemistry [10]	

5.4 Headspace Analysis

Headspace analysis is a method of choice for the determination of volatile compounds in polymers. The principle of the analysis is quite simple. The sample is placed in a container leaving a large headspace, which is filled with an inert gas (sometimes under pressure) that also serves as the GC carrier gas. Under the prevailing equilibrium conditions a proportion of the volatiles in the sample transfers to the gas-filled headspace, which is then withdrawn and analysed by GC.

5.4.1 Instrumentation

The HS-100/HS-101 automatic headspace analysers by Perkin Elmer employ a pneumatic pressure-balanced system. The HS-100 model is suitable for use with Perkin Elmer Sigma 2000 series of gas chromatograph while the HS-101 is designed for use with the 8000 series. With the advanced microprocessor system of the 8000 series and the moving-needle design the HS-101 offers application possibilities unavailable with other headspace injection systems. The HS-101 has a 100-sample storage magazine and is suitable for unattended and night operation.

5.4.2 Applications

Some of the applications of headspace analysis include the determination of vinyl chloride and other impurities in PVC, styrene monomer in PS, methyl methacrylate monomer in polyacrylates, ethylene in polyethylene, acrylonitrile monomer in ABS terpolymers, epichlorohydrin in epoxy resins, and residual solvents in polymers (see next).

A further apparatus for carrying out solid polymer headspace analysis of PS for styrene monomer and aromatic volatiles has been described by Crompton and co-workers [1, 2]. In this apparatus the polymer is heated to 300 °C in the absence of solvents, prior to its examination by GC. The apparatus is illustrated in **Figure 5.3**.

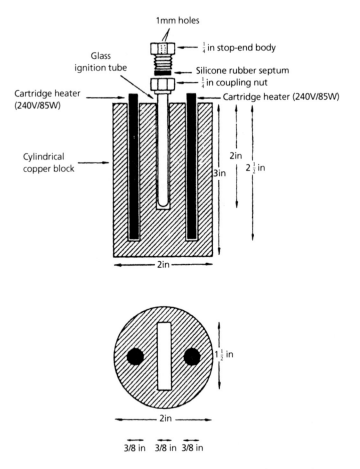

Figure 5.3 Apparatus for liberating volatiles from polymers.

A glass ignition tube is supported as shown in a Wade 0.8 cm diameter brass coupling unit, covered with a silicone rubber septum and sealed with a Wade 0.8 cm brass stop-end body. The stop-end body has two 1 mm diameter holes drilled through the cap. The whole unit is placed in a slot in a cylindrical copper block which is heated to 300 °C.

A sample of the polymer (0.25–0.50 g) is placed in an ignition tube and sealed with Wade fittings and a septum, as described previously. The tube is then heated in the copper block under the required conditions of time and temperature. A sample of the headspace gas is withdrawn from the ignition tube into a Hamilton gas-tight hypodermic syringe via the septum and injected into a GC.

Using this method several PS from different manufacturers were heated at 200 °C for 15 minutes under helium, and the liberated volatiles were examined by GC. All the samples liberated the same range of aromatic hydrocarbons, these differing only in their relative concentrations.

Steichen [12] has discussed a modified solution approach for the GC determination of monomers in polymers by headspace analysis. The more volatile monomers (vinyl chloride, butadiene, and acrylonitrile) are determined by dissolution of the polymer and analysis of the equilibrated headspace above the polymer solution. It is possible to determine vinyl chloride and butadiene at the 0.05 ppm level and acrylonitrile down to 0.5 ppm. The injection of water into polymer solutions containing styrene and 2-ethylhexyl acrylate monomers prior to headspace analysis greatly enhanced the detection capability for these monomers making it possible to determine styrene down to 1 ppm and 2-ethylhexyl acrylate at 5 ppm.

Steichen [12] also discussed the application of headspace analysis to solid polymer samples.

5.5 Headspace Gas Chromatography–Mass Spectrometry

5.5.1 Instrumentation

In instances where the chemical identity of some or all of the volatiles present in a polymer are not known it is possible to obtain the required information by the application of this technique.

Thus, workers at Hewlett Packard [13] using this technique identified six volatiles in styrene–acrylonitrile–maleic anhydride terpolymer (benzene, toluene, ethylbenzene, styrene, aromatics, and maleic anhydride).

- *Sample concentration in headspace analysis.* Some of the top-of-the-range headspace analysers feature a sample concentrator that can be used to increase the sensitivity of the technique. CDS supplies an Analytical 330 Sample Concentrator (Appendix 1) which may be attached to a headspace analyser, the concentrated volatiles being subsequently pulse injected into a GC.

5.6 Purge and Trap Analysis

This is an alternative technique to headspace analysis for the identification and determination of volatile organic compounds in polymers. The sample is swept with an inert gas for a fixed period of time. Volatile compounds from the sample are collected on a solid sorbent trap – usually activated carbon. The trap is then rapidly heated and the compounds collected and transferred as a plug under a reversed flow of inert gas to an external GC. Chromatographic techniques are then used to quantify and identify sample components.

5.6.1 Instrumentation

OIC Analytical Instrument (Appendix 1) supplies the 4460 A purge and trap concentrator. This is a microprocessor-based instrument with capillary column capability. It is supplied with an autosampler capable of handling 76 sample vials. Two automatic rinses of sample lines and vessel purge are carried out between sample analyses to minimise carry-over.

Tekmar are another supplier of purge and trap analysis equipment. Its LSC 2000 purge and trap concentrator features glass-lined stainless steel tubing, menu-driven programming with four-method storage, and a cryofocusing accessory. Cryofocusing is a technique in which only a short section of the column or a pre-column is cooled. In its simplest form a section of the column near the inlet is immersed in a flask of coolant during desorb. After desorb the coolant is removed and the column allowed to return to the oven temperature.

CDS Analytical also supplies a microprocessor-controlled purge and trap concentrator which features cryogenic trapping and cryogenic refocusing and thermal desorption to a trap or directly to a GC.

References

1. T.R. Crompton, L.W. Myers and D. Blair, *British Plastics*, 1965, **38**, 12, 740

2. T.R. Crompton and L.W. Myers, *European Polymers Journal*, 1968, **4**, 355.

3. F. Halmo, S. Surova, M. Balakova and A. Halmova in *Proceedings of Technical Rubber Goods: Part of our Everyday Life*, Puchov, Slovakia, 1996, p.233.

4. *Perkin Elmer Gas Chromatography Applications No. 3.*

5. L. Brown, *Analyst*, 1979, **104**, 1165.

6. N.F. Skelly and E.R. Husser, *Analytical Chemistry*, 1978, **50**, 1959.

7. E.R. Husser, R.H. Stehl and D.R. Price, *Analytical Chemistry*, 1977, **49**, 154.

8. A.G. Harrison, M.J. Taylor, J.H. Scrivens and H. Yates, *Polymer*, 1997, **38**, 2549.

9. G.C. Clover and E. Murphey, *Analytical Chemistry*, 1959, **31**, 1682.

10. T.R. Crompton and D. Buckley, *Analyst*, 1965, **90**, 76.

11. S.R. Betso and J.D. McLean, *Analytical Chemistry*, 1976, **48**, 766.

12. R.J. Steichen, *Analytical Chemistry*, 1976, **48**, 1398.

13. P.L. Wylie, P.D. Perkins and G.E. Kanahan, *Identification of Volatile Residues in Polymers by Headspace Gas Chromatography – Mass Spectroscopy*, Hewlett Packard Gas Chromatography Application Brief, 1986.

6 Fingerprinting Techniques

6.1 Glass Transition Temperature (T_g) and Melting Temperature (T_m)

The identification of polymers is frequently based on a comparison of glass transition or melting temperature, with literature values (**Table 6.1**).

Very often thermogravimetric curves are characteristic for particular polymers and can therefore be used for their identification. Because of the poor heat conductivity, temperature gradients occur within samples at high heating rates. To obtain reproducible results, a standardisation of heating rate should be used, e.g., 10 °C/min, when comparing an unknown polymer with a set of reference polymers.

Other thermal analysis techniques have been used to identify polymers by the fingerprint approach. These include differential thermal analysis (Chapter 9), which has been used to identify polymers including polyethylene (PE), polypropylene (PP), polymethyl methacrylate (PMMA), polyester, polyvinyl chloride (PVC), and polycarbonate [1], and pyrolysis or photolysis techniques as discussed next. Equipment for carrying out these pyrolysis techniques is reviewed in Appendix 1.

6.2 Pyrolysis Techniques

6.2.1 Conventional Pyrolysis Gas Chromatography

Pyrolysis is an analytical technique whereby complex involatile materials are broken down into smaller volatile constituent molecules by the use of very high temperatures. Polymeric materials lend themselves very readily to analysis by this technique. Providing that the pyrolysis conditions are kept constant, a sample should always degrade into the same constituent molecules. Therefore, if the degradation products are introduced into a gas chromatograph (GC), the resulting chromatogram should always be the same and a fingerprint uniquely characteristic of the original sample should be obtained.

In one technique a pyrolyser probe is inserted into a purpose-designed adaptor that is installed in the injector unit of a GC. Different types of adaptor are available for packed

Table 6.1 T$_g$ and T$_m$ values of polymers			
Polymer	Abbreviation	T$_g$, °C	T$_m$, °C
Polybutadiene		-86	(-20)
Polyisobutylene	PIB	-73	(44)
Polyethylene vinyl acetate copolymer	EVA	-20..20	40.100
Polyethylene, low density	LDPE	(-100)	120
Polybutene	PB		130
Polyethylene, high density	HDPE	(-70)	135
Polyoxymethylene copolymer	POM		164..168
Polypropylene	PP	(-30)	165
Polyvinyl chloride soft	PVC w	-40..10	
Polyvinylidene chloride	PVDC	-17	
Polyoxymethylene	POM		175..180
Polyvinylidene fluoride	PDF2		178
Polyamide 11	PA11		186
Polyvinyl acetate	PVAC	30	
Polyvinyl chloride	PVC	85	(190)
Polybutylene terephthalate	PBTB	65	220
Polyamide 6	PA6	(40)	220
Polyamide 6,10	PA6,10	(46)	226
Polyvinyl alcohol	PVA	85	
Polystyrene	PS	90..100	
Polymethyl methacrylate	PMMA	105	
Epoxy resin	EP	50..150	
Polyphenylene oxide	PPO		230
Polycarbonate	PC	155	(235)
Source: Author's own files			

or capillary column work. Correct selection of the appropriate adaptor is a prerequisite for optimum system performance.

The choice of packed or capillary pyrolysis gas chromatography (PGC) is generally a matter of personal requirements. The type and range of samples to be analysed, the

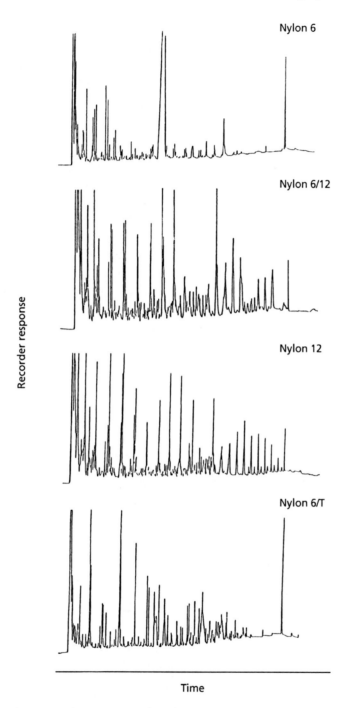

Figure 6.1 Pyrolysis-gas chromatography of Nylon 6, Nylon 6/12, Nylon 12 and Nylon 6/T. (*Source: Author's own files*)

complexity of pyrogram required, and the length of the analysis will all play a significant role in decision making. For the general fingerprinting or analysis of a wide range of routine rubbers or plastics, packed column pyrolysis is more than adequate. However, for comparison of one batch of rubber with another batch of the same rubber, the detail obtained from capillary column PGC will probably give the best results because minor details can be compared.

Applications for pyrolysis are vast and include all types of synthetic polymers, rubbers, and plastics, as well as latexes, paints, and varnishes, in fact, almost any sample that contains involatile organic material that can be contained in a tube or coated onto a platinum ribbon so that it may be pyrolysed. **Figure 6.1** illustrates the difference in the pyrograms obtained for four different Nylons.

PGC procedures generally follow one of two basic patterns. In the flash technique, a sample is deposited on an electrically heatable filament, or placed in a boat in a furnace, and the GC carrier stream is used to transport the pyrolysate directly to the column. The alternative procedure involves heating of the polymer in a separate enclosure, trapping the off-gases, and admitting the pyrolysate to the GC after a given collection interval. Each approach has certain advantages, and the type of information desired from PGC should therefore dictate the sampling method to be employed.

6.2.1.1 Filament Pyrolysis

For quantitative identification a thin film of sample is usually coated onto a Nichrome or platinum spiral or is placed in a small boat so that the weight of the residue remaining after heating can be determined. Although the filament may catalyse the degradation, with 20–30 µg samples the programs obtained with Nichrome, platinum, or gold-plated platinum filaments are identical. The power supply to the filament is generally controlled by a variable transformer and timed by a stopwatch, but exact measurement of the filament temperature is difficult. More elaborate automatic time and voltage controls have been suggested. If desired, the pyrolyser temperature can also be manually programmed to obtain a better equilibrium and to remove the pyrolysis products from the heated zone immediately after they are formed. For quantitative studies of the mechanism and the kinetics of polymer degradation where accurate analysis of the volatile and non-volatile reaction products obtained at a certain temperature and under closely controlled conditions is required, it is preferable to employ preheated tube furnaces than to refine the design of the filament-type pyrolyser. Despite the fact that the filament-type pyrolyser does not allow optimum control of degradation conditions, the programs are entirely satisfactory for identification purposes when polymers of known composition are available for comparison.

Advantages of this pyrolysis technique include the use of microgram samples and the rapid establishment of a reproducible pyrolysis temperature that is essentially dependent on the composition of the ferromagnetic conductor. However, the possibility of catalytic side reactions due to the heating element should not be overlooked. With 0.1 µg samples it may be possible to coat the wire with a monomolecular layer of polymer film which would greatly reduce the possibility of secondary reactions between components of a mixture.

Curie point filament pyrolysis has been used to produce PGC for various polymers used in paint production. Good inter-laboratory reproducibility is claimed for this procedure. The Curie point unit gives rapid and reproducible heating to the Curie point of the pyrolysis wire. If pyrolysis wires of a fixed composition are used in different pyrolysers, identical pyrolysis temperatures are obtained that give highly reproducible pyrograms. A temperature of 610 °C was chosen, as it is in the region in which most polymers give characteristic fragmentation patterns.

6.2.2 Laser Pyrolysis Gas Chromatography

The advantages claimed for this technique include rapid heating and cooling of the sample and relatively simple fragmentation patterns. Folmer and Azarranga [2] and Folmer [3] studied this technique in detail and applied it to a range of polymers.

Folmer [3] studied the effects of different operating conditions and methods of sample preparation on fragmentation patterns. Clear or translucent samples give reproducible results if mixed with carbon.

6.2.3 Photolysis Gas Chromatography

Juvet and co-workers [4] have used this technique for the preparation of fingerprint pyrograms of polymers. They claim that this technique yields considerably more simple and reproducible decomposition patterns than filament, furnace, and Curie point pyrolysers and claim that this is due to greater control of energy input and a more predictable manner in which the polymers decompose photolytically. Photolysis is carried out using a pure thin film of the polymer which is then irradiated using a medium-pressure mercury source. The photolysis products are swept onto a GC to produce a pattern characteristic of the polymer. The retention indexes of the photolysis products of some common polymers are given in **Table 6.2**.

PE primarily yields a series of products eluting at integral multiples of $I = 100$; these are undoubtedly *n*-alkanes and the corresponding alkenes. Products at intermediate I values are probably due to branched-chain hydrocarbon fragments.

213

Table 6.2 Photolysis products of common polymers		
Polymer	Retention index, I on SE-30	Relative peak area*
Polyethylene	<500	1.0
	604 ± 12	0.1
	680 ± 8	0.02
	702 ± 6	0.1
	765 ± 5	0.02
	800 ± 5	0.15
	876 ± 5	0.01
	900 ± 5	0.07
Polystyrene	<500	0.02
	660 ± 10 (benzene)	1.0
	784 ± 8 ⎫	
	830 ± 8 ⎬	ca. 0.5[b]
	886 ± 6 ⎭	
Poly(methylmethacrylate)	<500	1.0
	720 ± 8	0.5
	1525 ± 5	0.2
	1550 ± 5	0.2
Poly(tetrafluoroethylene)	<500	...
*Irradiation time, 30 minutes		
[b] Total of three incompletely resolved peaks		
Reprinted from R.S. Juvet, S.L.S. Smith and K. Panghi, *Analytical Chemistry*, 1972, **44**, 49, with permission from the American Chemical Society [4]		

Polystyrene (PS) yields benzene (I = 600) as the major photolysis product with apparently smaller amounts of toluene, ethylbenzene, and styrene formed. Quantitation of the last three materials is complicated by variable amounts of residual monomer in the samples available and a small amount of thermal decomposition occurring in the injection port.

UV absorption of a polymer influences the response to photolytic degradation. Polyolefins, which are relatively transparent in the UV, receive essentially constant radiation throughout the sample and thus yield photolysis products as a function of total sample weight. Highly UV absorbing polymers, such as PS, strongly attenuate the incident radiation. Measurements of the photolysis yield of benzene from PS films showed that photolysis is surface area controlled and products are formed within a thin surface layer.

Photolysis of PMMA proceeds primarily by formation of the monomer, corresponding to the product observed at $I = 720$. As in the case of PS, small amounts of residual monomer and monomer formed from thermal decomposition were observed in non-irrigated samples. Methanol and methyl formate, which have also been reported as photolysis products, are not separated on SE-30 but analysis on Carbowax 20 M gives two peaks corresponding to these substances.

6.2.4 Pyrolysis Mass Spectrometry

Hughes and co-workers [5] and Menzelaar and co-workers [6] have shown that pyrolysis mass spectrometry (PMS) has considerable potential for the characterisation and discrimination of natural and synthetic polymers. The equipment combined a Curie point pyrolyser and quadruple mass spectrometer operated in an optimum geometric arrangement. The pyrolysate was passed through an empty glass chromatograph column and jet separator before entering the mass spectrometer. Forty or more sequential mass spectral scans were integrated by computer processing to give a composite mass pyrogram. The system was used to study a wide range of polymeric materials. The sensitivity is sufficiently high to allow samples of 5 µg or less to give adequate electron impact spectra, but in the chemical ionisation mode larger samples are necessary. Mass pyrograms are usually characteristic of the sample type and frequently allow discrimination between samples of similar composition.

The advantages of PMS over PGC for generating information about polymeric materials are its speed, sensitivity, ease of producing data that can be processed by a computer, and the elimination of the variables associated with GC. A major disadvantage of PMS is that a complex mixture is produced by a combination of pyrolysis and electron impact fragmentation, which makes a mass pyrogram more difficult to interpret than the chromatogram produced in PGC, in which only a pyrolytic breakdown is involved.

Jackson and Walker [7] studied the applicability of pyrolysis combined with capillary column GC to the examination of phenyl polymers (e.g., styrene–isoprene copolymer) and phenyl ethers {e.g., bis[m-(m-phenoxy phenoxy)phenyl]ether}. In the procedure the polymer sample is dissolved in benzene. The pyrolysis Curie point temperature wire is dipped 6 mm into the polymer solution. The polymer-coated wires are then placed in a vacuum oven at 75–80 °C for 30 minutes to remove the solvent. **Figure 6.2** shows a characteristic pyrogram of the copolymer (isoprene–styrene) resulting from a 10-second pyrolysis at 601 °C. When the polyisoprene is pyrolysed, C_2, C_3, C_4, isoprene, and $C_{10}H_{16}$ dimers are produced. When PS is pyrolysed, styrene and aromatic hydrocarbons are the products. **Figure 6.2** shows that the copolymer product distribution and relative area basis resemble the two individual polymer product distributions.

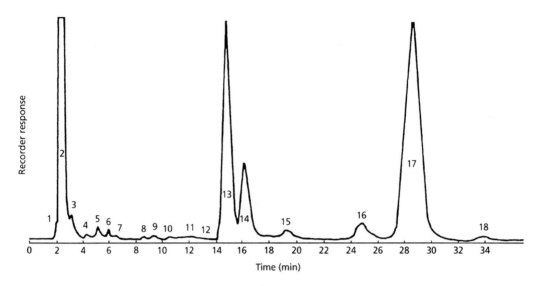

Figure 6.2 Pyrograms of isoprene-styrene copolymer at 600 °C for 10 seconds duration. (*Reprinted from M.T. Jackson and J.Q. Walker, Analytical Chemistry, 1971, 43, 74, with permission from the American Chemical Society [7]*)

Mattern *and co-workers* [8] carried out laser mass spectrometry on polytetrafluoro-ethylenes. They found a fragmentation mechanism common to each fluoropolymer yields structurally relevant ions indicative of the orientation of monomer units within the polymer chain. A unique set of structural fragments distinguished the positive ion spectra of each homopolymer, allowing identification.

The in-source direct PMS method discussed in Section 3.8.2 [9] when used in conjunction with a spectral library search is capable of providing rapid identification of some iso polymers.

6.3 Infrared Spectroscopy

Infrared (IR) spectra of thin films of a polymer in the region up to 4000 cm^{-1} are characteristic of the polymer. Computerised retrieval from data in a library of standard polymers has been used in the IR fingerprinting technique to facilitate polymer identification [10].

Alexander [11] has described a method for obtaining spectra of thin films of polymer that are free of interference fringes.

Various methods have been used to prepare polymers for IR spectroscopy.

6.3.1 Potassium Bromide Discs [12]

The spectra were run either as potassium bromide discs (1.5–2 µg polymer per 400 mg potassium bromide) or as polymer films of varying thickness up to 12 µm using a sodium chloride prism.

6.3.2 Hot Pressed Film [13]

Osland [14] has described a heated press for the preparation of plastic films for analysis by IR spectroscopy. The press can produce films of reproducible thickness as thick as 500 µm.

Alternatively, the sample material can be dissolved in a suitable organic solvent and a film cast onto glass or a cell window.

A new sample handling accessory with which films of constant thicknesses can be prepared has been introduced by Phillips Analytical. The plastic film press contains a thermostatically controlled oven unit that is calibrated up to 300 °C. It also contains a cooling facility which may be connected to a low-pressure compressed air supply to cool the prepared films rapidly. Reproducible thickness is ensured by using a set of brass dies that can be heated and cooled quickly. The dies can produce films of 20, 50, 100, 200, and 500 µm thickness.

Deschant [15] has discussed in detail the IR spectra of 35 polymers and Hippe and Kerste [16] developed an algorithm for the IR identification of vinyl polymers. Pyrolysis followed by IR spectroscopy is a particularly useful technique for application to polymers that are rendered opaque or completely non-transparent by the presence of pigments or fillers. Leukoth [17] used this technique to determine the chemical composition of plastics and rubbers containing high proportions of pigments or fillers.

Particular studies of the IR spectra of polymers include isotactic poly(1-pentane), poly(4-methyl-1-pentene), and atactic poly(4-methyl-pentene) [18], chlorinated PE [19], aromatic polymers including styrene, terephthalic acid, isophthalic acid [20], PS [21, 22], styrene–glycidyl-*p*-isopropenylphenyl ether copolymers [23], styrene–isobutylene copolymers [24], vinyl chloride–vinyl acetate–vinyl fluoride terpolymers [25], vinyl chloride–vinyl acetate copolymers [26], styrene copolymers [27], ethylene–vinyl acetate copolymers, graft copolymers, and butadiene–styrene [28] and acrylonitrile–styrene copolymers [29].

6.4 Pyrolysis Fourier Transform Infrared Spectroscopy

6.4.1 Theory

This technique has been studied by Washall and Wampler [30]. Pyrolysis of polymer samples enables IR spectra to be obtained from the vaporised polymer. Many polymers are difficult to analyse by more conventional Fourier transform infrared (FT-IR) sampling techniques due to additives that cause scattering. Carbon-filled rubbers are examples of this type of problem. By dissociating the bonds of the polymer backbone, polymer functionality and structure can be determined much like PGC. Using direct pyrolysis FT-IR in conjunction with other techniques gives a broad picture of the polymer structure and degradation mechanism. The latter is easily investigated using time-resolved pyrolysis in which the sample is heated at a slow rate over a long period of time. In this way degradation products can be monitored as they evolve from the sample.

While the coupling of pyrolysis with IR spectroscopy is not a new technique for analysing polymer samples, the interface is new. The traditional steps of condensing the pyrolysates on IR transmitting window materials and transferring the windows to the sample compartment are eliminated.

6.4.2 Instrumentation

The instrumentation is discussed in Appendix 1.

The system applied in the study mentioned above consisted of a CDS model 122 Pyroprobe with a ribbon filament as the heating surface (see Chapter 3 and Appendix 1). This pyrolyser heats by varying the resistance of the platinum element. Temperature rise times for flash pyrolysis are typically of the order of milliseconds. IR spectra were obtained with an FT-IR bench system equipped with a CDS pyrolysis/FT-IR interface. The data were collected at 8 cm^{-1} with a deuterated triglycine sulfate (DTGS) detector. The interface is cylindrical in shape with two potassium bromide windows for the IR beam to pass through.

When the beam enters the pyrolysis cell, it is focused directly above the platinum filament of the pyrolyser. An IBM PS2 computer was used to collect the data obtained. The spectra were baseline corrected using a polynomial baseline correction routine. Sample sizes ranged from 100 to 500 μm.

6.4.3 Applications

Some pyrolysis FT-IR spectra obtained by the technique discussed above are shown in **Figure 6.3**. Direct pyrolysis FT-IR of PS (**Figure 6.3(a)**) reveals a spectrum almost

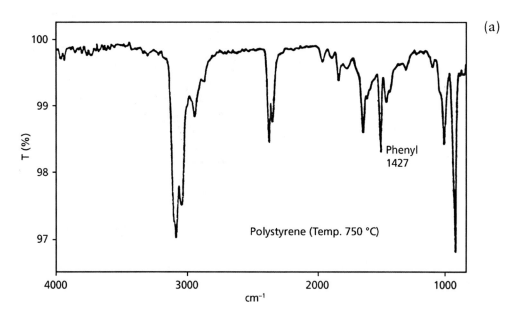

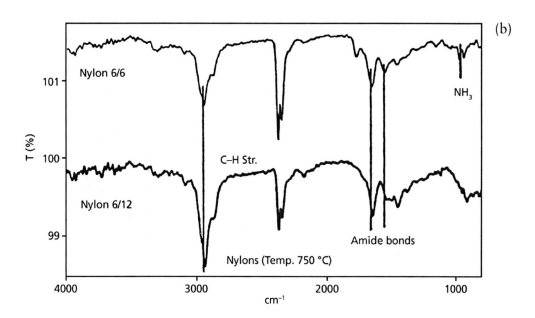

Figure 6.3 Pyrolysis-FTIR spectra of (a) polystyrene, (b) Nylons, (c) PVC and (d) polymethylmethacrylate – continued overleaf.

(*Source: Chemical Data Systems, Oxford, Pennsylvania, USA [30]*)

(c)

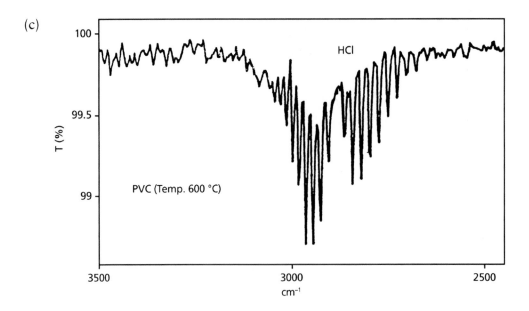

(d)

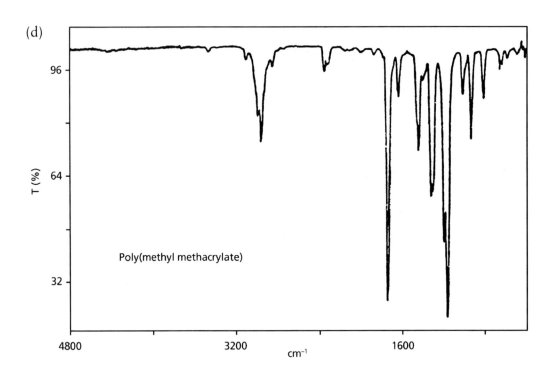

Figure 6.3 *Continued*

identical to that of the styrene monomer. Absorption bands indicating vinyl stretch (910 and 1000 cm^{-1}) and aromatic ring stretching at 1496 cm^{-1} are very evident. **Figure 6.3(b)** shows a comparison of Nylon 6,6 and Nylon 6,12 both pyrolysed at 750 °C. The difference between Nylon 6,6 and Nylon 6,12 lies in the length of the hydrocarbon chain. A relative measure of the CH$_2$ group content to the amide group content should be a good indication of the differences. The primary identifier is the intensity of the CH$_2$ stretching band compared to the amide stretch at 1650 cm^{-1}. This ratio gives a qualitative differentiation between the two polymers. Similar distinctions can be made for other Nylon samples.

A classic example of the verification of degradation mechanisms is the pyrolysis of PVC. This polymer undergoes side group elimination of hydrogen chloride in the initial step. The resulting polymer chain is a polyene. This polyene then undergoes several scission steps to produce aromatics. Of the aromatics produced, the more prevalent are benzene, toluene, and naphthalene. Since hydrogen chloride is strongly active in the IR region, pyrolysis FT-IR of PVC produces a spectrum of hydrogen chloride as well as some minor absorption bands indicating some aliphatic side products. **Figure 6.3(c)** shows the spectrum of PVC at 600 °C in the 3500–2400 cm^{-1} region. The unmistakable pattern of hydrogen chloride is present.

Figure 6.3(d) shows the spectrum of pyrolysed PMMA. A carbonyl stretching band is clearly present at 1745 cm^{-1}. Also evident is an ester CH$_3$–O stretch at 1018 cm^{-1}. The spectrum obtained in the pyrolysis experiment can be directly compared to a reference spectrum of methyl methacrylate.

6.5 Raman Spectroscopy

Yashino and Shinomiya [31] have published Raman spectra of solutions of various polymers. The technique has also found a limited application in structural studies of polymers [32–44]. For example, the three C=C stretching bands in polybutadiene corresponding to the three possible configurations can be observed by Raman spectroscopy. Raman spectroscopy has many applications in the identification of polymers in which additives obscure the polymer peaks obtained in the IR spectrum.

6.6 Fourier Transform Near-Infrared Raman Spectroscopy

The appearance of the Raman spectra is essentially independent of the physical form of the sample. The only obvious difference between spectra of pellets, powders, films, fibres, and foams is in the signal intensity. For constant laser power these intensities reflect the packing of the solid into the region being observed and the path of the exciting radiation. As would be expected powders give stronger spectra than solids. Fibres and films give poorer spectra, presumably because of inefficient packing. The spectra of foams vary.

The quality of the spectra obtained in two minutes is more than adequate for qualitative identification. Differences in crystallinity are readily seen for both PP and polyester samples. Some common inorganic fillers such as glass and talc are weak Raman scatterers and are not evident in the spectra. In other cases inorganic fillers are seen, e.g., $BaSO_4$ in voided polyethylene terephthalate film.

The Raman spectra shown in **Figure 6.4** were obtained with a modified Perkin Elmer 1720 FT-IR spectrometer using a Nd:YAG laser at 1.064 μm and a liquid nitrogen-cooled InGaAs detector. Excitation and collection of radiation were at 180°, the samples being held in a metal cup at one focus of an ellipsoidal mirror.

Most samples absorb 1.064 μm radiation only weakly. Even so there is sufficient sample heating to limit the laser power that can be used. The measurements were made using a focused laser which emphasises the heating effects. Samples were examined using laser powers of 200 and 750 mW. In most cases there was no change in the spectra but in some filled samples bands broadened at the higher power. Strongly absorbing samples, notably all carbon-filled materials, were damaged at low laser power and gave no spectra.

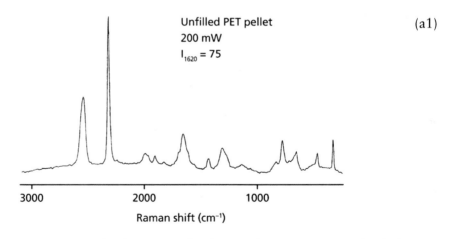

Figure 6.4 Raman spectra of polyesters and polypropylene. (a) PET = polyethylene-terephthalate; (b) polypropylene. (*Source: Author's own files*)

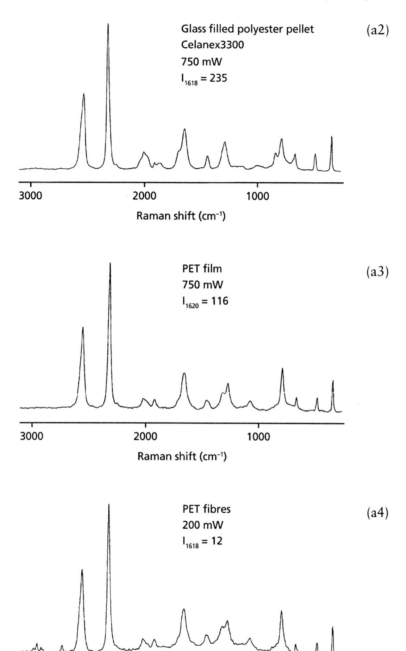

Figure 6.4 *Continued*

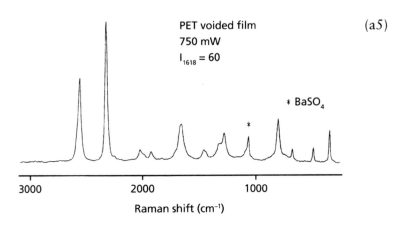

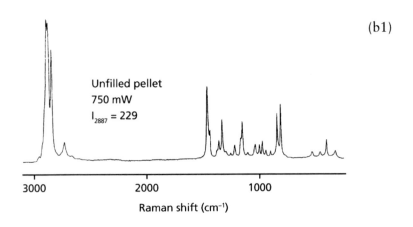

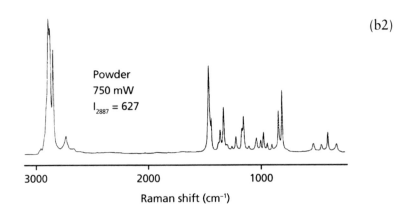

Figure 6.4 *Continued*

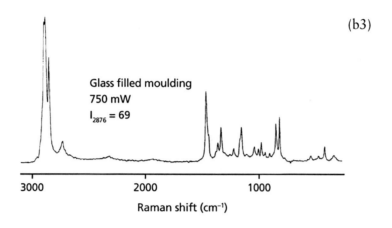

(b3)

Glass filled moulding
750 mW
$I_{2876} = 69$

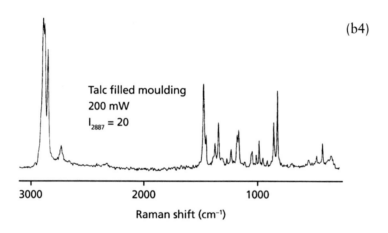

(b4)

Talc filled moulding
200 mW
$I_{2887} = 20$

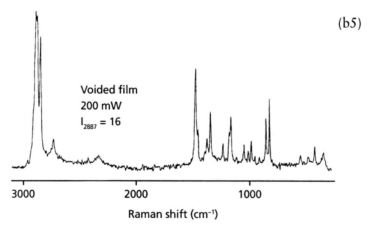

(b5)

Voided film
200 mW
$I_{2887} = 16$

Figure 6.4 *Continued*

6.7 Radio Frequency and Low Discharge Mass Spectrometry

This technique is discussed in Section 3.11.6 [25]. It is capable of rapidly characterising bulk polymers, copolymers, and terpolymers.

References

1. K. Bosch, *Beitrage Geschichte Medica*, 1975, **33**, 280.

2. O.F. Folmer and L.V. Azarranga, *Journal of Chromatographic Science*, 1969, **7**, 665.

3. O.F. Folmer, *Analytical Chemistry*, 1971, **43**, 1057.

4. R.S. Juvet, S.L.S. Smith and K. Panghi, *Analytical Chemistry*, 1972, **44**, 49.

5. J.C. Hughes, B.B. Wheals and M.J. Whitehouse, *Analyst*, 1977, **102**, 143.

6. H.L.C. Menzelaar, M.A. Posthumus, P.G. Kistemaker and J. Kiskinaker, *Analytical Chemistry*, 1973, **45**, 1546.

7. M.T. Jackson, Jr., and J.Q. Walker, *Analytical Chemistry*, 1971, **43**, 74.

8. D.E. Mattern, Lin Fu-Tyan and D.M. Hercules, *Analytical Chemistry*, 1984, **56**, 2762.

9. K. Qian, W.E. Killinger, M. Casey and G.R. Nicol, *Analytical Chemistry*, 1996, **68**, 1019.

10. M. Razinger, M. Penca and J. Zapan, *Analytical Chemistry*, 1981, **53**, 1107.

11. R. Alexander, *The Analytical Report*, No.7, Perkin Elmer Ltd., Beaconsfield, UK, 1985.

12. *Perkin Elmer Application & News Bulletin*, 1963, **1**, 1-8.

13. M. Bowden, P. Donaldson, D.J. Gardiner, J. Birnie and D.L. Gerrard, *Analytical Chemistry*, 1991, **63**, 2915.

14. R. Osland, *Laboratory Practice*, 1971, **37**, 73.

15. J. Deschant, *Ultraspektroskopische Untersuchungen an Polymerern*, Akadamie Verlag, Berlin, Germany, 1972, p.516.

16. Z. Hippe and A. Kerste, *Bulletin de l' Academie Polanaise des Sciences: Serie des Sciences Chimiques*, 1973, **21**, 395.

17. G. Leukoth, *Gummi Asbestos Kunstoffe*, 1974, **27**, 794.

18. S.M. Gabbay and S.S. Stivala, *Polymer*, 1976, **17**, 121.

19. B.M. Quernum, P. Berticat and G. Vallet, *Polymer Journal*, 1975, 7, 277.

20. G. Stanescu, *Revista de Chemie (Bucharest)*, 1963, **64**, 42.

21. H.J. Sloane, T. Johns, W.F. Ulrich and W.J. Cadman, *Applied Spectroscopy*, 1965, **19**, 130.

22. T. Yoshini and M.J. Shimoniya, *Polymer Science*, 1965, **A3**, 2811.

23. S.M. Aliev, M.R. Bairomov, A.G. Aziziv, S.A. Aliev and S.T. Akhmedov, *Azerbaidzhanskii Khimicheskii Zhurnal*, 1976, **3**, 70.

24. J.E. Forette and H.L. Pozek, *Journal of Applied Polymer Science*, 1974, **18**, 2973.

25. R. Aslanova, A.A. Yul'chivaev and K.U. Usmanov, *Uzbekskii Khimicheskii Zhurnal Z1*, 1973, **17**, 31.

26. K.S. Chia, G.M.S. Chen and J. Chin, *Chemistry (Taipei)*, 1973, **20**, 241.

27. F.M. Mirabella, E.M. Barrall and J.F. Johnson, *Polymer*, 1976, **17**, 17.

28. J.V. Dawkins and M.J. Hemming, *Applied Polymer Science*, 1975, **19**, 3107.

29. W.H. Littke, W. Flebec, R. Schmote and B.W. Kimmer, *Faserforschung und Textiltechnik*, 1975, **26**, 503.

30. J.W. Washall and T.P. Wampler, *A Pyrolysis/FT-IR System for Polymer Quality Control – A Feasibility Study*, Chemical Data Systems, Oxford, PA, USA, 1988.

31. T. Yashino and M. Shinomiya, *Journal of Polymer Science*, 1965, **A3**, 2811.

32. V.F. Gaylor, A.L. Conrad and J.H. Landerl, *Analytical Chemistry*, 1957, **29**, 228.

33. B. Schrader, *Angewandte Chemie - International Edition*, 1973, **12**, 11, 884.

34. F.J. Boerlo and J.L. Koenig in *Polymer Characterisation: Interdisciplinary Approaches*, Ed., C. Craver, Plenum Press, New York, NY, USA, 1971, p.1-13.

35. F.J. Boerio and J.L. Koenig, *Journal of Polymer Science: Polymer Symp*osia, 1973, **43**, 205.

36. J.L. Keonig, *Chemical Technology*, 1972, **2**, 411.

37. N.A. Slovokhotova, *Metody Ispyt Kontr Issled Machinostroit Mater*, 1973, **3**, 41.

38. C.J.H. Schutte, *Fortschritte der Chemischen Forschung*, 1972, **36**, 57.

39. G. Zerbi, *Chimica e Industria (Milan)*, 1973, **55**, 334.

40. I.V. Kumpanenko and K.S. Kazanskil, *Uspekhi Khimii Fizicheskikh Polimeri*, 1973, 64.

41. R.D. Andrews and T.R. Hart in Characterisation of *Metal and Polymer Surfaces*, Ed., L-H. Lee, Academic, Press, New York, NY, USA, 1977, Chapter 2, p.207-40.

42. M.G. Chauser, V.D. Ermakova, O.B. Mishenko, V.S. Gunova and M.I. Cherkashin, *Armyanskii Khimicheskii Zhurnal*, 1973, **26**, 608.

43. J.L. Koenig in *Proceedings of the Polymer Characterisation Conference*, 1975, Cleveland, OH, USA, p.73.

44. J.M. Chalmers, *Polymer*, 1977, **18**, 681.

45. C.R. Shick, P.A. de Palma and R.K. Marcus, *Analytical Chemistry*, 1996, **68**, 2113.

7 Polymer Additives

The analysis of additives in polymers has been reviewed by several workers [1, 2]. A variety of analytical techniques have been considered.

7.1 IR and Raman Spectroscopy

7.1.1 Instrumentation

Further details of instrumentation are given in Chapter 3 and Appendix 1.

7.1.2 Applications

Spell and Eddy [3] described infrared (IR) spectroscopic procedures for the determination of up to 500 ppm of various additives in polyethylene (PE) pellets following solvent extraction of the additives at room temperature. They showed Ionol (2,6-di-*tert*-butyl-*p*-cresol) and Santonox R (4,4'-thio-bis(6-*tert*-butyl-*m*-cresol)) are extracted quantitatively from PE pellets by carbon disulfide in 2–3 hours and by isooctane in 50–75 hours. The carbon disulfide extract is suitable for scanning in the IR region between 7.8 and 9.3 μm, while the isooctane extract is suitable for scanning in the UV region between 250 and 350 nm.

Miller and Willis [4] obtained IR spectra of antioxidants in polymer films. They compensated with additive-free polymer in the reference beam. IR spectroscopy is more specific than UV spectroscopy [5-7].

Patticini [8] has described an IR method for the determination of 1–8% of mineral oil in polystyrene (PS). In this method the PS sample is dissolved in carbon tetrachloride, together with known mineral oil standards. The solutions are evaluated by measurements made between 3100 and 3000 cm^{-1} using a spectral subtraction technique.

Fourier transform near-IR Raman spectroscopy (400–10,000 cm^{-1}) is useful for the examination of additives in polymer extracts [9].

An example of the application of Raman spectroscopy is the identification of additives in fire-retardant polypropylene (PP). When a sample of PP was examined by IR spectroscopy the strongest bands (9.8 and 14.9 μm) were due to a talc-type material and bands of medium intensity were assigned to PP and possibly antimony trioxide (13.4 μm). Additional weak bands in the 7.3–7.7 μm region were possibly due to decabromodiphenyl ether. In the Raman spectrum, however, the strongest bands (250 and 185 cm^{-1} shift) confirmed the presence of antimony trioxide and some bands of medium intensity confirmed the presence of decabromodiphenyl ether (doublet at 140, triplet at 220 cm^{-1} shift) and PP (800, 835, 1150, 1325, 1450, and 2900 cm^{-1} shift). The silicate bands that obscured the regions of the IR spectrum were not observed in the Raman spectrum.

Although both of these spectroscopic methods have a wide use in their own right, this example demonstrates well the complementary value of the two methods, taking advantage of the fact that elements of high atomic number, e.g., antimony and bromine, have relatively more intense Raman spectra but the lighter elements show up clearly in the IR spectra.

Pasch [10] in his recent review of the development in polyolefin characterisation discusses the use of high-temperature gas chromatography (GC) coupled with Fourier transform IR (FT-IR) spectroscopy for the determination of antioxidants in polyolefins.

7.2 Ultraviolet Spectroscopy

7.2.1 Instrumentation

Further details of instrumentation are given in Section 1.12 and Appendix 1.

7.2.2 Applications

Straightforward UV spectroscopy is liable to be in error owing to interference by other highly absorbing impurities that may be present in the sample [11-13]. Interference by such impurities in direct UV spectroscopy has been overcome or minimised by selective solvent extraction or by chromatography. However, within prescribed limits UV spectroscopy is of use and, as an example [14-18], procedures have been developed for the determination of Ionol and of Santonox R in polyolefins.

Organic and inorganic pigments are used for coloration of polymers, polymer films, and polymer coatings on metal containers. Vapour-phase UV absorption spectrometry at 200 nm has been used [19] to identify such pigments. In this method powdered samples are directly vaporised in a heated graphite atomiser. Thermal UV (TUV) profiles of organic pigments show absorption bands between 300 and 900 °C, while profiles of inorganic pigments are characterised by absorption bands at temperatures above 900 °C. Temperature, relative intensity, and width of the bands allow the identification of the pigments. The technique shows fast acquisition of thermal UV profiles (2–3 minutes for each run), good repeatability, and wide thermal range (from 150 to 2300 °C).

An example of the identification of pigments is given in **Figure 7.1.**

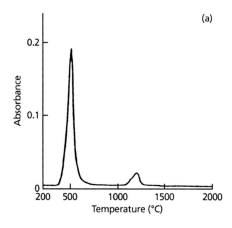

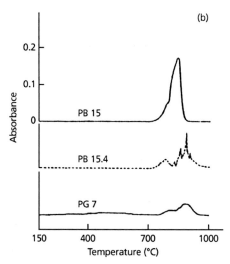

Thermal cycle	1	2	3
Temperature, °C	200	2000	2650
Ramp times, s	10	8	2
Hold temperature, °C	25	–10	5
Time constant		0.5	

Figure 7.1 TUV profile of (a) 1.1 mixture of organic pigment yellow (2-nitro *p*-toluidene coupled with acetoacetanilide and inorganic pigment PY 34 lead chromate. (b) PB 15.4 copper phthalocyonine (beta form); PG 7 polychlorinated copper phthalocyanine (14-16 chlorine atoms)

A 1:1 mixture of organic yellow pigment (2-nitro-*p*-toluidine coupled with acetoacetanilide) and inorganic PY 34 (lead chromate) was vaporised using the conditions quoted for **Figure 7.1(a)**. The thermal UV profile shows clearly two absorption bands at about 500 and 1250 °C. The first band is attributed to the vapours that originate from the decomposition and pyrolysis of the organic pigment. The second band corresponds to the decomposition and vaporisation of lead chromate at high temperature (mp 844 °C). It is possible therefore to determine by a rapid run whether the pigment is a mixture or belongs to the organic or inorganic group.

Lutzen and co-workers [20] describe an in-line monitoring UV method for the determination of polymer additives such as thermal and UV stabilisers and antioxidants.

7.3 Luminescence and Fluorescence Spectroscopy

7.3.1 Instrumentation

Perkin Elmer and Hamilton both supply luminescence instruments (see Appendix 1).

The Perkin Elmer LS-3B instrument is a fluorescence spectrometer with separate scanning monochromes for excitation and emission, and digital displays of both monochromator wavelengths and signal intensity. The LS-5B instrument is a rotating luminescence spectrometer with the capability of measuring fluorescence, phosphorescence, and bio- and chemiluminescence. Delay time (t_d) and gate width (t_g) are variable via the keyboard in 10 μs intervals. The instrument collects excitation and emission spectra.

Both instruments are equipped with a xenon discharge lamp source and have an excitation wavelength range of 230–720 nm and an emission wavelength range of 250–800 nm.

These instruments feature keyboard entry of instrument parameters which, combined with digital displays, simplifies instrument operation. A high-output pulsed xenon lamp, having low power consumption and minimal ozone production, is incorporated within the optical module.

Fluorescence spectrometers are equivalent in their performance to single-beam UV–visible spectrometers in that the spectra they produce are affected by solvent background and the optical characteristics of the instrument. These effects can be overcome using software built into the Perkin Elmer LS-5B instrument or using application software for use with the Perkin Elmer models 3700 and 7700 computers.

The Perkin Elmer LS-2B micro-filter fluorimeter is a low-cost, easy-to-operate filter fluorimeter that scans emission spectra over the wavelength range 390–700 nm (scanning) or 220–650 nm (individual interference filters).

232

The essentials of a filter fluorimeter are as follows:

- A source of UV/visible energy (pulsed xenon).
- A method of isolating the excitation wavelength.
- A means of discriminating between fluorescence and excitation energy.
- A sensitive detector and a digital display of the fluorescence intensity.

The model LS-2B has all these features arranged to optimise sensitivity for micro-samples. It can also be connected to a highly sensitive 7 µl liquid chromatographic detector for detecting the constituents in the column eluent. It has the capability of measuring fluorescence, time-resolved fluorescence, and bio- and chemiluminescence signals. A 40-portion autosampler is provided. An excitation filter kit containing six filters – 310, 340, 375, 400, 450, and 480 µm – is available.

7.3.2 Applications

Aromatic amines and phenols are among the few classes of compounds in which a large proportion of them exhibit useful fluorescence. Parker and Barnes [21] found that in solvent extracts of rubbers the strong absorption by pine tar and other constituents masks the absorption spectra of phenylnaphthylamines, whereas the fluorescence spectra of these amines are sufficiently unaffected for them to be determined directly in the unmodified extract by the fluorescence method. In a later paper Parker [22] discussed the possibility of using phosphorescence techniques for determining phenylnaphthylamines. Drushel and Sommers [7] have discussed the determination of Age Rite D (polymeric dihydroxy quinone) and phenyl-2-naphthylamine in polymer films by fluorescence methods and Santonox R and phenyl-2-naphthylamine by phosphorescence methods.

Uvitex OB UV stabiliser has an intense UV absorption at a wavelength of 378 nm, which is high enough to be outside the region many potentially interfering substances present in the polymer extract would be excited to fluorescence. This is illustrated by a fluorimetric procedure for the determination of Uvitex OB in PS down to 10 ppm. Antioxidants such as Ionol CP (2,6-di-*tert*-butyl-*p*-cresol), Ionox 330 (1,3,5-trimethyl-2,4,6-tri(3,5-di-*tert*-butyl-4-hydroxybenzyl)benzene), Polygard (tris(nonylated phenyl)phosphite), Wingstay T (a butylated cresol), Wingstay W, and many others do not interfere in this procedure.

UV stabilisers can be determined by direct analysis of polyolefins and polyvinyl chloride (PVC) (i.e., without solvent extraction) using the Perkin Elmer model LS-50 luminescence spectrometer (L255-0105) fitted with a front surface accessory. The relation between fluorescence emission and stabiliser content was found to be linear over the range obtained for both natural and extruded polymer samples [23].

7.4 Nuclear Magnetic Resonance Spectroscopy (NMR)

The technique discussed in Chapter 3 of the analysis of polymers has found limited applications in the analysis of polymer additives. Wide-line NMR spectroscopy has been used [24] for the determination of the diisooctylphthalate content of PVC. The principle of the method is that the narrow-line liquid-type NMR signal of the plasticiser is easily separated from the very broad signal due to the polymer - integration of the narrow-line signal permits determination of the plasticiser. A Newport Quantity Analyser Mk 1 low-resolution instrument, equipped with a 40 ml sample assembly and digital read-out, has been used to determine 20–50% of plasticiser in PVC.

7.5 Mass Spectrometry

7.5.1 Instrumentation

Further details of instrumentation are given in Appendix 1 and Section 3.11.

7.5.2 Applications

Lattimer and co-workers [25] have applied mass spectrometry (MS) to the determination of antioxidants and antiozonants in rubber vulcanisates. Direct thermal desorption was used with three different ionisation methods [electron impact (EI), chemical ionisation (CI), field ionisation (FI)]. The vulcanisates were also examined by direct fast atom bombardment mass spectrometry (FAB-MS) as a means for surface desorption/ionisation. Rubber extracts were examined directly by these four ionisation methods. Of the various vaporisation/ionisation methods, it appears that field ionisation is the most efficient for identifying organic additives in the rubber vulcanisates. Other ionisation methods may be required, however, for detection of specific types of additives. There was no clear advantage for direct analysis as compared to extract analysis. Antiozonants examined include aromatic amines and a hindered bisphenol. These compounds could be identified quite readily by either extraction or direct analysis and by use of any vaporisation/ionisation method.

Altenau and co-workers [26, 27] used MS to identify quantitatively volatile antioxidants in 0.02–0.03 inch thick samples of synthetic styrene–butadiene rubbers and rubber-type vulcanisates. They extracted the polymer with acetone in a Soxhlet apparatus, removed excess solvent, and dissolved the residue in benzene. Substances identified and determined by this procedure include N-phenyl-β-napthylamine, 6-dodecyl-2,2,4-trimethyl-1,2-dihydroquinolines, trisnonylphenylphosphate, isobutylene-bisphenol, 2-mercaptobenzothiazole sulfenamide (accelerator), N-cyclohexyl-2-benzothiazole sulfenamide, *N-tert*-butyl-2-benzothiazole sulfenamide, 2-(4-morpolinothio)benzothiaz ole, 2-(2,6-dimethyl-morphalinothio)benzothiazole, *N,N*-diisopropyl-2-benzothiazoles, 2-mercaptobenzothiazole, and *N,N'*-dicyclohexyl-2-benzothiazole sulfamide.

Laser desorption/ionisation FT-MS and FAB-MS have been used to determine non-volatile polymer additives such as thioester, phosphite, phosphonate, and hindered amine types [28].

Laser desorption/FT ion cyclotron resonance MS has been used to identify and determine the following types of polymer additives [29]: UV absorbents, e.g., Tinuvin [30], antioxidants, e.g., Irganox MD-1024, and amide wax anti-slip additives.

7.6 Gas Chromatography

7.6.1 Instrumentation

Further details of instrumentation are given in Chapter 3 and Appendix 1.

7.6.2 Applications

GC has been used extensively for the determination of more volatile components of polymers, e.g., monomers, residual solvents, and antioxidants, and when coupled with complementary techniques such as pyrolysis, photolysis, and MS for the elucidation of the structure of additives (**Table 7.1**) [31]. Halmo and co-workers [32] have used capillary GC for the determination of antioxidants in vulcanised rubbers.

7.7 High-Performance Liquid Chromatography

7.7.1 Theory

Modern high-performance liquid chromatography (HPLC) has been developed to a very high level of performance by the introduction of selective stationary phases of small particle sizes, resulting in efficient columns with large plate numbers per metre. There are several types of chromatographic columns used in HPLC.

The most commonly used chromatograph mode in HPLC is reversed-phase chromatography. Reversed-phase chromatography is used for the analysis of a wide range of neutral and

235

Table 7.1 Gas chromatographic determination of antioxidants in polymers		
Polymer	**Antioxidant**	**Reference**
Polystyrene	Ionol (2-di-*t*-butyl-*p*-cresol)	[324]
Polyethylene	2-*t*-butyl methyl phenol 2,6 di-*t*-butyl-4-methyl phenol *p*-*t*-butyl phenol	[325, 326]
Polypropylene	Butylated hydroxy anisole Butylated hydroxy toluene Ionox 330 (1,3,5-tri-methyl, 2,4,6-tri (3,5 di-*t*-butyl-4-hydroxy benzyl) benzene	[327, 328]
PVC Polyolefins	Dilauryl ββ′ thiodipropionate Dilauryl sulfonyl ββ′ dipropionate	[324-334]
PVC	Plasticisers	[329, 330]
Pyrolysis - gas chromatography - mass spectrometry		
Acrylic polymers	Unsaturated acids	[331]
Photolysis - gas chromatography - mass spectrometry		
Polyethylene	Dilaurylthiodipropionate	[258]
Gas chromatography - mass spectrometry		
Polypropylene crosslinked rubbers	Various additives	[315, 332]
Polystyrene	Isobutyronitrile Benzoyl peroxide Lauryl peroxide Tetramethyl succinonitrile	[333]
Gas chromatography - infrared spectroscopy		
Polyolefins	Various additives	[334]
Source: Author's own files		

polar organic compounds. Most commonly, reversed-phase chromatography is performed using bonded silica-based columns, thus inherently limiting the operating pH range to 2.0–7.5. The wide pH range (0–14) of some columns (e.g., Dionex Ion Pac NSI and NS 1–5 μm columns) removes this limitation, and consequently they are ideally suited for ion-pairing and ion-suppression reversed-phase chromatography: the two techniques that have helped extend reverse-phase chromatography to detection of ionisable compounds.

Typically, reversed-phase ion-pairing chromatography is carried out using the same stationary phase as reversed-phase chromatography. A hydrophobic ion of opposite charge to the solute of interest is added to the mobile phase. Samples that are determined by reverse-phase ion-packing chromatography are ionic and thus capable of forming an ion pair with the added counter ion. This form of reversed-phase chromatography can be used for anion and cation separations and for the separation of surfactants and other ionic types of organic molecules.

Ion suppression is a technique used to suppress the ionisation of compounds (such as carboxylic acids) so they will be retained exclusively by the reversed-phase retention mechanism and chromatographed as the neutral species.

7.7.2 Instrumentation

Four basic types of elution are used in HPLC. These are discussed next with reference to the systems offered by LKB, Sweden.

7.2.2.1 Isocratic System

This system (**Figure 7.2(a)**) consists of a solvent delivery for isocratic reversed-phase and gel-filtration chromatography. This isocratic system provides an economic first step into HPLC techniques. The system is built around a high-performance, dual-piston, pulse-free pump providing precision flow from 0.01 to 5 ml/min.

Any of the following detectors can be used with this system:

- fixed wavelength UV detector (LKB Unicord 2510)
- variable UV–visible (190–600 nm)
- wavelength monitor (LKB 2151)
- rapid diode array spectral detector (LKB 2140) (discussed later)
- refractive index detector (LKB 2142)
- electrochemical detector (LKB 2143)
- wavescan EG software (LKB 2146)

7.2.2.2 Basic Gradient System

This system (**Figure 7.2(b)**) is a simple upgrade of the isocratic system with the facility for gradient elution techniques and greater functionality. The basic system provides for manual

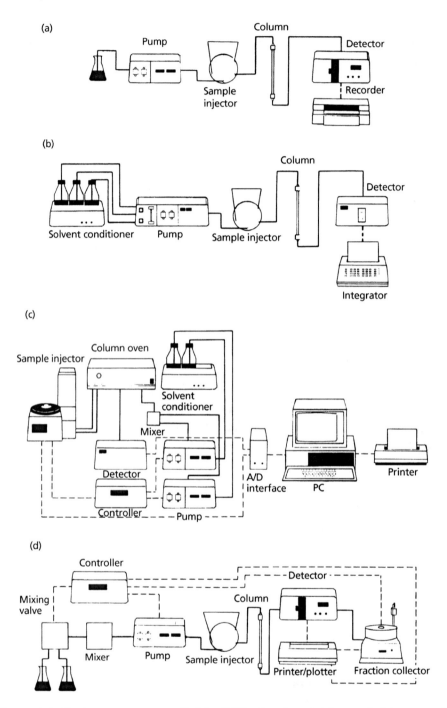

Figure 7.2 Elution systems supplied by LKB, Sweden: (a) isocratic bioseparation system; (b) basic system; (c) advanced chromatography system; (d) inert system.

operating gradient techniques such as reversed-phase, ion exchange, and hydrophobic interaction chromatography. Any of the detectors listed above for the isocratic system can be used.

7.2.2.3 Advanced Gradient System

For optimum functionality in automated systems designed primarily for reversed-phase chromatography and other gradient techniques, the LKB advanced gradient system (**Figure 7.2(c)**) is recommended. Key features include:

- A configuration that provides the highest possible reproducibility of results.

- A two-pump system for highly precise and accurate gradient formation for separation of complex samples.

- Full system control and advanced method developments provided from a liquid chromatography controller.

- Precise and accurate flows ranging from 0.01 to 5 ml/min.

This system is ideal for automatic method development and gradient optimisation.

7.2.2.4 Inert System

By a combination of the use of inert materials (glass, titanium, and inert polymers) this system (**Figure 7.2(d)**) offers totally inert fluidics. Primary features of the system include:

- The ability to perform isocratic or gradient elution by manual means.
- Full system control from a liquid chromatography controller.
- Precise and accurate flows from 0.01 to 5 ml/min.

This is the method of choice when corrosive buffers, e.g., those containing chloride or aggressive solvents, are used.

7.2.2.5 Chromatographic Detectors

Details concerning the types of detectors used in HPLC are given in **Table 7.2**. The most commonly used detectors are those based on spectrophotometry in the region 185–400

Table 7.2 Detectors used in HPLC

Type of detector		Supplier	Detection part no.	HPLC instrument part no.
Spectrophotometric (variable wavelength)	190-390 nm	Perkin Elmer	LC90	-
	195-350 nm	Kontron	735 LC	Series 400
	195-350 nm	Shimadzu	SPD-7A	LC-7A
	195-350 nm	Shimadzu	SPD-6A	LC-8A
	195-350 nm	Shimadzu	SPD-6A	LC-6A
	206-405 nm (fixed wavelength choice of 7 wavelengths between 206 and 405 nm)	LKB	2510 Uvicord SD	
	190-370 nm	Cecil Instruments		
	190-400 nm	Cecil Instruments		
Variable wavelength UV - visible	190-600 nm	Varian	2550	2500
	190-600 nm	LKB	2151	Uvicord SD
	190-700 nm	Kontron	432	Series 400
	190-800 nm	Kontron	430	Series 400
	185-900 nm	Kontron	720 LC	Series 400
	200-570 nm	Kontron	740 LC	Series 400
	190-800 nm	Dionex	VDM II	Series 400
	190-750 nm	Isco	V4	Microbo system
	214-660 nm (18 preset wavelengths)	Isco	UAS and 228	Microbo system
	195-700 nm	Shimadzu	SPD 7A	LC-7A
	195-700 nm	Shimadzu	SPD-6AV	LC-8A
	195-700 nm	Shimadzu	SPD-6AV	LC-6A
	193-350 nm	Shimadzu	SPD-6A and SPDM 6A	LC-9A
	196-700 nm	Shimadzu	SPD-6AV	LC-9A
	190-900 nm	Shimadzu	-	LC-10A

Table 7.2 Continued

Type of detector		Supplier	Detection part no.	HPLC instrument part no.
Variable wavelength (*continued*)	190-900 nm	Varian	LC Star Systems System for biomolecules	Star 9060
	190-900 nm	Pharmacia		-
	190-900 nm	ICI	-	LC/1210/1205
	190-600 nm	Hewlett Packard	Programmably variable wavelength detector	9050 series
	380-600 nm	Cecil Instruments	CE 1200	Series 1000
	190-800 nm	Applied Chromatography Systems	750/16 and 5750/11	-
Conductivity	-	Dionex	CDM 11	4500 i
		Roth Scientific	-	Chrom-A-Scope
		Shimadzu	CDD-6A	LC-9A
Electrochemical detector	-	Dionex	PAD-11	4500i
		LKB	2143	Wave-scan EG
		Roth Scientific	-	Chrom-A-Scope
		Cecil Instruments	CE 1500	-
		PSA Inc.	5100A	-
		Applied Chromatography Systems	650/350/06	-
		Shimadzu	L-ECD-6A	LC-9A
Refractive index detector	-	LKB	2142	Wavescan E.G.
		Roth Scientific	-	Chrom-A-Scope
		Cecil Instruments	CE 1400	Series 1000
		Shimadzu	RID-6A	LC-9A
			-	LC-1240

Table 7.2 Continued

Type of detector	Supplier	Detection part no.	HPLC instrument part no.
Fluorescence	Shimadzu	RF-551	LC-9A
220-900	Shimadzu	FLD-6A	LC-9A
-	Shimadzu	RF 335	LC-9A
220-650			
Differential viscosity mass detection (evaporative)	Roth Scientific	-	Chrom-A-Scope
	Applied Chromatography Systems	750/14	-
Diode array	Varian	9065	2000L and 5001 5500 series
	Perkin-Elmer	LC135, LC235 and LC480	-
	LKB	2140	-
	Hewlett Packard	Multiple wavelength detector	1050 series

Source: Author's own files

nm, visible–UV spectroscopy in the region 185–900 nm, post-column derivatisation with fluorescence detection (see next), conductivity, and multiple wavelength UV detectors using a diode array system detector (see next). Other types of detectors available are those based on electrochemical principles, refractive index, differential viscosity, and mass detection.

- Post-column derivatisation: fluorescence detectors

Modern column liquid chromatography has been developed to a very high level by the introduction of selective stationary phases of small particle sizes, resulting in efficient columns with large plate numbers per metre. The development of HPLC equipment has been built upon the achievements in column technology, but the weakest part is still the detection system. UV–visible and fluorescence detectors offer tremendous possibilities, but because of their specificity it is possible to detect components only at very low concentrations when using a specific chromophore or fluorophore. The lack of a sensitive all-purpose detector in liquid chromatography like the flame ionisation detector (FID) in GC is still disadvantageous for liquid chromatography for the detection of important groups of compounds. Consequently, chemical methods are increasingly used to enhance the sensitivity of detection. On-line post-column derivatisation started with the classic work of Spackman and co-workers [44] and has recently been of increasing interest and use [45–52]. With on-line post-column detection the complexity of the chromatographic equipment increases. An additional pump is required for the pulseless and constant delivery of the reagent.

- Diode array detectors

With the aid of an high-resolution UV diode array detector, the eluting components in a chromatogram can be characterised on the basis of their UV spectra. The detector features high spectral resolution (comparable to that of a high-performance UV spectrophotometer) and high spectral sensitivity. The high spectral sensitivity permits the identification of spectra near the detection limit, i.e., within the sub-milliabsorbance range. Several manufactures (Varian, Perkin Elmer, LKB, and Hewlett Packard, see **Table 7.2**) have developed diode array systems. In the polychromator incorporated in the Perkin Elmer LC 480 diode array system the light beam is dispersed within the range 190–430 nm onto a diode array consisting of 240 light-sensitive elements. This effects a digital resolution of 1 nm, which thus satisfies the spectral resolution determined by the entrance slit.

- Electrochemical detectors

These are available from several suppliers (see **Table 7.2**). ESA supply the model PS 100A coulochem multi-electrode electrochemical detector. Organics, anions, and cations can be detected by electrochemical means.

7.7.3 Applications

There are three main applications of HPLC in polymer analysis:

- Determination of polymer additives (or other low molecular weight, non-polymeric substances) in polymers, as discussed next

- Separation of polymers into molecular weight fractions, especially lower molecular weight polymers (oligomers) (Chapter 8)

- Molecular weight determination in polymers – gel permeation chromatography (GPC); see Chapter 8

Figure 7.3 shows a chromatogram, obtained using GPC, of a mixture of additives solvent extracted from a sample of PP. The mobile phase was water (channel A) and acetonitrile (channel B). The analysis, performed using a Hewlett Packard model HP109 Series M liquid chromatograph equipped with a diode array detector, shows the presence of methylated hydroxytoluene, methylated hydroxyethylbenzene, Amide E (erucamide), Irganox 1010, and Irganox 1076 (both sterically hindered phenols). Numerous other applications of HPLC to the determination of polymer additives have been discussed (**Table 7.3**).

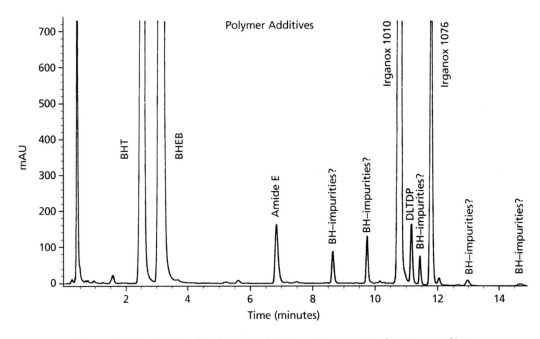

Figure 7.3 Analysis of polymer additives (*Source: Author's own files*)

Table 7.3 Application by high performance liquid chromatography to the determination of polymer additives		
Polymer	Determined	Reference
Polyethylene	Santonox R Ethanox 736 CA05 Irganox 1035 Iroganox 259 Topanol	[346]
Polyolefins	DLTDP DMTDP Irganox 1076 DSTDP	[347]
PVC	Plasticisers: Dibutyl phthalate Di-2-ethyl hexyl phthalate Di-iso decyl phthalate Tricresyl phosphate	[347]
Rubbers	Amine antioxidants Antiozonants	[349]
Polyethylene	Irganox 1076 Iroganox 1010 BHT	[348]
PVC	Didecyl phthalate Decyl benzyl phthalate Di-benzyl phthalate	[350]
Source: Author's own files		

7.8 Complementary Techniques

When HPLC is used in conjunction with other techniques such as MS and IR spectroscopy then identification and quantitation of polymer additives becomes more certain.

7.8.1 HPLC with Mass Spectrometry

One of the limitations of GC and consequently of GC–MS is that in many cases polymer additives are insufficiently volatile to be separated on GC columns operating at even the maximum of their temperature range. As a consequence of this, there has, in recent years, been a growing interest in applying HPLC, which is not subject to this temperature limitation, to the analysis of polymer extracts.

7.8.1.1 Instrumentation

Hewlett Packard, for example, supply the HP 5988A and HP 5987A mass-selective detectors for use with liquid chromatographs (LC) (see Appendix 1). The particle beam liquid chromatograph–mass spectrometer uses the same switchable EI-CI source and the same software and data systems that are used for a GC–MS system. Adding a GC creates a versatile particle beam LC–GC–MS system that can be switched from LC–MS to GC–MS in an instant.

EI spectra from the system are reproducible and can be searched against standard or custom libraries for positive compound identification. CI spectra can also be produced.

The particle beam system is a simple transport device, very similar to a two-stage jet separator. The solvent vapour is pumped away, while the analyte particles are concentrated in a beam and allowed to enter the mass spectrometric source. Here they are vaporised and ionised by electron impact.

The different ways a particle beam LC–MS can be configured reflect the versatility of the system in accommodating both the application and the availability of existing instrumentation. The system consists of the following elements:

- Particle beam interface mounted on the Hewlett Packard 5988A or 5987A mass spectrometer.

- Liquid chromatograph (either the integrated Hewlett Packard 1090 or modular Hewlett Packard 1050).

- Data system (either HP 59970C Chem Station for single instrument operation or the Hewlett Packard 100 RTE A-series for multi-instrument or multi-tasking, multi-user operation).

7.8.1.2 Applications

Like other forms of molecular spectroscopy, MS may be used as a 'fingerprint' technique to identify the components of additive systems extracted from polymer compositions. The strengths of MS are high sensitivity and the ability to distinguish between closely related compounds of differing relative molecular mass, e.g., the various alkyl thiodipropionates used as synergistic stabilisers in polyolefins and the UV-absorbing benzotriazole derivatives.

Often it is not necessary to separate the components before examination as some separation may be achieved by careful variation of the sample probe temperatures to produce, in effect, a fractional distillation of the components. The presence, however, of large amounts of low molecular weight polymers such as PE and PP can cause interference by producing a high hydrocarbon background extending to several hundred relative atomic mass units. In such instances thin-layer chromatographic separation can be used as a clean-up procedure.

Rudewicz and Munson [40] have described the use of secondary ion MS (SIMS) for the determination of additives in PP without prior separation by solvent extraction or precipitating techniques. The additives are vaporised from PP samples in a heatable glass probe under CI conditions using an ammonia (1.1%) in methane reagent gas mixture. The dominant ion in this mixture, NH_4^+, is a low-energy reagent ion that reacts with the additives to give very simple spectra of $(M + H)^+$ or $(M + NH_4)^+$ ions and little fragmentation. The detection of additives in a hydrocarbon matrix is very selective.

Using this technique Rudewicz and Munson [40] determined Ionox 330 and Irganox 168 antioxidants and the UV absorber, UV 531 in PP. The additives begin to vaporise rapidly at temperatures near the melting point of PP, about 176 °C. The additives, however, can clearly be distinguished by their different masses and their very different heating profiles. The lower molecular weight additive gives a relatively sharp peak with a maximum at approximately 210 °C, and the higher molecular weight additive gives a broader peak that has a maximum near 300 °C.

The information obtained by absorbance detection when coupled with LC is usually not specific enough to allow the qualitative identification of compounds present in complex polymer mixtures.

Vargo and Olsen [58] used MS detection in series with absorbance detection to identify or characterise antioxidants and UV-stabilising additives in acetonitrile extracts of PP, which were separated by LC. Nanogram quantities of additives could be detected. The procedure was used to characterise additives in PP samples of unknown composition. Butylated hydroxytoluene and Irganox 1076 were identified in the extract of an automotive component moulding. In addition, several other additives were identified (palmitic acid, dioctylphthalate, stearic acid, and octadecanol).

Yu and co-workers [59] discussed liquid chromatographic interfaces for bench-top single quadruple LC-MS. The two most popular interfaces are particle beam and atmospheric pressure ionisation types. The system was applied to the analysis of additives in PP. Dilts [60] used a photodiode array detector coupled with particle beam LC–MS to characterise the degradation of Isonox 129, Irganox 1010, Irganox 1076, and Irgafos 168 in polyolefins. Sidwell [61] examined extractables from plastic and rubber components of medical products by LC–MS and GC-MS.

7.8.2 HPLC with IR Spectroscopy

This technique is extremely useful. For example, it has been used to identify and determine plasticisers in PVC and antioxidants in PE.

Dwyer [62] used a combination of chromatography and IR spectroscopy to provide a versatile tool for the characterisation of polymers. The HPLC-FT-IR interface systems described, deposit the output of a chromatograph onto an IR optical medium, which is then scanned to provide data as a time-ordered set of spectra of the chromatogram. Polymer analysis applications described include the identification of polymer additives, the determination of composition/molecular weight distributions in copolymers, the mapping of components of polymer alloys and blends, the molecular configuration changes in polymers, and component identification in complex systems.

Thilen and Shishoo [63] optimised experimental parameters for the quantification of polymer additives (Irganox 1010 and Irgafos 168) using supercritical fluid extraction combined with HPLC. The experimental parameters of temperature, pressure, and modifiers were varied to find the best extraction conditions. The optimum temperature and pressure for extraction of these PP additives were found to be 12.0 and 38.4 MPa, with methanol as the modifier. The quantitative extractions were significantly faster than those reported previously in the literature.

7.9 Ion Chromatography

This technique, discussed in Chapter 1, can be applied to the separation of mixtures of organic substances and as such may have applications in the analysis of additives, additive breakdown products, or catalyst residues in polymer extracts. Phenols and carboxylic acids are possibilities in this respect.

7.10 Supercritical Fluid Chromatography

7.10.1 Theory

Until recently the chromatographer has had to rely on either GC or HPLC for separations, enduring the limitations of both. Lee Scientific has created a new dimension in chromatography, one that utilises the unusual properties of supercritical fluids. With the new technology of capillary supercritical fluid chromatography (SFC), the chromatographer benefits from the best of both worlds – the solubility behaviour of liquids and the diffusion and viscosity properties of gases. Consequently, capillary SFC offers unprecedented versatility in obtaining high-resolution separations of difficult compounds.

Beyond its critical point, a substance can no longer be condensed to a liquid, no matter how great the pressure. As the pressure increases, however, the fluid density approaches that of a liquid. Because solubility is closely related to density, the solvating strength of the fluid assumes liquid-like characteristics. Its diffusivity and viscosity, however, remain. SFC can use the widest range of detectors available to any chromatographic technique. As a result, capillary SFC has already demonstrated a great potential in applications to polymer additives.

SFC is now one of the fastest growing analytical techniques. The first paper on the technique was by Klesper and co-workers [64], but SFC did not catch analysts' attention until Novotny and co-workers [65] published the first paper on capillary SFC.

Most SFC use carbon dioxide as the supercritical eluent, as it has a convenient critical point of 31.3 °C and 7.3 MPa. Nitrous oxide, ammonia, and *n*-pentane have also been used. This allows easy control of density between 0.2 and 0.8 g/ml and the utilisation of almost any detector from LC or GC.

Wall [66] has discussed recent developments including timed split injection, extraction, and detection systems in SFC.

7.10.2 Instrumentation

Suppliers of this equipment are discussed in Appendix 1.

Capillary SFC utilises narrow 50 or 100 μm id columns of between 3 and 20 m in length. The internal volume of a 3 m × 50 μm id column is only 5.8 μl. SFC operates at pressures from 10.3 to beyond 41.4 MPa. To allow injections of about 10–50 μl to be introduced to a capillary column, an internal loop LC injector (Valco Instruments, Switzerland) has been used with a splitter (**Figure 7.4**), which was placed after the valve to ensure that a smaller volume was introduced onto the column. This method works well for compounds that are easily soluble in carbon dioxide at low pressures. Good reproducibility is attained for capillary SFC using a direct injection method without a split restrictor.

This method (**Figure 7.4(b)**) utilised a rapidly rotating internal loop injector (Valco Instruments, Switzerland) which remains in-line with the column for only a short period of time. This then gives a reproducible method of injecting a small fraction of the loop onto the column. For this method to be reproducible the valve must be able to switch very rapidly to put a small slug of sample into the column. To achieve this a method called timed-split injection was developed (Lee Scientific). For timed-split to operate it is essential that helium is used to switch the valve: air or nitrogen cannot provide sharp

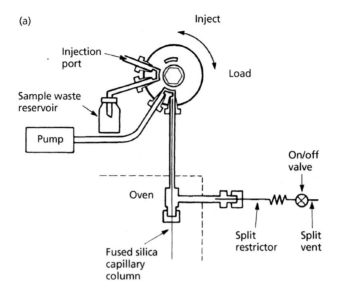

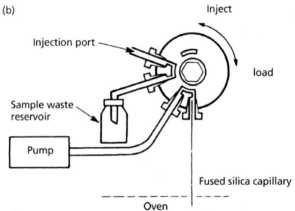

Figure 7.4 Sample injectors: (a) single split valve injector; (b) times split and direct valve injector. (Source: *Author's own files*)

enough switching pulses. The injection valve itself must have its internal dead volumes minimised. Dead volumes prior to the valve allow some on the sample to collect prior to the loop, effectively allowing a double slug of sample to be injected which appears at the detector as a very wide solvent peak.

7.10.2.1 Detection Systems

SFC uses detectors from both LC and GC. A summary of detection systems used in SFC has been documented [67].

One of the most commonly used detection systems is electron capture detector. A sensitivity to about 50 pg (minimum detection limit) on a column is obtainable [88].

A paper has been published showing the use of the photoionisation detector (PID) [89]. The PID is to a certain extent specific in that only compounds that can be ionised by a UV lamp will give a response. The solvents used were dichloromethane and acetonitrile, both of which should have little response in the photoionisation detector. However, a clear sharp solvent peak was observed.

The amount detected by this system (0.3 pg on column) was below the level that could have been determined using a FID. Initial indications show that the PID may be a very useful detector for cases requiring lower levels on the SFC and where concentration of samples is not possible.

A sulfur chemiluminescence detector (Sievers Research Inc., Colorado, USA) has been investigated. Good sensitivities and chromatograms have been shown for standards and real samples. This detector shows no response to carbon dioxide and gives low picogram sensitivities for a wide range of sulfur compounds.

7.10.3 Applications

Dionex Corporation, in an application note on its series 600-D SFC, describe a method for the determination of polymer additives. The possibilities of SFC for determining polymer additives have been recently demonstrated [90-101]. The low elution temperature and high resolution of capillary fluid chromatography makes this technique very attractive. Other advantages are that a FID can be used and that interfacing with spectroscopic detectors is somewhat easier than with HPLC. Quantitation in capillary SFC has been found to be more difficult than in HPLC and capillary gas chromatography, however, because of lack of precision in injection [97].

Polymer additives – mould release agents, plasticisers, antioxidants, and UV absorbers, with molecular weights extending beyond 1000 – are generally unsuitable for GC or LC analysis because of their low volatility, lack of chromophore, or thermal instability. SFC is now the method of choice for the analysis of such compounds. **Figure 7.5** shows the chromatogram of a polymer containing Tinuvin 1130.

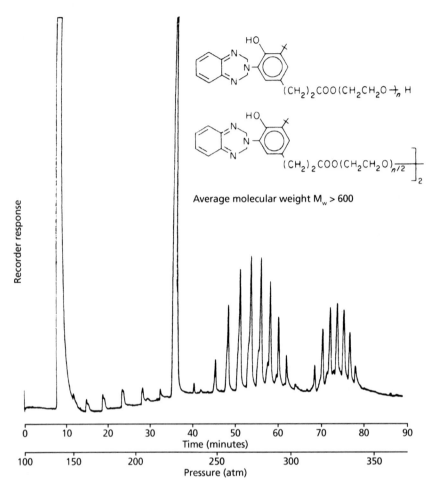

Figure 7.5 Supercritical fluid chromatogram of polymer additive, Tinuvin 1130. Conditions: column 10 m × 50 μm, SB-methyl; mobile phase, carbon dioxide at 150 °C with pressure programming detection; flame ionisation detector. (*Source: Author's own files*)

7.11 Thin-Layer Chromatography

7.11.1 Theory

Thin-layer chromatography (TLC), like GC, comes into its own when dealing with mixtures of substances. TLC using plates coated with 250 μm absorbent is an excellent technique for separating quantities of up to 20 mg of additive mixtures into their individual components and provides enough of each to prepare a recognisable IR or UV spectrum which can be compared with spectra of authentic known compounds. However, this technique does not conveniently handle larger quantities and in these cases separation on the 50–500 mg scale can be carried out with only a small loss of resolution by chromatography on a silica gel or aluminium packed column. The separated bands are marked under UV light, removed from the plate, and extracted with diethyl ether. Separations on this scale provide sufficient of each fraction for full characterisation by NMR spectroscopy and MS, as well as IR spectroscopic techniques.

7.11.2 Applications

Thin-layer liquid chromatography has been employed for the determination of alkylated cresols and amine antioxidants [102, 103] in polybutadiene, phenolic antioxidants in PE [104-106] and PP [106], dilauryl and distearyl thiodipropionate antioxidants in polyolefins, ethylene–vinyl acetate copolymer, acrylonitrile–styrene terpolymer and PS, UV absorbers and organotin stabilisers in polyolefins [102], and accelerators such as guanidines, thiazoles, thiurans, sulfenamides, diethiocarbamides, and morpholine disulfides in unvulcanised rubber compounds.

7.12 Polarography

7.12.1 Instrumentation

Further details of instrumentation are given in Appendix 1.

7.12.2 Applications

Some applications of polarography to the determination of polymer additives and catalyst remnants are reviewed in **Table 7.4**.

7.13 Pyrolysis Gas Chromatography–Mass Spectrometry

Applications of this technique have been reviewed by Geissler [107], Meyer-Dulheuer and co-workers [108], and workers at Shimadzu-Volkswagen [109, 110]. One of these papers [110] lists additive fragment spectra for 174 additives.

Table 7.4 Applications of polarography		
Polymer	Determinand	Reference
Polystyrene	*p-tert* butyl Perbenzoate	[375]
Polyester acrylates	Ionol (2,6 di-*t*-butyl-*p*-cresol)	[396]
Polystyrene	Ionol 4,4 Isopropylidene-diphenol	[377]
Polyolefins	Antioxidants	[378, 379]
Rubbers	Accelerators Phenolic antioxidants Amine antioxidants	[59, 76, 117, 123, 161, 284, 291, 380-387]
Source: Author's own files		

7.14 X-ray Photoelectron Spectroscopy

Pena and co-workers [111] discuss factors affecting the adsorption of organophosphorus polymer stabilisers onto carbon black.

7.15 Secondary Ion Mass Spectrometry

Time-of-flight secondary ion MS (ToFSIMS) with either gallium or indium primary beams has been evaluated as a method for measuring the homogeneity of distribution of Chimassorb 944 FD hindered amine antioxidant in low-density polyethylene. The parent ion for the oligomer at *m/z* 599 was so weak that it could not be used to map the distribution of the additive throughout its most commonly used concentration range (0.1–0.5% w/w) in PE. Instead a mass fragment at *m/z* 58 was found to be sufficiently clear of interferences for use as a surrogate for the parent ion. As a result, imaging of the antioxidant distribution was possible to concentrations as low as 0.1% and a linear concentration calibration curve was obtained. The use of an indium primary beam improved the correlation of the antioxidant. Furthermore, indium reduced the contribution from the PE background at *m/z* 58 in relation to the total counts acquired.

7.16 X-ray Fluorescence Spectroscopy

Lopez-Cervantes and co-workers [112] used this technique to identify oxygen absorbers in polymers.

7.17 Solvent Extraction Systems

Tecator supplies units for carrying out organic solvent extractions on polymers. The Soxtec HT2–HT6 systems are recommended for carrying out solvent extractions of additives in polymers and rubbers.

References

1. M.P. Thomas, *Journal of Vinyl and Additive Technology*, 1996, **2**, 330.

2. M.J. Forrest, *Analysis of Plastics*, Rapra Review Report No. 149, Rapra Technology Ltd., Shrewsbury, UK, 2002, **13**, 5, 146.

3. H.L. Spell and R.D. Eddy, *Analytical Chemistry*, 1960, **32**, 1811.

4. R.G.J. Miller and H.A. Willis, *Spectrochimica Acta*, 1959, **14**, 119.

5. G. Guichon and J. Henniker, *British Plastics*, 1964, **37**, 74.

6. J.P. Luongo, *Applied Spectroscopy*, 1965, **4**, 117.

7. H.V. Drushel and A.L. Sommers, *Analytical Chemistry*, 1964, **36**, 836.

8. S.C. Patticini, *Determination of Mineral Oil in Polystyrene, Perkin Elmer IR Bulletin No. 85*, 1981.

9. P. Hendra and H. Mould, *International Laboratory*, 1988, 30.

10. H. Pasch, *Analytica Chimica Acta*, 2001, **165**, 91.

11. P.J. Cornish, *Journal of Applied Polymer Science*, 1963, **7**, 727.

12. A. Fiorenza, G. Bonomi and A. Sareda, *Materie Plastiche ed Elastomeri*, 1965, **31**, 1045.

13. Z.H. Auler, *Gummi Asbesos Kunststoffe*, 1961, **14**, 1024.

14. T. Kawaguchi, K. Ueda and A. Koga, *Journal of the Society of Rubber Industry (Japan)*, 1955, **28**, 525.

15. L.H. Ruddle and J.R. Wilson, *Analyst*, 1969, 94, 105.

16. C.L. Hilton, *Rubber Age*, 1958, **84**, 263.

17. H.P. Burchfield and J.N. July, *Analytical Chemistry*, 1960, **19**, 383.

18. C. Cieleszby and F.Z. Nagy, *Lebensmittel-Forschung*, 1961, **114**, 13.

19. P. Tittarelli, T. Zerlia, A. Colli and G. Ferrari, *Analytical Chemistry*, 1983, **55**, 220.

20. J. Lutzen, K. van Veen and S.T. Balke in *Proceedings of 60th SPE Annual Technical Conference ANTEC 2002*, San Francisco, CA, 2002, Session T23, Paper No.346.

21. C.A. Parker and W.J. Barnes, *Analyst*, 1957, **82**, 606.

22. C.A. Parker, *Analytical Chemistry*, 1961, **37**, 140.

23. V.F. Gaylor, A.F. Conrad and J.H. Landerl, *Analytical Chemistry*, 1957, **29**, 224.

24. P.B. Mansfield, *Chemistry and Industry*, 1971, **28**, 792.

25. R.P. Lattimer, R.E. Harris, C.K. Rhoe and H.R. Shutlen, *Analytical Chemistry*, 1986, **58**, 3188.

26 A.S. Hilton and A.G. Altenau, *Dublin Chemistry and Technology*, 1973, **46**, 1035.

27. M.W. Hayes and A.G. *Altenau, Rubber Age*, 1970, **102**, 59.

28. C.L. Johlman, C.H. Wilkins, J.D. Hogan, T.L. Donovan, D.A. Laude and M.J. Youseffi, *Analytical Chemistry*, 1990, **62**, 1167.

29. B. Asamoto, J.R. Young and R.J. Citerin, *Analytical Chemistry*, 1990, **62**, 61.

30. E. Schroder and G. Rudolph, *Plaste und Kautschuk*, 1963, **1**, 22.

31. F. David, L. Vanderroost and P. Sandra, *Analysis of Polymer Additives Using the HP5890 Series 11, Cool on Column Inlet with Electronic Pressure Control*, Hewlett Packard Application Note No. 228-149, 1991.

32. F. Halmo, S. Surova, M. Balakova and A. Halmova in *Proceedings of Technical Rubber Goods: Part of our Everyday Life*, Puchov, Slovakia, 1996, p.233.

33. C.B. Roberts and J.D. Swank, *Analytical Chemistry*, 1964, **36**, 271.

34. R.E. Long and G.C. Guvernator, *Analytical Chemistry*, 1967, **39**, 1493.

35. E.L. Styskin, A. Ya Gurvich and S.T. Kumak, *Khimicheskaya Promyshlennost*, 1973, **5**, 359.

36. H.S. Knight and H. Siegel, *Analytical Chemistry*, 1966, **38**, 1221.

37. M.M. Robertson and R.M. Rowley, *British Plastics*, 1960, **33**, 1.

38. J. Haslam and W.W. Soppet, *Journal of the Chemical Society*, 1948, **67**, 33.

39. J.L. Sharp and G. Paterson, *Analyst*, 1980, **105**, 517.

40. P. Rudewicz and B. Munson, *Analytical Chemistry*, 1986, **58**, 358.

41. S.T. Likens and G.B. Nickerson, *Proceedings of the American Society of Brewing Chemistry*, 1964, 5.

42. D.M.W. Anderson, *Analyst*, 1959, **84**, 50.

43. R.S. Juvet, S.L.S. Smith and K. Panghi, *Analytical Chemistry*, 1972, **44**, 49.

44. D.H. Spackman, W.H. Stein and S. Moore, *Analytical Chemistry*, 1958, **30**, 1190.

45. *Chemical Derivatization in Analytical Chemistry, Volume 1*, Eds., R.W. Frei and J.F. Lawrence, Plenum Press, New York, NY, USA, 1981.

46. *Chemical Derivatization in Analytical Chemistry, Volume 2*, Eds., R.W. Frei and J.F. Lawrence, Plenum Press, New York, NY, USA, 1981.

47. *Reaction Detection Gas Chromatography*, Ed., L.S. Krull, Marcel Dekker, New York, NY, USA, 1986.

48. H. Englehardt and U.D. Neue, *Chromatographia*, 1982, **15**, 403.

49. H. Englehardt and B. Lillig, *Journal of High Resolution Chromatography (and Column Chromatography)*, 1985, **8**, 531.

50. H. Englehardt and B. Lillig, *Chromatographia*, 1982, **15**, 403.

51. H. Englehardt, R. Klinker and B. Lillig in *Kopplungsverfahren in der HPLC*, Eds., H. Englehardt and K.P. Hupe, GIT-Verlag, Darmstadt, Germany, 1985.

52. M. Uihlein and E. Schwab, *Chromatographia*, 1982, **15**, 140.

53. J.L. DiCesare, M.W. Dong and L.S. Ettre, *Chromatographia*, 1981, **14**, 257.

54. *Shimadzu Applications Data Book C190-E001*, Shimadzu Corporation, Tokyo, Japan.

56. J.F. Schabon and L.F. Fenska, *Analytical Chemistry*, 1980, **52**, 1411.

55. J. Protivova and S.J. Pospisil, *Chromatography*, 1974, **88**, 99.

57. R.E. Majors, *Journal of Chromatographic Science*, 1970, **8**, 338.

58. J.D. Vargo and K.L. Olsen, *Analytical Chemistry*, 1985, **57**, 672.

59. K. Yu, E. Block and M. Balogh, LC-GC International, 1999, **12**, 577.

60. M.J. Dilts in *Proceedings of an IUPAC Polymer Symposium: Functional and High Performance Polymers*, Taipei, Taiwan, 1994, p.445.

61. J. Sidwell, *Proceedings of a Rapra Conference on Medical Polymers 2003*, Dublin, Ireland, 2003, Paper No.8, p.51.

62. J.L. Dwyer in *Proceedings of Materials - Challenge, Diversification and the Future*, Anaheim, CA, USA, 1995, Volume 40, Book 1, p.673.

63. M. Thilen and R. Shishoo, *Journal of Applied Polymer Science*, 2000, **76**, 938.

64. E. Klesper, A. Corwin and D. Turner, *Journal of Organic Chemistry*, 1962, **27**, 700.

65. M. Novotny, P.S. Springston and M. Lee, *Analytical Chemistry*, 1981, **53**, 407A.

66. R.J. Wall, *Chromatography and Analysis*, 1988.

67. D.W Later, D.J. Bornhop, E.D. Lee, J.D. Henion and R.C. Wiebolt, LC-GC, 1987, **5**, 804.

68. E.J. Kuta and F.W. Quackenbush, *Analytical Chemistry*, 1960, **32**, 1669.

69. V.V. Budyina, V.G. Marinini, Yu V. Vodzinskii, A.I. Kalinina and I.A. Korschunov, *Zavodskaia Laboratoriia Diagnostika Materialov*, 1970, **36**, 1051.

70. A.A. Vasileva, Yu V. Vodzinskii and I.A. Korschunov, *Zavodskaia Laboratoriia Diagnostika Materialov*, 1968, **34**, 1304.

71. *Analytical Applications Report No. 637D*, Southern Analytical Ltd., Camberley, Surrey, UK.

72. *Analytical Applications Report No. 651/2*, Southern Analytical Ltd., Camberley, Surrey, UK.

73. E. Barendrecht, *Analytica Chimica Acta*, 1961, **24**, 498.

74. F. Mocker, *Kautchuk und Gummi Kunststoffe*, 1964, **11**, 1161.

75. W. Cooper, D.E. Eaves, M.E. Tunnicliffe and G. Vaughan, *European Polymer Journal*, 1965, **1**, 121.

76. G.W. Tindall, R.L. Perry and J.L. Little, *Analytical Chemistry*, 1991, **63**, 1251.

77. M.M. O'Mara, *Journal of Applied Polymer Science*, 1970, **13**, 1887.

78. V.F. Gaylor, A.L. Conrad and J.H. Landerl, *Analytical Chemistry*, 1957, **29**, 228.

79. G. Zerbi, *Chimica e l'Industria (Milan)*, 1973, **55**, 334.

80. F. Mocker, *Kautschuk und Gummi*, 1959, **12**, 155.

81. R. Hank, *Rubber Chemistry and Technology*, 1967, **40**, 936.

82. A. Zwerg, E. Lancaster, M.J. Neglea and W.H. Jura, *Journal of the American Chemical Society*, 1964, **86**, 413.

83. Yu Zodzinskii and G.S. Semchikova, *Po Khim i Khim Teknol*, 1963, 272.

84. V.F. Gaylor, P.J. Elving and A.L. Contrad, *Analytical Chemistry*, 1953, **25**, 1078.

85. J.F. Hadenberg and H. Freiser, *Analytical Chemistry*, 1953, **25**, 1355.

86. J.W. Hamilton and A.L. Tappel, *Journal of the American Oil Chemists' Society*, 1963, **40**, 52.

87. *Analytical Applications Report No. 637D*, Southern Analytical Ltd., Camberley, Surrey, UK.

88. S. Kennedy and R. Wall, LC-GC, 1988, **445**, 10.

89. P. Sim, C. Elson and M. Quillaim, *Journal of Chromatography*, 1988, **445**, 239.

90. D. Later, B. Richter and M. Anderson, LC-GC, 1986, **4**, 992.

91. J. Doehl, A. Farbrot, J. Greibrokk and B. Iverson, *Journal of Chromatography*, 1987, **392**, 175.

92. M.W. Raynor, K.D. Bartle, I.L. Davis, A. Williams, A.A. Clifford, J.M. Chalmers and B.W. Cook, *Analytical Chemistry*, 1988, **60**, 427.

93. R. Moulder, J.P. Kithinji, M.W. Raynor, K.D. Bartle and A.A. Clifford, *High Resolution Chromatography*, 1989, **12**, 688.

94. P.J. Arpino, D. Dilettato, Khoa Nguyen and A. Bruchet, *High Resolution Chromatography*, 1980, **13**, 5.

95. M. Ashraf-Khorassani and J.M. Levy, *High Resolution Chromatography*, 1990, **13**, 742.

96. N.J. Cotton, K.D. Bartle, A.A. Clifford and C.J. Dowle in *Proceedings of the 11th International Symposium on Capillary Chromatography*, Ed., P. Sandra, Monterey, CA, USA, 1990, p.665.

97. *Analysis of Polymers and Polymer Additives by Supercritical Fluid Chromatography and Supercritical Fluid Extraction*, Dionex Applications Note P/N TN 26LS, Dionex Corporation, Sunnydale, USA.

98. K.D. Bartle and R.M. Smith in *Supercritical Fluid Chromatography*, Ed., R.M. Smith, Royal Society of Chemistry, Cambridge, UK, 1988, p.1-25.

99. T.L. Chester, L.J. Burkes, T.E. Delaney, D.P. Innis, G.P. Owens and J.D. Pinkston in *Supercritical Extraction and Chromatography*, Eds., B.A. Charpentier and M.R. Sevenants, American Chemical Society Symposium Series No.336, ACS, Washington, DC, 1988, p.144-160.

100. M.W. Raynor, I.L. Davies, K.D. Bartle, J.D. Williams, J.D. Chalmers and B.W. Cook, *European Chromatography News*, 1987, **1**, 18.

101. F. David, L. Vanderroost and P. Sandra in *Proceedings of the 13th International Symposium on Capillary Chromatography*, Ed., P. Sandra, Riva del Garda, Italy, 1991, p.1539.

102. D. Simpson and B.R. Currell, *Analyst*, 1971, **96**, 515.

103. L.J. Gaeta, E.W. Schleuter and A.G. Altenau, *Rubber Age*, 1969, **47**, 101.

104. R.F. van der Heide and O. Wouters, *Lebensmittel-Forschung*, 1962, **117**, 129.

105. E. Schroder and E. Hagen, *Plaste und Kautschuk*, 1968, **15**, 625.

106. R.S. Delves, *Journal of Chromatography*, 1969, **40**, 110.

107. M. Geissler, *Kunststoffe Plast Europe*, 1997, **87**, 22.

108. T. Meyer-Dulheuer, H. Pasch and M. Geissler, *Kautschuk und Gummi Kunststoffe*, 2000, **53**, 574.

109. *Additives for Polymers*, 1998, **11**, 6-7.

110. *Adhesive Technology*, 1999, **16**, 29.

111. J.M. Pena, N.A. Allen, M. Edge, C.M. Laiuw and B. Valange, *Polymer Degradation and Stability*, 2001, **72**, 31.

112. J. Lopez-Cervantes, D.I. Sanchez-Machado, S. Pastorelli, R. Rijk and P. Paseiro-Losada, *Food Additives and Contaminants*, 2003, **20**, 291.

8 Polymer Fractionation and Molecular Weight

8.1 Introduction

Polymers do not normally consist of a particular molecule with a unique molecular weight, but rather are a mixture of molecules with a molecular weight range that follows a particular distribution. With some types of polymers the picture is further complicated by the appearance of what are known as crosslinks. These are chemical bonds that link one polymer chain to another. Crosslinking will, therefore, increase the molecular weight of a polymer and, incidentally, decrease its solubility in organic solvents. These are some of the features that make it possible to produce for a given polymer, say polypropylene (PP), a range of grades of the polymer, each with different physical properties and end uses and each characterised by a different molecular weight distribution (MWD) curve and degree of crosslinking. The factors that control these parameters in a polymer are complex, and are linked with the details of the manufacturing process used. They will not be discussed further here.

The measurement of the molecular weight is a task undertaken in its own right by polymer chemists, and is concerned with the development of new polymers and process control in the case of existing polymers. Additionally, however, it is necessary to separate a polymer not into unique molecules each with a particular molecular weight, but into a series of narrower MWD fractions. This is required in order to obtain a more detailed picture of the polymer composition and structure, and these separated fractions may be required for further analysis by a wide range of techniques.

In the simplest case it is required simply to separate, for example, the total gel fraction (i.e., crosslinked material which is insoluble in organic solvents) of a polymer from the total soluble fraction (non-crosslinked material). An example would be the separation of polystyrene (PS) into its gel and soluble fractions. In a more complicated case it may be required to carry out a separation of the original polymer into a series of fractions each with a narrower MWD than the parent polymer. These methods were formerly based on fractionation techniques, gradient elution, or precipitation.

Nowadays, such fractionations are carried out by a variety of techniques such as gel permeation chromatography (GPC)/size exclusion chromatography (SEC), high-performance liquid chromatography (HPLC), supercritical fluid chromatography (SFC), gas chromatography (GC) (for lower molecular weight polymers), liquid adsorption chromatography, field flow fractionation, thin-layer chromatography (TLC), capillary isotachoelectrophoresis, etc., as discussed next.

Various methods for the fractionation of polymers and the determination of molecular weight distributions and absolute molecular weight are discussed in this chapter under the following headings: high-performance GPC/SEC, HPLC, SFC, GC, TLC, nuclear magnetic resonance (NMR) spectroscopy, osmometry, light scattering, viscometry, ultracentrifugation, field desorption mass spectrometry, capillary electrophoresis, chromatography–MS, ion exchange chromatography, liquid adsorption chromatography, time-of-flight secondary ion mass spectrometry (ToFSIMS), matrix-assisted laser desorption/ionisation Fourier transform mass spectrometry, thermal field flow fractionation, desorption–chemical ionisation, and grazing emission X-ray fluorescence spectrometry.

8.2 High-Performance GPC and SEC

8.2.1 Theory

8.2.1.1 Fractionation of Copolymers

It is well known that copolymer properties are affected by composition in addition to molecular weight. Most copolymers have a chemical composition distribution and a MWD. Although the MWD of homopolymers can be measured by SEC rapidly and precisely, accurate information on the MWD of copolymers cannot be obtained by SEC alone. This is because separation in SEC is achieved according to the sizes of molecules in solution and the molecular weights of the copolymers are not proportional to molecular size unless the composition fluctuation is negligible and the chemical structure is the same across the whole range of molecular weights. Information on these distributions should be obtained by separating the copolymer by composition independently of molecular weight and then determining the MWD of each fraction. Alternatively, the MWD is determined first independently of composition and the chemical composition distribution of the same molecular weight species is measured. This is the principle of cross-fractionation, which can be performed by means of a combination of several chromatographic methods [1].

Some workers used SEC for the first fractionation, and TLC [2, 3] and high-performance precipitation liquid chromatography [4, 5] were applied as the second chromatographic method. Balke and Patel [6, 7] performed cross-fractionation by orthogonal chromatography (SEC–SEC) in which different mobile phases were used. Tanaka and co-workers [8] used the combination of column adsorption chromatography (first fractionation) and SEC (second fractionation). They separated styrene–methyl methacrylate graft copolymers using liquid adsorption chromatography with a mixture of ethyl acetate and benzene as the mobile phase. Styrene–methyl methacrylate random copolymers were also separated according to chemical composition by liquid adsorption chromatography on silica [9–12].

Mori [13] separated styrene–methyl methacrylate copolymers (54–85% methyl methacrylate) according to chemical composition and molecular size by liquid adsorption chromatography and SEC. They used a 1,2-dichloroethane–chloroform–silica gel system for liquid adsorption chromatography, i.e., chemical composition chromatography, and detection was achieved by UV absorption at 254 nm. Next, the MWD of the liquid adsorption chromatography fractions was analysed by SEC using a refractive index detector. Silica gel, having a pore size of 0.003 μm, was used as an adsorbent and mixtures of 1,2-dichlorethane and chloroform (including 1% ethanol as a stabiliser) were used as the mobile phase for liquid adsorption chromatography. The initial mobile phase for liquid adsorption chromatography was 1,2-dichloroethane and then the content of chloroform in the mobile phase was increased stepwise up to 100%. When 1,2-dichloroethane was used as the mobile phase, the copolymers adsorbed on the external surface of silica gel. With increasing chloroform content in the mobile phase, the copolymer was desorbed as a function of the composition. The early-eluted fraction in liquid adsorption chromatography had lower molecular weight averages and higher styrene content than those for the unfractionated copolymer. The late-eluted fraction had the opposite values.

8.2.1.2 Determination of Molecular Weight

Molecular weight together with its distribution is one of the basic descriptive characteristics of polymers. Molecular weight and its distribution exercise remarkable influences on thermal temperature, tackiness, tensile strength, shock resistance, fusing, fluidity, reactivity, hardness, weatherability, and other properties of polymers (**Table 8.1**). Therefore, measurements of the molecular weight and its distribution are extremely important in estimating the quality of raw materials and products, analysing quality in the production processes, and carrying out research and development.

Methods other than GPC that have been used to measure MWD of polymers include osmosis methods, viscosity methods, light scattering methods, ultracentrifugation, precipitation fractionation and column fractionation, and others, but these methods are complicated, and in some cases lengthy in operation.

Also, since a polymer is a mixture of molecules differing in molecular weight, the value of a measurement obtained by such methods is nothing but a statistical mean value assuming

Table 8.1a Effect of molecular weight or MWD on polymer properties (general correlations)		
Polymer property	Increasing mol wt.	Narrowing MWD
Tensile strength	+	+
Toughness	+	−
Brittleness	+	−
Hardness	+	−
Abrasion resistance	+	+
Softening temperature	+	+
Melt viscosity	+	+
Adhesion	−	−
Chemical resistance	+	+
Solubility	+	o
+ *property increases* − *property decreases* o *little change* *Reprinted from Gel Permeation Chromatography with an FT-IR Detector, 1988, with permission from Perkin Elmer, Inc. [1]*		

Table 8.1b Effect of molecular weight or MWD on polymer properties (specific correlations)		
Polymer	Property	Correlation
Nylon 66	Fibre tenacity	Increases with $\bar{M}_n$
Polymethyl methacrylate	Sensitivity as an electron resist	Increases with $\bar{M}_n$ and narrow MWD
Polystyrene	Solution viscosity and shear stability index	Decreases with a decrease of $\bar{M}_w$ caused by shearing
Polyethylene	Strength, toughness Melt fluidity	Increase with $\bar{M}_n$ Decreases with increasing Mn and decreasing $\bar{M}_w / \bar{M}_n$
Epoxy resin	'Acceptance quality' of circuit boards	Overall GPC curve or MWD profile
Cellulose triacetate	Density Shrinkage of film	Increases with MWD Decreases with MWD

the polymer to be of a single molecular composition. Accordingly, if the same average molecular weight is shown for two samples, the MWD between them may nevertheless be very different, and properties of the polymers may then be different. Therefore, determination of MWD (by GPC) is very important for better understanding the true nature of a polymer.

Of the many separation modes of HPLC, GPC is the simplest. This is now most commonly called SEC. In this mode, the separation of substances proceeds according to the hydrodynamic radius, or effective 'size', of each solute molecule in solution, and is relatively independent of the chemical or electronic nature of the substances and their mobile phase carrier. This independence includes the idea that the column packing, usually composed of porous particles of controlled pore size, is also relatively inert toward the solutes and solvents employed so that adsorption/desorption phenomena do not enter into account.

A generalised description of how SEC works on an essentially mechanical basis (without resort to chemical, van der Waal, and thermodynamic effects, and other considerations) is as follows. Separation takes place according to the effective size of molecules (conveniently represented by molecular weight or chain length) relative to the pore diameter and volume of the porous stationary phase of the column. For a packing with a given mean pore diameter, molecules larger than a particular size (called the exclusion limit) are excluded from entering the pores. Their passage through the column is hindered only by the tortuosity of the route between and around particles, with the result that they elute first and do so completely with the first interparticle volume of the mobile phase. (Of course, molecules, or more properly particulates, that are too large to pass at all do not elute since the packing acts as a simple filter). Conversely, molecules of a size below a specific value (called the permeation limit) are equally free to enter a pore and diffuse into the smallest of confinements within the pore, and are thus equally likely to be retained by a molecular sieve effect, the result being that these molecules are retained longest and cannot be separated from one another by this mechanism. Assuming they are not trapped or adsorbed, these small molecules elute completely only with the total mobile phase volume of the column.

For molecules of sizes between the two limits described previously, separation proceeds by a combination of size exclusion and diffusion permeation with the larger molecules spending less time within pores so that they are eluted ahead of smaller molecules. The shorter residence times for the larger molecules are due mostly to the steric hindrance they exhibit between each other and with the surfaces of the substrate; the smaller molecules being able to approach the walls of the pores much more readily, thus resulting in their having longer total free paths within pores and equivalently longer intrapore residence times (other factors being equal).

- Average Molecular Weight

The average molecular weight can be calculated from the results of measurements of MWD. As regards polymer properties, the number-average molecular weight ($\bar{M}_n$) is said to correlate with tensile strength and impact resistance, the weight-average molecular weight ($\bar{M}_w$) with brittleness and flow properties, and the Z-average molecular weight ($\bar{M}_z$) with flexibility and stiffness. $\bar{M}_n$ and $\bar{M}_w$ determined by GPC are relative values. Absolute values are obtained by the osmotic pressure or vapour pressure methods for $\bar{M}_n$, and the scattering or viscosity methods for $\bar{M}_w$. The viscosity-average molecular weight ($\bar{M}_v$) controls extrudability and moulding properties. The molecular weight of the peak apex in GPC is nearly at the middle between $\bar{M}_v$ and $\bar{M}_n$. Among these average molecular weights, the following relationship is established:

$$\bar{M}_n < \bar{M}_w < \bar{M}_z \tag{8.1}$$

The degree of dispersion (dispersity) of the molecular weight of a sample is related to the ratio of the weight-average molecular weight to the number-average molecular weight ($\bar{M}_w / \bar{M}_n$). The closer the dispersity value is to 1, the narrower is the dispersion of molecular weight.

The average molecular weight is calculated as follows:

$$\bar{M}_n = \frac{\sum H_i}{\sum \left(H_i / M_i \right)} \times QF \tag{8.2}$$

$$\bar{M}_w = \frac{\sum M_i H_i}{\sum H_i} \times QF \tag{8.3}$$

$$\bar{M}_z = \frac{M_i^2 H_i}{\sum M_i H_i} \times QF \tag{8.4}$$

$$\text{Polydispersivity} = \frac{\bar{M}_n}{\bar{M}_w} \tag{8.5}$$

where H_i = peak height, M_i = molecular weight or molecular chain length, and QF = molecular weight per unit chain length of polymer, which is 1 when the molecular weight is used as the unit of M_i.

Measurement and analysis of molecular weight and its distribution in polymers by GPC consists of the following four steps:

1. Preparing samples to be measured.

2. Preparing calibration curves.

3. Measuring GPC elution curves.

4. Analysing GPC data.

A calibration curve is obtained by analysing several standard samples on a particular GPC column, each of which has a small MWD and the average molecular weight of which is known. In many cases monodisperse PS samples are used as the standard. The logarithms of molecular weight are plotted on the ordinate and the retention volume on the abscissa. The average molecular weight of an actual sample is obtained from this calibration curve.

There are many reports on high-speed GPC of high polymers. Most of the samples that are soluble in the previously mentioned solvents can be analysed. **Table 8.2** shows some examples of molecular weight determination that have been achieved on mixed-bed columns. **Table 8.3** gives samples that can be analysed using various solvents by high-speed GPC to obtain the average molecular weight or the MWD.

The stationary phase or filter of a GPC column must meet a number of requirements including a relatively large pore volume, controlled pore diameter, inertness toward sample and mobile phase, and rigidity necessary to prevent collapse under high pressure. Both synthetic polymeric gels and porous silica gels are employed, with the synthetic polymers being most common since selective pore size and pore size distribution are fairly easy to achieve. A typical example is a PS gel (styrene–divinylbenzene copolymer). GPC columns HS910 to HS960, packed with highly crosslinked PS gels, cover a variety of pore sizes and volumes. These columns can be used to separate uncrosslinked PS when a suitable mobile phase is chosen. The calibrations are mobile phase dependent since a mobile phase that solvates the uncrosslinked PS sample will also swell the packing material. Thus, the calibrations given assume a mobile phase that has caused swelling. For a given packing under different analytical conditions, a different calibration will result. The chromatographer must choose conditions carefully in order to gain accurate and repeatable results; however, the relative ease with which this is done makes GPC a valuable tool in the separation of a wide range of polymers.

Modern gel permeation chromatographs present molecular weight data in a variety of ways, as shown in **Table 8.4**.

Table 8.2 Mixed bed columns available from PL Laboratories

Gel	Molecular weight range	Applications	Examples
Traditional PL gel 50 Å - 106 Å e.g. 50, 100 and 500°	<20,000	–	
PL 20 µM gel mixed A	$1000 - 4 \times 10^7$	Ultra high MW polymer distributions	Polystyrene standard calibration, polypropylene analysis, high density polyethylene analysis, polybutadiene analysis
PL 10 µM gel mixed B	$500 - 10^7$	Polymer MW distributions	Polystyrene standard separation, polystyrene analysis, polymethyl methacrylate analysis polyethylene terephthalate, polymide, polyvinylchloride, pyrrolidone, fluoropolymer, butyl rubber, polyethylene, copolymers
PL 5 µM gel mixed C	$200 - 3 \times 10^6$	Rapid polymer MW distributions	Polystyrene standard separation, polycarbonates, hydroxyethyl cellulose, polyether sulfone, polyurethane
PL 5 µM gel mixed D	$200 - 0.4 \times 10^6$	Resins, condensation polymers	Polystyrene standard separation, resins, epoxy resin, polybutadiene, polysiloxane, polycarbonate
PL gel 3 µM mixed D	up to 30,000	Low MW resins	Polystyrene standard separation, epoxy resins, prepolymers, novolak, polyesters, phenolic resins
PL aquagel - OH 50, 10 µM	200 - 100,000	Water soluble polymers	Polyvinyl alcohol, polyacrylic acid, poly 2-vinyl pyridine
PL aquagel - OH 50, 10 µM	$20,000 - 1 \times 10^6$	Water soluble polymers	
PL aquagel - OH 60, 10 µM	$100,000 - 2 \times 10^7$	Water soluble polymers	Polyacrylamides, polyethylene oxides

Source: Author's own files

Table 8.3 Samples and suitable mobile phase	
Sample	Mobile phase
Alkyd resins	1, 2, 3
Acrylic resins	1, 2, 3
Alkyl phenol resins	1
Epoxy resins	1, 2, 3
Ethylene-propylene copolymer	1, 3
Natural rubbers	1, 2, 3
Nitro-cellulose	1, 2, 3
Butyl rubbers	1, 2, 3
Phenolic resins	1
Butadiene rubbers	1, 2, 3
Butadiene-styrene rubbers	1, 2, 3
Silicones	1, 2, 3, 4
Neoprene rubbers	2, 3
Polyisobutylene	1, 2, 3
Polyisoprene	1, 2, 3
Unsaturated polyesters	1
Polyethers	1, 2, 3
Polyethylene glycols	1
Polypropylene glycols	1
Polyvinyl acetate	1
Polystyrene	1, 2, 3, 4
Polyvinyl carbazole	1
Polymethylmethacrylate	1
Melamine resin	1
Polyethylene oxide	1
Polysulfone	1
Acrylonitrile styrene copolymer	1
Polyisobutyl ether	1
1: Tetrahydrofuran (THF) 2: Benzene 3: Toluene 4: Chloroform *Source: Author's own files.*	

Table 8.4 Molecular weight data for a variety of polymers, obtained by GPC						
	$\bar{M}_n$	$\bar{M}_w$	$\bar{M}_z$	$\bar{M}_p$	$\bar{M}_v$	Polydispersity ($\bar{M}_w$ / $\bar{M}_n$)
Styrene acrylate copolymers [14]	9195	80,455	374,788	888	0	8.75
Polyacrylates [15]	379	2366	9015	1251	0	6.23
Polyester [16]	959	2617	13 088	481	0	2.73
Polycarbonate [17]	14,132	26,363	38,099	29,283	0	1.86

8.2.2 Applications

8.2.2.1 Polymer Fractionation Studies

GPC is used for polymer profiling, i.e., when MWD is not required - examples are production quality control, checking lot-to-lot variations, and analysis of competitors' products. The technique has been applied to numerous polymers and resins, and to oligomers, e.g., tetrahydrofuran (THF) solutions of a mixture of PS oligomers, epoxy resins, or mixed dimethyl formamide tetrahydrofuran solutions of styrene butadiene latex.

Some particular applications are now reviewed. Prokai and Simonsick [18] coupled GPC with Fourier transform - mass spectrometry (FT-MS) through a Finnigan UltraSource interface. Sodiated molecular ions of the GPC effluent are produced using electrospray ionisation (ESI). GPC/ESI/FT-MS combines the size separation-based technique of GPC with one of the most powerful mass spectrometric techniques of FT-MS offering high mass accuracy (ppm), ultra-high resolving power (greater than 10^6), and the capability to perform tandem MS. The utility of GPC/ESI/FT-MS for the characterisation of oligomeric mixed polyesters is demonstrated.

Other applications include the fractionation of high molecular weight Ryaluronan [19], methyl methacrylate, *n*-butyl acrylate, styrene, and maleic anhydride copolymers [20, 21], partially crosslinked polyethylenes (PE) [22], star PS [23], high-density polyethylene [24], polyglycerols [25], and stereoregular polyethyl methacrylate [26].

8.2.2.2 Molecular Weight Distributions

Materials studied include *cis*-1,4-polybutadiene [27-44], ethylene–propylene copolymers [28], PS [30-39, 44], isotactic polystyrene [40], styrene–butadiene copolymers [41–43,

45, 46], butadiene and methylstyrene copolymers [45, 46], sulfated PS [47], polyvinyl chloride (PVC) [30, 48–51], vinyl acetate–vinyl chloride copolymers [52], vinyl chloride–vinylidine chloride copolymers [53], vinylidine chloride–methyl methacrylate copolymers [53], polyacrylonitrile [54, 55], acrylonitrile–butadiene–styrene terpolymers [34], polyethylene terephthalate (PET) [56-59], poly(2-methoxyethyl methacrylate) [30], polycarbonates (PC) [61, 62], PMMA [34, 60, 63-68], polyethylene glycol (PEG) oligomers [69], polypropylene glycol [69, 70], polyvinyl pyrrolidone [55], polyvinyl acetate [34], epoxy resins [34], polyamides [71, 72], PE and polymethyl methacrylate (PMMA) [26].

8.2.2.3 Absolute Molecular Weights

Materials studied include polyesters and polyethers [73], PMMA [74], star-shaped styrene block copolymers [75, 76], PC [77], isoprene–styrene graft copolymers [76], polyethylene oxide–maleic anhydride–vinyl methyl ether copolymers [78], polybutylene terephthalate [79], PE [80], (phenoxymethyl)thizrane [81], polytrimethylene carbonate [82], polybutylene succinate [83], polymethyldiundecenylsilane [84], poly(N-vinyl carbazole) [85], titanium-containing copolymers [86], styrene-macro zwitterion polymers [87], and poly(octadecene-alt-maleic anhydride) [88].

8.2.2.4 Arms in Star-Shaped Polystyrene

Various workers [89–91] have attempted to determine the number of arms in these polymers using ion exclusion methods. Sugimoto and co-workers [89] found seven arms.

Several workers reviewed the application of GPC/SEC to the characterisation of polymers [91–96]. Chen and co-workers [97] analysed polyamides by SEC and laser light scattering. The choice of suitable mobile phases is discussed.

8.3 High-Performance Liquid Chromatography

8.3.1 Instrumentation

Further details of the instrumentation are given in Appendix 1.

8.3.2 Applications

8.3.2.1 Chromatography of Polymers on Activated Silica Gel

This is another technique for achieving fractionation of polymers. It is based on functionality distribution and has been applied to the fractionation of polyoxyethylene and polyoxypropylene [98], polytetrahydrofuran [99], polyepichlorohydrin, oligobutadienes [98], and carboxy- and hydroxy-terminated polybutadiene [99, 100]. Isotactic PMMA has been separated from PMMA using competitive absorption on silica gel from chloroform solution. This method is unsatisfactory for unsaturated polymers due to poor recovery from the column, apparently caused by polymerisation on the column. Law [101, 102] circumvented this problem by using partially deactivated silica gel, and recycling the unfractionated portion through additional columns of increasing activity to obtain the desired resolution. Using this technique he separated carboxy-polybutadienes and hydroxy-polybutadienes according to functionality using stepwise elution from silica gel. Recoveries in the 95–199% range were achieved. Fractions obtained from the silica gel separation were subjected to analysis via GPC of chloroform solutions and infrared or near-infrared spectroscopy which yielded not only functionality distribution data, but also provided the relationship between MWD and functional type.

The accurate determination of the chemical composition distribution for copolymers is very important for their characterisation. Among several techniques to measure chemical composition distribution, HPLC holds great promise because of its high efficiency. The following separations on a silica gel column have been reported: styrene–methyl acrylate copolymers on a silica gel column [103], styrene–acrylonitrile copolymers by precipitation liquid chromatography [104], styrene–butadiene copolymers on a polyacrylonitrile gel column [105], styrene–methyl methacrylate copolymers on a silica gel column [106-108], styrene–methyl methacrylate block copolymers by column adsorption chromatography using a 50 mm id cylindrical column [109], and styrene-*n*-butyl methacrylate copolymers by orthogonal chromatography [110]. HPLC on unmodified or activated silica gel has been used to separate prepolymer oligomers (defined as polymers of molecular weight less than 10,000) of PS [111, 112], *p*-alkylphenol-formaldehyde [113], and PET [110].

The separation of PS homopolymers on a C_4 bimodal pore diameter reversed phase column has been studied. Separation of PS on a molecular weight basis was achieved using a THF–acrylonitrile system [114].

Various types of detector have been employed for the detection of polymers separated by liquid adsorption chromatography, including conductimetric detectors for polyoxyethylenes [115] and UV detectors for ethyl methacrylate–butyl methacrylate copolymers [116]. Hancock and Synovec [117] carried out a rapid characterisation of linear and star branched polymers, particularly PS by a gradient detection method. This method measures average molecular weight in methylene dichloride solutions of the polymers and gives more specific results than those obtained using a refractive index detector. Petro and co-workers [118] used a moulded monolithic rod of macro-porous poly(styrene-*co*-divinylbenzene) as a separation medium for the HPLC of synthetic polymers. 'On-column' precipitation–redissolution chromatography was used as an alternative to SEC for the examination of styrene oligomers and polymers. The solvent gradient starts with a high proportion of a poor solvent for the polymer, e.g., water, methanol, or acetonitrile, and as the fractionation proceeds an increasing proportion of a good solvent for the polymer, e.g., THF, is used. Excellent separations were achieved by this procedure which is a suitable alternative to SEC. The technique can be used to separate copolymers and determine molecular weights of copolymers such as styrene-2-naphthol methacrylate.

8.3.2.2 Chromatography of Oligomers

Mourey and co-workers [111] separated *n*-butyl lithium-polymerised PS standards (general structure $CH_3(CH_2)_2(CH_2CHPh)_nCH_2CHPh$) with molecular weight values of 800, 2100, and 4800 anionically on silica gel with 3:1 *v/v* *n*-hexane–THF or *n*-hexane–ethyl acetate or *n*-hexane–dichloromethane (**Figure 8.1**). THF and ethyl acetate gave separations according to the number of oligomer units, and dichloromethane separated the stereoisomers of individual oligomers.

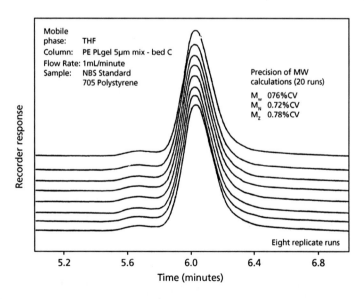

Figure 8.1 CIPC curves of polystyrene standards on Perkin Elmer PL-Gel 5C mixed gel column. (*Reprinted with permission from Perkin Elmer*)

Samples were injected into a column packed with either LiChrosorb Si60 silica (E. Merck, 5 µm particle diameter) or Hibar II LiChrospher Si500 (10 µm particle diameter). PS oligomers were separated by gradient elution by use of either of two Waters Associates M6000A pumps with a 720 system controller or a Varian 5060 liquid chromatograph. UV absorption of the eluant was monitored at 265 nm with a Perkin Elmer LC-55 variable-wavelength detector. Oligomer retention times with ethyl acetate gradient are shorter than those obtained with THF, but the distribution profiles obtained from both solvents are similar, and in both examples individual oligomer peaks are sharp and symmetrical. As many as 40 oligomers were separated in a 2100 weight average molecular weight PS using the THF gradient.

In further work Mourey and co-workers [111, 112] investigated the adsorption chromatography of anionically and cationically prepared PS oligomers on a 6 mm Lichrosorb Si60 silica (E. Merck) column with *n*-hexane–dichloromethane eluents. End-group differences between the cationically prepared (PhCH=CH(CHPhCH$_2$)nCHPhCH$_3$) and anionically prepared (CH$_3$(CH$_2$)$_3$(CH$_2$CHPh)$_n$CH$_2$CHPh) PS produced significant differences in the retention of oligomers that are equivalent in length.

GC methods for determining unreacted terephthalic acid, ethylene glycol, and mono(2-hydroxyethyl) terephthalate (MHET) (**Figure 8.2**) are capable of determining only about 25% of the total prepolymer sample, as the higher molecular weight oligomers are not resolved. HPLC on a DuPont model 530 liquid chromatograph with (100:1 *v/v*)

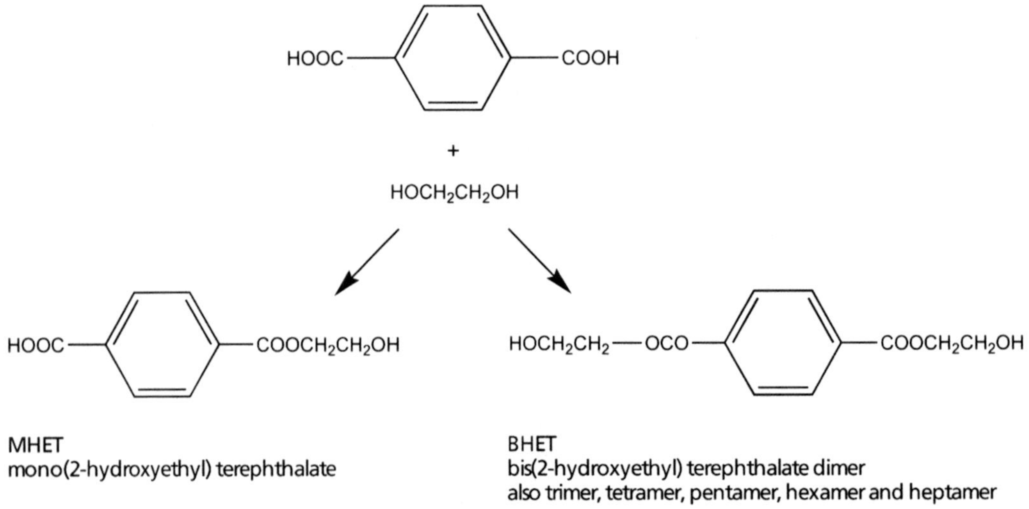

Figure 8.2 Formation of MHET and BHET from ethylene glycol and terephthalic acid

chloroform:reagent alcohol eluant using a DuPont Zorbax SIL column and a 254 mm UV detector gives good separations of the seven PET prepolymer oligomers. The bis(2-hydroxyethyl) terephthalate (BHET) content determined by GC is used as the known standard for the HPLC separation. All chromatograms were run at room temperature. The solvent system was chloroform (1250 parts by volume) and reagent alcohol (12 parts by volume). The reagent alcohol consisted of 90 parts ethanol and 5 parts each of methanol and isopropanol. Approximately 15 mg of sample were placed in a three-dram vial, and 5 ml chloroform/alcohol (9:1 *v/v*) were added. The sample was stirred on a magnetic stirrer until dissolved.

Chromophoric complexes of polyester monomer, dimer, and trimer have approximately the same absorbance at a given wavelength per terephthalyl equivalent. Since the oligomer equivalent per terephthalyl group is also approximately the same, the equal UV weight response assumption appears to be valid.

Table 8.5 lists the HPLC results with precision data for a prepolymer sample. These results show the greatly increased assay of the prepolymer that is possible using the HPLC method. The prepolymer chromatogram shows the presence of oligomers higher than heptamer which are not well enough resolved for quantitation.

8.3.2.3 Oligomer Functionality in Oligo(epichlorohydrins)

Bektashi and co-workers [120] studied the effect of Lewis acids and of the medium on the MWD and oligomer functionality type distribution obtained in epichlorohydrin polymerisation by means of HPLC, which was applied to separation of oligo(epichlorohydrins) under critical conditions, in combination with GPC. Chromatography under critical and exclusive separation modes enabled seven types of functionalities, including cyclic and six linear types, to be detected. The low molecular weight fractions of the oligomers corresponded to cyclic and linear tetra-, hexa-, hepta-, and octamers. The maximum content of cyclic oligomers was 50%.

8.4 Supercritical Fluid Chromatography

8.4.1 Theory

Supercritical fluids can dissolve a variety of solutes such as polymers with high molecular weight and low volatility. The solubility may be easily varied by changing the density via the applied pressure. The low viscosity means that the pressure drop across the column is small, and the consequent permeability allows capillary columns with high efficiencies to

Table 8.5 Polyester prepolymer HPLC analytical results		
Component		Lot No. 1
BHET	[a]	10.15
Dimer	X	14.52
	SD	±0.35
	RSD	2.4%
Trimer	X	13.30
	SD	±0.63
	RSD	4.7%
Tetramer	X	10.42
	SD	±0.76
	RSD	7.3%
Pentamer	X	6.90
	SD	±0.59
	RSD	8.6%
Hexamer	X	5.00
	SD	±0.25
	RSD	5.0%
Heptamer	X	3.57
	SD	±0.51
	RSD	14.3%
HPLC total (less BHET)		53.71
GC total		19.38
Total assay		73.09
Number of determinations		5
[a] *Results from GC analysis* SD: *Standard deviation* RSD: *Relative standard deviation* Source: *Author's own files.*		

be used. Solubility can be altered by increasing the temperature - increased temperatures also increase solute diffusion coefficients in the mobile phase with a consequent increase in resolution.

The narrow columns necessary in SFC make small injection and detector volumes necessary, but SFC is compatible with both GC and HPLC detectors. Coupling to spectrometric detectors such as those of mass, Fourier transform infrared (FT-IR), and even NMR spectrometers can also be carried out.

The technology for the preparation of capillary columns for SFC is now well advanced. Column diameters below 100 μm are necessary because the solute diffusion coefficients are smaller than in GC. Practical efficiencies of up to 5000 plates per metre are possible for 50 μm id columns.

Retention of SFC is decreased by increasing the mobile phase density. Density programming is made possible via software for pump pressure control which incorporates the appropriate pressure–density isotherms. The resolution of homologues decreases with increasing density and therefore asymptotic rather than linear density programmes may be required. Simultaneous density–temperature programming is carried out to achieve variations in selectivity – indeed all the four significant operating parameters in chromatography may be varied.

8.4.2 Instrumentation

See Section 7.10 for further details of the instrumentation.

8.4.3 Applications

The high dissolving power of supercritical fluids for some polymers and the high efficiencies of capillary columns make SFC particularly suitable for analyses of oligomeric and polymeric materials. The separation of 42 styrene oligomers has been reported [121, 122]. The separation of every oligomer in a mixture is routinely possible, in contrast to SEC in which resolution is much lower. For example, **Figure 8.3(a)** shows the separation of a polydimethysiloxane mixture. Not only are the individual linear oligomers resolved, but also a series of small peaks is assigned to branched-chain oligomers. The analysis of numerous other polymers and surface-active agents by SFC had been reported, including methylphenylsiloxanes, styrene and other vinyl aromatic polymers, polyolefins and waxes, polyethers, polyglycols (underivatised, since analysis temperatures are well below the decomposition temperature), polyesters, and more polar polymers such as epoxies (**Figure 8.3(b)**) and isocyanates.

Such analyses generally involve density programming to bring out oligomers at fairly regular intervals. Simultaneous temperature programming also extends the range of compounds eluted and gives greater chromatographic efficiencies by increasing solute diffusion coefficients in the mobile phase.

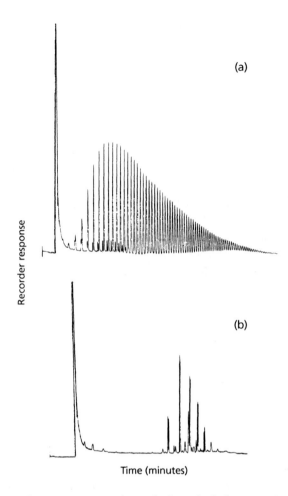

Figure 8.3 (a) SFC chromatogram of a polydimethylsiloxane. Conditions: column, 20 m × 50 μm id SB-methyl; mobile phase, CO_2 at 120 °C with asymptotic density programming, detector FID. (b) SFC chromatogram of epoxy acrylate oligomers. Conditions: column 20 m × 50 μm id SB-biphenyl-100; mobile phase, CO_2 at 70 °C with linear density programming; detector FID. (*Source: Author's own files*)

For very high molecular weight polymers, hydrocarbon (e.g., pentane) mobile phases modified with polar additives such as alcohols and ethers are required. Many such separations have been reported by SFC using packed columns, often with gradient elution, but such an approach poses special problems in capillary SFC because of the low flow rates. Capillary SFC with solvent programming is a likely future growth area.

A number of advantages accrue from combining the separating power of capillary SFC with the explicit structural information of MS. Most of the common ionisation modes have been shown to be compatible with SFC, including electron impact, chemical ionisation, and charge exchange [123]. The variety of structural data available from SFC with negative-ion chemical ionisation detection has been demonstrated: while methane reagent gas gave mainly the M + I ion, both methane with carbon dioxide and ammonia with carbon dioxide as reagent gases gave many more fragment ions.

By changing the density of the supercritical fluid, different fractions may be selectively extracted from the complex mixture or sample matrix. On decompression, the extracted solutes are precipitated and may be collected for injection into a GC or SFC for analysis. **Figure 8.4** shows a simple apparatus for on-line SFE/SFC: solutes extracted from the sample matrix are deposited from the end of a restrictor into the internal loop of the microinjection valve of the capillary SFC. The valve loop contents are subsequently switched into the SFC column by means of liquid or supercritical carbon dioxide.

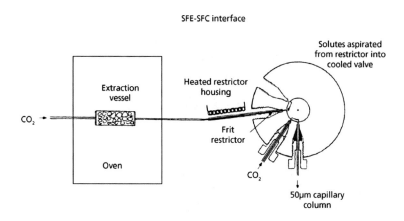

Figure 8.4 Apparatus for coupled supercritical fluid extraction/SFC

SFC has been used to determine oligomers in PET [124] and low molecular weight, high-density polyethylene wax [125]. Carbon dioxide, propane, and propane-modified carbon dioxide have been studied as eluents.

SFC coupled to ToFSIMS has been applied to the fractionation of polydimethylsiloxane oligomers in the molecular weight range of 1,000–10,000 [126].

8.5 Gas Chromatography

GC is, of course, limited to lower molecular weight polymers that can be volatilised.

Mikkelson [127] reported a GC analysis of PEG 400; the original sample was injected directly into the GC, a technique also employed by Puschmann [128]. Celedes Pasquot [129] converted the PEG into methyl ethers in a reaction with dimethylsulfate before injection into the GC.

The rapid formation of trimethylsilyl ether derivatives of polyhydroxy compounds followed by their separation and estimation by GC has been described by Sweeley and co-workers [130]. This most useful technique has been applied to the liquid PEG 200, 300, and 400. Fletcher and Persinger [131] studied this application using PEG 200, 300, and 400 and reported the determination of response factors relative to ethylene glycol for conversion of peak areas to weight percent when a thermal conductivity detector was used.

p-Toluenesulfonic acid-catalysed acetylation has been used to produce volatile derivatives of propylene oxide–glycerol condensates which are suitable for GC [132]. Graphs relating initial retention distance on the gas chromatogram with molecular weight give straight-line relationships.

Uttenbach and co-workers [133] characterised and determined formaldehyde oligomers by capillary column GC. The homologous samples of polyoxymethylene glycols and the corresponding monomethyl ethers were derivatised with ammonia prior to GC.

Temperature-programmed GC has been employed for the fractionation of polyoxymethylene oligomers [134].

8.6 Thin-Layer Chromatography

TLC is a useful technique for separating polymers into molecular weight fractions on a fairly small scale. It has been used to fractionate PET [135, 136], styrene–butadiene copolymers [137], styrene acrylonitrile copolymers [138], polyoxypropylene glycols [136], Nylon–styrene graft copolymers and PMMA [139–141], styrene–methacrylate copolymers [142], poly-α-methylstyrene [143], polyvinyl acetate–styrene copolymers, and polyvinyl alcohol–styrene copolymers [144].

8.7 NMR Spectroscopy

The use of NMR spectroscopy in MWD studies of polymers has received limited attention in recent years [20, 21, 79, 82, 89].

High-resolution magnetic resonance spectroscopy has been used for the determination of composition and number-average molecular weight of polyester urethanes [145].

Sionim and co-workers [146] described NMR methods to determine chain structure, composition, and number-average molecular weight of unsaturated polyesters.

8.8 Osmometry

Osmotic pressure measurements of solutions of a polymer in a solvent can be employed to determine the molecular weight of the polymer. After approximately one hour the osmotic pressure of solutions of poly-α-methylstyrene in toluene becomes practically constant. From this osmotic pressure and the concentration of the polymer in solution it is possible to calculate the number-average molecular weight of the polymer from the following relationship:

$M = aRT/PV$

where M = molecular weight, R = gas constant, and T = temperature (K).

Osmometry has been used to measure the molecular weight of a wide range of polymers including PVC [147], PET [148–150], Nylon [149], PS [148], and linear polyesters [150], including polytetramethylene terephthalate, polypentamethylene terephthalate, polyhexamethylene terephthalate and polytetramethylene isophthalate.

8.9 Light Scattering Methods

Two principal light scattering techniques have been used to determine the MWD of polymers. The first of these is the temperature gradient method. This method has been applied to the determination of the MWD of PE [151, 152]. In this method a solution of PE in α-chloronaphthalene solvent and a non-solvent containing a very low concentration of PE is slowly cooled.

The high molecular weight species become insoluble and separate out, causing a small amount of turbidity. As the temperature continues to decrease, increasing amounts of polymer are precipitated out according to their molecular weight. Finally, a point is reached at which even the lowest molecular weight species become insoluble in the solution. At this point the turbidity is greatest, and ideally all the polymer is precipitated but remains in suspension as very fine particles. If the increase in turbidity is plotted against the decreasing temperature, a cumulative plot is obtained which is similar to a cumulative wt% *versus* molecular weight plot. The increase in turbidity is related to the cumulative wt% and the molecular weight is related to the decrease in temperature. The ordinate is in terms of ΔE, the decrease in millivolts in phototube output from the starting value. Taylor and Tung [151, 152] discuss the application to their results of procedures described by Morey and Tamblyn [153] and Claesson [154] to convert the experimental turbidity into MWD data and the difficulties they encountered in this work.

Gamble and co-workers [155] used a photoelectric turbidimeter for measuring MWD of poly-α-olefins and ethylene–propylene copolymers. This instrument measures changes in turbidity as a function of temperature.

The second method of determining MWD is turbidimetric titration. Beattie [156] developed an absolute turbidimetric titration method for determining solubility distribution (which is closely related to MWD) of PS. In this method a polymer is precipitated from its solution in methyl ethyl ketone by addition of a non-solvent (isopropanol) of the same refractive index as that of the solvent. He showed, by the use of light scattering theory, that under these conditions the concentration of polymer that is precipitated can be calculated from the maximum turbidity on an absolute basis. Beattie concluded that with PS, under the specified conditions, the reproducibility of turbidimetric precipitation curves is very good and that the method is accurate. Gooberman [157] also studied PS but dissolved the polymer in butanone and titrated with isopropanol.

Tanaka and co-workers [158] have used the high-temperature turbidimetric titration procedure originally described by Morey and Tamblyn [153] for determining the MWD of cellulose esters. This method has been applied to the measurement of the MWD of PP. They found that the type of MWD of this polymer is a log-normal distribution function in a range of $I(M)$ (cumulative wt%) between 5 and 90%. The effect of heterogeneity in the MWD of PP on the viscosity–molecular weight equation was examined experimentally - the results agreed with those calculated from theory. Strict temperature control ($\pm 0.15\ °C$) is necessary in these determinations [159].

Taylor and Graham [160] have described a dual-beam turbidimetric photometer which they claim has distinct advantages over the earlier single-beam instruments [161, 162]. They have applied it to the determination of the MWD of PE, PP, ethylene–propylene rubbers, and other polymers [163].

Light scattering measurements based on turbidimetric titration have been applied to a wide range of polymers including PE [164], PP [165], PS [166–170], polymethylstyrene [171], PVC [172], chlorinated PVC–styrene [165], polyethylene oxide [173], polyphenylene oxide [174], PET [175, 176], PEG [166], PS [177–179], styrene–methyl methacrylate copolymer [180], water-soluble acrylic acid and acryloamide copolymers [181], poly-α-methylstyrene [182], PS-block-polyisoprene [183], and poly(esterurethane) [184].

Thielking and Kuliche [185] used on-line coupling of field flow fractionation and multi-laser light scattering for the characterisation of sulfonated PS macromolecules in aqueous solution. Seven sulfonated PS standard polymers ($1.8 - 300 \times 10^4$ g/mol) were dissolved in model solutions for macromolar polyelectrolytes.

Various workers have reviewed the use of osmometry to determine molecular weight and other characteristics in polymers [186–188]. Suitable commercially available instrumentation has been discussed [186, 188].

8.10 Viscometry

The viscosity-average molecular weight of a polymer sample may be computed from the chromatogram hi of the polymer by the formula:

$$M_v = (\textstyle\sum h_i M_i^a / \sum h_i)^{1/\mu}$$

Substituting the Mark–Houwink equation gives:

$$[\eta] = K(\textstyle\sum h_i M_i^a / \sum h_i)$$

where a and K are Mark–Houwink parameters and $[\eta]$ is the intrinsic viscosity of the polymer. It is thus possible to calculate average molecular weights from viscosity data on the weighed fractions obtained by chromatography, or on the original unfractionated polymer provided appropriate calibration data are available for the calculation of constants in the above equation.

Mao and co-workers [189] have developed a new method for the determination of polymer viscosity-average molecular weights using flow piezoelectric quartz crystal (PQC) viscosity sensing. Experimental apparatus with a 9 MHz AT-cut quartz crystal and a flow detection cell was constructed and shown to give highly reproducible data at a temperature of 25 ± 0.1 °C and fluid flow rate of 1.3–1.6 ml/min. A response model for the PQC in contact with the dilute polymer solutions (concentration less than 0.01 g/ml) was proposed in which the frequency change from the pure solvents, Δf_s, follows $\Delta f_s = k_6 \eta_1^{1/2} + k_7$, where η_1 is the absolute viscosity of dilute polymer solution and k_6 and k_7 are the proportionality constants. This model was investigated with PEG samples (PEG-20000 and PEG-10000) under the aforementioned experimental conditions using water as solvent. Based on this model, the method was developed and tested with an unknown polyvinyl alcohol sample against the conventional capillary viscosity method. There was good agreement between this method and the conventional method. The method proposed by Mao and co-workers [189] has a number of advantages over the conventional viscosity method: it is simpler and more rapid, the instruments required are cheaper and portable, the sample quantity required is smaller, and the experimental apparatus can be used in continuous measurement. Consequently, it is concluded that this method offers an attractive and promising alternative to the conventional capillary viscosity method for determination of polymer molecular weights.

Minigawa and co-workers [190] measured intrinsic viscosities of 10.9–0.1 dl/g corresponding to viscosity averaged molecular weight (viscosity-average molecular weight M_v) of 1,590,000 to 3,000 for polyacetonitriles. The molecular weight dependence of the glass transition temperature is discussed.

Lukhovitskii and Karpo [191] established that, in the concentration range accessible for viscometry, the efflux time of a polymer solution is a linear function of the polymer concentration. This is consistent with the Einstein–Simha equation. In the general case, the efflux time of a solution with a polymer concentration tending to zero, T_e, does not coincide with the efflux time of the pure solvent, T_0. When the efflux time of a polymer solution is reduced by T_e rather than by T_0 (as is done in the standard method), the reduced viscosity becomes independent of the polymer concentration and equal to the intrinsic viscosity. The advantages of the proposed method are especially important for the determination of the intrinsic viscosity (M_w) of ultra-high molecular weight polymers.

Intrinsic viscosity measurements have been used to determine the molecular weight of PP [192-195], PET [196, 197], PE [198–204], PVC [203], polyacrylamide [205, 206], polyvinyl pyrrolidone [207], and styrene–methyl methacrylate copolymers [208].

8.11 Ultracentrifugation

Some 30 years ago the analytical ultracentrifuge played an important part in characterising weight-average MWD, size distribution, and density distribution in polymers. This technique was displaced by SEC. However, with improvements in instrumentation, ultracentrifuge methods are to some extent making a comeback.

Machtle and Klodwig [209] have reported on improvements in instrumentation in the analytical centrifuge technique, including an eight-cell interference optics multiplexer and an eight-cell Schlieren optics multiplexer, both based on the same light source, a modulable laser, and a titanium rotor with a maximum speed of 60,000 rpm.

Sedimentation methods have also been applied to PS [210], polyglycols [211], and polyesters [212].

8.12 Field Desorption Mass Spectrometry

Field desorption mass spectrometry has been shown to be a method of choice for determining molecular weights of non-volatile and higher molecular weight chemicals [212–224]. Field desorption provides a gentle means of ionisation that imparts little excess energy to the molecule. Consequently, molecular ions (or at times protonated or cation-adduct molecular ions) are frequently the strongest ions observed. Fragment ions and ions due to thermal decomposition products are often absent or else of relatively low intensity. Field desorption mass spectrometry can be used to obtain good qualitative distributions of oligomers of low molecular weight polymers. Oligomeric mixtures studied in this regard include *n*-paraffins [215], polypivalolactone [216, 217], poly(2,2,4-trimethyl-1,2-dihydroquinoline) [221–223], PS [218–220], and polypropylene glycol [221–223]. Components of several oligomeric polymer chemical mixtures have also been qualitatively characterised by field desorption mass spectrometry [221–223].

In none of the work described previously was an attempt made to determine accurate molecular weight averages $\bar{M}_n$ and $\bar{M}_w$. Lattimer and Hermon [223] applied the technique to the determination of molecular weight averages of narrow MWD butyl lithium-catalysed PS oligomers with molecular weight up to 5300 amu. Molecular ion spectra were obtained by the latter method for PS oligomers. Field desorption-derived molecular weight averages ($\bar{M}_n$ and $\bar{M}_w$) compared favourably to values obtained by conventional techniques (vapour pressure osmometry, intrinsic viscosity, kinetic data, and SEC). This agreement indicates that the relative intensities of the field desorption mass spectrometry oligomer molecular ions (with appropriate corrections for isotopic abundances) could be used directly to give reasonably accurate relative molar concentrations of the oligomer.

8.13 Capillary Electrophoresis

Poli and Schure have used this technique to fractionate PS sulfonate polymers [225]. Wallingford [226] used capillary gel electrophoresis to separate ionic and non-ionic ethoxylated polymers. The oligomeric distributions of several sulfated and phosphated alkylphenol ethoxylated surfactants were resolved on commercial crosslinked polyacrylamide gel columns. Non-ionic detergents and PEG oligomers reacted with phthalic anhydride to provide determination of PEG oligomers ranging from ethylene glycol up to species containing 120 monomer units. There was a linear relationship between molecular weight and migration time. Auriola and co-workers [227] have shown that capillary electrophoresis in PEG solution-filled uncoated silica capillaries is suitable for the separation of oligonucleotides in 100 mM ammonium formate buffers. Reversed polarity was used because the electroosmotic flow at 4.5 ml/min was significantly lower than the mobility of the anionic analytes. On-column transient isotachophoresis was developed for pre-concentration of oligonucleotide samples in PEG solution-filled capillaries. The isotachoelectrophoresis focusing step allowed injection of 20% of column length without loss of peak resolution. The detection limit of the method was 100 ng/ml of 16-mer oligonucleotide using UV detection at a wavelength of 254 nm (signal-to-noise ratio 10).

8.14 Liquid Chromatography–Mass Spectrometry

Jones and co-workers [228] used particle beam liquid chromatography coupled with a MS detector to determine styrene oligomers up to the *n*-18 oligomer of PS. Bryant and Semlyen [229] analysed cyclic oligomers of polybutylene terephthalate using column chromatography GPC (size exclusion), fast atom bombardment mass spectrometry, and tandem mass spectrometry (MS-MS). They obtained molecular weight distributions.

Aaserud and co-workers [73] used GPC coupled to electrospray ionisation Fourier transform mass spectrometry (GPC/ESI/FT-MS) to establish chain microstructure and provide on-line absolute molecular weight of polyesters and polyethers. This approach reduces the number of components and thus the complexity entering the ion source of the MS. Ionisation is accomplished using electrospray ionisation with cation doping (NaI) and ions are seen as their sodium adducts. Detection of the fractionated polymer was accomplished by FT-MS, which is capable of providing high resolution and mass accuracy, thus facilitating the direct determination of the ion's charge state, monomer repeat unit, and end-groups to about 10,000 daltons. Only 0.5% of the GPC effluent is required for the FT-MS. The relationship between the GPC elution times (hydrodynamic volume) and the absolute molecular weight of the individual polymer changes are used to generate GPC calibration curves which do not rely upon unrelated calibrants such as PS, thus affording accurate molecular weight data. The application of GPC/ESI/FT-MS for the characterisation of oligomeric mixed polyesters, hyperbranched dimethylol propionic acid, and propylene oxide oligomers produced from single-site organometallic catalysts is demonstrated. GPC/ESI/FT-MS reduces the complexity for MS analysis and yields architecture-dependent calibration curves that can be explained.

Dwyer [96] investigated the combination of SEC and matrix-assisted laser desorption/ ionisation mass spectrometry (MALDI-MS). He showed that there are a number of factors that can influence results. Raghaven and Egwim [230] used LC-MS in degradation studies of polyester films in alkali solution.

8.15 Ion Exchange Chromatography

Up to 30 oligomers of polyoxyethylene have been separated on a K^+ form cation exchange resin [231].

8.16 Liquid Adsorption Chromatography

Mori [232] separated ethyl methacrylate–butyl methacrylate copolymers by liquid adsorption chromatography using an UV detector.

8.17 Time-of-Flight Secondary Ion Mass Spectrometry (ToF SIMS)

This technique has been used to determine MWD in polyesters, PS, and PEG oligomers [233], and in perfluorinated ether polymers [234].

8.18 MALDI-MS

This technique coupled to HPLC has been used to determine the molecular weight of *tert*-octylphenol-ethoxylate surfactant polymers [235]. Schriemer and Li [236] used MALDI-MS to detect high molecular weight narrow polydisperse polymers up to 1.5×10^6 daltons.

In this rapid method of analysis MALDI results agree very well with those obtained by classic methods. Retinoic acid was used as the organic matrix for the laser desorption procedure and the samples were analysed as their silver anion adducts.

Dwyer [96] combined SEC and MALDI-MS to provide an insight into the structure and MWD of polymers. He showed that there are a number of factors that can influence results, particularly quantitative results, when utilising MALDI-MS. Preparation of samples can introduce significant artefacts of measurement. An important application of MALDI-MS is the generation of absolute molecular weights and MWD of polymer samples. The technique works well with samples of narrow polydispersity, but yields error bias with samples in which polydispersity is greater than 1.2. A possible way to circumvent polydispersity limitations is to perform GPC/SEC on a sample and collect a series of fractions (which will be of narrow polydispersity). While successful, the technique is labour intensive and long. A technique is described that integrates GPC and MALDI-MS to provide data rapidly across the polymer MWD. The technique utilises a continuous spray deposition column eluant onto a preformed matrix-coated plate. At the completion of chromatography, the matrix plate is directly inserted into the sample compartment of the spectrometer and profiled to provide a series of discrete time-ordered spectra.

Kricheldorf and co-workers [237] investigated cyclic PC with molecular weights up to 106 daltons using MALDI-MS.

8.19 Thermal Field Flow Fractionation

Kirkland and co-workers have used this technique to determine the MWD of water-soluble polymers including polyethylene oxide in the 10^4–2×10^6 molecular weight range, including sodium PS sulfates and dextrans [238]. Also, they applied the techniques using Mark–Houwink constants to PS, polyisoprene, poly-α-methylstyrene, polyacrylates, polyvinyl pyrrolidone, and PVC [238].

This technique has also been applied to the fractionation and determination of number-average MWD of anionic and cationic water-soluble polymers such as PS sulfonate (molecular weight 6500–690,000) and polyvinyl pyridine (molecular weight 28,000–240,000) [239]. Schuch and co-workers [240] carried out a field flow fractionation of polyvinyl formamide.

8.20 Desorption Chemical Ionisation Mass Spectrometry

Vincenti and co-workers [241] used desorption chemical ionisation mass spectrometry to determine MWD in PS, PEG, polysiloxanes, and polynorbornene.

8.21 Grazing Emission X-ray Fluorescence Spectrometry

Blockhuys and co-workers [242] analysed spray-coated thin films of E-poly[(3-methoxy-2,5-thiophenediyl-1″,2″-ethenediyl) (4″-methoxy-2″,5″-thiophenediyl-1″,2″-ethyenediyl)] conducting polymer by grazing emission X-ray fluorescence spectrometry. Measurement of the S/Cl ratio in this sulfur-containing polymer with one terminal chlorine allowed determination of the chain length and, hence, the molecular weight.

References

1. *Gel Permeation Chromatography with an FT-IR Detector*, Perkin Elmer Infrared Spectroscopy Applications, No.3.3.2, Perkin Elmer Corporation, Norwalk, CT, USA, 1988.

2. B.G. Belenkil and E.S. Gankina, *Journal of Chromatography*, 1977, **141**, 19.

3. S. Teramachi, A. Hasegawa and S. Yoshida, *Makromolecules*, 1983, **16**, 542.

4. G. Glockner, J.H.M. van den Berg, N.L.J. Meijerink, T.G. Scholte and R. Koningsveld, *Macromolecules*, 1984, **17**, 962.

5. G. Glöckner, J.H.M. van den Berg, N.L.J. Meijerink, T.G. Scholte and R. Koningsveld, *Journal of Chromatography*, 1984, **317**, 615.

6. S.T. Balke and R.D. Patel, *Journal of Polymer Science: Polymer Letters Edition*, 1980, **18**, 453.

7. S.T. Balke and R.D. Patel in *Polymer Charcacterisation: Spectroscopic, Chromatographic and Physical Instrumental Methods*, Ed., C.D. Craver, *Advances in Chemistry Series*, No.203, 1983, p.281.

8. T. Tanaka, M. Omoto, M. Donkal and H. Inagaki, *Journal of Macromolecular Science, Part B: Physics*, 1980, **17**, 211.

9. M. Danielewicz and M. Kubin, *Journal of Applied Polymer Science*, 1981, **26**, 951.

10. S. Vlori, Y. Uno and M. Suzuki, *Analytical Chemistry*, 1986, **58**, 303.

11. S. Shirno, K.A. Ishimuva and J. Enomoto, *Journal of Chromatography*, 1980, **193**, 243.

12. J.L. DiCesare, M.W. Dong and F.L. Vandemark, *American Laboratory*, 1981, **13**, 52.

13. S. Mori, *Analytical Chemistry*, 1988, **60**, 1125.

14. *Simultaneous Characterization of Styrene Acrylate Copolymers Using UV and RI Detection*, Hewlett Packard HPLC Application 87-6, Gel Permeation Chromatography. 1987.

15. *Characterization of Polyacrylates with GPC and Reverse Phase Chromatography*, Hewlett Packard HPLC Application 87-10, Gel Permeation Chromatography, 1987.

16. *Simultaneous Determination of Molecular Weight Data and Monomers of a Polyester*, Hewlett Packard HPLC Application 87-5, Gel Permeation Chromatography 1987.

17. *Analysis of Multiple Detection Data in Analysis of Polycarbonates*, Hewlett Packard HPLC Application 87-4, Gel Permeation Chromatography, 1987.

18. L. Prokai and W.J. Simonsick, *Polymer Preprints*, 2000, **41**, 1, 665.

19. R. Mendichi, A.G. Schieroni, *Polymer*, 2002, **43**, 6115.

20. M.S. Montaudo, *Polymer*, 2002, **43**, 1587.

21. M.S. Montaudo, *Polymer News*, 2002, **27**, 115.

22. M. Parth, N. Aust and K. Lederer, *International Journal of Polymer Analysis and Characterization*, 2003, **8**, 175.

23. T. Kakuchi, A. Narumi, T. Matsuda, Y. Miura, N. Sugimoto, T. Satoh and H. Kaga, *Macromolecules*, 2003, **36**, 3914.

24. P.M. Cotts, *Polymer Preprints*, 2002, **43**, 1, 297.

25. H. Kautz, R. Hanselmann, H. Frey, E. Schwab and S. Mecking in ACS Polymeric Materials: Science and Engineering, 2001, **84**, Paper No.525, p.943.

26. Donghyun Cho, Insun Park, Taihyun Chang, K. Ute, I. Fukuda and T. Kitayama, *Macromolecules*, 2002, **35**, 6067.

27. J. Maley, *Journal of Polymer Science*, 1965, C-8, 253.

28. T. Ogawa and T. Inaba, *Journal of Polymer Science*, 1977, **21**, 2979.

29. M.F. Vaughan in *Proceedings of Industrial Polymer Characterization by Molecular Weight Conference*, 1973, p.111.

30. S. Mori, *Analytical Chemistry*, 1981, **53**, 1813.

31. E.P. Otaka, *Journal of Chromatography*, 1973, **76**, 149.

32. C.V. Uglea, *Makromolekulare Chemie*, 1973, **166**, 275.

33. B.G. Beden'kii, E.S. Gankina, P.D. Nefedov, M.A. Kuznetsova and M.D. Valchikhina, *Journal of Chromatography*, 1973, 77, 209.

34. D. Kranz, H. Rahl and H. Baumann, *Angewandte Makromolekulare Chemie*, 1972, **26**, 67.

35. L.H. Tung and J.R. Runyon, *Journal of Applied Polymer Science*, 1973, **17**, 1589.

36. J.C. Moore and J.G. Hendrickson, *Journal of Polymer Science*, 1965, **C-8**, 233.

37. S.P. Zhdanov, B.G. Belenkii, P.P. Nekdov and E.V. Kormal'di, *Journal of Chromatography*, 1973, 77, 149.

38. J.G. Moore, *Journal of Polymer Science*, 1964, A-2, 835.

39. B.G. Belen'kii, E.S. Gankina, P.P. Nefedov, M.A. Kuznetsova and M.D. Valchikhina, *Journal of Chromatography*, 1973, 77, 209.

40. J.M. Guenet, Z. Gallot, C. Pilot and D. Bennett, *Journal of Polymer Science*, 1977, **21**, 2181.

41. F.S.C. Chang in *Polymer Molecular Weight Methods*, Ed., M. Ezrin, Advances in Chemistry Series No.125, ACS, Washington, DC, USA, 1973, p.154.

42. M. Hoffmann and H. Urban, *Makromolekulare Chemie*, 1977, **178**, 268.

43. M.R. Amber, R.D. Mate and J.R. Durdon, *Journal of Polymer Science*, 1974, **12**, 1771.

44. A.M. Winns and S.J. Swarin, *Journal of Applied Polymer Science*, 1975, **19**, 1243.

45. K. Stojanov, Z.H. Shirazi and T.O.K. Audo, *Chromatographia*, 1978, **11**, 274.

46. K. Stojanov, Z.H. Shirazi and T.O.K. Audo, *Berichte der Bunseng-Gesellschaft-Physical Chemistry*, 1976, **81**, 767.

47. F.E. Regnier and R. Noel, *Journal of Chromatographic Science*, 1976, **14**, 316.

48. J. Janca and M. Kolinsky, *Plaste und Kautschuk*, 1976, 13, 138.

49. L.E. Daley, *Journal of Polymer Science*, 1965, **C8**, 253.

50. F.T. Lin, L.X. Tung, F.Y. Liu and W.W. Hsu, *J Chim Inst Chem Eng*, 1973, **4**, 43.

51. L. Nakao and K. Kuramoto, *Nippon Secchaku Kyokaishi*, 1972, 8, 4, 186.

52. J. Janca and M. Kolinsky, *Journal of Applied Polymer Science*, 1977, **21**, 83.

53. A. Revillon, R. Dumont and A. Guyot, *Journal of Polymer Science, Polymer Chemistry Edition*, 1976, **14**, 2263.

54. *Fire Retardant Specifications*, DoE/PSA, London, UK.

55. S. Mori, *Analytical Chemistry*, 1983, 55, 2414.

56. D.R. Cooper and J.A. Semylen, *Polymer*, 1973, **14**, 185.

57. E.E. Paschke, B.A. Bidingmeyer and J.C. Bergman, *Journal of Polymer Science: Polymer Chemistry Edition*, 1977, **15**, 983.

58. M. Miarzk, Z. Sir and J. Coupek, *Angewandte Makromolekulare Chemie*, 1977, **64**, 147.

59. S. Shino, *Analytical Chemistry*, 1979, **51**, 2398.

60. J. Janca, P. Vlcek, J. Trekoval and M. Kolinsky, *Journal of Polymer Science: Polymer Chemistry Edition*, 1975, **13**, 1471.

61. A.P. Robertson, J.A. Cook and J.Y. Gregory in *Proceedings of a Kinetics Symposium*, Boston, MA, USA, 1972, p.258.

62. J. Moore and D.E. Hillman, *British Polymer Journal*, 1971, **3**, 259.

63. G. Meyerhoff, *Journal of Polymer Science*, 1971, **9**, 596.

64. Y. Miyamoto, S. Tomoshige and H. Inagak, *Polymer Journal*, 1974, **6**, 564.

65. Z. Grubisic, P. Rempp and H.J. Benoit, *Polymer Science B: Polymer Letters*, 1967, **5**, 753.

66. A.R. Weiss and E. Cohn-Ginsberg, *Journal of Polymer Science*, 1970, A-2, **8**, 148.

67. B.G. Belinski and P.P. Nefedov, *Vysokomolekulyarnye Soedineniya Seriya A*, 1972, **14**, 1568.

68. M. Kolinsky and J. Janca, *Journal of Polymer Science*, 1974, A-1, **12**, 1181.

69. D. Berck and L. Nova, *Chemicky Prumsyl*, 1973, **23**, 91.

70. I.T. Vakhtina and O.G. Tarakanov, *Plaste und Kautschuk*, 1974, **21**, 28.

71. A. Meyer, Y. Lin, L. Hoipkemeier, M. Khoshbin, S. Leff, N. Jones and D. Kranbuehl, *Polymer Preprints*, 2000, **43**, 1, 58.

72. J. Chen, W. Radke and H. Pasch, *Macromolecular Symposia*, 2003, **193**, 107.

73. D.J. Aaserud, W. Zhong and W.J. Simonsick, *Polymer Preprints*, 2000, **41**, 1, 657.

74. Y. Kojima, T. Matsuoka and H. Takahashi, *Journal of Materials Science Letters*, 2002, **21**, 473.

75. I-C. Liu and R.C-C. Tsiang, *Journal of Polymer Science, Polymer Chemistry Edition*, 2003, **41**, 976.

76. K. Se, R. Sakakibara and E. Ogawa, *Polymer*, 2002, **43**, 5447.

77. R. Walkenhorst, *LC-GC Europe*, 2002, **15**, 40.

78. A.M. Alta and K.F. Arndt, *Journal of Applied Polymer Science*, 2002, **86**, 1138.

79. J.J.L. Bryant and J.A. Semlyen, *Polymer*, 1997, **38**, 4531.

80. B.G. Millar, G.M. McNally and W.R. Murphy in *Proceedings of 60th SPE Annual Technical Conference, ANTEC 2002*, San Francisco, CA, USA, 2002, Session W36, Paper No.723.

81. T. Imai, T. Satoh and T. Kakuchi, *Polymer Preprints*, 2002, 43, 1, 574.

82. H.R. Kricheldorf and A. Stricker, *Polymer*, 2000, **41**, 7311.

83. E.K. Kim, J.S. Bae, S.S. Im, B.C. Kim and Y.K. Han, *Journal of Applied Polymer Science*, 2001, 80, 1388.

84. C. Drohmann, O.B. Gorbatsevich and A.M. Muzafarov, *Polymer Preprints*, 2000, **41**, 1, 959.

85. M.T.R. Laguna, J. Gallego, F. Mendicuti, E. Saiz and M.P. Jarazona, *Macromolecules*, 2002, 35, 7782.

86. K.E. Branham, H. Byrd, R. Cook, J.W. Mays and G.M. Gray, *Journal of Applied Polymer Science*, 2000, **78**, 190.

87. A. Bozanko, W.D. Carswell, L.R. Hutchings and R. Richards, *Polymer*, 2000, **41**, 8175.

88. M.C. Davies, J.V. Dawkins, D.J. Hourston and E. Meehan, *Polymer*, 2002, **43**, 4311.

89. N. Sugimoto, T. Matsuda, Y. Miura, A. Narumi and H. Kaga, *Polymer Preprints*, 2002, **43**, 1, 664.

90. T. Huang and D.M. Knauss, *Polymer Bulletin*, 2002, **49**, 143.

91. S.T. Balke, T.H. Mouray, D.R. Rabbelo, T.A. Davis, A. Kraus and K. Skoniecznyik, *Journal of Applied Polymer Science*, 2002, **85**, 552.

92. *Rapid GPC Polymer Screening*, Polymer Laboratories Ltd., Church Stretton, UK, 2002.

93. R. Walkenhorst, *LC-GC Europe*, 2001, **14**, 676.

94. M. Netopilik, *International Journal of Polymer Analysis and Characterization*, 2001, **6**, 349.

95. L. Prokai and W.J. Simonsick, *Polymer Preprints*, 2000, **41**, 1, 663.

96. J.L. Dwyer, *Polymer Preprints*, 2000, **41**, 1, 659.

97. J. Chen, W. Radke and H. Pasch, *Macromolecular Symposia*, 2003, **193**, 107.

98. A.I. Kuzaev, E.N. Suslova and S.G. Entelis, *Doklady Akademii Nauk SSR*, 1973, **208**, 142.

99. M.S. Chang, D.M. French and P.L. Rogers, *Journal of Macromolecular Science A*, 1973, 7, 1727.

100. R.D. Law, *Journal of Polymer Science, Part A-1: Polymer Chemistry*, 1973, **11**, 175.

101. R.D. Law, *Journal of Polymer Science, Part A-1: Polymer Chemistry*, 1971, 9, 589.

102. R.D. Law, *Journal of Polymer Science, Part A-1: Polymer Chemistry* 1971, 7, 2097.

103. S. Teramachi, A. Hasegawa, Y. Shima, M. Akatsuka and M. Nakajima, *Macromolecules*, 1979, **12**, 992.

104. G. Glockner, J.H.M. van den Berg, N.L.J. Meijerink, T.G. Scholte and R. Koningsveld, *Macromolecules*, 1984, **17**, 962.

105. H. Sato, H. Takeuchi, S. Suzuki and Y. Tanaka, *Makromolekulare Chemie, Rapid Communications*, 1984, 5, 719.

106. M. Danielewicz and M. Kuhin, *Journal of Applied Polymer Science*, 1981, **26**, 951.

107. S. Mori and Y. Uno, *Analytical Chemistry*, 1987, **59**, 90.

108. S. Mori and Y. Uno, *Analytical Chemistry*, 1986, **58**, 303.

109. T. Tanaka, M. Omoto, N. Donkai and H. Inagaki, *Journal of Makromolecular Science*, 1980, **B-17**, 211.

110. L.M. Zaborski, *Analytical Chemistry*, 1977, **49**, 1166.

111. T.H. Mourey, G. Smith and L.R. Snyder, *Analytical Chemistry*, 1984, 56, 1773.

112. T.H. Mourey, *Analytical Chemistry*, 1984, 56, 1177.

113. F.J. Ludwig, *Analytical Chemistry*, 1984, 56, 2081.

114. D.M. Northop, D.E. Martire and R.P.W. Scott, *Analytical Chemistry*, 1992, **64**, 16.

115. T. Okada, *Analytical Chemistry*, 1990, **62**, 734.

116. S. Mori, *Analytical Chemistry*, 1990, **62**, 1902.

117. D.O. Hancock and R.E. Synovec, *Analytical Chemistry*, 1988, 60, 2812.

118. M. Petro, F. Svec, I. Gitsov and J.M.J. Frechet, *Analytical Chemistry*, 1996, **68**, 315.

119. E.R. Atkinson, Jr., and S.I. Calouche, *Analytical Chemistry*, 1971, **43**, 460.

120. N.R. Bektashi, D.H. Alieva, R.A. Ozhalilovi and A.V. Ragimov, *Polymer Science, Series B*, 2000, **42**, 276.

121. E. Klesper and W. Hartmann, *Journal of Polymer Science, Polymer Letters Edition*, 1977, **15**, 707.

122. E. Klesper and W. Hartmann, *Journal of Polymer Science, Polymer Letters Edition*, 1977, **15**, 9.

123. N.M. Frew, C.G. Johnson and R.M. Bromind in *Supercritical Fluid Extraction and Chromatography*, Eds., B.A. Charpentier and M.R. Sevenants, American Chemical Society Symposium Series No.366, ACS, Washington, DC, USA, 1988, p.208.

124. K.D. Bartle, T. Boddington and A.A. Clifford, *Analytical Chemistry*, 1991, **63**, 2371.

125. J.C. Via, C.L. Brane and L.R. Taylor, *Analytical Chemistry*, 1994, **66**, 603.

126. O. Hagenhoff, A. Benninghaven and H. Barthel, *Analytical Chemistry*, 1991, 63, 2466.

127. L. Mikkelson in *Characterization of High Molecular Weight Substances*, Pittsburgh Conference on Analytical Chemistry, Pittsburgh, PA, USA, 1962.

128. H. Puschmann, *Fette, Seifen, Anstrichmittel*, 1963, **65**.

129. C. Celedes Pasquot, *Revue Francaise Corps Gras*, 1962, **9**, 145.

130. C.C. Sweeley, R. Bentley, M. Makita and W.W. Wells, *Journal of the American Chemical Society*, 1963, **85**, 2497.

131. J.P. Fletcher and H. Persinger, *Journal of Polymer Science*, 1968, A-1, **6**, 1025.

132. L.W. Myers, Shell Research Ltd., Carrington, Cheshire, UK, private communication.

133. D.F. Uttenbach, D.S. Millington and A. Gold, *Analytical Chemistry*, 1984, **56**, 470.

134. T. Okada, *Analytical Chemistry*, 1991, **63**, 1043.

135. S. Teramachi and T. Fukao, *Polymer Journal*, 1974, **6**, 532.

136. F. Hori, Y. Ikada and J.J. Sakurada, *Journal of Polymer Science, Polymer Chemistry Edition*, 1975, **13**, 755.

137. T. Katada and J.L. White, *Makromolecules*, 1974, **7**, 1, 106.

138. I.A. Vachtina, R.F. Khrenova and O.G. Tarnkov, *Zhurnal Analiticheskoi Khimii*, 1973, **28**, 1625.

139. R. Buter, Y.Y. Tan and G. Challa, *Polymer*, 1973, **14**, 171.

140. B.G. Belenskii, E.S. Gankina, P.P. Nefedov, M.A. Lazareva, T.S. Savitskaya and M.D. Voichikhina, *Journal of Chromatography*, 1975, **108**, 61.

141. H. Inagaki and F. Kamiama, *Macromolecules*, 1973, **6**, 107.

142. T. Kotaka, T. Uda, T. Tanaka and H. Inagaku, *Makromolekulare Chemie*, 1975, **176**, 1273.

143. D.O. Geymer, Shell Chemical Company, Emeryville, CA, USA, private communications.

144. H. Inagaki, T. Miyamoto and F. Kamiyama, *Journal of Polymer Science*, 1969, B-7, 329.

145. F.W. Yeager and J.W. Becker, *Analytical Chemistry*, 1977, **49**, 722.

146. I.Y. Sionim, Y.G. Urman and U.N. Klyuchnikov, *Plasticheskie Massy (USSR)*, 1973, **4**, 68.

147. M.R. Ambler and R.D. Mate, *Journal of Polymer Science*, 1972, A-1, **10**, 2677.

148. I. Sakurada, Y. Ikada and T. Kawahara, *Journal of Polymer Science*, 1973, **11**, 2329.

149. A.R. Berens, *Polymer Preprints*, 1974, **15**, 2, 197.

150. M. Gilbert and F.J. Hylart, *Journal of Polymer Science*, 1971, A-1, **9**, 227.

151. W.C. Taylor and L.H. Tung in *Proceedings of the 140th American Chemical Society Meeting*, Chicago, IL, USA, 1961.

152. W.C. Taylor and L.H. Tung, *SPE Transactions*, 1962, April, p.119.

153. D.R. Morey and J.W. Tamblyn, *Journal of Applied Physics*, 1945, **16**, 419.

154. S. Claesson, *Journal of Polymer Science*, 1955, **16**, 193.

155. L.W. Gamble, W.T. Nipke and T.L. Lane, *Journal of Applied Polymer Science*, 1965, **9**, 1503.

156. W.H. Beattie, private communication.

157. G. Gooberman, *Journal of Polymer Science*, 1959, **40**, 469.

158. S. Tanaka, A. Nakamura and H. Morikawa, *Die Makromolekulare Chemie*, 1965, **85**, 164.

159. U. Gruber and H.G. Elias, *Die Makromolekulare Chemie*, 1964, **78**, 58.

160. W.C. Taylor and J.P. Graham, *Polymer Letters*, 1964, **2**, 169.

161. W.C. Taylor and L.H. Tung, *SPE Transactions*, 1963, **2**, 119.

162. F.H. Pager and H. Zorgmann, *Kunststoffe im Bau*, 1979, **14**, 57.

163. H. Werslau, *Makromolekulare Chemie*, 1956, **20**, 111.

164. W.H. Beattie, private communication.

165. S. Tanaka, A. Nakamura and H. Morikawa, *Die Makromolekulare Chemie*, 1965, **85**, 164.

166. U. Gruber and H.G. Elias, *Die Makromolekulare Chemie*, 1964, **78**, 58.

167. G. Gooberman, *Journal of Polymer Science*, 1969, **40**, 469.

168. V.I. Klenin and S. Yu Schogolov, *Journal of Polymer Science, Polymer Symposium*, 1973, **42**, Part 2, 965.

169. O. Hall in *Techniques of Polymer Characterisation*, Ed., P.W. Allen, Academic Press, New York, NY, USA, 1959, Chapter 11.

170. W.H. Beattie, *Journal of Polymer Science A*, 1965, **3**, 527.

171. V.I. Klenin, A.F. Padol'skii, S. Yu Shchegolev, B.I. Shvortsburd and N.E. Petrova, *Vysokomolekulyarnye Soedineniya Seriya A*, 1974, **16**, 974.

172. H.H. Cantow, M. Kowalski and C. Krozer, *Angewandte Chemie - International English Edition* 1972, **11**, 336.

173. V.A. Grechanovskii, *Vysokomolekulyarnye Soedineniya*, 1975, **17**, 2721.

174. J. Lanikova and H. Hiensek, *Chemicky Prumsyl*, 1973, **23**, 10.

175. E. Banicka and V.S. Ciganekova, and P. Nasek, *Macrotest*, 1973, **2**, 64.

176. C. Vasile, A. Onu, O. Popa and T. Matei, *Materiale Plastice (Bucharest)*, 1973, **10**, 631.

177. K. Takashima, K. Nakae and M. Shibata, *Makromolecules*, 1974, **7**, 641.

178. O.V. Kallistov, *Zavodskaia Laboratoriia*, 1972, **38**, 711.

179. T. Kamata and T. Nakahara, *Journal of Colloid Interface Science*, 1973, **43**, 89.

180. H. Dautzenberg, *Journal of Polymer Science C*, 1972, **39**, 123.

181. K.V. Pogorel'skii, A. Asanov and K.S. Akhedov, *Doklady Akademii Nauk SSR*, 1970, **27**, 28.

182. J.H. Bradley, *Journal of Polymer Science C*, 1965, **8**, 305.

183. G. Liu, X. Yan and S. Duncan, *Macromolecules*, 2002, **35**, 9788.

184. D.A. Wrobleski, E.B. Orler and M.E. Smith, *Polymer Preprints*, 2001, **2**, 1, 669.

185. H. Thielking and W.M. Kuliche, *Analytical Chemistry*, 1996, **68**, 1169.

186. *British Plastics and Rubbers*, 2002, February, 14.

187. H. Lindner and O. Glatter, *Macromolecular Symposia*, 2000, **162**, 81.

188. *B1-MWA (Molecular Weight Analyser), On-Line Monitoring of Polymerisation Reactions*, Brookhaven Instruments Corporation, Hottsville, NY, USA, 2002.

189. Y. Mao, W. Wei, J. Zhang and Y. Li, *Journal of Applied Polymer Science*, 2001, **82**, 63.

190. M. Minigawa, H. Kanoh, S. Tanno and M. Satoh, *Macromolecular Chemistry and Physics*, 2002, **203**, 2481.

191. V.I. Lukhovitskii and A.I. Karpo, *Polymer Science, Series B*, 2001, **43**, 195.

192. A. Nakajima and H. Fujiwara, *High Polymers Japan*, 1950.

193. A. Nakajima and H. Fujiwara, *High Polymers Japan*, 1964, **37**, 909.

194. T. Walsh, W. Engewald and E. Kowash, *Plaste und Kautschuk*, 1976, **23**, 584.

195. T. Wozniak, *Pezegl Wiok*, 1976, **30**, 18.

196. M.I. Simonova and E.M. Alzenshteini, *Zavodskaia Laboratoriia*, 1974, **40**, 435.

197. E.O. Shemalz, *Faserforsh Textiltech*, 1970, **21**, 209.

198. J.B. Platonov, V.M. Belyaev and F.P. Grigor'eva, *Plasticheskie Massy (USSR)*, 1975, **4**, 77.

199. ISO 834-1, *Fire Resistance Tests – Elements of Building Construction – Part 1: General Requirements*, 1999.

200. J. Lanikova and M. Hlousek, *Chemicky Prumysl*, 1977, **27**, 628.

201. J.H. Ross and R.L. Shamk in *Polymer Molecular Weight Methods*, Ed., M.Ezrin, Advances in Chemistry Series, No.125, ACS, Washington, DC, USA, 1971, p.108.

202. R. Chiang, *Journal of Polymer Science*, 1965, A-3, 3679.

203. F.W. Cuesta de la Billmeyer, *Journal of Polymer Science*, 1963, A-1, 1721.

204. N. Das and S.R. Palit, *Journal of Polymer Science*, 1973, A-1, **11**, 1025.

205. M.A. Langhorst, F.M. Stanley, S.S. Cutie, J.H. Sugarman, L.R. Wilson, D.A. Hoagland and R.K. Prud'homme, *Analytical Chemistry*, 1986, **58**, 2242.

206. W. Machtle, *Makromolekulare Chemie*, 1982, **183**, 2215.

207. C.S. Wu, L. Senak, J. Bonilla and J. Cullen, *Journal of Applied Polymer Science*, 2002, **86**, 1312.

208. P.N. Songkhla and J. Wootthikanokkhan, *Journal of Polymer Science, Part B: Polymer Physics*, 2002, **40**, 6, 562.

209. W. Machtle and U. Klodwig, *Makromolekulare Chemie*, 1979, **180**, 2507.

210. W.E. Blair, *Journal of Polymer Science, Part C: Polymer Symposia*, 1965, **8**, 287.

211. H. Suzuki and C.G. Leonis, *British Polymer Journal*, 1973, 5, 6, 485.

212. H-D. Beckey, *Principles of Field Ionization and Field Desorption Mass Spectroscopy*, Pergamon Press, New York, NY, USA, 1977.

213. H.R. Schulten, *Advances in Mass Spectrometry*, 1977, 7, 83.

214. H.R. Schulten, *Methods Biochemical Analysis*, 1977, 24, 313.

215. W.L. Mead, *Analytical Chemistry*, 1968, 40, 743.

216. R.H. Wiley and J.C. Cook Jr., *Journal of Makromolecular Science - Chemistry*, 1976, A10, 5, 811.

217. R.H. Wiley, *Macromolecular Review*, 1979, 14, 379.

218. R.P. Lattimer, K.R. Welch, J.B. Pausch and U. Rapp, *Varian MAT 311A Application Note No. 27*, Varian MAT Mass Spectrometry, Florham Park, NJ, USA, 1978.

219. R.P. Lattimer, D.J. Harmon and K.R. Welch, *Analytical Chemistry*, 1979, 51, 1293.

220. T. Matsuo, H. Matsuda and I. Katakuse, *Analytical Chemistry*, 1979, 51, 1329.

221. R.P. Lattimer and K.L. Welch, *Rubber Chemistry and Technology*, 1978, 51, 925.

222. R.P. Lattimer and K.L. Welch, *Rubber Chemistry and Technology*, 1980, 51, 151.

223. R.P. Lattimer and D.J. Hermon, *Analytical Chemistry*, 1980, 52, 1808.

224. P.J. Florey, *Principles of Polymer Chemistry*, Cornell University Press, New York, NY, USA, 1960, Chapter 7.

225. J.B. Poli and M.R. Schure, *Analytical Chemistry*, 1992, 64, 896.

226. R.A. Wallingford, *Analytical Chemistry*, 1996, 68, 2541.

227. S. Auriola, I. Jaaskelainen, M. Regina and A. Urtti, *Analytical Chemistry*, 1996, 68, 3907.

228. G.G. Jones, R.E. Pauls and R.C. Willoughby, *Analytical Chemistry*, 1991, 63, 460.

229. J.J.L. Bryant and J.A. Semlyen, *Polymer*, 1997, 38, 4531.

230. D. Raghavan and K. Egwim, *Journal of Applied Polymer Science*, 2000, 78, 2454.

231. T. Okada, *Analytical Chemistry*, 1990, **62**, 327.

232. S. Mori, *Analytical Chemistry*, 1990, **62**, 1902.

233. I.V. Bletsos, D.M. Hercules, D. van Leyen, B. Hagenhoff, E. Niehuis and A. Benninghoven, *Analytical Chemistry*, 1991, **63**, 1953.

234. D.E. Fowler, R.D. Johnson, D. van Leyen and A. Benninghoven, *Analytical Chemistry*, 1990, **62**, 2088.

235. Z. Liang, A.G. Marshall and D.G. Westmoreland, *Analytical Chemistry*, 1991, **63**, 815.

236. D.C. Schriemer and L. Li, *Analytical Chemistry*, 1996, **68**, 2721.

237. H.R. Kricheldorf, G. Schwarz, S. Boehme and C.L. Schultx, *Journal of Polymer Science, Polymer Chemistry Edition*, 2003, **41**, 890.

238. J.J. Kirkland and S.W. Rementer, *Analytical Chemistry*, 1992, **64**, 904.

239. M.A. Benincasa and J.C. Giddings, *Analytical Chemistry*, 1992, **64**, 790.

240. H. Schuch, S. Frenzel and F. Runge, *Progress in Colloid and Polymer Science*, 2002, **121**, 43.

241. M. Vincenti, E. Pelizzetti, A. Guarini and S. Costanzi, *Analytical Chemistry*, 1992, **64**, 1879.

242. F. Blockhuys, M. Claes, R. Van Grieken and H.J. Geise, *Analytical Chemistry*, 2000, **72**, 3366.

9 Thermal and Chemical Stability

9.1 Introduction

A wide range of techniques have been applied to the elucidation of the thermal and chemical stability of polymers. Theses include:

- Thermogravimetric analysis (TGA)

- Differential thermal analysis (DTA)

- Differential scanning calorimetry (DSC)

- Thermal volatilisation analysis (TVA)

- Evolved gas analysis (EGA)

- Mass spectroscopic methods (MS)

Thermal analysis methods can be broadly defined as analytical techniques that study the behaviour of materials as a function of temperature [1]. These are rapidly expanding in both breadth (number of thermal analysis-associated techniques) and in depth (increased applications). Conventional thermal analysis techniques include DSC, DTA, TGA, thermomechanical analysis, and dynamic mechanical analysis (DMA). Thermal analysis of a material can be either destructive or non-destructive, but in almost all cases subtle and dramatic changes accompany the introduction of thermal energy. Thermal analysis can offer advantages over other analytical techniques including: variability with respect to application of thermal energy (step-wise, cyclic, continuous, etc.), small sample size, the material can be in any 'solid' form – gel, liquid, glass, solid, ease of variability and control of sample preparation, ease and variability of atmosphere, it is relatively rapid, and instrumentation is moderately priced. Most often, thermal analysis data are used in conjunction with results from other techniques.

Cerrada [2] has reviewed current trends in the thermal testing of polymers.

Some of the thermal and chemical stability applications to which these various techniques have been applied are summarised in **Table 9.1**.

Table 9.1 Techniques used in thermal and chemical stability studies		
Analysis method	**Thermal stability**	**Chemical stability**
TGA	Weight loss measurement Moisture content Thermal stability QC testing	Chemical stability Chemical composition Decomposition kinetics Catalyst activity
DTA	Thermal stability QC testing	Chemical stability Chemical composition Decomposition kinetics Extent of rate of resin cure Catalyst activity
DSC	Thermal stability QC testing	Chemical stability Chemical composition Decomposition kinetics Extent of resin cure Resin cure kinetics Catalyst activity
TVA		Polymer decomposition Volatiles identification
EGA		Polymer decomposition Volatiles identification
Source: Author's own files		

9.2 Theory

The principles of the various techniques are discussed next.

9.2.1 Thermogravimetric Analysis

Weight changes in a material as it is heated, cooled, or held at a constant temperature in an inert atmosphere are measured to determine composition and thermal stability (weight *versus* temperature plots). The technique is used primarily to determine the composition of polymers and to predict their thermal stability at temperatures up to 1000 °C. The technique can characterise materials that exhibit weight loss or gain due to decomposition, oxidation, or dehydration. It is especially useful for studying polymers including composites. It is used to obtain estimates of polymer lifetimes under specific test conditions (i.e., a kinetic approach).

Temperature *versus* weight plots provide information on:

- Polymer decomposition temperatures.

- Thermal stability in inert atmospheres.

- Oxidative stability (see Chapter 11).

- Compositional analysis of rubbers and elastomers.

- Ash, inert filler, moisture, residual solvent, and plasticiser contents of polymers.

A combination of TGA analysis with Fourier transform IR (FT-IR) spectroscopy with periodic analysis of the evolved gases is a very useful means of studying polymer decomposition.

9.2.2 Differential Thermal Analysis

This technique measures heat-related phenomena that are associated with transitions in materials. In this technique the polymer sample is temperature programmed at a controlled rate and, instead of determining weight changes as in TGA, the temperature of the sample is continually monitored. Just as a phase change from ice to water or *vice versa* is accompanied by a latent heat effect, so when a polymer undergoes a phase change from, for example, crystalline to an amorphous form, heat is either evolved or absorbed.

This technique has been used to characterise polymers at temperatures up to 150 °C. Under a reactive gas (oxygen) or an inert gas (nitrogen) plots of the applied temperature *versus* the temperature of specimens (or calories per second) detect positive (i.e., exothermic) or negative (i.e., endothermic) temperature changes (ΔH) in the reactions or phase changes occurring upon heating the polymer. The theory of this technique has been discussed by Earnest [3].

9.2.3 Differential Scanning Calorimetry

DSC measures the amount of energy absorbed or released by a sample as it is heated, cooled, or held at a constant temperature. The precise measurement of sample temperature is also made with DSC. The theory of the technique has been discussed by McNaughton and Mortimer [4].

DSC measures the temperature and the heat flow associated with transitions in materials as a function of the time and temperature, i.e., heat flow–sample temperature plots. Such measurements provide quantitative and qualitative information about physical or chemical changes that involve exothermic or endothermic processes or changes in heat capacity.

DSC instruments can be used in the DSC mode, i.e., heat flow–temperature of sample, or the DTA mode, i.e., sample temperature–temperature of sample. DSC can provide information on:

- Heat of transition, e.g., glass transition (T_g).
- T_g (temperature–heat-flow).
- Melting point T_m (temperature–heat-flow).
- Crystallisation time and temperature.
- Crystallinity (%) (temperature–heat-flow).
- Specific heat.
- Oxidative stability.
- Rate of cure – heat of cure (temperature–heat-flow rate).
- Degree of cure – onset of cure (temperature–heat-flow rate).
- Reaction kinetics.
- Thermal stability.
- Compatibility of polymer blends.
- Resin substrate uniformity in composites.
- Study of post-photoinitiated curing reactions.
- Polymer heat history studies (temperature–heat-flow).

Since polymerisation processes are mainly exothermic reactions in which each additional chain formation step generates a defined amount of heat, the reaction process can be monitored using DSC directly and continuously. Combining DSC with an irradiation unit enables curing studies to be carried out in photoinitiated polymerisations. The somewhat indiscriminate use of the terms DSC and DTA has made it necessary for the IUPAC to offer definitions of these processes in a communication of nomenclature [5, 6].

The purpose of differential thermal systems is to record the difference between the enthalpy change that occurs in a sample and that in some inert reference material, when they are both heated. These systems may be classified into three types as follows:

1. Classic DTA
2. Boersma DTA
3. DSC

In the classic and Boersma DTA systems, both sample and reference are heated by a single heat source. Temperatures are measured by sensors embedded in the sample and reference materials (classic) or attached to the pans that contain the materials (Boersma). A plot is made, usually by means of a recorder, of the temperature difference $\Delta T = T_S - T_0$ between sample and reference as ordinate against time as abscissa. The magnitude of ΔT at a given time is proportional to (a) the enthalpy change, (b) the heat capacities, and (c) the total thermal resistance to heat flow, R. High sensitivity requires a large value of R, but unfortunately the value of R depends on the nature of the sample, the way it is packed into the sample pan, and the extent of thermal contact between sample pan and holder; also, R varies with temperature. Attachment of the temperature sensors to the pans in the Boersma method is made in an attempt to reduce the effect of variations in the thermal resistance caused by the sample itself.

It is not possible with either of these DTA systems to make a simple conversion of the peak area, from a plot of ΔT against time, into energy units. This is because of (a) the need to know the heat capacities and (b) the variation of R, and hence the calibration constant, with temperature. Consequently DTA systems are not very suitable for calorimetric measurements.

The technique referred to as DSC is specifically that described next and any other thermal analysis method, in particular Boersma-type DTA, which may have been described as 'differential scanning calorimetry' in the literature will be referred to as 'indirect DSC'.

The important difference between the DTA and DSC systems is that in the latter the sample and reference are each provided with individual heaters. This makes it possible to use a 'null-balance' principle. It is convenient to think of the system as divided into two control loops. One is for average temperature control, so that the temperature, T_0, of the sample and reference may be increased at a predetermined rate, which is recorded. The second loop ensures that if a temperature difference develops between the sample and reference (because of exothermic or endothermic reaction in the sample), the power input is adjusted to remove this difference. This is the null-balance principle. Thus, the temperature of the sample holder is always kept the same as that of the reference holder by continuous and automatic adjustment of the heater power. A signal, proportional to the difference between the heat input to the sample and that to the reference, dH/dt, is fed into a recorder. In practice this recorder is also used to register the average temperature of the sample and reference.

In an idealised thermogram or record of the differential heat input, dH/dt is plotted against temperature T (or time t on the same axis). A dH/dt peak corresponding to the endotherm is obtained in such plots. The most advanced commercial instrument, the Perkin Elmer DSC-2, provides a maximum sensitivity of 0.1 mcal/s for a full-scale deflection and is normally used with milligram quantities of sample. The operational temperature range is from −175 to +725 °C.

Details of the theory and design of DSC instruments are discussed next.

In DSC instruments the thermal mass of the sample and reference holders is kept to a minimum, thermal resistances are reduced as much as possible, and a high 'loop gain' in the closed loop of the differential power control circuit is used. These measures ensure that the response of the system is short. Consequently, the assumption that the sample and reference holders are always the same temperature, T_p, is valid. The response of the system depends on the thermal resistance between the holders and surroundings, R, but this is unaffected by a change in the sample.

9.2.4 Thermal Volatilisation Analysis

In this technique, in a continuously evacuated system the volatile products are passed from a heated sample to the cold surface of a trap some distance away. A small pressure develops which varies with the rate of volatilisation of the sample. If this pressure is recorded as the sample temperature is increased in a linear manner, a thermal volatilisation analysis thermogram showing one or more peaks is produced. The trace obtained is somewhat dependent on heating rate, which therefore should be standardised.

9.2.5 Evolved Gas Analysis

The various techniques discussed previously measure heat changes, phase changes, pressure changes, and weight loss under controlled conditions. Although in many instances these measurements are very meaningful, they are limited in the information they can provide. TGA, for example, measures the weight of a sample as a function of temperature and time as the temperature is varied. This gives valuable quantitative information but tells us nothing about the chemistry of the processes occurring. To bridge this gap a field has opened up called EGA in which it is possible to monitor the by-products of reactions associated with heat. In the EGA technique the sample is heated at a controlled rate under controlled conditions and the weight changes monitored (i.e., TGA). At the same time the reaction products are led into a suitable instrument for identification, and in some cases quantitation. Many variants of this approach have been developed based on three methods for thermally breaking down samples: pyrolysis, linear programmed thermal degradation (i.e., without recording weight change), and the thermogravimetric approach (i.e., sample weight continuously recorded).

- Pyrolysis with gas chromatography (GC).

- Pyrolysis with mass spectrometry (MS).

- Linear programmed thermal degradation with computerised GC.

- TGA analysis with FT-IR spectroscopy or IR spectroscopy.

- Linear programmed thermal degradation with MS.

EGA is the generic name for a wide range of processes in which the polymer is heated under controlled conditions of temperature and a chosen atmosphere ranging from inert (e.g., nitrogen or helium) to reactive (e.g., oxygen) and the breakdown products produced are examined by any one of a wide variety of analytical techniques, either qualitatively or quantitatively, or both. A classic example of EGA is the pyrolysis–gas chromatography (PGC) technique described in Section 3.7. Some other examples of the application of the technique include polymer weight change–degradation product studies, the production of polymer thermochromatograms, and polymer additive degradation studies (Chapter 7).

EGA is a technique in thermal analysis for characterising compounds evolved during sample heating. It is, in a sense, a reverse technique of TGA and if only one compound is evolved during sample heating, then EGA and TGA should principally give identical information. Usually a number of components evolve during heating of a sample according to complicated degradation kinetics and this valuable information is poorly reflected on a TGA curve. In view of this last fact, it is surprising that the value of EGA has frequently been overlooked, and thus used less often than TGA.

9.3 Instrumentation

As discussed earlier, instrumentation is useful in studies of thermal and chemical stability of polymers, including the following (see **Table 9.2**): TGA, DTA, DSC, TVA, and EGA (including TGA–GC, TGA–FT-IR spectroscopy, TGA–GC–MS).

The availability of such instrumentation is discussed next.

9.3.1 Instrumentation for TGA, DTA, and DSC

Various suppliers are listed in **Table 9.2**; see also Appendix 1.

311

Table 9.2 Thermal measurement possible with commercial analysers (excluding evolved gas analysis and thermal volatilisation analysis)

	Perkin Elmer Series 7 system	Dupont 9900 Thermal Analysis Redcroft Omnitherm	PL Thermal Sciences/Stanton 210 system	TA Instruments Thermal analyst Scientific	Sectron (Clayton Scientific)
TGA	TGA7	951	PL-TGA and PL-STA	TGA 2950	
DTA	1700 (high temp.)	DTA analyser	PL-DTA and PL-STA		
DSC	DSC-7	910 912 dual sample cell DSC-2910	PL-DSC and PL-STA	DSC	Micro DSC calorimeter
Pressure DSC (PDSC)		Dual sample PDSC cell			
TMA	TMA7	943 (discussed further in Chapter 19)	PL-TMA	TMA	
Data handling and software options	(a) 1020 series for DSC-7 and TGA-7 (b) PC series for DSC-7, TMA-7, and high temperature DTA-1700 (c) 7 series personal workstation selected options		PL-ETA software supports DHTA, DSC, TGA modules	IBM personal computer systems/2*-model 70386 computer	

9.3.2 Instrumentation for TVA and EGA

9.3.2.1 TGA–Fourier Transform Spectroscopy

Stanton Redcroft (now PL Thermal Sciences Ltd) was a pioneer in developing techniques that coupled a Stanton Redcroft 1000 °C thermobalance thermogravimetric analyser with FT-IR spectroscopy (Digilab FTS40 FT-IR spectrometer with 3200 data system).

For certain applications, compared to a MS combination, FT-IR instrumentation has the advantages that it is inherently more simple – no vacuum involved – and it is research-grade quality, whereas a MS is likely to be a basic quadrupole device. While the MS could be more sensitive and would possibly provide more information as an evolved gas analyser, it would also have a much greater need for maintenance than an FT-IR instrument.

Perkin Elmer supply a system 2000 TGA–IR interface. This is a single supplier fully optimised system with a powerful range of advanced graphics and post-run processing software including the ability to produce stack plots showing small changes and trends in the concentrations of evolved gases.

9.3.2.2 TGA–Mass Spectrometry

PL Thermal Sciences produce the PL STA MS. In this instrument a PL-STA 1500 analyser capable of making TGA and DTA measurements is coupled with an MS, thereby augmenting weight loss and the differential temperature response data with quadrupole mass spectral data of the evolved gases such as oxygen, sulfur dioxide, and carbon dioxide. With a transfer time of evolved gases to the MS of about 100 ms, the MS-EGA is virtually simultaneous.

9.3.2.3 Gas Chromatography–Mass Spectrometry

Shimadzu produce a range of instruments combining GC with MS (the GCMS-QP1000EX system). This system covers a range of evolved organics in the mass range 1–50,000. It can be used in the following modes:

1. The TGA-GC-MS mode in which gases evolved in thermogravimetry are trapped for a given length of time and then the collecting tube is rapidly heated to introduce the gases into the gas chromatograph column. The separated gases then pass into the MS to permit separate determination and identification.

2. The TG-MS mode. Gases evolved in TGA are directly introduced into the MS in real time. This mode allows the concentrations of the evolved gases to be plotted against the weight loss of sample under heating.

9.4 Applications

9.4.1 Thermogravimetric Analysis

This technique involves continuous weighing of a polymer as it is subjected to a temperature programme. This technique can provide quantitative information about the kinetics of the thermal decomposition of polymeric materials from which the thermal stability can be evaluated. It is used to study the influence of factors such as effect of crystallinity, molecular weight, orientation, tacticity, substitution of hydrogen atoms, grafting, copolymerisation, and addition of stabilisers on polymer degradation. **Figure 9.1** shows decomposition profiles for polytetrafluoroethylene (PTFE) and fibre glass-reinforced Nylon.

The lifetime or shelf life of a polymer can be estimated from the kinetic data. Ozawa [7] observed that the activation energy of a thermal event could be determined from a series of thermogravimetric runs performed at different heating rates. As the heating rate increased, the thermogravimetric changes occurred at higher temperatures. A linear correlation was obtained by plotting the logarithm of the heating rate or scan speed against the reciprocal of the absolute temperature at the same conversion or weight loss percentage. The slope was directly proportional to the activation energy and known constants. To minimise errors in calculation, approximations were used to calculate the exponential integral [8-12]. It was assumed that the initial thermogravimetric decomposition curve (2–20% conversion) obeyed first-order kinetics. Rate constants and pre-exponential factors could then be calculated and used to examine relationships between temperature and conversion levels. The thermogravimetric decomposition kinetics can be used to calculate:

1. The lifetime of a sample at selected temperatures.

2. The temperature that will give a selected lifetime.

3. The lifetimes at all temperatures at known percentage conversion.

Thermograms (percentage weight *versus* temperature) can be prepared for specimens of polymer obtained at four different heating rates, 2.5, 5, 10, and 20 °C/min, in a dynamic air atmosphere. From these data the rate of decomposition of the polymer, the activation energy, and the relationship between the rate constant or half-life and temperature can be calculated .

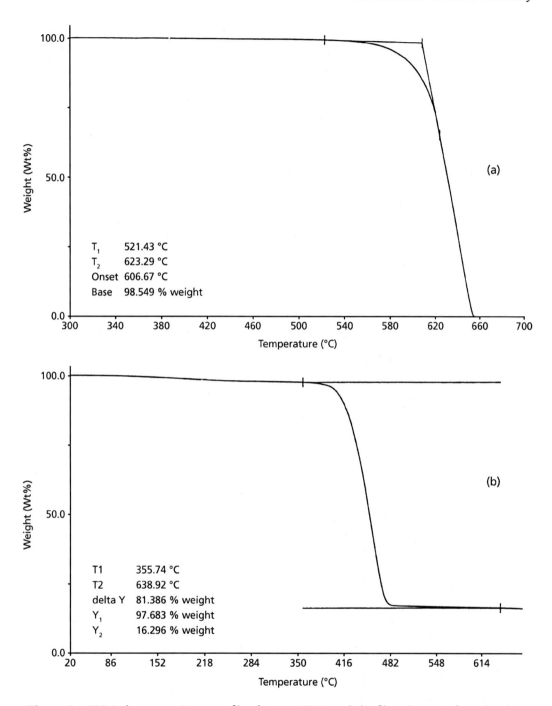

Figure 9.1 TGA decomposition profiles for (a) PTFE and (b) fibreglass reinforced nylon.
(*Source: Author's own files*)

The results agree with the theoretical prediction that, as the rate of heating increases the thermograms are displaced to higher temperatures. Activation energies at selected percentage conversion levels can be calculated using the results. The activation energy is calculated from the slope of the graph of scan time against the inverse of the absolute temperature. After the activation energy has been determined, the rate constants, half-lives, and percentage conversions can be calculated for certain temperatures. The rate constant is plotted against temperature to provide information on the stability of the sample from ambient temperature up to 800 °C. The half-life can be calculated from these kinetic data and a graph of half-life *versus* temperature plotted. The temperature equivalent to a half-life of 60 minutes is also determined. These data can be presented in tabular form (**Table 9.3**) to facilitate quantitative comparisons. The rate of percentage conversion with respect to time can be computed for any selected temperature. The results calculated using the thermogravimetric decomposition kinetics can be used for comparative studies. However, if the lifetimes at specific temperature and percentage conversions are known for the sample, adjusted lifetimes can be calculated. These types of data can be obtained by elevated temperature measurements on 'bulk' samples.

Table 9.3 Thermogravimetric analysed PTFE half-life table							
Temp. (°C)	Temp. (K)	1000/T	$K(T)$ (l/min)	Years	Days	Hours	Minutes
400	673.16	1.485	1.667E-06	0	288	14	26
410	683.16	1.463	4.154E-06	0	115	20	57
420	693.16	1.442	1.007E-05	0	47	18	21
430	703.16	1.422	2.383E-05	0	20	4	36
440	713.16	1.402	5.504E-05	0	8	17	51
450	723.16	1.382	1.242E-04	9	3	21	0
460	733.16	1.363	2.740E-04	0	1	18	9
470	743.16	1.345	5.920E-04	0	0	19	30
480	753.16	1.327	1.253E-03	0	0	9	13
490	763.16	1.310	2.600E-03	0	0	4	26
500	773.16	1.293	5.296E-03	0	0	2	10
510	783.16	1.276	1.059E-02	0	0	1	5
520	793.16	1.260	2.081E-02	0	0	0	33
530	803.16	1.245	4.022E-02	0	0	0	17
540	813.16	1.229	7.647E-02	0	0	0	9
Activation energy: 348.9 kJ/mole, pre-exponential factor: 1.98 E+ 2 l/min, conversion: 2.5–10%. Reprinted from T. Ozawa, Bulletin of Chemical Society of Japan, 1965, 38, 1881, with permission from Chemical Society of Japan [7]							

TGA has been used to study degradation kinetics and various factors affecting thermal stability of polymers, such as crystallinity, molecular weight, orientation, tacticity, substitution of hydrogen atoms, grafting, copolymerisation, and addition of stabilisers. More important systems investigated include cellulose [13–18], polystyrene (PS) [14], ethylene–styrene copolymers [19, 20], styrene–divinyl benzene-based ion exchangers [21], vinyl chloride–acrylonitrile copolymers [22], polyethylene terephthalate (PET) [23], polyesters such as polyisopropylene carboxylate [22], polyglycollide [24-26], Nylon 6 [27], polypyromellitimides, poly-N-naphthylmaleimides [28, 29], polybenzo-bis(amino imino pyrolenes) [30], polyvinyl chloride (PVC) [31-34], acrylamide–acrylate copolymers and polyacrylic anhydride [35-37], polyamides [38], amine-based polybenzoxazines [39], polyester hydrazides [40], poly-α-methylstyrene tricarbonyl chromium [41], polytetrahydrofuran [42], polyhexylisocyanate [43], polyurethanes [44], and ethylene–vinyl acetate copolymer [45].

One of the most common uses of TGA is the quantitative determination of components in a mixture. Because of differences in thermal stability, the selection of the appropriate temperature programme and sample atmosphere results in the quantitative separation of the major components of a polymer or elastomer.

Thermogravimetry of the following rubber blends has been studied in detail by Lockmuller and co-workers [46] from the point of view of controlling decomposition mechanisms and minimisation of variance: chloroprene rubber blends, butadiene–acrylonitrile rubbers, and rubber adhesives. These workers carried out a factor analysis in the TGA of rubber blends and butadiene–acrylonitrile copolymers using singular value decomposition and variance minimisation.

High-resolution TGA has been applied to decomposition studies on polymethyl methacrylate (PMMA), ethylene–vinyl acetate copolymer and acrylonitrile–butadiene–styrene terpolymer [47–50]. The results obtained on a supposedly pure sample of PMMA homopolymer indicated that a small quantity of impurity, possibly unreacted methyl monomer or even polyethylene methacrylate, is present. Conventional TGA does not resolve this impurity.

9.4.2 TGA–FT-IR Spectroscopy and DSC–FT-IR Spectroscopy

These techniques have been discussed by Castle and McClure [51] who applied the Perkin Elmer system 2000 TGA-FT-IR interface system to thermal decomposition studies on PVC–polyvinyl acetate copolymers. They identified acetic acid and hydrogen chloride as two of the decomposition products.

The techniques have also been applied to isocyanate- and ester-based foams [52]. See also Section 9.4.6.

A combination of TGA and/or DSC and FT-IR spectroscopy has been used to study the thermal behaviour of some polymethacrylates [53] and polyamides [54], chlorinated elastomers [55], polyamides and polyimides [56], poly(3-hydroxybutyrate) and poly(3-hydroxybutyrate-*co*-3-hydroxyvalerate) [57], polypropylene carbonate [58], polyacrylonitrile [59], polyimides [54], polyurethanes [60], styrene–nitrostyrene copolymers [61], bisphenol A-based polycarbonates [62], and polymethacrylates [53].

9.4.3 Differential Thermal Analysis

This technique has been used in polymer stabilisation studies [63, 64], polymer pyrolysis kinetic studies [65, 66], and examination of the effects of thermal history on polyethylene (PE) [67].

9.4.4 Differential Scanning Calorimetry

The applications of this technique have been reviewed [68]. In addition to chemical reactions associated with thermal and oxidative decomposition, DSC can also be used to monitor the curing of thermosets (see Section 10.3). Examples are the polymerisation of unsaturated polyester resins, poly-addition of epoxy resins with curing agents, and isocyanates with polyols.

An isothermal DSC curve shows at a glance whether a reaction proceeds normally: in other words, the rate of reaction and thus the heat flow reaches a maximum upon the reaction mixture's attainment of the reaction temperature. To locate a suitable isothermal reaction temperature, a dynamic experiment is carried out at 10 °C/min. The optimum isothermal temperature will lie between the start of reaction (at 20% of the peak height) and the peak maximum temperature. For example, an epoxy resin used for powder coating gives values of 180 °C to 220 °C. Conversely, an autocatalytic reaction shows an increasing reaction rate after an induction period.

To determine the extent of reaction as a function of reaction time, it is assumed that the area under the curve increases proportionally to the conversion, i.e., the conversion at a time *t* is equal to the partial area at the time *t* divided by the total area. The graph of the extent of reaction *versus* reaction time is constructed by taking, for example, five calculated values.

Figure 9.2 shows a typical curing curve from a thermoset material showing the exothermic peak produced as it is released during the curing process. A particular example is the study of curing kinetics in diallyphthalate moulding compounds [69–71]. DSC has also

318

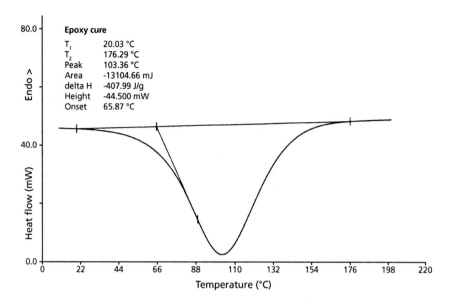

Figure 9.2 Use of differential scanning calorimetry in curing studies
(*Source: Author's own files*)

been used to monitor the crosslinking [72] rate of PE–carbon black systems, and to study the decomposition of polyoxypropylene glycols [72] and the thermal stability of polyvinyl alcohol modified at low levels by various reagents and by grafting with other vinyl monomers [73]. The technique has been used in kinetic studies and as a check on experimentally determined activation energies and Arrhenius frequency factors. A kinetic study of isothermal cure of epoxy resin has been carried out [74, 75]. Kinetic parameters associated with the crosslinking process of formaldehyde–phenol and formaldehyde-melamine copolymers have been obtained from the exotherm of a single DSC temperature scan [76]. Using this technique, kinetic studies have been made of the polymerisation of styrene [77, 78], methylacrylate [79, 80], vinyl acetate [80], bis-maleinimides [81], and phenolformaldehyde [82], and also curing of epoxide resins [83-85], polyesters [86], poly(N-isopropyl acrylamide) [87], unsaturated polyesters [88], polyvinylidene fluoride [89], poly(aniline-*co*-fluoroaniline) [90], styrene–butadiene terpolymers [91], polysiloxanes [92], poly(ester carbonate) [93], ethylene–norborneol copolymers [94], polyacrylonitrile [95], polyamide 6,66 [96-97], butylacrylate–styrene copolymers [98], polyoxyethylene [97], *tert*-butyl methacrylate–PS block copolymers [99], ethylene–α-olefin copolymers [100], polyhydroxyurethanes [101], poly(butylene succinates), *co*-butylene adipates [102], soy protein thermoplastics [103], PVC–polytaconite blends [104], polyamide 6,6 [105], atactic PS [106], epoxy resins [107], PE, and polypropylene (PP) and PET [108].

A combination of programmed isothermal techniques has been used for characterising unresolved multi-step reactions in polymers [109].

9.4.5 Thermal Volatilisation Analysis

TVA thermograms for various PMMA are illustrated in **Figure 9.3**. As in the case of TGA, the trace obtained is somewhat dependent on the heating rate. With PMMA the two stages in the degradation are clearly distinguished. The first peak above 200 °C represents the reaction initiated at unsaturated ends formed in the termination step of the polymerisation. The second, larger peak corresponds to the reaction at higher temperatures initiated by random scission of the main chain. It is apparent that as the proportion of chain ends in the sample increases, the size of the first peak also increases. Such TVA thermograms illustrate very clearly the conclusion drawn by Macallum [110] in a general consideration of the mechanism of degradation of this polymer. The peaks occurring below 200 °C can be attributed to trapped solvent, precipitation, etc. These show up very clearly, indicating the usefulness of TVA as a method for testing polymers for freedom from this type of impurity.

The technique has been applied to a range of polymers including PS [111, 112], styrene–butadiene copolymers [111], PVC [111], polyisobutene, butyl rubber and chlorobutyl rubber, and poly-α-methylstyrene [111].

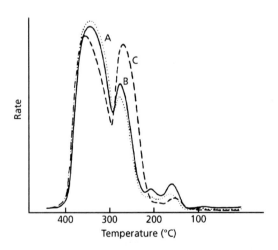

Figure 9.3 Thermograms (10 °C/min) for samples of PMMA of various molecular weight: (A) 820,000; (B) 250,000; (C) approximately 20,000
(*Source: Author's own files*)

320

9.4.6 EGA–TGA–Gas Chromatogravimetry and TGA–Gas Chromatography–Mass Spectrometry

To date, most significant work in the EGA field has been done by MS [113–115] – a valuable tool in identifying the evolved components. However, EGA can be applied where identification is trivial or an exact knowledge of the nature of the evolving components is not necessary. The number of components and their rate of evolution with temperature also give much information about the thermal behaviour of a sample. Then, instead of complicated and expensive MS, a more simple, straightforward, and less expensive technique – GC, can be used.

The problem with GC as an analysis tool for the evolved component is that it takes much more time to separate the evolved components than to analyse their mass in the MS. If the sample is heated by a linear thermal programme (as is the case in thermal analysis), then many interesting kinetic phenomena may occur during separation of components in the GC. This makes on-line GC analysis of the evolved components difficult. The evolved components could be trapped at moments of interest [116], thus overcoming the sampling rate problem. However, this is not the best solution. With the introduction of computer control and some modification of the conventional GC apparatus, the speed of GC analysis makes on-line evolved gas analysis possible.

Modern high-speed GC systems are able to separate some light hydrocarbons within a second [117, 118]. This approach requires special sampling valves, narrow-bore columns (diameter = 0.005 cm), and detectors with a fast response. However, in EGA these fast separations are not necessary. Taking samples on-line from the reactor and separating them within 1 minute is frequently satisfactory in order to better understand degradation kinetics. The apparatus for this approach can be constructed of commercially available parts.

Denq and co-workers [119] studied the evolved gases and the residual protective char yield in the thermal degradation of epoxy resins blended with propyl ester phosphazines by different methods, e.g., TGA, FT-IR analysis, pyrolysis, GC, and MS. The results showed that thermal degradation of the epoxy resin and its blends was a single-stage process, that the chief degradation temperature of the blends was below that for pure epoxy resins, that the residual char yield of epoxy blends at 550 °C was higher than for pure epoxy resins, that propyl ester phosphazine accelerated the production of small molecules such as acetone, phenol, and isopropyl phenol during thermal degradation, and that honeycomb structure compounds were formed in the residual char of the epoxy resin and its blends. Elemental analysis of the residual char indicated the presence of phosphorus, while other elements were mostly diffused into the evolved gases.

Kullik and co-workers [120] have described an evolved gas analyser based on thermogravimetry and GC in which the sample under study is heated in the reactor. An inert gas flows through the reactor and carries the evolved products through one path of a sampling valve into the air. The pure carrier gas flows through the other path of the sampling valve into the GC column. By a command from the computer the valve reverses for a short time and the sample flows into the column. The detector signal is recorded digitally and stored on a floppy disc for further use. For experimental control a computer was used.

The Apple IIe has several slots where application-dependent interface cards can be inserted. Also, it is relatively easy to programme the Apple IIe to control an experiment using a 6502 uP (microprocessor) code and Applesoft Basic programming language. Two built-in interface cards are used – one for communication between the Apple IIe and an HP 7225 A plotter and the other for interfacing the Apple IIe and a standard crate which contains a 24-bit analogue-to-digital converter (ADC) (Institute of Cybernetics, Estonian Academy of Sciences). The ADC records the GC detector signal. The analogue signal counting frequency was 4 points per second. Two 1-bit outputs of the Apple IIe are used for controlling the solenoid valve status. The solenoid valve controls the Deans-type flow switch located in the gas chromatographic oven. This sampling system is capable of performing injections of gas samples with a precision of 0.3% and duration as short as 0.1 second.

The sample is in a quartz test tube in the flow-through reactor. The rector temperature is programmable linearly from 50 to 500 °C at a speed of up to 20 °C/min. For reactor temperature control a separate temperature programme is used. The amount of evolved gas depends on the sample weight which varies in the range 1–100 mg.

The evolved products are separated in two short, open tube columns: a 10 m × 0.125 mm column coated with OV-43 and a 45 m × 0.5 mm column coated with SE-30. The apparatus is built on the basis of a Fractovac 200 GC (Carlo Erba Instrument 10).

Reaction temperature and chromatogram running time information are obtained from the apparatus. Therefore it is possible to produce a thermogravimetogram. There are three plotting forms possible.

1. Chromatograms are produced in a sequence as they appear on the chart recorder. As the chromatograms are stored on a floppy disc, post-processing is possible. Chromatograms can be plotted with a different attenuation and the time axes can be expanded or contracted to obtain a better knowledge about a particular feature of a chromatogram. Also, the presentation facilitates quantification and obtaining kinetic data (see **Figure 9.4(a)**).

2. A thermochromatogram can be considered as a surface. The isometric projection of this surface is called a stack plot. In this plot chromatograms are plotted one above the other and the upper chromatogram is shifted relative to the lower one to form a

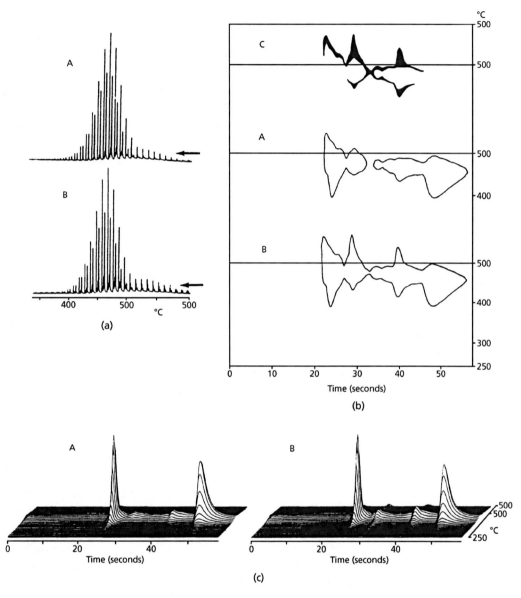

Figure 9.4 Plotting of thermograms. (a) Sequences of rubber thermogram slices. The arrows mark cutting levels for (A) *cis*-1,4-polyisoprene and (B) *cis*-1,4-polyisoprene with 5% *cis*-3,4-polyisoprene. (b) Stack plots of rubber thermochromatograms. (c) Contour plots of rubber thermochromatograms. Contour pattern derived by subtracting B from A. Sample heating rate 10 °C/min; GC column SE-30; column temperature 150 °C

(Reprinted from E. Kullik, M. Kalyurand and M. Lamburg, with permission from Laboratory Practice, 1987, Jan/Feb. p.73 [120])

323

plotting angle. To control the 'up' and 'down' portions of the plotter pen, a reference function is formed. If a particular chromatogram point value exceeds this function value at this point, then the plotter pen gets the command 'down' and the reference function gets a new value that equals the value of the chromatogram point. After that the plotter moves towards this point, drawing a line on the chart. If the chromatogram point value is lower than that of the reference function value, then the pen gets the command 'up', the reference function value remains unchanged, and the line will not be drawn on the plotter chart – this simple rule enables the hidden line problem to be solved very easily. The stack plot gives a very good qualitative survey of the process occurring during sample heating, but any kind of measurement is difficult from this plot. If the thermochromatogram is considered to be a surface, the above sequence plot of chromatograms can be regarded as a set of thermochromatogram slices made by planes parallel to the plane formed by the time and evolving rate axes. The stack plots are shown in **Figure 9.4(b)**.

3. The third plot frequently used in two-dimensional data presentation is the contour plot. The contour plot is obtained by cutting the thermochromatogram surface by a plane that is parallel to the plane formed by the temperature and time axes, and by plotting the cutting line on the temperature–time plane. Between two adjacent chromatograms a bilinear surface for approximation of the thermochromatogram is used. The contour plot is very useful for comparing retention times of different thermochromatograms. An example of the contour plot is shown in **Figure 9.4(c)**.

To identify the separated evolved products emerging via the GC, Kullik and co-workers [121] either linked the apparatus to a MS or collected the evolved compounds in traps for subsequent MS analysis.

They applied the procedure to thermoresist rubbers. **Figure 9.4(a)** shows thermochromatograms and **Figure 9.4(b)** the 'stack plot' for two rubber samples. Thermochromatogram A in **Figure 9.4(a)** represents synthetic *cis*-1,4-polyisoprene while thermochromatogram B represents *cis*-1,4-polyisoprene with 5% *cis*-3,4-polyisoprene. Differences are evident, and at first glance a conclusion can be drawn that rubber B is more thermostable than A, whose degradation begins at lower temperatures and continues at higher temperature.

From an analytical point of view, thermochromatography should be a useful tool for differentiating samples for manufacture control. Differences between two similar thermochromatograms can be made clearly evident by using the contour plot (**Figure 9.4(c)**). In this figure A and B are contour plots for 1,4-*cis*-polyisoprene and 1,4-*cis*-polyisoprene containing 5% *cis*-3,4-isomer. C shows a contour pattern that characterises the deflection between the two samples, i.e., C = (B – A).

In **Figure 9.5(a)**, A is a stack plot for butadiene–styrene rubber and in **Figure 9.5(c)** B is a thermochromatogram for a butadiene–α-methylstyrene polymer which, as can be seen, degrades in a more complicated way than the butadiene–styrene rubber.

The effect of polymer ageing is reflected in thermochromatograms. **Figures 9.5(b) and (c)** depict TGA stack plots of the same PE, with the only difference being in the ageing of one polymer for a month at 100 °C. The changes due to ageing are marked by the arrows. This points to the conclusion that continuous heating of PE makes it more unstable. In **Figures 9.5(b) and (c)** the advantage of computerised data handling is seen. In both figures, the thermochromatograms are presented from a different point of view. This technique allows one to discover the features from the hidden areas and obtain more valuable information about degradation reactions.

Thermochromatography is useful in studying flame-resistant materials, e.g., cellulose fibres [122]. Phosphorus-containing antipyrenes increase the amount of water present in the degradation products, which, in turn, increases flame resistance of a polymer. The amount and composition of the burning products should also be taken into account in characterising the flame resistance of cellulose fibres [123]. In the thermochromatograms

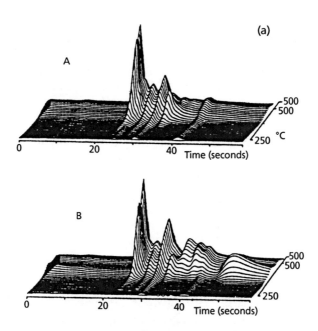

Figure 9.5 (a) Stack plots of the butadiene styrene rubber thermochromatograms.
A - poly butadiene styrene copolymer; B - polybutadiene α-methyl styrene copolymer.
Sample heating rate 10 °C/min; GC column SE-30; column temperature 150 °C.

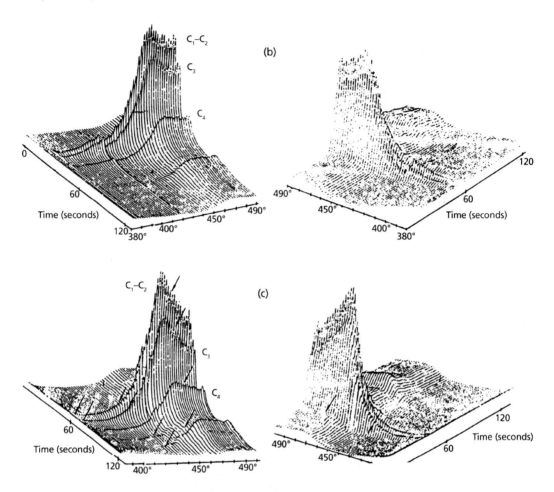

Figure 9.5 (*Continued*) Effects of ageing polyethylene: (b) Stack plot of unaged crosslinked low-density polyethylene; (c) Stack plot of crosslinked low-density polyethylene after ageing for 1 month at 100 °C. C_1, C_2, C_3, C_4 possible hydrocarbons, sample heating rate 3 °C/min, GC column SE30, column temperature 60 °C (*Reprinted from E. Kullik, M. Kalyurand and M. Lamburg, with permission from Laboratory Practice, 1987, Jan/Feb. p.73 [120]*)

shown in **Figure 9.6**, two fibres containing phosphorus antipyrenes are presented. One sample is cellulose fibre crosslinked with phosphonitrilamide (sample B); the second sample is a graft copolymer of cellulose with polymethylvinylpyridine phosphate (sample A). Both samples have the same oxygen index (~40) indicating similar flame resistance. However, as follows from **Figure 9.6**, the amount of the evolved gaseous burning product as a function of temperature is quite different for these fibres. Although identification of

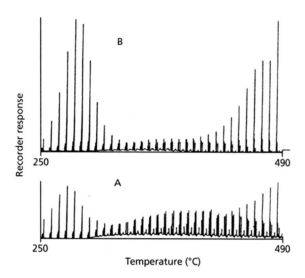

Figure 9.6 Sequence of slices of thermochromatograms. A - Cellulose pyridine phoshate; B - Containing phosphonitrilamide. Sample heating rate 7 °C/min, GC column OV-43, column temperature 60 °C
(*Reprinted from E. Heinsoo, A. Kogerman, O. Kirret, J. Coupek and S. Vilkova, Analytical Applied Pyrolysis, 1980, 2, 131, with permission from Elsevier [123]*)

the degradation products is complicated in this case, the components evolved are probably light hydrocarbons and oxygen compounds because a low gas chromatographic column temperature (60 °C) was used in separation.

Thermochromatography was used to compare the usefulness of different catalysts in the polymerisation processes. **Figure 9.7** shows thermochromatograms of the degradation products of urea melamine formaldehyde resins cured with equal amounts of $AlCl_3$, NH_4Cl, and $FeCl_3$ as catalysts. The thermal destruction of the polymers gives two main products: formaldehyde and methanol.

A combination of TGA and IR spectroscopy or FT-IR spectroscopy has also been used to carry out EGA in polymer and polymer additive degradation studies. The Stanton Redcroft apparatus for combining thermogravimetric apparatus with FT-IR spectroscopy is described in Section 9.3.2. This equipment combines the following features:

• TGA and FT-IR spectroscopy data stored on one computer using the same time basis.

• Data storage sufficient for long data collections, e.g., over 4 hours with a 5 second resolution.

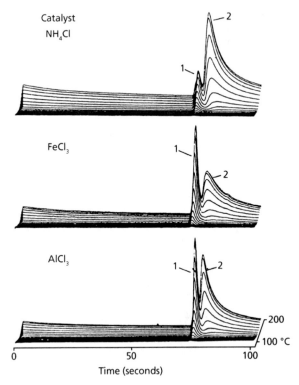

Figure 9.7 Thermochromatograms of the urea formaldehyde resins. 1 - formaldehyde; 2 - methanol. Sample heating rate 5 °C/min, GC column SE-30, column temperature 130 °C (*Reprinted from E. Heinsoo, A. Kogerman, O. Kirret, J. Coupek and S. Vilkova, Analytical Applied Pyrolysis, 1980, 2, 131, with permission from Elsevier [123]*)

- Spectral resolution of at least 2 cm⁻¹ to handle gaseous products often of low molecular weight.

- High sensitivity for trace component detection and identification.

- Ability to plot gas evolution profiles.

This equipment has, to date, been used for thermal degradation studies on PTFE and PVC, and filler, monomer, and plasticiser analyses in polymers. It is worth noting that while quantification from FT-IR spectroscopy results is feasible, there could be a discrepancy between the calculated weight loss of evolved gases and that found by TGA. This discrepancy can be accounted for by the evolution of a gas such as nitrogen which FT-IR spectroscopy is unable to detect.

There is another approach to the use of thermal degradation. In EGA the production of organic volatiles is monitored by continuously weighing the sample and identifying and determining the volatiles produced. In the alternative approach the sample is heated under controlled conditions in a sealed ampoule and the degradation products in the residue are identified and determined. This approach has been applied in studies of the thermal degradation of 0.5% of the phenolic antioxidant Santonox R in low-density polyethylene. Preliminary TGA on neat Santonox R indicates that it degrades at temperatures as low as 160 °C in air and degrades and/or volatilises faster at higher temperatures. In the absence of air, degradation and volatilisation occur at 250 °C and probably to some extent at lower temperatures. To establish the fate of 0.5% Santonox R which has been milled at temperatures between 160 and 250 °C into low-density polyethylene, the polymer was extracted with diethyl ether and examined by thin-layer chromatography followed by IR spectroscopy of the compounds separated on the thin-layer plate. In this procedure 1 ml of a 1% diethyl ether solution of the PE extract was applied along the side of a 20 cm × 20 cm plate coated with a layer of GR254 fluorescent silica gel. The plate was developed under a 254 nm ultraviolet (UV) light source. This revealed the presence of bands containing UV-absorbing substances at R_f = 0.0, 0.15, 0.25, and 0.30 (these four bands are also revealed by the 2:6 dibromophenoquinone-4-chlorimine detection reagent). The edge of the plate was treated with 20% sulfuric acid and heated to 160 °C and this revealed the presence of two further bands at R_f = 0.75 and 1.0.

The silica gel corresponding to each of the six observed bands was then separately scraped off from the plate and reheated with absolute ethyl alcohol to desorb the organic matter. The six extracts thus obtained were then prepared into potassium bromide micro-discs and examined by IR spectroscopy in the 2.5–15 μm range. The results suggested that upon milling into PE at 260 °C, Santonox R degrades into at least two products, i.e., a phenolic breakdown product (band 2) and non-phenolic hydrocarbon (band 4). The phenolic breakdown product (band 2) has some resemblance to the material produced upon heating Santonox R in air for 30 minutes at 250 °C.

Excess ferric ions react with Santonox R but not with the oxidation products to produce ferrous ions which can then be determined by the spectrophotometric 2,2′-dipyridyl method, giving an estimate of the unoxidised Santonox R content of the extract:

$$\text{Santonox (reduced)} + Fe^{3+} = \text{Santonox (oxidised)} + Fe^{2+}$$

Obviously, any oxidised Santonox in the polymer extract would not be included in this estimation. Now it is likely that some or all of the Santonox degradation products in PE exist in the oxidised form which would not be included in the result obtained by the dipyridyl method. Therefore, lower analytical results should be obtained for polymers in which the Santonox is more degraded. The method was applied to the PE in which 0.5%

Santonox R has been incorporated by milling at various temperatures. The results obtained showed that perceptible Santonox degradation occurred in the case of the polymer milled at 260 °C compared to polymers milled at lower temperatures.

Other polymers that have been examined by EGA include PVC [124–126], PS [127–130], styrene–acrylonitrile copolymers [131-133], PE and PP [134–139], polyacrylates and copolymers [140–143], and PET, polyphenylenes, and polyphenylene oxides and sulfides [143–148]. Studies involving the use of chromatography include the thermal degradation of PVC [149], vinyl plastics [150], and polysulfone [151].

9.4.7 Mass Spectrometric Methods

MS has been used as a means of obtaining accurate information regarding breakdown products produced upon pyrolysis of polymers. This includes applications to PS [152, 153], PVC [154], polyethers [155], PVC–polycarboxy piperadine polyurethanes [156], phenolics [157], PTFE [158], polybenzimidazole epoxies [159], ethylene–vinyl acetate copolymers [160], ethylene–vinyl alcohol copolymers [161], polybenzoxazines [162], polyxylyene sulfides [163], trimethoxysiloxy-substituted polyoxadisilpentanylenes [164], chlorinated natural rubber [165], and polyacrylonitrile [166].

Mass spectrometric thermal analysis, the determination of total ion current as a function of time and temperature in combination with DTA [167], has proved to be very useful for distinguishing between energy changes caused by phase transformations and those caused by decomposition reactions. Determination of total ion current alone is not particularly informative in most cases of polymer degradation, when several products may be formed simultaneously or sequentially. Shulman and co-workers [158] studied this by repeatedly scanning spectra as the temperature was raised on a linear programme, then plotting peak height as a function of temperature. Mass spectrometric thermal analysis determinations conducted in this way proved helpful in several polymer investigations. They present a preliminary discussion of this technique, including several suggested applications and new methods of data treatment which permit determination of kinetic parameters from mass spectrometric thermal analysis.

For qualitative analysis of polymers, it is not necessary to know the product of the reaction, since identification can be based on temperatures and relative heights at the maximum of several of the more prominent fragments once these have been established for known materials. Additional information about the degradation chemistry can be secured if one chooses peaks characteristic of specific products.

Since a Knudsen cell inlet system and MS are being evacuated continuously, the pressure in the vicinity of the ionising filament is dependent on the rate of production of the effusing gases. Peak height is therefore proportional to the rate of production of a given species. For competitive first-order reactions, the amount of each product is proportional to its rate of production, so that the composition of the product mixture at a particular temperature may be derived from a single spectrum. Also, since peak height is proportional to the rate at the test temperature, a semi-logarithmic graph of the peak height *versus* reciprocal absolute temperature (Arrhenius plot) will give a straight line with a slope of E_a/R (E_a = activation energy) over a range in which the amount of polymer (or of a functional group) does not change significantly. At heating rates of 15–50 °C/min, it is possible to obtain linear plots over a 50–200 °C temperature increment, the increment varying inversely as the activation energy. In this way, activation energies for reactions leading to each product can be obtained easily. A composite activation energy comparable to that derived from TGA or isothermal kinetic studies can be obtained by taking a weighted average of the individual activation energies. Furthermore, from areas under rate of formation *versus* time graphs, the relative yield of each product can be determined. Unfortunately, identification of products of polymer degradation from the mass spectrum of an unfractionated effluent cannot be considered reliable unless confirmation is secured using other analytical methods, such as GC.

Polymer samples were placed in a tungsten crucible in a Knudsen cell inlet system of a time-of-flight MS. The system was evacuated. Samples were then heated at a linear rate. Spectra were determined at 1 minute intervals. The thermocouple was welded to a support rod 3 mm below the crucible and had been calibrated against a thermocouple in the bottom of the crucible at each heating rate. Details are summarised in **Tables 9.4 and 9.5** for three polymers.

Table 9.4 Experimental details of mass spectrometric thermal analysis			
Polymer	Form	Weight, mg	Heating rate °C/min
Phenolic	Powder	14.9	29
Polytetrafluoroethylene (a), (b)	Chips	12.7	14.5
Polybenzimidazole (c)	Powder	15.5	20

(a) DuPont Teflon

(b) E_a = 93 kcal/mole for C_2F_4 formation

(c) Poly-2,2'-m-phenylene)-5,5'-bibenzimidazole (cured under 1.4 MPa of nitrogen to 400 °C)

Reprinted from G.P. Shulman, Journal of Polymer Science: Part B, Polymer Letters, 1965, 3, 911 [158], with permission from Wiley.

Table 9.5 Products of poly-2,2'-(*m*-phenylene)-5,5'-bibenzimidazole pyrolysis		
Product	Yield, % of volatiles	Activation energy, kcal/mole
H_2	13	38
CH_4	1.5	–
NH_3	13	54
H_2O	38	22
HCN	30	48
CO	4	31
CO_2	0.3	–
C_6H_5OH	0.9	–
Reprinted from G.P. Shulman, Journal of Polymer Science: Part B, Polymer Letters, 1965, 3, 911 [158], with permission from Wiley.		

An Arrhenius plot for polybenzimidazole shows that the average activation energy (E_a) is temperature dependent, varying from 26 to 40 kcal/mol, compared to a value of 44 ± 11 kcal/mol derived from isothermal kinetics. Similar data for the phenolic resin [168] give a range of 16–48 kcal/mol.

Risby and Yergey [154] have described the design of a temperature programmable probe computer and present mechanisms of thermal desorption decomposition energies and pre-exponential factors in the linear programmed thermal degradation mass spectrometry (time resolved pyrolysis) of PS and PVC. Their results show a stepwise thermal degradation process. Linear programmed thermal degradation–mass spectrometry is based on a collection of sequential mass spectra during the programmed heating of the sample. The data, with temporal dependence, allow detection of subtle differences in structure which would not be apparent with data from PGC or pyrolysis–mass spectrometry (PMS). In addition, activation energies and pre-exponential factors for the decomposition processes can be obtained from linear programmed and thermal degradation–mass spectrometry by changing the rate of sample heating; such results cannot be obtained by PGC or PMS.

Figures 9.8(a) and (b) show mass spectra as a function of temperature for the thermal degradations of uncrosslinked PS (M = 1,000,000) and 12% divinyl benzene crosslinked PS. The difference between the samples is clearly shown, as the heavily crosslinked PS starts to degrade at a lower temperature and the degradation occurs over a wider temperature range. These differences appear even more dramatic in **Figures 9.8(c) and (d)** which show the profiles for specific ions (105, 117, 209, and 221 *m/z*) as a function of scan number (sample temperatures). In **Figure 9.8(c)** more than one kinetic process for the evolution of the monomer fragment (105 *m/z*) is indicated since the profile is clearly broader than the profiles for the specific ions at 117, 209, and 221 *m/z* – these differences in evolution profiles may perhaps be explained by different types of microstructures within the uncrosslinked PS which can lose a styrene fragment. The activation energies for the removal of the monomer from these microstructures would be expected to differ. The hypothesis of different microstructures within the PS is supported by further evidence.

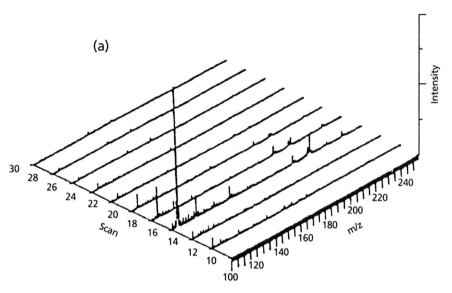

Figure 9.8 (a) Three-dimensional plot for thermal degradation at 16 °C/s of uncrosslinked polystyrene (M, 100,000) shown as *m/z versus* intensity *versus* scan number (10 = 73 °C, 30 = 563 °C). (b) Three-dimensional plot for thermal degradation at 15.5 °C/s of polystyrene - 12% divinylbenzene, shown as *m/z versus* intensity *versus* scan number (10 = 97 °C, 30 = 564 °C). (c) Specific non profiles for thermal degradation at 16.0 °C/s of uncrosslinked polystyrene (M, 100,000) shown as intensity *versus* scan number: (A) Mass 105 *m/z*, temperature of evolution maximum 423 °C; (B) Mass 117 *m/z*, temperature of evolution maximum 423 °C; (C) Mass 209 *m/z*, temperature of evolution maximum 423 °C; (D) Mass 221 *m/z*, temperature of evolution maximum 423 °C. (d) Specific ion particles for thermal degradation at 15.5 °C/s of polystyrene - 12% divinyl benzene shown as *versus* scan number: (A) Mass 105 *m/z*, temperature of evolution maximum 423 °C; (B) Mass 117 *m/z*, temperature of evolution maximum 423 °C; (C) Mass 209 *m/z*, temperature of evolution maximum 423 °C; (D) Mass 221 *m/z*, temperature of evolution maximum 423 °C
(Reprinted from T.H. Risby, J.A. Yergey and J.J. Scocca, Analytical Chemistry, 1982, 54, 2228, with permission from the American Chemical Society [154])

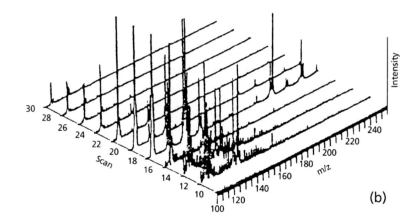

(b)

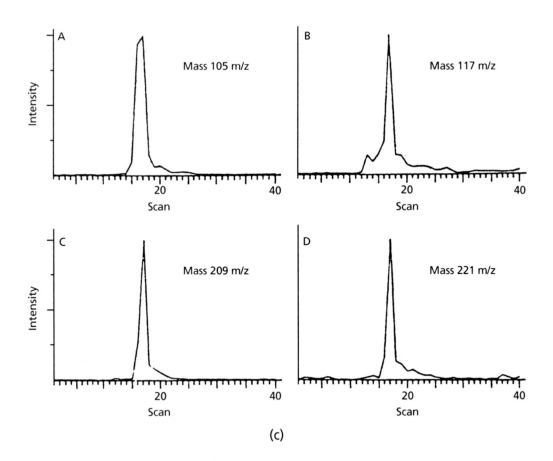

(c)

Figure 9.8 (*Continued*)

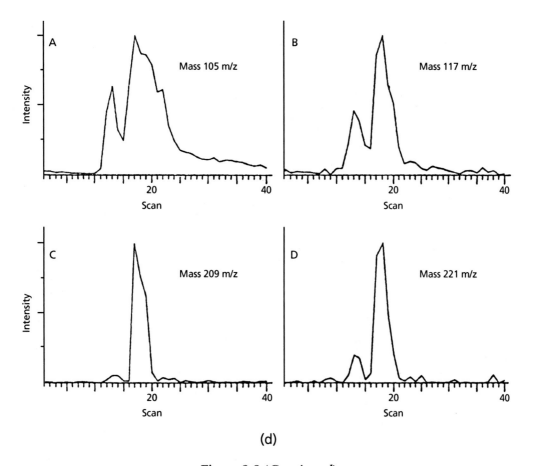

(d)

Figure 9.8 (*Continued*)

The evolution profiles for larger fragments are sharper, suggesting less variety in kinetic processes of evolution, i.e., less variance in microstructure. In **Figure 9.8(d)** (crosslinked PS) profiles for all the fragments are broad suggesting that each ion may be produced by a variety of kinetic processes. Crosslinking would be expected to produce different types of microstructures. (However, differences could be explained by the fragments having arisen from either styrene or divinylbenzene). PS with a lower degree of crosslinking has a more random structure and would be expected to have many different microstructures. This is indeed supported by the complex profiles obtained for this polymer (**Figure 9.8**).

The temperatures that correspond to the maxima of the evolution profiles are also shown in **Figures 9.8(c) and (d)**. There are temperature maxima variations for the evolution of different fragments from the sample and for the same fragment from different samples. Also crosslinked polymers begin pyrolysing at temperatures 200 °C lower than the maximum evolution temperature. This is not observed with the uncrosslinked polymers.

Since it was impossible to separate and identify the kinetic processes corresponding to the loss of specific fragments from the sample of crosslinked PS, and difficult to ensure that the same chemical species were involved, Risby and Yergey [154] did not attempt to obtain values for the activation energies by varying the rate of sample heating. In addition, evolution profiles for specific fragments from uncrosslinked PS samples were complex, and values for the activation energies given in **Table 9.6** are based on 235 *m/z*. There is good agreement between the values for the activation energies and the pre-exponential factors for the two uncrosslinked PS, as would be expected since they differ only in molecular weight. Also, there is reasonable agreement of the value of the activation energy with a literature value of 48.04 kcal obtained with a much larger sample size.

The major fragment ions observed for the PS samples are listed in **Table 9.7**. These fragments have been observed by other workers using PGC and/or MS and using chemical ionisation mass spectrometry [169].

Udseth and Friedman [170] carried out a MS study of PS with a molecular weight of 2,100. The PS was evaporated from a probe filament heated at 1000 °C/s under both electron impact and methane and argon chemical ionisation conditions. Smooth rhenium ribbon surface direct insertion probes were used with both the electron impact and

Table 9.6 Temperature data, activation energies and pre-exponential factors for un-crosslinked polystyrene			
M_r 390,000		M_r 100,000	
Ramp rate, °C/s	Temp of maxima, °C	Ramp rate, °C/s	Temp of maxima, °C
15.9	390	16.0	388
7.79	385	7.77	386
3.92	363	3.87	365
1.97	352	1.99	349
0.99	345	0.99	343
Activation energy	42 kcal	40 kcal	
Pre-exponential factor	3.6×10^{13}/s	6.1×10^{12}/s	
M_r: Relative molecular mass Reprinted from T.H. Risby, J.A. Yergey and J.J. Scocca, *Analytical Chemistry*, 1982, 54, 2228, with permission from the American Chemical Society [154]			

Table 9.7 Polystyrene pyrolysis products		
Basic structure of fragment	Mass of ion with max intensity in the region of mol wt	% total ion current at max evolution
$HC\!\!=\!\!CH_2$ with Ph	104	4.0
$HC\!\!=\!\!CHCH$ with Ph	117	4.7
$HC\!\!=\!\!CHCHCH$ with Ph	130	1.8
$HC\!\!=\!\!CHCH_2CH\!\!=\!\!CH_2$ with Ph	144	2.1
$CHCH_2CH$ with Ph / Ph	194	1.9
CH_2CHCH_2CH with Ph Ph	208	9.7
$CH_2CHCHCHCH$ with Ph Ph	222	2.4
$HC\!\!=\!\!CHCH_2CH\!\!=\!\!CHCH_2$ with Ph Ph	234	6.0

*Uncrosslinked Mr 390,000
Reprinted from T.H. Risby, J.A. Yergey and J.J. Scocca, Analytical Chemistry, 1982, 54, 2228, with permission from the American Chemical Society [154]*

chemical ionisation sources. Under electron impact conditions extensive fragmentation and depolymerisation were observed but oligomers up to (C_8H_8) were detected. Under chemical ionisation conditions oligomers up to (C_8H_8)$_{27}$ were detected and spectra that approximately reproduced the oligomers' distribution were obtained. These workers also measured quantitatively temperature dependencies of rates of desorption and activation energies of evaporation rates for many of the ionic species.

Evaporation studies were carried out for both a chemical ionisation and an electron impact ion source and in each case the products were extracted and analysed by a computer-controlled quadrupole MS.

Figure 9.9 shows electron impact mass spectra. Numbers above selected ions indicate the temperature at which counting rates of 550 counts/ms were observed. The numbers inside the parentheses give the number of the styrene monomer units and the respective masses of the ion. Temperatures of desorption were arbitrarily defined in terms of 550 counts/ms which corresponds to approximately two-thirds of the maximum intensity of the relatively low-abundance ions in **Figure 9.9(a)** curve II. The electron impact data in **Figure 9.9(a)** are divided into two molecular weight regions which overlap at the trimer ion at $m/z = 370$. The solid lines in the figure are used to indicate ions that consist of styrene monomer units and the butyl initiator, with masses given by the relationship $m = 104n + 58$. These ions are designated as the A oligomer sequence. Ions with values of n ranging from 1 to 11 are found in the electron impact mass spectrum. The A sequence constitutes the only ions observed above $m/z = 780$. The dashed lines represent ions with masses that do not fit the A sequence, produced by thermal surface reactions and/or electron impact decompositions. Sequential mass distributions can also be found in the dashed line spectra. For example, there is a set of ions with masses 291, 395, 499, 603, and 707, which differ by the mass of a styrene monomer unit.

Temperatures of desorption shown in **Table 9.8** show a gradual increase for A oligomer ions with increasing molecular weight. The observation of lower molecular weight ions at lower temperatures and during earlier stages in the desorption is inconsistent with the hypothesis that these ions are electron impact fragmentation products. Temperatures of desorption establish the lower molecular weight ions as primarily products of thermal depolymerisation reactions. Low molecular weight fragment ions not part of the A sequence with masses at 325 and 351 are included in **Table 9.8**.

These ions, indicated by dashed lines in **Figure 9.9(a)** appear late in the desorption process and at high enough temperatures to be either products of pyrolysis or electron impact decompositions of the highest molecular weight A sequence oligomers detected. With the exception of these fragment ions, the correlation of increasing mass and temperature of desorption of the A sequence is clearly shown in **Table 9.8**.

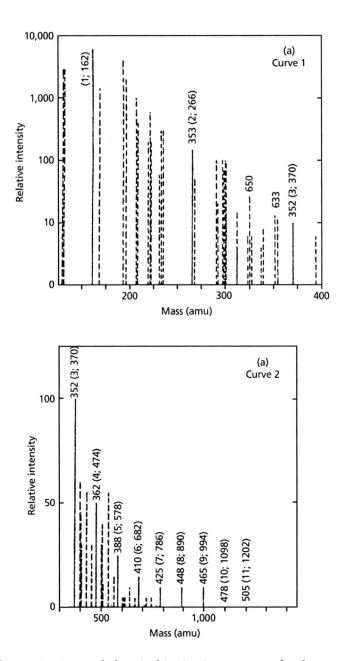

Figure 9.9 Electronisation and chemical ionisation spectra of polystyrene. (a) Partial EI spectrum of polystyrene. The solid lines are the trimer spectrum and the numbers in parenthesis give first the number of monomer units present and second the assigned monoisotropic mass. The numbers not in parethesis give the evaporation temperature in kelvin. The dashed lines are fragment sequence peaks.

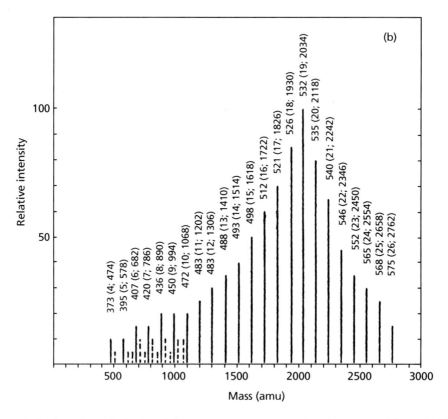

Figure 9.9 (*Continued*) (b) Partial argon CI spectrum of polystyrene. The solid lines are the monomer spectrum and the dashed lines are intermediate fragment peaks. The format for the numbers in the sample are as those used in Figure 9.9(a)

The second column in **Table 9.8** gives activation energies for the rate-limiting processes responsible for the generation of the respective ions in the electron impact spectrum. These activation energies are calculated from slopes of linear plots of the logarithm of relative ion intensity *versus* reciprocal absolute temperature.

Activation energies in **Table 9.8** for A sequence oligomers are higher for the $n = 2$ and $n = 3$ oligomers, relatively constant for $n = 4$–8, and lowest for $n = 9$–11. If the respective rate-limiting processes were desorption in this homologous series of oligomers, a gradual increase in activation energy with molecular weight would be expected. If oligomer ions were formed exclusively by electron impact decomposition of the higher molecular weight polymers, then both activation energies and temperatures of desorption of the respective ions would be expected to be almost identical. The dimer and trimer ions have virtually the

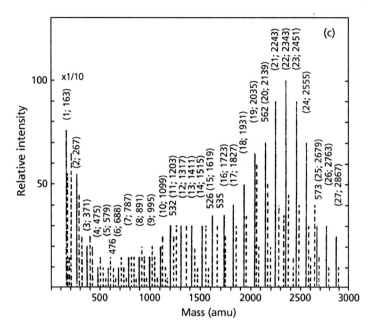

Figure 9.9 (*Continued*) (c) Methane CI spectrum of polystyrene. The solid lines are the number of monomer sequence and the dashed lines are the solvation or fragment peaks. The format of the numbers is the same as used in Figure 9.9(a)

(*Reprinted from H.R. Udseth and N. Friedman, Analytical Chemistry, 1981, 53, 29, with permission from American Chemical Society [170]*)

same temperatures of desorption, but the 10 kcal/mol difference in activation energy supports the conclusion that these ions are formed in surface thermal depolymerisation reactions at different rates. Differences in temperature of desorption for oligomer ions with n = 3–8 support the argument that, in spite of similar activation energies, these oligomers are not for the most part produced from the same parent species by electron impact dissociation.

Sample plots of log(relative intensity) *versus* $1/T$ (T = absolute temperature) are given in **Figure 9.10**. A complex mechanism for the production of neutral species which are parents of the trimer ions is shown by the segmented plot of the temperature dependence of the relative intensity of the trimer ion. There is at lower temperatures a slope that gives a 41 kcal/mol activation energy and a higher temperature process that gives an activation energy lower than 27 kcal/mol or lower than that of the pentamer ion.

Chemical ionisation mass spectra obtained with argon and methane reagent gases are presented in **Figures 9.9(b) and (c)**, respectively. These spectra were obtained to

Table 9.8 Evaporation temperatures and activation energies for some of the peaks in the EI spectrum		
Mass, amu	Activation energy, kcal/mol*	Temperature, °C
266	51	80
325	17	377
351	17	360
370	41	79
474	36	89
578	27	115
682	26	137
786	27	152
890	29	175
994	21	192
1098	18	205
1202	18	232

*Estimated uncertainty ± 3 kcal/mol
Reprinted from H.R. Udseth and N. Friedman, Analytical Chemistry, 1981, 53, 29, with permission from American Chemical Society [170]

determine the effect of the contact of a weakly ionised plasma with heated sample surface on the competitive depolymerisation and desorption processes. The possibility of chemical interaction of neutral reagent gases with the sample is, in the case of methane, remote. With argon, neutral molecule–surface chemical interactions can be eliminated from consideration. The methane chemical ionisation spectrum was scanned from the mass of the monomer in the A sequence of ions up to mass 3000. With argon the lower molecular weight region was not investigated. The lowest oligomer shown in the argon chemical ionisation spectrum is the tetramer, m/z = 473. The chemical ionisation spectra were taken with relatively low resolution of ±4 mass units, in the mass range between 500 and 3000 amu. Resolution limitations are reflected by presentation of data as monoisotopic mass spectra. For ions containing of the order of 100 or more carbon atoms with approximately 1.1% ^{13}C, ions having one or more ^{13}C atoms would be the most abundant species in a polyisotopic spectrum. Higher molecular weight ions were not sufficiently well characterised with respect to mass, to establish the A sequence ions in the methane chemical ionisation as protonated species. Nor were the solvated A sequence ions, indicated by the dashed lines in the higher molecular weight region of the methane chemical ionisation spectrum identified precisely. The addition of a two- or three-carbon,

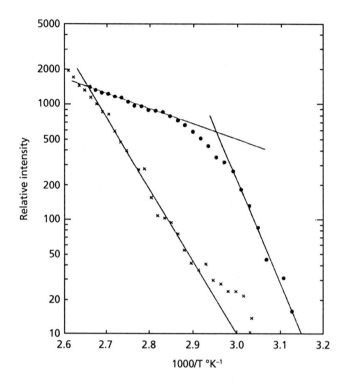

Figure 9.10 Evaporation plots from rapidly heated polystyrene obtained under EI conditions. x are the plot of the pentamer and • a plot of evaporation of the trimer (*Reprinted from H.R. Udseth and N. Friedman, Analytical Chemistry, 1981, 53, 29, with permission from American Chemical Society [170]*)

hydrocarbon, fragment was established. The graphical data in **Figure 9.10** accurately reflect relative intensities of oligomer and fragment ion peaks but are schematic in that they are presented as monoisotopic.

The methane and argon chemical ionisation spectra both show higher molecular weight species not seen under similar conditions of sample heating with the electron impact source. Some small differences may be noted but there is no conclusive evidence for volatility enhancement in which the argon chemical ionisation plasma in contact with the sample gives rise to lower temperature desorption. Temperatures of desorption for selected ions in the methane chemical ionisation spectrum are slightly higher than those found for electron impact or argon chemical ionisation. In both chemical ionisation spectra the correlation of increasing temperature of desorption with increasing ion molecular weight for the A sequence oligomer ions is maintained. Here again the lower molecular weight ions appear earlier in the sample process and cannot be taken as products of exothermic gas-phase ionic decomposition reactions. Such reactions might be expected with argon because of the relatively large differences between the recombination energy of argon ions and estimated ionisation potential of styrene polymers (approximately 4 or 5 eV).

In view of this potential exothermicity, the argon chemical ionisation spectrum is unusually clean and free of fragmentation products. These results may be explained by a very efficient collisional de-excitation of oligomer ions by neutral argon in the chemical ionisation source or by de-excitation of products of heterogeneous ion-molecule reactions on the surface of the sample probe. The only evidence in the mass spectrum for a somewhat more gentle ionisation with methane than argon is the shift to lower molecular weight in the maximum of the argon spectrum. With argon chemical ionisation, the $n = 19$ oligomer is the most abundant high-molecular-weight ion. In the methane chemical ionisation spectrum the most abundant ion in this mass region has 22 styrene monomer units.

In contrast to the values found in the electron impact mass spectrum, activation energies for the chemical ionisation spectra tend to increase with increasing molecular weights. Unfortunately, comparisons are limited because many of the ions detected in the chemical ionisation spectra are not seen under electron impact conditions. With argon chemical ionisation one can compare the n-mers with $n = 4-8$ and see a systematic 6–9 kcal increase in activation energy with argon chemical ionisation. However, with higher molecular weight oligomers with $n = 9-11$, the argon activation energies are 16–32 kcal/mol larger than those observed with the same mass ions under electron impact ionisation conditions. Temperatures of desorption and activation energies of PS are presented for chemical ionisation experiments using oxygen and methane in **Tables 9.9 and 9.10**.

Two sets of activation energies are given for the argon chemical ionisation studies, because with the higher molecular weight ions, plots of log intensity *versus* $1/T$ gave segmented lines which were resolvable into low- and high-temperature processes. This phenomenon was not observed in the methane chemical ionisation studies but was reproducible in the argon experiments. Comparison of the data on selected A sequence oligomers shows differences in activation energy of rate-limiting processes leading to desorption in the argon and methane chemical ionisation spectra. For example, for the $n = 5$ ion, the activation energies are 62 kcal/mol with argon chemical ionisation and 44 kcal/mol with methane chemical ionisation. No difference was found for the activation energy of the $n = 20$ oligomers in the methane spectra and during the high-temperature part of the argon spectra. An 18 kcal/mol difference is noted for the $n = 11$ oligomer ion at mass 1202 in the argon spectrum and the assigned mass of 1203 in the methane chemical ionisation spectrum.

These data all indicate a significant interaction of reagent ions with molecules on the surface of the sample probe during the sample heating process. There is evidence

Mass, amu	Activation energy, kcal/mol*		Temperature, °C
	Initial	Final	
474		32	100
578		32	122
682		34	134
786		36	147
890		35	163
994		37	177
1098		45	199
1202		50	210
1306		49	210
1410		49	215
1514		62	220
1618		62	225
1722	48	49	234
1826	55	47	248
1930	41	44	253
2034	42	63	259
2138	40	52	262
2242	42	36	267
2346	56	38	273
2450	45	43	279
2554	52	38	292
2658	65	28	295
2762	64	35	302
2866	64	37	304

Table 9.9 Evaporation temperature and activation energies for some of the peaks in the argon C1 spectrum of polystyrene

*Estimated uncertainty ± 3 kcal/mol through mass 1618 and + 6 kcal/mol for higher masses.

Reprinted from H.R. Udseth and N. Friedman, *Analytical Chemistry*, 1981, 53, 29, with permission from American Chemical Society [170]

Table 9.10 Evaporation temperatures and activation energies for some of the peaks in the methane C1 spectrum of polystyrene		
Mass, amu	Activation energy, kcal*	Evaporation temperature, °C
579	22	173
607	24	203
633	20	203
1203	31	259
1224	32	248
1253	33	259
1619	44	253
1665	42	262
2139	51	289
2659	51	300
Estimated uncertainty ± 3 kcal/mol. *Reprinted from H.R. Udseth and N. Friedman, Analytical Chemistry, 1981, 53, 29, with permission from American Chemical Society [170]*		

in the lower molecular weight portion of the methane chemical ionisation spectrum of production of relatively large yields of lower molecular weight depolymerisation products. The energies found in the electron impact and chemical ionisation studies suggest that sample exhaustion via low-temperature depolymerisation reactions was in part inhibited by interaction of gaseous ions with species on the surface of the sample probe. This inhibition could have been an ionic catalysis of recombination or polymerisation reactions that reduce the concentration of lower molecular weight oligomers which desorb at lower temperatures.

The lower activation energies in the electron spectra, particularly for *n*-mers with *n* = 9–11, may be the result of the failure of the basic assumption of zero-order kinetics for their desorption. These oligomers are detected at temperatures well above those required for desorption of 99% of the sample and are produced in the desorption of layers of sample. Under these circumstances sample depletion can play an important role in giving activation energies that are too low. Correction for this type of error is difficult. Suffice to say that activation energies for ions produced after most of the sample has been desorbed are probably the lower limits of values of activation energies of desorption and are useful only to establish the presence or absence of genetic relationships between these ions and lower molecular weight species.

The result of the interaction of the gaseous ions in the chemical ionisation source with the solid sample surface is to facilitate observation of spectra that can be best correlated with the structure of the solid polymer. If one assumes that the lower molecular weight oligomers of the A sequence, with values of $n = 1–3$, are primarily thermal depolymerisation products, then the remaining mass spectrum can be used to calculate average molecular weights of the polymer on the probe surface. Number- and weight-average molecular weights are calculated from the methane chemical ionisation spectrum to be 1987 and 2143, respectively, on the basis of the latter assumption. Values of number average molecular weight (M_n) and weight average molecular weight (M_w) calculated from the argon chemical ionisation spectrum are 1800 and 1970, respectively. The ratios of M_n/M_w in the argon and methane chemical ionisation spectra are 1.10 and 1.08, respectively, which gives good agreement with the nominal value for this sample of PS (2100).

Risby and Yergey [154] used PVC as an example of a polymer that undergoes a bimodal pyrolysis [171–177], in which initial evolution of hydrogen chloride leads to a conjugated polyene which subsequently pyrolyses to hydrocarbon fragments. The ability to observe the bimodal evolution is dependent upon the sample's heating rate. Since these evolution profiles involve fragments with different polarities, namely hydrogen chloride and hydrocarbons, both negative and positive ion, methane chemical ionisation MS methods were used to follow the progress of the pyrolysis. Negative ion detection was used in order to selectively detect halogenated species since there is some controversy as to whether halogenated hydrocarbons are also evolved with hydrogen chloride. Positive ion detection was used to detect the evolution of the hydrocarbon fragments.

Negative ions. PVC showed a bimodal evolution profile in the negative ion detection mode, but the first evolution was due to the release of the residual solvent, tetrahydrofuran. The second evolution was due to the following major species: $[Cl]^-$ (35, 37 m/z), $[HCl_2]^-$ (71, 73, 75 m/z), and $[PhCHCl]^-$ (125, 126, 127, 128 m/z). Presence of the ions $[Cl]^-$ and $[HCl_2]^-$ indicates the loss of hydrogen chloride from the polymer. The differences in the widths of the evolution profiles can be rationalised: dehydrochlorination produces allylic groups that will cyclise and the subsequent losses of hydrogen chloride will have different kinetics.

Examination of specific ion profiles shows that the onset of the evolution of hydrogen chloride (loss of Cl^-) occurs prior to the onset of the evolution of $[HCl_2]^-$ or $[PhCHCl]^-$ suggesting that the latter ions may be products of secondary gas-phase reactions. Although the evolution profiles were broad, the average activation energy could be obtained by the measurement of the temperature maxima as a function of sample heating rates. The results of these studies are given in **Table 9.11**.

The values for activation energy obtained by Risby and Yergey [154] (26–33 kcal) fall within the range of previously reported values (20–33 kcal) [178–180].

Table 9.11 Temperature data, activation energies and pre-exponential factors for polyvinyl chloride			
Ramp rate, °C/s	Temp. of max., °C	Ramp rate, °C/s	Temp. of max., °C
15.5	326	15.65	319
7.80	306	7.99	294, 354
3.96	306	4.00	283, 349
1.98	286	2.01	275, 340
1.00	276	1.02	254, 334
Activation energy		33 kcal	25, 72 kcal
Pre-exponential factor		6.3×10^{11}/s	6.3×10^{8}/s 6.3×10^{24}/s
Reprinted from T.H. Risby, J.A. Yergey and J.J. Scocca, Analytical Chemistry, 1982, 54, 2228, with permission from the American Chemical Society [154]			

Positive ions. Since the background ion current of the positive ion methane chemical ionisation mass spectrum in the region of 10–60 *m/z* was intense, no attempt was made to monitor the evolution of hydrogen chloride ($^{+}H_2Cl$, *m/z* 36). Protonated molecular ions were observed for the major fragments with the exception of methylnaphthalene, methylanthracene, and ethylanthracene. The nominal structures for the fragments that result from the thermal degradation of the polyene are shown in **Table 9.12**.

The specific ion current decreased with increasing molecular weight as would be expected. There was some evidence of the evolution of aliphatic hydrocarbons, but they were not characterised. The evolution profiles for some of the aromatic hydrocarbon fragments were bimodal and the first evolution maximum occurred concomitantly with the dehydrochlorination step. These results confirm that the ion (PhCHCl)$^{-}$ was formed in secondary gas-phase reactions since it was not observed in the positive ion spectra. It is interesting to note that not all the ions have this bimodal evolution profile.

Although some of the evolution profiles for the loss of these aromatic hydrocarbon fragments were bimodal and were clearly due to multiple kinetic processes, attempts were made to measure average activation energies. The activation energies for dehydrochlorination and polyene degradation are presented in **Table 9.11**. The activation energy of the first evolution maxima was lower than the average activation energy for the dehydrochlorination. This is difficult to rationalise since dehydrochlorination must precede or occur concomitantly with the formation of the polyene. However, this difference may not be significant and may reflect errors in the measurement of the activation energies by this method. If this difference is significant, it may be because this

Table 9.12 Hydrocarbon pyrolysis products from polyvinyl chloride (7.80 °C/s)			
Hydrocarbon fragment	Mass of ion with max intensity in region of mol.wt.	% total ion current at max	Temperature of evolution max, °C
Benzene	79	5.8	266-294
Toluene	93	1.0	321-348
Styrene	105	2.0	294
Ethylbenzene	107	3.0	266-294
Propylene-benzene	119	3.4	294
Propyl-benzene	121	1.4	294
Naphthalene	129	4.2	294
Methyl-naphthalene	142	0.9	321-348
Ethyl-naphthalene	156	1.4	294
Propylene-naphthalene	169	0.9	294
Propyl-naphthalene	171	0.8	294
Anthracene	179	0.7	294, 321
Methyl-anthracene	192	0.6	294, 321
Ethyl-anthracene	206	0.5	294
Propylene-anthracene	219	0.2	294, 321
Propyl-anthracene	221	0.2	294

Reprinted from T.H. Risby, J.A. Yergey and J.J. Scocca, Analytical Chemistry, 1982, 54, 2228, with permission from the American Chemical Society [154]

technique measures the average activation energy for the loss of the particular fragment - the loss of hydrogen chloride has a much greater variation in activation energy than does the loss of the hydrocarbons. This may explain the wide variation in published activation energies for the same processes. The second evolution profile has a much higher activation energy, similar to the value found for the thermolysis of isoprene (56–63.05 kcal), which has a similar degree of unsaturation [181].

In a classic example of thermochemical analysis, O'Mara [173] studied the thermolysis of a PVC resin containing 54.4% chlorine by two techniques. The first method involved the heating of resin in the heated (325 °C) inlet of a MS in order to obtain a mass spectrum of the total pyrolysate. The second, more detailed, method consisted of degrading the resin in a PGC interfaced with a MS through a molecule enricher.

Table 9.13 Pyrolytic analysis of HCl from PVC compounds and chlorinated PVC		
Sample type	Percentage HCl	
	Determined	Theoretical
PVC plastisol	30.1	29.1
PVC plastisol	37.0	37.3
PVC plastisol	44.7	44.5
PVC-vinyl acetate copolymer	53.5	52.9
PVC-vinyl acetate copolymer	46.3	46.3
PVC compound	38.5	38.6
PVC compound	47.5	47.7
Chlorinated polyethylene	34.9	34.3
Chlorinated polyethylene	40.7	41.1
Chlorinated poly(vinylchloride)*	63.9	62.2
Chlorinated poly(vinylchloride)	68.5	71.5

*Impure material, theoretical recovery not expected
Reprinted from M.M. O'Mara, Journal of Polymer Science: Polymer Chemistry, 1970, 8, 1887, with permission fom Wiley [173]

Samples of PVC resin and plastisols (10–20 mg) were pyrolysed at 600 °C in a helium carrier gas flow. Since a stoichiometric amount of hydrogen chloride is released (58.3%) from PVC when heated at 600 °C, over half of the degradation products, by weight, is hydrogen chloride. The major components resulting from the pyrolysis of PVC are hydrogen chloride, benzene, toluene, and naphthalene. In addition to these major products, a homologous series of aliphatic and olefinic hydrocarbons ranging from C_1 to C_4 are formed.

O'Mara [173, 182] obtained a linear correlation between the weight of PVC pyrolysed and the weight of hydrogen chloride obtained by GC. Good agreement is obtained between the expected and determined hydrogen chloride contents by the procedure (**Table 9.13**).

The technique was then applied to various PVC into which different inorganic fillers had been incorporated ($CaCO_3$, CaO, $Al(OH)_3$, Na_2CO_3, Al_2O_3, LiOH [183, 184], TiO_2, SnO_2, and ZnO_2) [183]. This work provided valuable information regarding reaction mechanisms that occur upon heating filled and unfilled PVC to high temperatures.

Chang and Mead [160] coupled a TGA-GC with a high-resolution MS (HRMS) in tandem in order to study the thermal stability of a ethylene–PS foam (75:25 by weight). They used a stainless steel system consisting of two valves and two traps to join a TGA to a GC. This arrangement has the capability of making multiple trapping of desired portions of a thermogram. The TGA effluent gas was collected in a cold trap and then injected directly into the GC for separation. The separated components were individually introduced into the HRMS for unequivocal identification. Results indicated excellent separation between the contents of the two traps. Gas chromatographic resolution and peak shape were comparable to conventional injection systems. Sixteen effluent compounds from the thermal degradation of PS foam were identified. The elemental compositions of the molecular ions of 31 components were obtained from the pyrolysis of ethylene–vinyl acetate copolymer.

This system consisted of a TGA, GC, HRMS, GC to HRMS interface unit, and a GC interface unit. The TGA was a DuPont model 950 of the DuPont Thermal Analysis System (EI DuPont de Nemours & Co., Wilmington, DE, USA). The heating rate was 10–15 °C per minute. The sample size was about 2–20 mg. Helium was used as the TGA carrier gas at a flow rate of 60–80 cm^3 per minute.

The GC was a Perkin Elmer model 881 (Perkin Elmer Co., Norwalk, CT, USA) with a flame ionisation detector. Two GC columns were employed: a 1.8 m × 0.3 cm stainless steel column containing 5% SE-30 on Chromosorb W and a 3 m × 0.5 cm stainless steel column containing 15% SE-30 Chromosorb W. The GC carrier gas was helium and the flow rate was approximately 60 cm^3 per minute.

The HRMS was a GEC model 21-110B (EI DuPont de Nemours & Co., Wilmington, DE, USA). Photoplates with 30-exposure capacity were used to record the MS of the GC peaks. These photoplates were processed via a real-time shared comparator computer system [185].

The GC to MS interface unit included a Biemann–Watson molecular separator [186], a GC stream splitter, and two Hoke needle valves (Hoke Inc., Cresskill, NJ, USA). One needle valve was placed in between the stream splitter and the molecular separator and the MS. The ion source pressure was maintained at approximately $1065\text{-}1335 \times 10^{-6}$ Pa during the operation. The temperatures of the connection lines were maintained at 220–250 °C.

The TGA and GC interface unit was constructed from an eight-way valve (valve A), a six-way valve, and two collecting traps. All parts were made from stainless steel except the Viton O-rings used in the valves. These multi-direction linear valves were products of Loenco Inc., Altadena, CA, USA (Cat. No. L-20606 and L-208-8). The traps were made from 20 cm × 0.6 cm stainless steel tubing with a capacity of approximately 3 ml. The connection tubes were of 0.3 cm stainless steel tubing. All joints were of Swagelok fittings. The ball joint on the standard (~50 ml capacity) quartz furnace tube of the DuPont

TGA was removed, and to this furnace tube an approximately 25 mm length of 6 mm borosilicate glass tubing was joined through a graded seal. Through a glass-to-metal seal on this borosilicate glass tube, the remaining interface was joined. The interface unit was enclosed in an oven except the traps. The oven temperature was maintained at 150 °C. The oven temperature was limited to the thermal stability of the Viton O-rings which start to soften at approximately 220 °C. The temperature of the connection lines was maintained at 200–250 °C. The traps were cooled with liquid nitrogen. A simply constructed oval-shaped sleeve heater consisting of heater tape covered with asbestos was used to warm up the traps to 200–250 °C. An external cold trap (vent trap) was connected to the exit of the TGA effluent to prevent possible backflow of moisture into the traps.

Figures 9.11(a) and (b) show, the thermogravimetric thermogram and the gas chromatogram obtained for PS foam, respectively. It is seen in **Figure 9.11(b)** that the thermally degraded products consist of a complex mixture of about 20 components. Sixteen of these components were readily identified by the HRMS, as listed in **Table 9.14**.

One component was identified as a styrene dimer. Both low- and high-resolution mass spectra are obtained from the computer printout of the photoplate data. No effort was made to identify the minor components.

It is interesting to note the presence of two oxygenated compounds: C_8H_8O (peak 11) and C_9H_8O (peak 12). Peak 11 may arise from oxygenation of ethylbenzene, and peak 12 may arise from oxygenation of α-methylstyrene.

Ethylene–vinyl acetate (75:25 by weight) showed two distinct TGA breaks. The thermogravimetric thermogram was obtained from pyrolysis of a 9.2 mg sample in a helium atmosphere at a 15 °C per minute heating rate. Two traps were employed to collect the effluents evolved that corresponded to the two breaks. Gas chromatograms were obtained from the contents of traps 1 and 2. Both were acquired by maintaining the column temperatures at ambient for 6 minutes, and then programming the column temperatures from 50 to 225 °C at 10 °C per minute. Only negligible amounts of trap 1 contents were observed in trap 2. The exact mass of the molecular ion, its corresponding elemental composition, deviation in millimass units (observed – theoretical mass), and its probable molecular structure are listed in **Table 9.15**. Trap 1 contents arise from the vinyl acetate portion of the polymer, whereas trap 2 contents, with the exception of acetone, consist of aliphatic and olefinic hydrocarbons from the ethylene portion.

Peltonen [159] applied TGA followed by GC and MS to the determination of volatile compounds arising from epoxy powder paints during the curing process carried out at 200 °C. The volatile compounds may be residues from the synthesis of epoxy resin or they can be constituents of the paint. Two types of epoxy powder paint were involved in this study. They differ from each other in the form of hardener used.

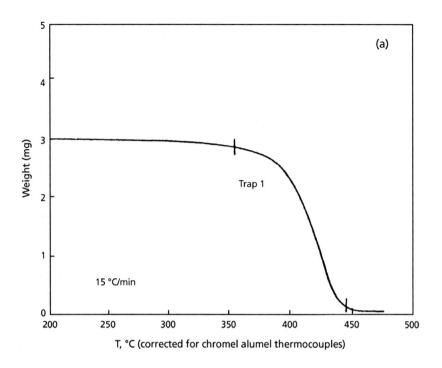

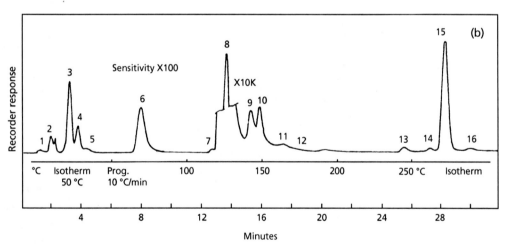

Figure 9.11 (a) A TGA thermogram of polystyrene foam, 3 mg sample, TGA heating rate 15 °C/min, helium carrier gas; (b) GC chromatogram of effluent compounds from pyrolysis of polystyrene foam. Column temperature 50 °C for 7 min, then programmed at 24 °C/min at 250 °C

(*Reprinted from T.L. Chang and T.E. Mead, Analytical Chemistry, 1971, 43, 534, with permission from American Chemical Society, [160]*)

Table 9.14 Components from decomposition of polystyrene foam				
Peak No.	Molecular ion found	Elemental composition	Deviation, mmu*	Probable structure
1	43.9901	CO_2	−0.30	Carbon dioxide
2	72.0950	C_5H_{12}	−1.10	Pentane
3	86.1105	C_6H_{14}	−1.36	Hexane
4	84.1105	C_6H_{12}	1.86	Hexene
5	78.1477	C_6H_6	−0.80	Benzene
6	92.0622	C_7H_8	0.38	Toluene
7	106.0785	C_8H_{10}	−0.29	Ethylbenzene
8	104.0633	C_8H_8O	−0.81	Styrene
9	118.0773	C_9H_{10}	0.86	α-Methylstyrene
10	132.0946	$C_{10}H_{12}$	−0.77	$Ar-C_4H_7$
11	120.0582	C_8H_8O	−0.76	ArC_2H_3O
12	132.0584	C_9H_8O	−0.94	$Ar-C_3H_3O$
13	182.1088	$C_{14}H_{14}$	0.66	$Ar-C_2H_4-Ar$
14	196.1269	$C_{15}H_{16}$	1.76	ArC_3H_6-Ar
15	208.1245	$C_{16}H_{16}$	0.60	Styrene dimer
16	220.1242	$C_{17}H_{16}$	0.90	$(Ar)_2-C_5H_6$

*Deviation = (theoretical mass − observed mass) × 1000.
Reprinted from T.L. Chang and T.E. Mead, Analytical Chemistry, 1971, 43, 534, with permission from American Chemical Society, [160]

Dicyandiamide-hardened paints have a glossy surface, whereas with all acid anhydride hardener the result is semi-glossy. In the powder, the amount of hardener is about 4% and that of the accelerator 1%.

The solid resins in powder paints are produced in two different ways. In the conventional method, the resin is produced in isobutyl methyl ketone. In the fusion method, bisphenol-A is condensed catalytically without solvents to give a liquid epoxy resin. In this study, the powders were cured in a device permitting precise control of temperature, type of atmosphere, rate of gas flow, and the collection of the volatile compounds. The compounds were studied with a high-resolution GC and a GC–MS. The mass loss of the powders during curing was studied by thermogravimetry.

Table 9.15 Components from decomposition of ethylene-vinyl acetate copolymer				
Peak No.	Molecular ion found	Elemental composition	Deviation, mmu[a]	Probable structure
1	58.0146	C_3H_6O	0.24	Acetone
2	60.0207	$C_2H_4O_3$	0.36	Acetic acid
3	42.0470	C_3H_6	−0.15	Propene
	44.0630	C_3H_8	−0.49	Propane
4	56.619	C_4H_8	0.63	Butene
	58.0777	C_4H_{10}	0.53	Butane
5	58.0407	C_3H_6O	1.08	Acetone
	70.0780	C_5H_{10}	0.15	Pentene
	72.0944	C_5H_{12}	−0.59	Pentane
6	84.0939	C_6H_{12}	−0.01	Hexene
	86.1098	C_6H_{14}	−0.26	Hexane
7	98.1086	C_7H_{14}	0.88	Heptene
	100.1244	C_7H_{16}	0.79	Heptane
8	112.1259	C_8H_{16}	−0.75	Octene
	114.1417	C_8H_{18}	−0.88	Octane
9	126.1402	C_8H_{18}	0.58	Nonene
	128.1556	C_9H_{20}	0.83	Nonane
10	140.1557	$C_{10}H_{20}$	0.79	Decene
	142.1707	$C_{10}H_{22}$	1.38	Decane
11	154.1719	$C_{11}H_{22}$	0.15	Undecene
	156.1869	$C_{11}H_{24}$	0.82	Undecane
12	168.1863	$C_{12}H_{24}$	1.40	Dodecene
	170.2034	$C_{12}H_{26}$	−0.05	Dodecane
13	180.2109	$C_{13}H_{26}$	1.53	Tridecene
	182.2019	$C_{13}H_{28}$	−1.37	Tridecane
14	196.2201	$C_{14}H_{28}$	−1.09	Teradecene
15	198.2371	$C_{14}H_{30}$	−2.41	Teradecane
	210.2362	$C_{15}H_{30}$	−1.50	Pentadecane
16	224.2517	$C_{16}H_{32}$	−1.36	Hexadecene
	226.2666	$C_{16}H_{34}$	−0.64	Hexadecane

Peak numbers 1 and 2 from Trap 2. [a]Deviation = (theoretical mass − observed mass × 100) Reprinted from T.L. Chang and T.E. Mead, Analytical Chemistry, 1971, 43, 534, with permission from American Chemical Society, [160]

355

The TGA was carried out with a Stanton Redcroft 770 apparatus (Stanton Redcroft, UK). In the analyses, the temperature was raised from 20 to 200 °C at a rate of 20 °C/min and the sample size was 5 mg.

In the curing process, temperature was adjusted to 200 °C. The air flow rate was 500 ml/min (synthetic air: 20% O_2, 80% N_2). The 10 g sample was spread over a distance of 800 mm along a glass tube (1500 × 17 mm). The trap (1000 × 4 mm) was cooled to –78 °C with acetone and dry ice. The volatiles were washed out from the spiral with 1 ml of acetone.

The GC was a Hewlett-Packard 5790-A (Hewlett-Packard, USA) equipped with flame ionisation and nitrogen–phosphorus detection systems. The chromatograms were recorded with a Hewlett-Packard 3392-A integrator. A BP-5 fused-silica column (25 m × 0.2 mm id) from SGE (Australia) was used and the carrier gas was helium at a flow rate of 1 ml/min. The oven temperature was kept at 35 °C for 1 minute after injection, then raised to 240 °C for 20 minutes. The injector temperature was 220 °C, the detector temperature 280 °C, and the volume injected 0.1 μl.

Mass spectral data (chemical ionisation and high-resolution electron impact) were obtained with a Finnigan-MAT 8230 double-focusing system with an Incross data system and a Varian 3800 GC. In chemical ionisation the reactant gas was isobutane, the ionisation energy 160 eV, the emission current 0.05 mA, the resolution 1000, and the scan range 65–400 *m/z* at a rate of 1 scan/s. In high-resolution electron impact mass spectrometry the ionising energy was 70 eV, the emission current 1 mA, the resolution 3000, and the scan range 50–350 *m/z* at a rate of 0.55 scans/s. The GC conditions were as previously with the exception of the column, which was an SE-54 (25 m × 0.32 mm id) (Orion Analytica, Finland). A Hewlett-Packard 5970 mass selective detector coupled to the GC (Hewlett-Packard 5890) was used for low-resolution electron impact mass spectrometry. The scan range was 45–350 *m/z*. The GC conditions were the same as previously, but here the column was a Hewlett-Packard Ultra 2 (25 m × 0.2 mm id).

The thermogravimetric mass losses were 0.3–0.4% and did not depend on the manufacturing process of the resin, or on paint colour (**Table 9.16**). The mass losses were similar in air and in nitrogen. Most of the mass loss may have been due to condensed moisture released during heating. No further mass loss occurred after 20 minutes.

The volatiles emitted from white amide-cured epoxy powder paints, cured and manufactured by the solution method are listed in **Table 9.17**. The most abundant compounds detected were isobutyl methyl ketone (4-methylpentan-2-one) and the various isomers of xylene which are typical solvents used in the production of resins. The volatiles characteristic of all three amide-cured epoxy powders studied were an unidentified compound and melamine.

Table 9.16 Mass loss of epoxy powder paints during curing		
Production method and colour	Mass loss in air, %	Mass loss in nitrogen, %
Conventional:		
White 1	0.3	0.3
White 2	0.4	0.3
Black 4	0.4	0.4
Black 5	0.3	0.3
Fusion:		
White 3	0.3	0.3
Black	0.4	0.3
Reprinted from K. Peltonen, Analyst, 1986, 111, 819, with permission from The Royal Society of Chemistry [159]		

The source of the melamine is not clear but it has been suggested that dicyandiamide decomposes to cyanide during curing. Cyanide can react with dicyanamide to form melamine. However, cyanamide was not observed among the volatilised products.

Gas chromatographs for a black paint with a resin made by the solvent method are shown in **Figure 9.12**. **Table 9.18** lists the compounds identified. Even in this example, the most abundant compounds were isobutyl methyl ketone and xylenes. The origin of 2-phenyl-4-H-imidazole is not certain, but it might have been added as a catalyst or it may arise from the anhydride adduct during the curing process.

Most of the compounds identified gave observable molecular ions in electron impact mass spectrometry. Chemical ionisation mass spectroscopy with isobutane was carried out to determine the relative molecular masses. In all instances, the quasi-molecular ion was M + 1 with a relative abundance of 10%; no M + 57 was observed. Some compounds were identified on the basis of their mass spectra only because no reference compounds were obtainable.

High-resolution mass spectra were obtained for unidentified compounds a, b, c, d, and f. The molecular formula of compounds a, b, c, and d is $C_{12}H_{22}O$. Proposed structures are isomers of ketones, e.g., a and b might be 2- or 4-isomers of ethylisopropyl cyclohexyl ketone and c and d might be cyclohex-4-ylhexan-2-one or cyclohex-4-yl-4-methylpentan-2-one, respectively. An unidentified compound (e) was found also for the quasi-molecular ion M + 1 105. The only fragment was M – 18 in chemical ionisation mass spectroscopy. The molecular formula of unidentified compound f is $C_{15}H_{14}O$. The proposed structure is

357

No. in Figure 9.13	Compound	Paint			Method of identification
		1	2	3	
1	Solvent				
2	4-Methylpentan-2-one	X	X		GC+GC-MS
3	Chlorobenzene	X			GC+GC-MS
4	1,2-Dimethylbenzene	X	X		GC+GC-MS
5	Ethylbenzene	X	X		GC+GC-MS
6	1,4-Dimethylbenzene	X	X		GC+GC-MS
7	1,3-Dimethylbenzene	X	X		GC+GC-MS
8	Trimethylbenzene	X	X		GC+GC-MS
9	3-Isobutylbut-3-en-2-one	X	X		GC-MS
10	α-Methylbenzyl-alcohol	X	X		GC+GC-MS
11	Acetophenone	X	X		GC+GC-MS
12	Unidentified[a]	X			
13	Unidentified[b]	X			
14	Unidentified[c]	X			
15	Unidentified[d]	X			
16	Unidentified[e]	X	X	X	
17	Melamine	X	X	X	GC+GC-MS
	Phenylglycidyl ether		X		GC+GC-MS
	Unidentified[f]			X	

Table 9.17 Volatile compounds emitted from amide-hardened epoxy powder paints during curing

[a] EI: 53(8), 55(24), 57(100), 58(120), 67(8), 85(60). CI: M+1183(100)

[b] EI: 53(8), 55(26), 57(100), 67(10), 69(10), 85(80), 97(8), 125(6). CI: M+1183(100)

[c] EI: 55(32), 57(28), 68(32), 83(93), 107(25), 123(100), 167(20), 182(6). CI: M+1183(100)

[d] EI: 55(30), 67(54), 82(16), 83(30), 10(12), 125(100), 140(10), 182(10). CI: M+1183(100)

[e] EI: 53(1), 55(100), 56(98), 57(24), 71(15), 73(52), 74(8), 86(6). CI: M+1105(100)

[f] EI: 85(8), 126(16), 139(10), 152(8), 165(20), 195(100), 196(16), 210(10). CI: M+1211

Reprinted from K. Peltonen, Analyst, 1986, 111, 819, with permission from The Royal Society of Chemistry [159]

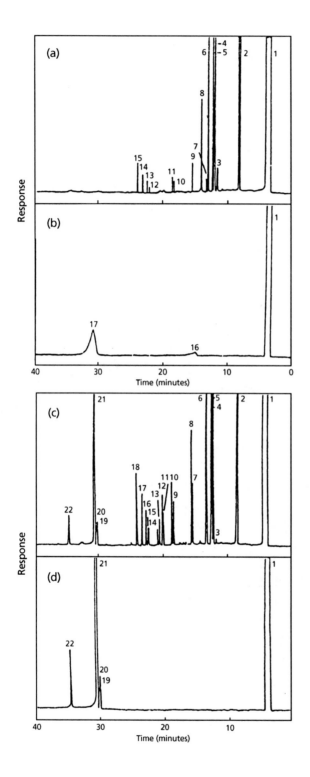

Figure 9.12 Gas chromatograms of compounds volatilised during curing of amide paint. (a) flame ionisation detector; (b) Nitrogen specific detector. Compounds as tested in Table 9.17. Gas chromatograms of compounds volatilised during curing of anhydride paint; (c) Flame ionisation detector; (d) Nitrogen specific detector. Compounds are listed in Table 9.18.

(Reprinted from K. Peltonen, Analyst, 1986, 111, 819, with permission from The Royal Society of Chemistry [159])

Table 9.18 Volatile compounds emitted from anhydride-hardened epoxy powder paints during curing					
No. in Figure 9.13	Compound	Paint			Method of identification
		4	5	6	
1	Solvent				
2	4-Methylpentan-2-one	X	X		GC+GC-MS
3	Chlorobenzene	X			GC+GC-MS
4	1,2-Dimethylbenzene	X	X		GC+GC-MS
5	Ethylbenzene	X	X		GC+GC-MS
6	1,4-Dimethylbenzene	X	X		GC+GC-MS
7	3-Isobut-3-en-2-one	X			GC-MS
8	Benzaldehyde	X			GC+GC-MS
9	α-Methylbenzyl-alcohol	X	X		GC+GC-MS
10	Acetophenone	X	X		GC+GC-MS
11	Dec-6-en-3-one	X			GC-MS
12	Dec-6-en-5-one	X			GC-MS
13	3-Methylbenzyl alcohol	X	X		GC+GC-MS
14	4-Methylbenzyl alcohol	X	X		GC+GC-MS
15	Unidentified[a]	X			
16	Unidentified[b]	X			
17	Unidentified[c]	X			
18	Unidentified[d]	X			
19	Phenylimidazoline	X			GC-MS
20	Phenylimidazoline	X			GC-MS
21	2-Phenyl-4H-imidazoline	X		X	GC+GC-MS
22	Diphenylimidazoline	X			GC-MS
–	Phenol		X		GC+GC-MS
–	4-Methyl-2,6-di-*tert*-butylphenol		X		GC+GC-MS
–	Phenyl glycidyl ether			X	GC+GC-MS

[a] EI: 53(8), 55(24), 57(100), 58(120), 67(8), 85(60). CI: M+1183(100)

[b] EI: 53(8), 55(26), 57(100), 67(10), 69(10), 85(80), 97(8), 125(6). CI: M+1183(100)

[c] EI: 55(32), 57(28), 68(32), 83(93), 107(25), 123(100), 167(20), 182(6). CI: M+1183(100)

[d] EI: 55(30), 67(54), 82(16), 83(30), 10(12), 125(100), 140(10), 182(10). CI: M+1183(100)

Reprinted from K. Peltonen, Analyst, 1986, 111, 819, with permission from The Royal Society of Chemistry [159]

2-biphenyl-2,3-epoxypropane. The eight most intense ions of the unidentified compounds and their relative abundances, together with the M + 1 ions in their chemical ionisation spectra, are given in **Tables 9.17 and 9.18**.

9.4.7.1 Gas Chromatography–Mass Spectrometry

Hemvichian and Ishida [162] and Ziatdinov and co-workers [164] have applied this technique, to products produced in a thermogravimetric analyser during the decomposition of polybenzo-oxazines and siloxy-substituted polyoxadisilapentanylenes, respectively.

9.4.7.2 Pyrolysis Gas Chromatography–Mass Spectrometry

Various workers have applied this combination of techniques to studies carried out on ethylene–vinyl acetate copolymers [161], chlorinated natural rubber latex [165], and polyacrylonitrile [166].

9.4.7.3 Thermal Degradation of Poly(Xylylene Sulfide) and Poly(Xylylene Disulfide) [163]

Thermodegradative data showed that these polymers decompose with two separate steps in the temperature ranges 250–280 °C and 600–650 °C, leaving a large amount of residue (about 50% at 800 °C). The pyrolysis products detected by direct pyrolysis mass spectrometry in the first degradation step of poly(xylylene sulfide) and poly(xylylene disulfide) were terminated by three types of end groups, namely methyl, primary thiol, and thioaldehyde, originating from thermal cleavage reactions involving a series of homolytic chain scissions followed by hydrogen transfer reactions, generating several oligomers containing some intact xylylene sulfide repeating units. The presence of pyrolysis compounds containing some stilbene-like units in the first degradation step was also observed. Their formation is accounted for by a parallel cleavage involving the elimination of hydrogen sulfide from the poly(xylylene sulfide) main chains. These unsaturated units can undergo crosslinking at higher temperatures, producing the large amount of char residue observed. The thermal degradation compounds detected by direct pyrolysis mass spectrometry in the second decomposition step at about 600–650 °C were aromatic molecules containing dihydrofenanthrene and fenanthrene units. A mechanism of their formation is proposed. The information produced by the direct pyrolysis mass spectrometry and the flash PGC/MS techniques were compared [163].

361

9.5 Examination of Thermal Stability by a Variety of Techniques

Frequently studies of the thermal decomposition of polymers do not involve the application of a single technique but of several techniques in order to obtain more detailed information.

Some examples are as follows: styrene–nitrostyrene copolymers (DTA, TGA, FT-IR) [61], polyurethane (DSC, TGA) [60], bisphenol A-based PS (DSC, TGA, FT-IR) [62], styrene–isoprene copolymer (TGA–MS) [187], poly(l-lactide) (PGC-MS, TGA) [188], polymethacrylates (TGA–FT-IR, ^{13}C NMR) [53], polyacrylonitrile (FT-IR, NMR, PGC) [59], polypropylene carbonate, (PGC–MS, TGA–FT-IR) [58], polyimides (TGA–FT-IR–MS, PGC-MS, TGA–MS) [54], poly(3-hydroxybutyrate) and poly(3-hydroxyvalerate) (PGC-MS, FT-IR, TGA) [57], chlorinated elastomers (DSC, TGA, FT-IR) [55], polyamides and polyimides (NMR, FT-IR, DSC, TGA) [56], sulfonated maleated ethylene–propylene–diene rubber (DSC, TGA, DMA, dielectric thermal analysis) [189], bismaleimides (FT-IR, NMR, DSC, TGA) [189], styrene–butadiene copolymers (DSC, thermal field flow fractionation) [190], poly(allylazide) (FT-IR, NMR, DSC, TGA) [191], 3-alkoxy-substituted polythiophene (TGA, DSC) [193], fluorinated epoxies (DSC, TGA, dielectric thermal analysis) [194], polyimides (NMR, DTA, TMA, TGA) [195], and polyamides (TGA, flash PGC–MS) [196].

9.6 Heat Stability of Polypropylene

One of the most favourable features of PP is its high softening temperature, and consequently the possibility of its use for applications previously closed to low-cost thermoplastics because of high-temperature service requirements. However, the particular chemical structure of the polymer, (i.e., the presence of tertiary carbon atoms in the molecule) renders it particularly liable to oxidation at high temperatures, a feature that must be minimised by incorporating suitable antioxidants. In fact this thermal degradation may be accelerated by the presence of certain metals, and it may then be desirable to employ special stabiliser systems.

In addition, the possibility of interactions between the heat stability additives and other additives employed cannot be ignored, and such possibilities have been taken into account. For example, as the majority of applications require pigmented material it was considered necessary to study the effect, if any, of pigment systems used in the PP range on the heat stability of PP. As the majority of pigments used to colour PP are based on inorganic compounds, e.g., cadmium sulfides, zinc sulfides, and sulfoselenides, and since certain metals are known to catalyse the thermal degradation of PP, the possibility exists that some pigments may accelerate this degradation at elevated temperature.

Finally, there are a number of applications where resistance to both heat and light is required, and it is necessary to study the influence of UV stabilisers on the heat stability of pigmented PP.

A variety of tests have been employed in order to follow the progress of thermal oxidative degradation. These include the rate of oxygen uptake by the specimen, change in molecular weight, change in impact strength, and change in flexural endurance (i.e., flex life). The rate of onset of thermal degradation is of course influenced by the rate of diffusion of oxygen into the specimen via the surface, as well as by the chemical kinetics of the reactions between oxygen and the polymer/additive combination. Consequently, section thickness must be carefully standardised, and the appropriate reservations introduced in any attempt to predict the heat ageing life of sections other than the standard thickness chosen. The tests do, however, give a valid comparison of the relative effects of various additives and environmental conditions. Specimens 0.25 cm long × 0.8 cm wide × 0.64 mm thick, cut from compression moulded sheets, have been employed by some workers who selected flexure endurance as a convenient monitoring test. The effective heat ageing life is taken as the time at which 50% of the exposed specimen population fails at a flex life of one cycle of 180° amplitude.

The exposed strip of specimens from each sample of material were placed in glass test tubes which were suspended in an air circulating oven. Specimens were normally exposed at temperatures in the range 155 °C down to 115 °C.

Where the effect of direct contact with metal is being investigated, it has been found convenient to surround the specimens by finely divided metal to within half an inch of the top of the specimen. Various metals and their particle forms are as follows:

- Copper – turnings
- Brass – filings
- Iron – filings
- Zinc – filings
- Nickel – fine powder
- Chromium – technical powder
- Lead – small shot
- Tin – coarse powder
- Aluminium – coarse powder

Grades of PP of melt index 3 were employed in these studies of general-purpose polymers and special high heat ageing stabilisation. Additional studies are discussed involving pigmented materials incorporating various UV stabilisers. In these latter studies these PP had a lower melt index (1.5). The test results are conveniently presented as a graph of exposure temperature on a linear scale *versus* the time for 50% failure on a logarithmic scale. Reasonably straight-line plots are obtained, and facilitate the extrapolation down to temperatures of 100 °C or lower.

Figure 9.13 shows a typical plot for the stabilising systems designated KM61 and KM81, which has been extrapolated down to 100 °C. It is seen that, at least in the absence of deleterious metals, the advantage of the 81 stabilising system only appears to show up at the higher temperatures. It may be concluded with reasonable confidence that at temperatures of 100 °C or below, negligible advantage is normally gained from the use of the 81 stabilisation system.

The metals having the most deleterious effect on the heat ageing life of PP are copper and brass, and it should be noted that although brass appears to be more severe in its effect at the highest temperatures examined, the differing slopes of the plots for brass and copper lead to a position where copper is the more deleterious at lower temperatures such as 100 °C. Nickel and chromium are next in order of severity, reducing the heat ageing life of a 61 system a matter of three-fold and two-fold, respectively, while iron, lead, and zinc are somewhat less severe. Tin and aluminium have a barely detectable effect on PP heat ageing life, and may be considered inert.

Use of the 81 system, which was specifically developed to counter these metal effects, gives a most marked improvement in terms of resistance to the effect of brass and copper, and a distinct improvement in the case of the less deleterious metals. **Figures 9.14–9.17** illustrate these points, which are summarised in **Table 9.19**.

9.6.1 Influence of Pigmentation and UV Stabilisation on Heat Ageing Life

The heat ageing results of PP pigmented with the 'Researched Colour' formulations are shown in **Table 9.20**. The data covering four of the worst pigmentations in terms of heat ageing effects are plotted, together with the natural material, in **Figure 9.18**. Here it is seen that the deleterious influence of these pigments is most marked at temperatures in excess of 100–120 °C, but appears of relatively small practical importance at lower temperatures.

The results indicate that the influence of UV stabilisers on heat ageing stability, when used in conjunction with pigments and antioxidants, is extremely difficult to predict in general terms. The results presented for four commercially employed UV stabilisers in **Table 9.21** and **Figure 9.19** indicate that in some instances UV stabiliser actually appears to enhance the heat ageing resistance, whereas in other instances antagonistic effects are shown, notably when such pigments as ferric oxide and cadmium sulfoselenide are employed.

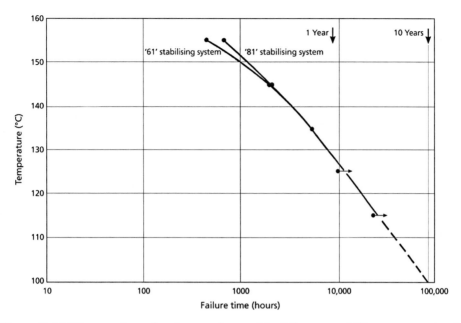

Figure 9.13 Heat ageing of polypropylene with different stabiliser systems in air. *(Source: Author's own files)*

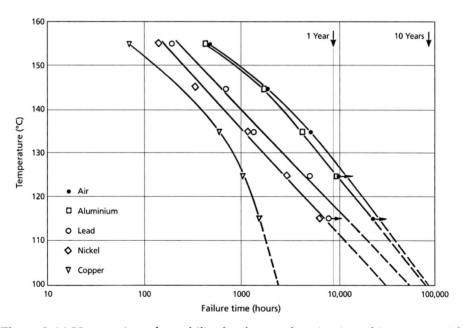

Figure 9.14 Heat ageing of a stabilised polypropylene in air and in contact with aluminium, lead, nickel and copper *(Source: Author's own files)*

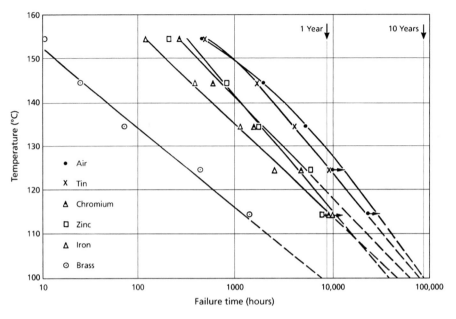

Figure 9.15 Heat ageing of stabilised polypropylene in air and in contact with tin, chromium, zinc, iron and brass (*Source: Author's own files*)

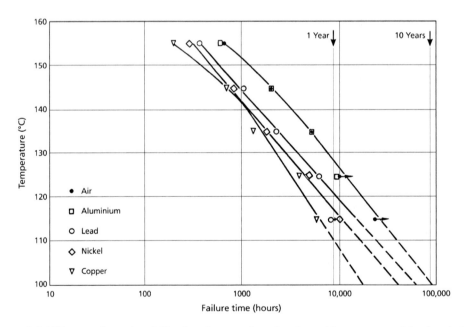

Figure 9.16 Heat ageing of stabilised polypropylene in air and in contact with aluminium, lead, nickel and copper (different stabilising system to that used in Figure 9.14)

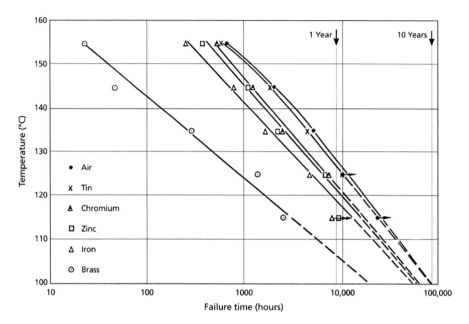

Figure 9.17 Heat ageing of stabilised polypropylene in air and in contact with tin, chromium, zinc, iron and brass (different stabilising system to that used in Figure 9.15). (*Source: Author's own files*)

Table 9.19 Effect of metal contact on life of polypropylene at 100 °C (based on Figure 9.15)		
Environment	Extrapolated life at 100 °C (F_{50})	
	'61' System	'81' System
Air	>87,600 hours (10 years)	>87,600 hours (10 years)
Aluminium	82,000 hours	>87,600 hours (10 years)
Tin	82,000 hours	87,600 hours (10 years)
Zinc	65,000 hours	62,000 hours
Lead	55,000 hours	62,000 hours
Iron	48,000 hours	56,000 hours
Chromium	40,000 hours	66,000 hours
Nickel	30,000 hours	40,000 hours
Brass	8,000 hours	18,000 hours
Copper	2,500 hours	20,000 hours

367

Table 9.20 Heat ageing of polypropylene pigmented with colour formulations				
Colour	Failure time (F_{50}), days			
	155 °C	145 °C	135 °C	115 °C
Natural control (1)	33	74	145	>940
Ivory	38	74	138	>940
Cream	42	86	155	>940
Primrose yellow	38	86	151	>940
Soft yellow	14	72	175	>940
Bright	10	72	147	>940
Golden rod	40	72	160	>940
Antique gold	12	82	160	>940
Sea blue	10	66	152	>940
Aqua blue	6	22	61	>940
Cadet blue	7	24	107	>940
Glacier blue	25	66	172	>940
Sapphire blue	38	71	148	>940
Turquoise	16	65	125	>940
Navy blue	9	61	125	785
Sandlewood	30	74	147	>940
Purple	43	73	151	>940
Peach bloom	60	82	159	>940
Orchid	38	64	180	>940
Coral	14	60	125	>940
Rose	50	78	148	>940
Sage green	56	74	151	>940
Fern green	14	52	115	>940
Sprout green	40	67	148	>940
Lime green	20	74	138	>940
Bright green	49	74	155	>940
Spring green	9	64	117	>940
Forest green	3	15	84	553
Emerald green	3	8	94	801
Pearl grey	38	86	125	>940

Colour	Failure time (F_{50}), days			
	155 °C	145 °C	135 °C	115 °C
Granite grey	20	80	138	>940
Charcoal	15	51	115	670
Dusk grey	27	64	110	710
Natural control (2)	30	70	145	>940
Ultra blue	22	82	155	>940
Lilac	66	86	152	>940
Pink	56	82	152	>940
Flame red	66	86	159	>940
Red/orange	53	74	143	>940
Orange	64	82	159	>940
Carmine red	13	64	111	855
Maroon	6	52	106	872
Red	42	86	172	>940
Vermillion	61	92	145	>940
Scarlet	40	86	18	>940
Mushroom	43	74	152	>940
Brown	14	44	84	565
Woodtone tan	28	52	103	855

Table 9.20 Continued ...

Source: Author's own files

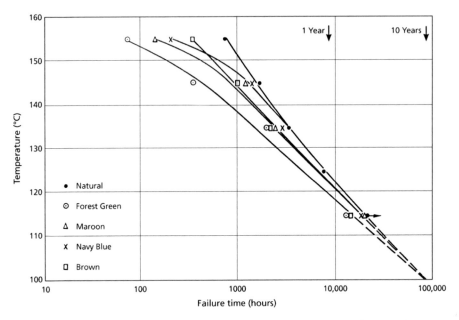

Figure 9.18 Heat ageing of natural and pigmented stabilised polypropylene in air. (*Source: Author's own files*)

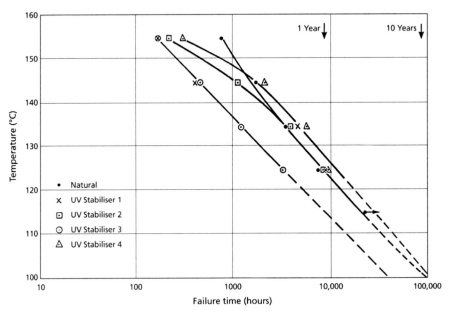

Figure 9.19 Heat ageing of natural stabilised polypropylene in air in conjunction with an ultraviolet stabiliser. (*Source: Author's own files*)

Table 9.21 Heat ageing of polypropylene containing single pigments and UV stabilisers						
Chemical nature of pigment (1% concentration)	UV stabiliser (0.5% concentration)	Failure time (F_{50}), days				
		155 °C	145 °C	135 °C	125 °C	115 °C
Natural	None	33	74	145	333	940
	No. 1	7	17	193	370	–
	No. 2	9	46	166	344	–
	No. 3	7	19	50	134	–
	No. 4	13	90	259	370	–
Titanium dioxide	None	37	86	148	–	–
	No. 1	2	18	69	249	–
	No. 2	2	7	12	82	–
	No. 3	6	22	40	134	–
	No. 4	15	92	210	370	–
Cadmium zinc sulfide	None	38	92	159	–	–
	No. 1	8	56	175	393	–
	No. 2	6	29	160	333	–
	No. 3	6	22	32	134	–
	No. 4	17	92	182	393	–
Copper phthalo-cyanine blue	None	2	8	34	19	–
	No. 1	1	4	39	85	–
	No. 2	1	5	25	134	–
	No. 3	7	22	32	134	–
Ultamarine (Lapis lazuli) Sodium aluminium silicate	None	6	26	145	–	–
	No. 1	2	11	130	292	–
	No. 2	2	4	22	107	–
	No. 3	5	21	50	179	–
	No. 4	17	74	193	365	–
Chlorinated ultramarine	None	22	64	145	–	–
	No. 1	4	33	207	431	–
	No. 2	2	3	60	194	–
	No. 3	8	21	50	–	–
Chlorinated phthalocyanine	None	2	6	84	–	–
	No. 1	<1	6	43	134	–
	No. 2	<1	2	7	79	–

Table 9.21 Heat ageing of polypropylene containing single pigments and UV stabilisers (*Continued*)						
Chemical nature of pigment (1% concentration)	UV stabiliser (0.5% concentration)	Failure time (F_{50}), days				
		155 °C	145 °C	135 °C	125 °C	115 °C
Chlorinated phthalocyanine	No. 3	5	15	50	134	–
	No. 4	7	29	50	190	–
Cobalt chlorine zinc aluminate, cadmium zinc sulfate	None	20	72	125	–	–
	No. 1	3.5	37	189	365	–
	No. 2	4	15	32	134	–
	No. 3	2.5	23	128	198	–
Cadmium sulfo-selenide barium sulfate	None	30	74	170	–	–
	No. 1	6	47	177	–	–
	No. 2	6	18	50	134	–
	No. 3	5	48	50	393	–
	No. 4	21	91	200	416	–
Cadmium sulfo-selenide ferrous oxide	None	2	7	19	–	–
	No. 1	1	4	18	71	–
	No. 2	1	2	7	60	–
	No. 3	6	17	50	168	–
	No. 4	3	7	33	155	–
Ferric oxide	None	1	4	25	–	–
	No. 1	<1	1	2	7	–
	No. 2	<1	1	2	3	–
	No. 3	8	16	50	150	–
	No. 4	7	25	150	190	–
Polyazo derivative	None	35	86	125	–	–
	No. 1	11	37	168	314	–
	No. 2	12	30	156	321	–
	No. 3	6	22	50	155	–
	No. 4	17	85	170	333	–
Cadmium sulfo-selenide	None	33	74	160	–	–
	No. 1	1	1	4	7	–
	No. 2	1	1	2	3	–
	No. 3	8	16	50	150	–
	No. 4	8	25	50	176	–

References

1. P. Cebe, M. Jaffe and C.E. Carraher in *Proceedings of the ACS Polymeric Materials Science and Engineering*, Dallas, TX, USA, Spring 1998, 78, p. 96.

2. M.L. Cerrada, *Revista de Plásticos Modernos*, 2002, 83, 501.

3. C.H. Earnest, *Thermal Analysis of Clays, Minerals and Coal*, Perkin Elmer Corporation, Norwalk, CT, USA, 1984.

4. J.L. McNaughton and C.T. Mortimer, *Differential Scanning Calorimetry*, Perkin Elmer Corporation, Norwalk, CT, USA, 1975.

5. W. Becker in *Brandverhalten von Kunststoffen*, Ed., J. Troitzsch, Carl Hanser Verlag, Munich, Germany, 1982.

6. G. Schreyer, *Konstruieren mit Kunststoffen*, Carl Hanser Verlag, Munich, Germany, 1972.

7. T. Ozawa, *Bulletin of the Chemical Society of Japan*, 1965, 38, 1881.

8. J.H. Flynn and L.A. Wall, *Polymer Letters*, 1966, 4, 323.

9. C.D. Doyle, *Journal of Applied Polymer Science*, 1961, 5, 285.

10. C.D. Zasko, *Journal of Applied Polymer Science*, 1980, 19, 333.

11. J. Krisonovsky, *Journal of Thermal Analysis*, 1980, 13, 571.

12. B.J. Toop, *Translating Electrical Industries*, EI-6 2, Institution of Electrical Engineers, London, UK, 1971.

13. A. Basch and N. Lwein, *Journal of Polymer Science, Polymer Chemistry Edition*, 1973, 11, 3071.

14. A. Basch and N. Lwein, *Journal of Polymer Science, Polymer Chemistry Edition*, 1973, 11, 3095.

15. A. Basch and N. Lwein, *Journal of Polymer Science, Polymer Chemistry Edition*, 1973, 11, 1707.

16. D. Dollimore and B. Holt, *Journal of Polymer Science, Polymer Physics Edition*, 1973, 11, 1703.

17. D.S. Varma and V. Narasimhan, *Journal of Applied Polymer Science*, 1972, **16**, 3325.

18. J.M. Funt and J.H. Magill, *Journal of Polymer Science, Polymer Physics Edition*, 1974, **12**, 217.

19. B.V. Kikta, J.L. Valade and W.N. Martin, *Journal of Applied Polymer Science*, 1973, **17**, 1.

20. M.D. Judd and A.C. Norris, *Journal of Thermal Analysis*, 1973, **5**, 179.

21. P.E. Tulupov and O.N. Karpov, *Zhurnal Fizicheskoi Khimii*, 1973, **47**, 1420.

22. B.L. Joesten and N.W. Johnston, *Journal of Macromolecular Science A*, 1974, **8**, 83.

23. V. Halip, V. Stan, A. Biro and R. Radovici, *Materiale Plastice (Bucharest)*, 1973, **10**, 601.

24. D.R. Cooper, G.J. Sutton and B.J. Tighe, *Journal of Polymer Science, Polymer Chemistry Edition*, 1973, **11**, 2045.

25. A. Patterson, G.J. Sutton and B.J. Tighe, *Journal of Polymer Science, Polymer Chemistry Edition*, 1973, **11**, 2343.

26. G.J. Sutton and B.J. Tighe, *Journal of Polymer Science, Polymer Chemistry Edition*, 1973, **11**, 1069.

27. D.S. Varma and S. Ravisankar, *Die Angewandte Makromolekulare Chemie*, 1973, **28**, 191.

28. J.M. Barrales-Rienda and J.G. Ramos, *Journal of Polymer Science: Polymer Symposium*, 1973, **42**, 1249.

29. K.A. Hodd and W.A. Holmes-Walker, *Journal of Polymer Science: Polymer Symposium*, 1973, **42**, 1435.

30. R. Kromalte, N.D. Malegina, B.V. Kotov, L.A. Oksent'evich and A.N. Prauednikov, *Vysokomolekulyarnye Soedineriya Seriya A*, 1972, **14**, 2148.

31. T.J. Gedemer, *Journal of Macromolecular Science*, 1974, **8**, 95.

32. E. Kiran, J.K. Gillham and E.J. Gipstein, *Journal of Makromolecular Science, B*, 1974, **9**, 341.

33. S.A. Liebman, D.H. Ahlstrom and C.R. Foltz, *Journal of Polymer Science*, 1978, **16**, 3139.

34. S.A. Leibman, D.H. Ahlstrom and C.R. Foltz, *Journal of Polymer Science, Polymer Chemistry*, 1978, **16**, 3139.

35. NEN 6065, *Determination of the Contribution to Fire Propagation of Building Products*, 2001. [In Dutch]

36. NEN 6066, *Determination of the Smoke Production During Fire of Building Products*, 1997. [In Dutch]

37. N.L. Dassanayake and R.W. Philips, *Analytical Chemistry*, 1984, **56**, 1753.

38. M.J. Turk, S.A. Ansari, W.B. Alston, G.S. Gahn, A.A. Frimer and D.A. Scheiman, *Journal of Polymer Science and Polymer Chemistry*, 1999, **37**, 3943.

39. K. Hemvichian and H. Ishida, *Polymer*, 2002, **43**, 4391.

40. D. Capatini, D. De Prisco, P. Laurienzo, M. Malinconico, P. Proietti and A. Roviello, *Polymer Journal (Japan)*, 2001, **33**, 575.

41. D.T. Grishin, L.L. Semenycheva, E.V. Telegina and V.K. Cherkasov, *Polymer Science*, Series A, 2003, **45**, 104.

42. T. Kojima, E. Inoue, M. Tsuchiya and K. Ishimura, *Journal of Thermal Analysis and Calorimetry*, 2003, **72** 737.

43. T.A.P. Seery, P. Dhar, D.H. Huber and F. Vatansever in *Proceedings of the ACS Polymeric Materials Science and Engineering*, Dallas, TX, Fall 1997, **77**, 634.

44. P-L. Kuo, J-M. Chang and T-L. Wang, *Journal of Applied Polymer Science*, 1998, **69**, 1635.

45. N.S. Allen, M. Edge, M. Rodríguez, D.M. Laiuw and E. Fontan, *Polymer Degradation and Stability*, 2000, **68**, 363.

46. O.H. Lockmuller, S.J. Breiner, M.W. Koch and M.A. Elomaa, *Analytical Chemistry*, 1991, **63**, 1685.

47. *High Resolution Thermogravimetric Analysis*, TA Leaflet TA-023TA Instruments Inc., New Castle, DE, USA.

48. J. Rouquerol, *Bulletin of the Society of Chemistry*, 1964, **31**.

49. E. Paulik and J. Paulik, *Analytica Chimica Acta*, 1971, **56**, 328.

50. S. Sorenson, *Journal of Thermal Analysis*, 1978, **13**, 429.

51. B. Castle and G. McClure, *American Laboratory*, 1989.

52. D.R. Clark and K.J. Gray, *Laboratory Practice*, 1987, **40**, No. 6.

53. Z. Ilter, M. Coskun, I. Erol, A. Unal and M. Ahmedzade, *Polymer Degradation and Stability*, 2002, **75**, 501

54. W. Xie, W.P. Pan and K.C. Chuang, *Journal of Thermal Analysis and Calorimetry*, 2001, **64**, 477.

55. A.R. Tripathy, P.H. Patra, J.K. Sinha and M.S. Banerji, *Journal of Applied Polymer Science*, 2002, **83**, 937.

56. H. Yagci and L.J. Mathias, *Polymer*, 1998, **38**, 3779.

57. S-D. Li, J-D. He, P.H. Yu and M.K. Cheung, *Journal of Applied Polymer Science*, 2003, **89**, 1530.

58. X.H. Li, Y.Z. Meng, Q. Zhu and S.C. Tjong, *Polymer Degradation and Stability*, 2003, **81**, 157.

59. M. Surianarayanan, R. Vijayaraghavan and K.V. Raghavan, *Journal of Polymer Science, Polymer Chemistry Edition*, 1998, **36**, 2503.

60. S. Vessot, J. Andrieu, P. Laurent, J. Galy and J.F. Gerard, *Journal of Coatings Technology*, 1998, **70**, 67.

61. M.J. Fernandez and M.D. Fernandez, *Polymer Degradation and Stability*, 1998, **60**, 257.

62. M.C. Delpech, F.M.P. Coutinho and M.E.S. Habibe, *Polymer Testing*, 2002, **21**, 156.

63. N. Dick and C.J. Westerberg, *Journal of Macromolecular Science A*, 1978, **12**, 455.

64. G. Poschet, *Kunststoffe Plastics*, 1978, **25**, 24.

65. L. Reich, *Thermochimica Acta*, 1973, **5**, 433.

66. T. Kotoyori, *Thermochimica Acta*, 1973, **5**, 51.

67. H.W. Holden, *Journal of Polymer Science: Part C, Polymer Symposia,* 1964, **6**, 53.

68. A. Panarotto and M. Bertucci, *Macplas International,* 1999, **10**, 64.

69. R. Slysh, A.C.A. Hettinger and K.E. Guyler, *Polymer Engineering and Science,* 1974, **14**, 264.

70. P.E. Willard, *Journal of Macromolecular Science A,* 1974, **8**, 33.

71. P.E. Willard, private communication, 1973.

72. S.L. Malhorta and L.P. Blanchard, *Journal of Macromolecular Science A,* 1974, **8**, 843.

73. N.G. Radhakrishnan and M.R. Padbye, *Angewandte Makromolekulare Chemie,* 1975, **43**, 177.

74. J. Sickfield and B. Heinze, *Journal of Thermal Analysis,* 1974, **6**, 689.

75. S. Sourouur and M.R. Kamal, *Thermochimica Acta,* 1976, **14**, 41.

76. R. Kay and A.R.A. Westwood, *European Polymer Journal,* 1975, **11**, 25.

77. J.R. Ebdon and B.J. Hunt, *Analytical Chemistry,* 1973, **45**, 1410.

78. P. Godard and J.P. Mercier, *Journal of Applied Polymer Science,* 1974, **18**, 1493.

79. T. Malavasic, I. Vizovisek, S. Lapanje and A. Moze, *Makromolekulare Chemie,* 1974, **175**, 873.

80. A. Moze, I. Vizovisek, T. Malavasic, F. Cernec and S. Lapanje, *Makromolekulare Chemie,* 1974, **175**, 1507.

81. K.U. Heinen and D.O. Hummel, *Kolloid-Zeitschrift und Zeitschrift für Polymer,* 1973, **251**, 901.

82. A. Sebenik, I. Vizovisek and S. Lapanje, *European Polymer Journal,* 1974, **10**, 273.

83. J.M. Barton, *Journal of Macromolecular Science A,* 1974, **8**, 25.

84. J.M. Barton, *Makromolekulare Chemie,* 1973, **171**, 247.

85. L.W. Crane, P.J. Dynes and D.H. Kaelble, *Journal of Polymer Science, Polymer Letters Edition,* 1973, **11**, 533.

86. E. Sacher, *Polymer*, 1973, **14**, 91.

87. V.I. Lozinsky, E.V. Kalinina, O.I. Putilina, V.K. Kulakova, E.A. Kurskaya, A.S. Dubovik and V.Y. Grimberg, *Polymer Science, Series A*, 2002, **44**, 1122.

88. J. Grenet, S. Marais, M.T. Legras, P. Chevalier and J.M. Saiter, *Journal of Thermal Analysis and Calorimetry*, 2000, **61**, 719.

89. M.W. Barique and H. Ohigashi, *Polymer*, 2001, **42**, 4981.

90. A.L. Sharma, V. Saxena, S. Annapoorni and B.D. Malhotra, *Journal of Applied Polymer Science*, 2001, **81**, 1460.

91. R.P. Quirk, K-C. Hua, L. Zhu and E.S.A. Moctezuma in *Proceedings of the 163rd ACS Rubber Division Meeting*, San Francisco, CA, USA, Spring 2003, Paper No. 5.

92. J. Sun, H. Tang, J. Jiang, X. Zhou, P. Xie, R. Zhang and P.F. Fu, *Journal of Polymer Science, Polymer Chemistry Edition*, 2003, **41**, 636.

93. R-S. Lee, T-F. Lin and J-M. Yang, *Journal of Polymer Science, Polymer Chemistry Edition*, 2003, **41**, 1435.

94. S. Mecking, F.M. Bauers and R.R. Thomann in in *Proceedings of the ACS Polymeric Materials Science and Engineering*, San Diego, CA, USA, Spring 2001, **84**, Paper No.580, p.1049.

95. M. Minigawa, H. Kanoh, S. Tanno and Y. Nishimoto, *Macromolecular Chemistry and Physics*, 2002, **203**, 475.

96. C.N. Kartalis, J.G. Poulakis and C.D. Papaspyrides, *Journal of Applied Polymer Science*, 2002, **86**, 1924.

97. H.Q. Xie and Y. Liu, *Journal of Applied Polymer Science*, 2001, **80**, 903.

98. C. Farcet, B. Charleux and R. Pirrin, *Macromolecular Symposia*, 2002, **182**, 249.

99. S.H. Qin and K.Y. Qiu, *Journal of Polymer Science, Polymer Chemistry Edition*, 2001, **39**, 1450.

100. M.A. Villar, M.D. Failla, R. Quijada, R.S. Mauler, E.M. Valles, M.B. Galland and L.M. Quinzani, *Polymer*, 2001, **42**, 9269.

101. M-R. Kim, H.S. Kim, C-S. Ha, D-W. Park and J-K. Li, *Journal of Applied Polymer Science*, 2001, **81**, 2735.

102. M.S. Nikolic and J. Djonlagic, *Polymer Degradation and Stability*, 2001, **74**, 263.

103. Q. Wu and L. Zhang, *Journal of Applied Polymer Science*, 2001, **82**, 3373.

104. O. Karal-Yilmaz, S. Tasevska, T. Grchev, M. Cvetkovska and B.M. Baysal, *Macromolecular Chemistry and Physics*, 2001, **202**, 388.

105. P.A. Eriksson, A.C. Albertsson, K. Eriksson and J.A. Manson, *Journal of Thermal Analysis and Colorimetry*, 1998, **53**, 19.

106. T. Sasaki, M. Tanaka and T. Takahashi, *Polymer*, 1998, **39**, 3853.

107. B.C. Chern, T.J. Moon, J.R. Howell and W. Tan, *Journal of Composite Materials*, 2002, **36**, 2061.

108. E.P. Soares, E. de Cássia D. Nunes, M. Saiki and H. Wiebeck, *Polimeros: Ciencia e Tecnologia*, 2002, **12**, 206.

109. A.A. Duwalt, *Thermochimica Acta*, 1974, **57**, 8.

110. J.R. MacCallum, *Die Makromolekulare Chemie*, 1965, **83**, 137.

111. I.C. McNeill, *Journal of Polymer Science: Part A-1: Polymer Chemistry*, 1966, **4**, 2479.

112. Y. Mehmet and R.S. Roche, *Journal of Applied Polymer Science,* 1976, **20**, 1955.

113. R.A. Hwelwitskii, I.M. Lubashenko, E.S. Brodskii in *Pyrolysis Mass Spectroscopy of Macromolecules*, Khimja, Moscow, Russia, 1980.

114. H.L.C. Muezelaar, J. Haverkamp and F.D. Hileman, *Pyrolysis Mass Spectroscopy of Recent and Fossil Biomaterials*, Elsevier, Amsterdam, The Netherlands, 1982.

115. T.H. Risby, I.A. Yergey and I.L. Socca, *Analytical Chemistry*, 1982, **54**, 2228.

116. K. Yamada, T. Oura and T. Haruki in *Proceedings of the International Conference of Thermal Analysis*, Budapest, Hungary, 1974, Part 3, p.1029.

117. G. Gaspar, *Analytical Chemistry*, 1978, **50**, 1572.

118. G. Schutjes, *Journal of Chromatography*, 1983, **279**, 269.

119. B-L. Denq, W-Y. Chiu, K-F. Lin and M-R.S. Fuh, *Journal of Applied Polymer Science*, 2001, **81**, 1161.

120. E. Kullik, M. Kalyurand and M. Lamburg, *Laboratory Practice*, 1987, January/February, p.73.

121. E. Kullik, M. Kalyurand and M. Lamburg, *Laboratory Practice*, 1987, January/February, p.74.

122. J.R. Deans, *Journal of Chromatography*, 1984, **289**, 43.

123. E. Heinsoo, A. Kogerman, O. Kirret, J. Coupek and S. Vilkova, *Analytical Applied Pyrolysis*, 1980, **2**, 131.

124. G. Ayrey, B.C. Head and R.C. Poller, *Journal of Polymer Science, Macromolecular Reviews*, 1974, **8**, 1.

125. J.D. Danforth and T.J. Takeuchi, *Journal of Polymer Science A-1*, 1973, **11**, 2091.

126. A. Guyot, A. Bert and R. Spitz, *Journal of Polymer Science A-1*, 1970, **8**, 1596.

127. Y. Shibazaki and H. Kamebe, *Kabishi Kogaku*, 1964, **21**, 65.

128. F. Beckwitz and H. Housinger, *Angewandte Makromolekulare Chemie*, 1975, **45**, 143.

129. L.A. Wall, *Elastoplast*, 1973, 5, 36.

130. J. Mitera, V. Kubelka, J. Novak and J. Mostecky, *Plasty e Kauchuk.*, 1977, **14** 18.

131. M. Chaigneau, *Analysis*, 1977, 5, 223.

132. N. Grassie and D.R. Bain, *Journal of Polymer Science A-1*, 1970, **8**, 2683.

133. N. Grassie and D.R. Bain, *Journal of Polymer Science A-1*, 1970, **8**, 2669.

134. C.R. Scmitt, *Journal of Fire Flammability*, 1972, **3**, 303.

135. C. Beachell and D.L. Beck, *Journal of Polymer Science Part A: General Papers*, 1965, **3**, 457.

136. M. Seegar and R.J. Gritter, *Journal of Polymer Science A-1*, 1977, **15**, 1393.

137. A. Tsuchiya and K. Sumi, *Journal of Polymer Science A-1*, 1969, **7**, 1599.

138. U.D. Moiseeva and M.H. Nieman, *Vysokomolekuliarnye Soedineniia*, 1961, **3**, 1383.

139. L.A. Wall, *Society of Petroleum Engineers*, 1960, **16**, 1.

140. M.C. McGaugh and S. Kottle, *Journal of Polymer Science A-1*, 1968, **6**, 1243.

141. N. Grassie and B.D.J. Torrance, *Journal of Polymer Science A-1*, 1968, **6**, 3303.

142. N. Grassie and B.D.J. Torrance, *Journal of Polymer Science A-1*, 1968, **6**, 3315.

143. M. Nagasawa and A. Holtzer, *Journal of Analytical Chemical Society*, 1964, **86**, 538.

144. B.B. Troitskii, V.A. Varyukhin and L.V. Khokhlova, *Trudy Khimz Khim Teknol*, 1974, **2**, 115.

145. J.C. Gilland and J.S. Lewis, *Angewandte Makromolekulare Chemie*, 1976, **54**, 49.

146. G.F.C. Ehlers, K.R. Fisch and W.R. Powell, *Journal of Polymer Science A-1*, 1969, **7**, 2969.

147. G.F.C. Ehlers, K.R. Fisch and W.R. Powell, *Journal of Polymer Science A-1*, 1969, **7**, 2955.

148. G.F.C. Ehlers, K.R. Fisch and W.R. Powell, *Journal of Polymer Science A-1*, 1969, **7**, 2931.

149. E.A. Boettner and B. Weiss, *American Industrial Hygiene Association Journal*, 1967, **28**, 535.

150. E.A. Boettner, G. Ball and B. Weiss, *Journal of Applied Polymer Science*, 1969, **13**, 377.

151. E.F. Hale, A.G. Farnham, R.N. Johnson and R.A. Glendinning, *Journal of Polymer Science A-1*, 1967, **5**, 2399.

152. S.G. Coloff and N.E. Vandenbergh, *Analytical Chemistry*, 1973, **45**, 1507.

153. S. Foti, A. Liguori, P. Moravigra and G. Montaudo, *Analytical Chemistry*, 1982, **54**, 647.

154. T.H. Risby, J.A. Yergey and J.J. Scocca, *Analytical Chemistry*, 1982, **54**, 2228.

155. J.C. Hughes, B.B. Wheals and M.J. Whitehouse, *Analyst*, 1977, **102**, 143.

156. K.T. Joseph and R.F. Browner, *Analytical Chemistry*, 1980, **52**, 1083.

157. G.J. Moe, *Thermochimica Acta*, 1974, **10**, 259.

158. G.P. Shulman, *Journal of Polymer Science, Part B: Polymer Letters*, 1965, **3**, 11, 911.

159. K. Peltonen, *Analyst*, 1986, **111**, 819.

160. T.L. Chang and T.E. Mead, *Analytical Chemistry*, 1971, **43**, 534.

161. N. Matsuda, H. Shirasaka, K. Takayama, T. Ishikawa and K. Takeda, *Polymer Degradation and Stability*, 2002, **79**, 13.

162. K. Hemvichian and H. Ishida, *Polymer*, 2002, **43**, 4391.

163. G. Montaudo, C. Puglisi, J.W. de Leeuw, W. Hartgers, K. Kishore and K. Ganesh, *Macromolecules*, 1996, **29**, 6466.

164. V.R. Ziatdinov, G. Cai and W.P. Weber, *Macromolecules*, 2002, **35**, 2892.

165. D. Yang, S-D. Li, J.P. Zhong and D-M. Jia, *China Synthetic Rubber Industry*, 2003, **26**, 47.

166. M.A. Aviles, J.M. Gines, J.C. del Rio, J. Pascual, J.L. Perez-Rodríguez and P.J. Sanchez-Soto, *Journal of Thermal Analysis and Calorimetry*, 2002, **67**, 177.

167. H.G. Langer, R.S. Guhlke and S.D. Smith, *Analytical Chemistry*, 1965, **37**, 433.

168. G.P.G. Shulman and H.W. Lochte, *Polymer Preprints*, 1965, **6**, 36.

169. J.L. Koenig in *Proceedings of the Polymer Characterisation Conference*, Cleveland, OH, USA, 1974, p.73.

170. H.R. Udseth and N. Friedman, *Analytical Chemistry*, 1981, **53**, 29.

171. Y. Shimzu and M.J. Munson, *Journal of Polymer Science, Polymer Chemistry Edition*, 1979, **17**, 1991.

172. P. Burille, M. Bert, A. Michel and A. Guyot, *Polymer Science, Polymer Letters Edition*, 1978, **16**, 181.

173. M.M. O'Mara, *Journal of Polymer Science, Polymer Chemistry Edition*, 1970, **8**, 1887.

174. E.P. Change and R. Salovey, *Journal of Polymer Science, Polymer Chemistry Edition*, 1976, **12**, 2927.

175. D.H. Ahlstrom, S.A. Liebman and K.B. Abbas, *Journal of Polymer Science, Polymer Chemistry Edition*, 1976, **14**, 2479.

176. A. Alajberg, P. Arpino, D. Deursiftar and G. Ginochow, *Journal of Analytical Applied Pyrolysis*, 1980, **1**, 203.

177. A. Ballisteri, S. Foti, G. Montaudo and E. Scamporrino, *Journal of Polymer Science, Polymer Chemistry Edition*, 1980, **18**, 1147.

178. A. Guyot, J.P. Benevise and Y. Trambouze, *Journal of Applied Polymer Science*, 1962, **6**, 103.

179. G. Talamini and G. Pezzin, *Die Makromolekulare Chemie*, 1960, **39**, 26.

180. R.R. Stromber, S. Straus and B.G. Achhammer, *Journal of Polymer Science*, 1959, **35**, 355.

181. S. Straus and S.L. Madorsky, *Industrial and Engineering Chemistry*, 1956, **48**, 1212.

182. M.M. O'Mara, *Journal of Polymer Science A-1*, 1971, **9**, 1387.

183. D.T. Watson and K. Bimann, *Analytical Chemistry*, 1964, **36**, 1135.

184. T. Iida, M. Nakanishi and K. Goto, *Journal of Polymer Science*, 1974, **12**, 737.

185. D.M. Desiderio, Jr., and T.E. Mead, *Analytical Chemistry*, 1968, **40**, 2090.

186. J.T. Watson and K. Biemann, *Analytical Chemistry*, 1965, **37**, 844.

187. M. Statheropoulos, K. Mikedi, N. Tzamtzis and A. Pappa, *Analytica Chimica Acta*, 2002, **461**, 215.

188. Y. Fan, H. Nishida, S. Hoshihara, Y. Shirai and Y. Tokiwa, *Polymer Degradation and Stability*, 2003, **79**, 547.

189. S.K. Ghosh, P.P. De, D. Khastgir and S.K. De in *Proceedings of the 151st ACS Rubber Division Conference*, Anaheim, CA, USA, Spring 1997, Paper No.93, p.33.

190. C. Hulubei and C. Gaina, *High Performance Polymers*, 2000, **12**, 247.

191. M. Sibbald, L. Lewandowski, M. Mallamaci and E. Johnson, *Macromolecular Symposia*, 2000, **155**, 213,

192. B. Gaur, B. Lochab, V. Choudhary and I.K. Varma, *Journal of Thermal Analysis and Calorimetry*, 2003, **71**, 467.

193. X. Hu and L. Xu, *Polymer*, 2000, **41**, 9147.

194. E.J. Nelson, S.H. Foulger and D.W. Smith, *High Performance Polymers*, 2001, **13**, 101.

195. J.A. Mikrayannidis, *Journal of Polymer Science, Polymer Chemistry Edition*, 1997, **35**, 1353.

196. G. Di Pasquale, A.D. La Rosa, A. Recca, S. Di Carlo, M.R. Bassani and S. Facchetti, *Journal of Materials Science*, 1997, **32**, 3021.

10 Monitoring of Resin Cure

Two techniques, dynamic mechanical thermal analysis (DMTA) and dielectric thermal analysis (DETA), have been used for the study of resin cure. Differential scanning calorimetry (DSC) has also been employed (for a discussion of the theory and instrumentation of DSC, see Chapter 9). The application of differential photocalorimetry to the measurement of cure rates of photocurable resins is discussed in Chapter 12.

10.1 Dynamic Mechanical Thermal Analysis

10.1.1 Theory

Although DMTA has found its major applications in the mechanical testing of plastics (see Chapter 15), it has also found application in degradation studies, chemical reaction studies such as resin curing measurements (i.e., degree and rate of cure), and studies of temperature properties of elastomers.

In resin cure studies the technique characterises the rheological changes in resins before, during, and after cure. Plots of temperature *versus* permittivity pinpoint the glass transition temperature (T_g) for the resin during cure. Plots of time *versus* logarithm of loss factor enable determinations of vitrification of resins during cure to be carried out.

The rheological changes in a polymer during complex thermal histories can provide information about polymer processing, chemical structure, and end-use performance (time or temperature versus logarithm of loss factor, time versus logarithm of conductivity, and temperature–permittivity plots).

10.1.2 Instrumentation

Some available instrumentation and software for DMTA is listed next (see also Appendix 1):

- Perkin Elmer: Series 7, Model DMA-7
- DuPont: 9000 Thermal Analysis System, Model 983
- PL Thermal Sciences: Model PL DMA
- TA Instruments: Model DMA 983

10.1.3 Applications

Viscous thermosetting materials such as prepregs, adhesives, or coatings can be evaluated by dynamic mechanical analysis (DMA) using a supported structure such as a glass cloth. In these experiments, information about the curing properties can be obtained as the thermoset progresses from a liquid to a rigid solid. **Figure 10.1** illustrates the results for an epoxy prepreg. Initially the prepreg has a G* (shear loss modulus) which is higher than G′ (shear storage modulus) because the low molecular weight liquid polymer exhibits low elasticity and high damping as it 'flows' during oscillatory testing. Both G′ and G* increase as molecular weight (cure) advances. G′ eventually exceeds G* as the polymer system gains elasticity due to molecular entanglement and network formation. The crossover of the G′ and G* curves has been related to the gel point for these materials.

Using DMA the curing reactions of phenol-formaldehyde resins have been followed [1]. The evolution of various rheological parameters was recorded for samples of the resins on cloth. A third-order phenomenological equation described the curing reaction. The influences of the structure, composition, and physical treatment on the curing kinetics were evaluated.

Schoff [2] has described the use of dynamic mechanical techniques in the characterisation of commercial organic coatings. Such techniques have become valuable methods for the determination of basic viscoelastic properties, following and measuring cure, and

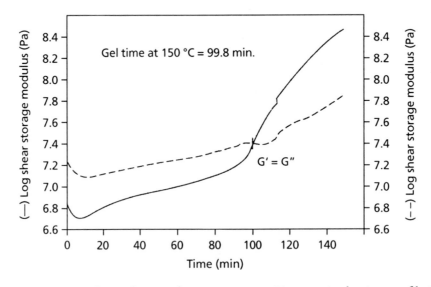

Figure 10.1 Isothermal cure of epoxy prepreg (*Source: Author's own files*)

for determining structure–property relationships. Disadvantages are also considered, including the requirement for free films or supported specimens and the problems of using results from small strain measurements to predict ultimate failures such as cracking, stone chipping, mar and scratch, and delamination.

Pawlowski and Dick [3] have discussed the application of the recently introduced Alpha Technologies RPA 2000 dynamic mechanical rheological tester to isothermal cure studies.

10.2 Dielectric Thermal Analysis

10.2.1 Theory

Insulating materials possess a dielectric constant (E′) characterising the extent of electrical polarisation that can be induced in the material by an electric field. If an alternating electric field is applied, the polarisation lags behind the field by a phase angle, δ. This results in partial dissipation of stored energy. The dissipation energy is proportional to the dielectric loss (ε″) and the stored energy to the dielectric constant (ε′).

The DETA technique normally obtains data from thermal scans at constant impressed frequency. The T_g at which molecular motions become faster than the impressed timescale are recorded as peaks in ε″ and tan δ.

It is a simple matter to multiplex frequencies over the whole frequency range 20–100 kHz and under such conditions the peaks in ε″ are shifted to higher temperatures as frequency is increased. A further option allows data to be obtained in the frequency plane under isothermal conditions.

This technique is used principally for the rheological characterisation of polymers (Sections 18.1.3 and 18.1.4) and measurement of dielectric constant (Section 18.2). In the field of thermal and mechanical stability it has also found application in the fields of resin cure kinetics and resin cure monitoring.

The technique measures changes in the properties of a polymer as it is subjected to a periodic electrical field. This produces quantitative data from which can be determined the capacitive and conductive nature of materials. Molecular relaxations (Section 13.9.1) can be characterised and flow and cure of resins monitored (Section 10.2.1).

While the theory of dielectric analysis is well known, its use has long been frustrated by lack of effective instrumentation, particularly the lack of adequate sensors and experimental controls, difficulties with sample handling, and the lack of analysis software for easy conversion of raw data into meaningful results. Modern dielectric thermal analyses make the technique a practical reality.

10.2.2 Instrumentation

Available instrumentation is listed below (see also Appendix 1):

- DuPont: Model DEA-2970
- PL Thermal Sciences: Model PL-DETA
- TA Instruments: Model DETA

10.2.3 Applications

DETA can be used to monitor the cure of resins. The technique can characterise dramatic rheological changes in resins before, during, and after cure. This information can be used to identify the appropriate storage temperature and processing conditions for thermosets, elastomers, adhesives, coatings, and many other polymeric materials.

The curves in **Figures 10.2(a) and (b)** show such rheological changes for an epoxy–amine mixture. This very viscous mixture was applied to a ceramic single-surface sensor with a spatula. The sensor and sample were then placed in a furnace, cooled to –75 °C, and then heated at 3 °C/min. The curves describe the effect on the material. Below 0 °C, the mixture is solid because main-chain molecular motion in the resin is characteristically, greatly restricted below the T_g. This suggests that such a mixture should be stored at a temperature below 0 °C in order to minimise the chemical reaction between the epoxy resin and the amine crosslinker. As the resin is heated from 0 °C through its T_g to about 50 °C, it is transformed into a viscous liquid. Above 50 °C, the viscosity decreases, as indicated by the increase in loss factor, which is due to the increased mobility of free ions. Above 150 °C, temperature-induced fluidity is overshadowed by an increase in molecular weight and network formation due to the epoxy–amine reaction. Also in this temperature range, the loss factor decreases as the increased molecular weight restricts ionic mobility. The cure reaction is complete at about 200 °C.

Curing studies such as this can be combined with isothermal experiments to generate time–temperature transformation diagrams for a material. The resulting information is critical for optimising storage, shipping, and processing conditions.

Dielectric analysis is a very sensitive technique for determining the vitrification of a chemically reactive resin. A plot of log ε'' (loss factor) *versus* time shows two log ε''

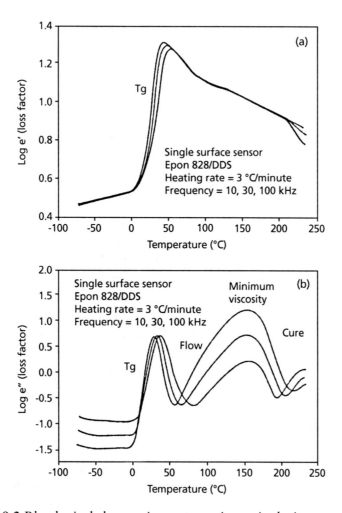

Figure 10.2 Rheological changes in epoxy-amine resin during cure (a) TG;
(b) Transitions as defined by ε″ curves (*Source: Author's own files*)

peaks corresponding to the onset of cure and to vitrification. Vitrification is the process
by which the chemical reaction is quenched during the curing cycle. It prevents the resin
from achieving a fully cured state. This is of critical concern because full cure is required
for the development of many desired properties, such as mechanical strength and solvent
resistance. Vitrification during processing can also result in problems with a product's
long-term performance, especially those required to operate at high temperatures. Such
exposure results in residual curing, which can lead to warpage, residual stress, loss of
paint adhesion, and many other problems.

In some applications, however, vitrification can be used to advantage. For example, limited crosslink density can be a desirable means of sustaining impact resistance. Thus the ability to identify proper processing conditions for control of vitrification can be critical to the success of many products.

DETA is an ideal technique for monitoring the cure of silicone potting compounds because the unreacted material (with its low viscosity) can be directly applied to the dielectric sensor and continuous measurement made as the material transforms from a low molecular weight liquid to a high molecular weight crosslinked rubber.

Figure 10.3(a) shows the results obtained for the uncured silicone sealant Lot A. This plot shows the loss factor, ε'', as a function of time. The temperature profile is also contained in this figure. The sealant initially shows an increase in ε'' as the resin is heated. This increase reflects a decrease in the viscosity of the unreacted material as the temperature increases. After 9.9 minutes, a maximum is obtained in the loss factor data which reflects the point of minimum resin viscosity. After this point, the material undergoes crosslinking, and the loss factor begins to decrease (as the molecular weight of the sealant begins to increase). A second maximum is observed in the loss factor at 16.7 minutes (at the beginning of the isothermal period). The appearance of the two maxima in the loss factor data indicates that Lot A cures in two discreet steps.

The results obtained for the second sealant sample (Lot B) show significant differences, as displayed in **Figure 10.3(b)**. The loss factor data for this sample yields only a single maximum at 17.2 minutes. This behaviour indicates that this particular sealant undergoes curing only after the isothermal temperature of 100 °C is reached. These results demonstrate that dielectric analysis yields excellent results for these potting compounds and can clearly distinguish between curing processes associated with these two materials. These differences should result in significantly different long-term properties for the cured sealants.

McIlhagger and co-workers [4] used an on-line parallel plate dielectric analyser to identify the key cure stages in liquid moulding processes. This technique appears to offer the greatest potential for determining the through-thickness cure state of a resin during cure. A laboratory dielectric instrument is utilised to simulate resin transfer moulding and autoclave cure cycles for composite structures continuing non-conductive and conductive fibres and for different resin systems used in the aerospace industry. Key resin cure stages are identified by an appropriate dielectric signal and correlated with data from other thermal and mechanical techniques.

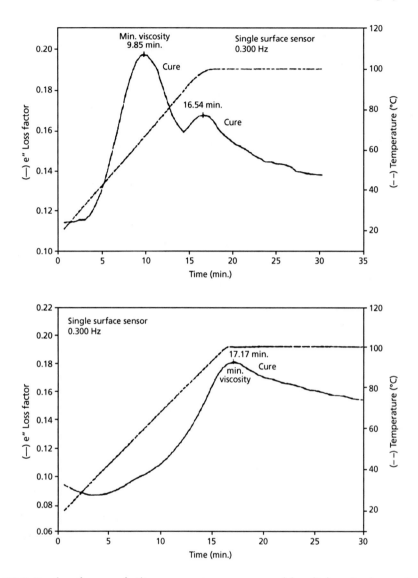

Figure 10.3 Study of cure of silicone potting compound by dielectric thermal analysis (a) Lot A; (b) Lot B (*Source: Author's own files*)

10.3 Differential Scanning Calorimetry

The theory of this technique is discussed in Section 9.2.3 and instrumentation is discussed in Chapter 9 and Appendix 1.

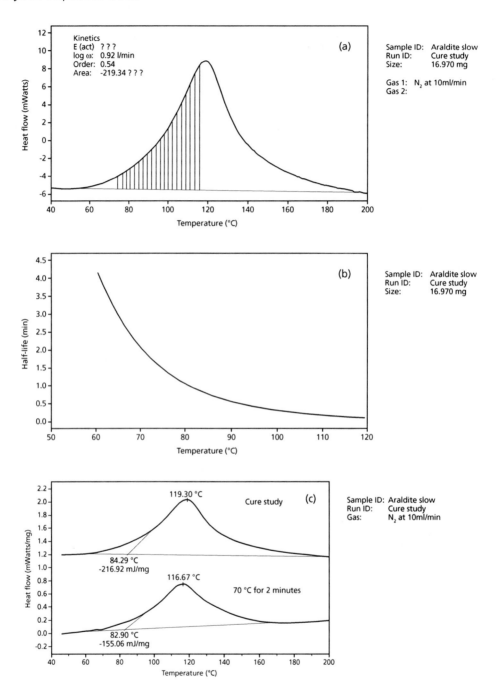

Figure 10.4 Application of differential scanning calorimetry to study cure reaction of Araldite. (a) Curing exotherm, (b) Half life curve, (c) Comparison of partially cured resin and uncured resin (*Source: Author's own files*)

DSC has been used in studies of cure reactions in thermosetting resins. It enables the extent of cure, the kinetics of cure including evaluation of the half-life of a cure reaction, and the heats of curing all to be obtained from the DSC curve (**Figure 10.4**).

Grenet and co-workers [5] have given details of a study of the curing kinetics of unsaturated polyesters using DSC and Fourier transform infrared spectroscopy.

10.4 Fibre Optic Sensor to Monitor Resin Cure

Lenhart and co-workers [6] have described a fibre optic sensor to monitor the curing of resins and interphase formation. The sensor was made by tethering a dimethylamino-nitrostilbene fluorophore to a triethoxy silane coupling agent, to give a fluorescently labelled silane coupling agent which was covalently grafted to the fibre surface. The sensor was used to study the curing of an epoxy resin, the position of the fluorescence maximum being used to detect a difference between the bulk resin and the interphase. The change in fluorescent shape during curing was quantified by subtracting the intensity at 635 nm from that at 580 nm, the intensity difference being plotted against time to establish a curing profile.

References

1. S. Markovic, B. Dunjic, A. Zlatanic and J. Djonlagic, *Journal of Applied Polymer Science*, 2001, **81**, 1902.

2. C.K. Schoff in *Proceedings of ACS Polymeric Materials Science and Engineering Conference*, Orlando, FL, USA, Fall 1996, **75**, 162.

3. H. Pawlowski and J.S. Dick, *Rubber and Plastics News*, 1997, **26**, 12.

4. A.T. McIlhagger, S.T. Matthews, D. Brown and B. Hill in *Proceedings of ICAC '99*, Bristol, UK, 1999, p.133.

5. J. Grenet, S. Marais, M.T. Legras, P. Chevalier and J.M. Saiter, *Journal of Thermal Analysis and Calorimetry*, 2000, **61**, 719.

6. J.L. Lenhart, J.H. van Zanten, J.P. Dunkers and R.S. Parnas in *Proceedings of ANTEC 2000 Conference*, Orlando, FL, USA, 2000, Paper No. 163.

11 Oxidative Stability

Seven different techniques have been applied to the study of the oxidative stability of polymers:

- Thermogravimetric analysis (TGA)

- Differential scanning calorimetry (DSC)

- Pressure differential scanning calorimetry (PDSC)

- Evolved gas analysis (EGA)

- Infrared spectroscopy

- Electron spin resonance spectroscopy (ESR)

- Matrix-assisted laser desorption/ionisation mass spectrometry (MALDI)

11.1 Theory and Instrumentation

The theory and instrumentation of these techniques has been discussed in earlier chapters. See also Appendix 1.

11.2 Applications

Some examples of the applications of these techniques are discussed next.

In most cases thermal oxidation studies in polymers would be conducted under an inert gas such as nitrogen or helium in order to avoid the formation of secondary oxidation products that would complicate interpretation. In some cases, however, information is required on the stability to oxidation of polymeric materials and in these instances the thermal experiment would be carried out in an atmosphere of oxygen or air. Applications of the various techniques listed above to the determination of oxidative stability are discussed next.

11.2.1 Thermogravimetric Analysis

In this technique the sample is heated in air or oxygen at a constant heating rate until complete oxidation or combustion occurs and weight changes are continuously recorded as a weight *versus* temperature plot. Often the shape of the derivative thermogram is typical for a certain polymer and can be used for characterisation.

Thizon and co-workers [1] have studied the effect of low concentrations of oxygen and water in nitrogen on the rate of thermal degradation of ethylene–propylene copolymers. Using TGA they are able to obtain estimates of the rate and change of polymer weight with time (dm/dt), hence activation energy at a range of polymer temperatures (240–350 °C) and oxygen contents in the head space gas (0–90 ppm) and water contents in the head space gas (0–600 ppm):

$$\frac{dm}{dt} = K \exp\left(\frac{E}{RT}\right)$$

It is known that oxygen reacts with hydrocarbon polymers even at moderate temperatures [2]. The resulting peroxides or hydroperoxides which are thermally unstable can induce a chain degradation through reactions involving the RO_2 radical. Thus, oxygen contamination of the head space gas in a thermogravimetric experiment is likely.

The oxygen effect is severe, particularly at low temperatures. The degradation rate can be multiplied by a factor of as large as six for an oxygen concentration of 90 ppm at 240 °C; it is multiplied by 1.2 at 280 °C and 2 at 240 °C for a concentration of 20 ppm. The water content of the carrier gas is not as critical at least for ethylene–propylene copolymers and does not appreciably affect the thermal degradation.

11.2.2 Differential Scanning Calorimetry

To examine oxidative breakdown, a sample is tested in flowing air or oxygen. Since oxidation is highly exothermic, onset of oxidation is clearly visible on a DSC trace. The time to the extrapolated onset of the exothermic reaction under isothermal conditions is called the induction period. Dynamic determination of oxidation stability (heating the sample at 10 °C/min, for example) is much quicker than the isothermal procedure and permits simultaneous measurement of glass transition or melting temperatures. This type of analysis is shown in **Figure 11.1** by the comparative performance between polyethylene (PE) and polypropylene (PP) films. As with the isothermal induction period, the resulting onset temperature is a measure of polymer stability.

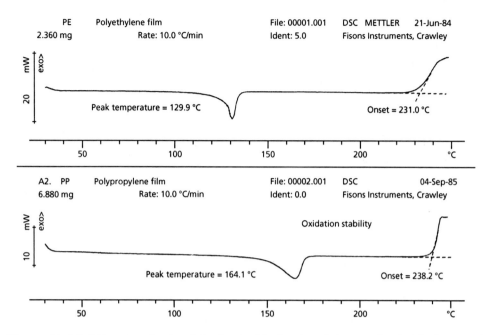

Figure 11.1 DSC curve obtained during oxidative breakdown of polyethylene and polypropylene. (*Source: Author's own files*)

An alternative technique is pressure DSC (PDSC). This technique measures heat flow and temperatures of transitions as a function of temperature, time, and pressure (elevated pressure or vacuum). The ability to vary pressure from 1.3 Pa to 6.8 MPa makes PDSC ideal for the determination of oxidative polymer stability and for other pressure-sensitive reactions.

High pressures of oxygen cause polymers to oxidise much faster than they would at atmospheric pressure. The oxidation process is observed as an exothermic peak. Thus, the oxidative stability of a polymer can be measured as the time to the onset of oxidation at a predetermined temperature and pressure.

DuPont supply a dual sample pressure differential calorimetry cell suitable for use in their 9900 thermal analysis system.

Antioxidants are common additives for polymers. Oven ageing and thermal analysis techniques such as DSC and TGA have been used with varying degrees of success to measure both the concentration and effectiveness of antioxidants. In most cases, the failure of these tests to correlate with actual end-use performance is due to the volatility of the antioxidants and other components at test temperatures. Even at a relatively modest test temperature (150 °C) the volatility of stabilisers varies greatly, e.g., butylated hydroxytoluene antioxidant loses 30% of its weight in 15 minutes while Irganox 1010 loses only 5% in the same time.

If an isothermal test were performed under these conditions, results would be greatly affected by the volatility of the antioxidant and not accurately reflect end-use performance.

PDSC eliminates the volatility problem in two ways. First, high pressure decreases the volatility by increasing the boiling point, and second, high pressure increases the concentration of the reacting gas, oxygen. This allows lower test temperatures or provides significantly shorter test times at equivalent temperatures.

A DSC heating experiment in air on an impact-modified PP sample shows the melting of two components followed by oxidative decomposition at 237.7 °C. When tested by TGA in nitrogen the same material shows thermal degradation starting near 300 °C. However, an expanded view of the top 5% weight loss shows a component (0.66%) of the sample is lost between 126 and 184 °C at ambient pressure. The identity of the component is unknown, but the mere fact that the weight loss occurs makes oxidative stability results suspect if they are generated at a temperature above 126 °C and at ambient pressure.

The following advantages of PDSC over DSC and oven ageing for determination of oxidative stability have been identified:

1. At 4.1 MPa oxygen pressure and a specimen size of 4–5 mg, the maximum oxidation rate is maintained throughout the reaction and the results are not dependent on oxygen concentration or diffusion rate.

2. Temperature and pressure can be precisely controlled with commercial thermal analysis equipment. This eliminates errors caused by temperature variations commonly encountered in oven testing.

3. PDSC oxidation exotherms are sharper and better defined than the usual differential scanning calorimetric tests at atmospheric pressure, making extrapolation of the onset of autoxidation easier and more precise.

4. The test temperature may be varied to permit analysis of samples in the molten, semi-molten, or non-molten state, or to permit changing the duration of the test. Conversely, studies may be conducted at relatively low temperatures if a longer test period can be tolerated.

5. Because high oxygen pressures accelerate autoxidation, the test can be carried out at sufficiently low enough temperatures to minimise volatilisation of stabilisers.

DSC has been used in conjunction with differential thermal analysis to investigate the kinetics of the oxidation of isotactic polypropylene (iPP) [3].

DSC and oxygen uptake experiments have been used to measure the oxidative stability of gamma-irradiated ethylene–propylene elastomers [4]. The oxidative irradiation environment generated peroxy radicals that were involved in the air-degraded samples. The specific heat capacity dependences on temperature determined for the two methods of irradiation were dissimilar.

Woo and co-workers [5] carried out polyolefin durability estimates by oxidative stability testing using DSC.

Groening and Hakkarainen [6] report results of an investigation that has established a correlation between the degradation product pattern and changes in the mechanical properties during thermooxidative degradation of in-plant recycled polyamide 6,6 with the aid of headspace solid-phase microextraction with gas chromatography (GC)/mass spectroscopy (MS), tensile testing, DSC, and Fourier transform infrared spectroscopy (FT-IR).

Ding and co-workers [7] found that the commonly used oxidative stability measurement of oxidative induction time (OIT) at high temperatures, typically employing molten polymers, grossly overestimates stability at lower temperatures in the solid state. The continuity and suitability of OIT data at lower temperatures and under widely different experimental parameters were examined for high-density polyethylenes (HDPE) and PP of various formulations. Upon raising the oxygen pressure from about 0.02 to 4 MPa, for example, an approximately 20-fold acceleration was observed. For a hindered phenol-stabilised PP, a normal 1/OIT *versus* the square root of oxygen pressure was observed, while a hindered amine-stabilised PP showed marked departure from this behaviour at lower pressures. For HDPE, a marked loss in OIT was observed when airflow rate through the DSC cell was reduced, along with a reduction in apparent activation energy. These effects were interpreted from experimental parameters, when possible, with polymer degradation mechanisms.

11.2.3 Evolved Gas Analysis

Schole and co-workers [8] have reported on the characterisation of polymers using an oxidative degradation technique. In this method the oxidation products of the polymers are produced in a short pre-column maintained at 100–600 °C just ahead of the separation column in a gas chromatograph. The oxidation products are swept on to the separation column and detected in the normal manner.

Chien and Kiang [9] oxidised PP at temperatures between 240 and 289 °C using a DuPont 900 thermogravimetric analyser. The products were separated on a gas chromatograph and identified on-line by an interfaced MS-GC identification system. The major products were carbon dioxide, water, acetaldehyde, acetone, butanol, formaldehyde, methanol, and other ketones and aldehydes. Rate constants and activation energies were calculated.

Wampler and Levy [10] have applied similar principles to the study of the oxidative degradation of PE and PP. The resulting chromatograms showed that the predominant materials produced depended on the rate at which the sample was heated. Oxidation products identified in the cure of PE were decyl aldehyde, dodecyl aldehyde, and decanol.

11.2.4 Infrared Spectroscopy of Oxidised Polymers

Extensive infrared studies have been carried out on oxidised PE [11–17]. Infrared absorption spectroscopy has been widely used to determine the oxidation products and the rate of formation of these products during the thermal or photo-oxidation of PE [13, 18, 19]. Acids, ketones, and aldehydes, the end products reported from these oxidations, have similar spectra in the 5.50–6.00 µm region. It is only in this carbonyl stretching region that the products have suitable absorptivity to give quantitative data. The absorption band of the acid (5.84 µm), ketone (5.81 µm), and aldehyde (5.77 µm) groups present in oxidised PE are so overlapped that they only give a broad band on relatively low-resolution infrared spectrometers. Interpretation of these data, based on the increase in total carbonyl rather than on a single chemical moiety, could lead to incorrect conclusions because of the large differences in the absorptivity of the various oxidation products. Acid absorptivity has been reported [20] to be 2.4 times greater than that of ketones and 3.1 times greater than that of aldehydes. Rugg and co-workers [12] using a grating spectrometer for increased resolution have demonstrated that the carbonyl groups formed by heat oxidation are mainly ketonic, while in highly photo-oxidised PE the amounts of aldehyde, ketone, and acid are approximately equal. This procedure is adequate for qualitative, but not suitable for accurate quantitative, data.

An infrared study of oxidative crystallisation of PE was made from examination of the 5.28 µm 'crystallinity' band and 7.67 µm 'amorphous' band and carbonyl absorption at 5.83 µm [21]. Miller and co-workers [22] used FT-IR infrared to study the effect of irradiation on PE. Aldehydic carbonyl and vinyl groups decreased and the ketonic carbonyl and *trans*-vinylene double bonds increased on irradiation.

Infrared reflection was used in studies of oxidation of PE at a copper surface in the presence and absence of an inhibitor, *N,N*-diphenyl-oxamide [23].

Cooper and Prober [17] have used alcoholic sodium hydroxide to convert the acid groups to sodium carboxylate (6.40 μm) to analyse PE oxidised with a corona discharge in the presence of oxygen and ozone. This procedure requires five days, and has been found to extract the low molecular weight acids from the film.

Heacock [24] has described a method for the determination of carboxyl groups in oxidised polyolefins without interference by carbonyl groups. This procedure is based upon the relative reactivities of the various carbonyl groups present in oxidised PE film to sulfur tetrafluoride gas. The quantity of the carboxyl groups in the film is then measured as a function of the absorption at 5.45 μm.

The appearance in the spectra of irradiated (gamma rays from ^{60}Co) PP specimens, after storage in air, of a strong band in the region of 5.85 μm, corresponding to carbonyl groups, must be explained by reaction of oxygen with the long-lived allyl radical, with formation of peroxide radicals which form carbonyl groups by decomposition [25].

The intensity of the 5.85 μm band of PP (and consequently the degree of oxidation) increases sharply with time of storage of specimens in air. Irradiated amorphous specimens oxidise to a considerably lesser extent than iPP specimens. The degree of oxidation of specimens at –196 °C increases more rapidly than when the specimens are irradiated at 25 °C. All these facts indicate that the lifetime of the allyl radicals is longer in crystalline PP, and that the concentration of these radicals is higher in specimens irradiated at low temperature. The free radicals are destroyed only after heat treatment of the specimens in an inert atmosphere at 150 °C. After this heat treatment the intensity of the 5.85 μm band ceases to increase on storage, i.e., no further oxidation occurs.

Adams and Goodrich [26] have compared on a qualitative basis the non-volatile oxidation products obtained by photo- and thermo-oxidation of PP. They used infrared spectroscopy and chemical reactions. The major functional group obtained by a photodecomposition is followed by vinyl alkene, then acid. In comparison, thermally oxidised PP contains relatively more aldehyde, ketone, and γ-lactone, and much less ester and vinyl alkene. Photodegraded PE contains mostly vinyl alkene followed by carboxylic acid. Gel permeation chromatography determined the decrease in PP molecular weights with exposure time. Adams and Goodrich determined that in photochemical oxidation there is one functional group formed per chain scission; in thermal oxidation there are two groups formed per scission. Adams and Goodrich [26] make the following comments regarding the infrared spectrum of oxidised PP.

Hydroxyl region. The hydroxyl absorption in the infrared spectrum of PP has a broad band centred at 2.90 μm (associated alcohols) with a definite shoulder at 2.77 μm (unassociated alcohols). At a similar extent of degradation, thermally oxidised polyolefins show hydroxyl bands of roughly half the absorbance values of the photo-oxidised polyolefins. Thus thermal oxidation produces about half as many hydroxyl groups as photo-oxidation in polyolefins.

A portion of the PP hydroxyl absorption could be due to hydroperoxides. If so, then an exposed sheet, with the volatiles removed, heated in a nitrogen atmosphere for two days at 140 °C, should show a decrease in the hydroxyl infrared band and an increase in the carbonyl band due to the decomposition of hydroperoxides under such treatment. The infrared spectrum of the photodegraded PP sheet subjected to the thermal treatment showed at 20% decrease in the hydroxyl band. However, the broad carbonyl band at 5.75 μm did not increase but showed a 5% decrease. However, the small γ-lactone (5.75 μm) and vinyl alkene (6.08 μm) bands did show a slight increase. Thus, these results are due not to hydroperoxide decomposition but to some carboxylic acids converting to γ-lactones and some terminal alcohols dehydrating to vinyl alkenes at the high temperature. While hydroperoxides are undoubtedly an intermediate in the photo-oxidation process, they decompose too rapidly under ultraviolet light to build up any significant concentration.

Carbonyl region. The PP carbonyl band after 335 hours exposure is broad, with few discernible features except for the vinyl alkene band at 6.08 μm. The broadness of the carbonyl band indicates a large variety of functional groups, and makes accurate quantitative analysis difficult. The large vinyl alkene band at 6.08 μm stands out clearly and distinct carboxylic acid (5.83 μm) and γ-lactone (5.58 μm) spikes can be readily identified.

After the volatile products are removed by the vacuum oven, the carbonyl band for PP decreases. Isopropanol extraction removes about 40% of the PP carbonyl. The carbonyl band is then narrow and appears to centre at the ester absorption at 5.75 μm.

Treatment with base converts lactones, esters, and acids to carboxylates (6.33 μm), leaving only a small band at 5.81 μm, which is due to aldehyde and ketone.

Upon reacidification of the PP, some of the original esters at 5.75 μm do not reform but become carboxylic acids and γ-lactones. Curiously, the vinyl alkene band becomes less intense with each step and broader, shifting down to 6.10–6.25 μm, the vinyl groups may be isomerised into internal alkenes or become conjugated during the various treatments, although no such change occurs with either the PP vinyl alkene or with the process-degraded PP vinyl alkene. Wood and Statton [27] developed a new technique to study molecular mechanics of oriented PP during creep and stress relaxation based on use of the stress-sensitive 10.25 μm band and the orientation-sensitive 11.12 μm band. The far infrared spectrum of iPP was obtained from 400 to 10 cm^{-1} and several band assignments were made [28]. The isotacticity of PP has been measured from infrared spectra and pyrolysis gas chromatography following calibration from standard mixtures of iPP and atactic polypropylene. The infrared spectrum of oxidised PP indicated small amounts of OOH groups plus larger concentrations of stable cyclic peroxides or epoxides in the PP chain [28].

Grassie and Weir [29] described an apparatus for the measurement of the uptake of small amounts of oxygen by polystyrene (PS) with a high degree of precision. Grassie and Weir [30] also investigated the application of ultraviolet and infrared spectroscopy to the assessment of PS films after vacuum photolysis in the presence of 253.7 nm radiation using the apparatus mentioned previously. During irradiation there is a general increase in absorption in the range 230–350 nm. Rates of increase are relatively much greater in the 240 and 290–300 nm regions. Absorption in the 240 nm region is characteristic of compounds having a carbon–carbon double bond in conjunction with a benzene ring. Styrene, for example, has an absorption band at 244 nm.

Schole and co-workers [31] have applied an oxidative degradation technique to the study of PS. In this technique the PS sample is mixed with a support in a pre-column which is mounted at the inlet to a GC column.

Shaw and Marshall [32] have carried out an infrared spectroscopic examination of emulsifier-free PS which, had been oxidised during polymerisation. Evidence was found for the presence of surface carboxyl groups bound to the polymer chains, presumably formed by oxidation during polymerisation. The band at 5.86 μm was assigned in part to the carbonyl stretching mode of dimeric carboxylic acid, formed by oxidation, in the PS chains. Absorption at 5.65 μm, which was very weak, was tentatively attributed to the carbonyl stretching mode of the monomeric form of this acid. The structure of the acid end group was not established but the results obtained suggested that it was possibly a phenylacetic acid residue or a residue of standard (unoxidised) and of oxidised emulsion polymerised PS in the region 12.5–25.0 μm.

An alternative technique involves infrared spectroscopic examination of films of polymers that have been subjected to oxidative degradation at elevated temperatures [26, 33-36]. Changes in concentrations of carbonyl and other oxygenated functional groups can be obtained by this method.

Ahlblad and co-workers [37] determined the oxidation profiles of polyamide 6,6 using FT-IR spectroscopy and imaging chemiluminescence. FT-IR analysis provides carbonyl index depth profiles and imaging chemiluminescence gives a peroxide depth profile.

Kaczmarek and co-workers [38] investigated the course of photo-oxidative degradation of polyacrylic acid, polymethacrylic acid, and polyvinyl pyrrolidone using infrared and ultraviolet spectroscopy.

11.2.5 Electron Spin Resonance Spectroscopy

Ohnishi and co-workers [39] have carried out an ESR study of the radiation oxidation of polyvinyl chloride (PVC). This technique has also been applied to studies of PE [40], polymethyl methacrylate [41], PVC [42], polymethacrylic acid [43], and polycarbonate [44].

11.2.6 Matrix-Assisted Laser Desorption/Ionisation Mass Spectrometry

Gallet and co-workers [45] studied the oxidative thermal degradation of polyethylene oxide–ethylene oxide triblock copolymer. It was found by MALDI that degradation starts after 21 days at 80 °C.

11.2.7 Imaging Chemiluminescence

Forsstram and co-workers [46] studied the thermo-oxidative stability of polyamide 6 in the temperature range 100–140 °C using chemiluminescence and imaging chemiluminescence. Tests were carried out on unstabilised polyamides and polyamides stabilised with the phenolic antioxidant Irganox 1098. The oxidation rate is discussed in terms of chemiluminescence intensity *versus* the rate of oxygen uptake along with associated changes in chemical and mechanical properties.

References

1. M. Thizon, C. Eon, P. Valentin and G. Guiochon, *Analytical Chemistry*, 1976, 48, 1861.

2. P.M. Norling and A.V. Taboesky in *Thermal Stability of Polymers*, Ed., R.T. Conley, Dekker, New York, NY, USA, 1970, Volume 1, Chapter 5.

3. A.A. Duiwalt, *Thermochimica Acta*, 1974, 57, 8.

4. T. Zaharescu, V. Meltzer and R. Vilcu, *Polymer Degradation and Stability*, 1999, 64, 101.

5. L. Woo, C.L. Sandford and S.Y. Ding, *Polymer Preprints*, 2001, 42, 1, 394.

6. M. Groening and M. Hakkarainen, *Journal of Applied Polymer Science*, 2002, 86, 3396.

7. S. Ding, M.K.T. Ling, A.R. Khare, C.L. Sandford and L. Woo, *Journal of Applied Medical Polymers*, 2000, **4**, 28.

8. R.G. Schole, J. Bednarczyk and T. Yamanichi, *Analytical Chemistry*, 1966, **38**, 331.

9. J.C.W. Chien and F.J.Y. Kiang, *Makromolekulare Chemie*, 1980, **181**, 47.

10. T.P. Wampler and E.J. Levy, *Journal of Analytical and Applied Pyrolysis*, 1985, **8**, 153.

11. L.H. Cross, R.B. Richards and H.A. Willis, *Discussions of the Faraday Society*, 1950, **9**, 235.

12. R.M. Rugg, J.J. Smith and R.C. Bacon, *Journal of Polymer Science*, 1954, **13**, 535.

13. A.W. Pross and R.M. Black, *Journal of the Society of the Chemical Industry (London)*, 1950, **69**, 115.

14. A.C. Beachell and S.P. Nemphos, *Journal of Polymer Science*, 1956, **21**, 113.

15. A.C. Beachell and G.W. Tarbet, *Journal of Polymer Science*, 1960, **45**, 451.

16. J.P. Luongo, *Journal of Polymer Science*, 1960, **42**, 139.

17. G.D. Cooper and X. Prober, *Journal of Polymer Science*, 1960, **44**, 397.

18. G. Kimmerle, *Combustion Toxicology*, 1974, **1**, 4.

19. H. Kveder and G. Ungar, *Nafta (Zagreb)*, 1973, **24**, 85.

20. Yu A. Pentin, B.N. Tarasevich and B.N. El'tsefon, *Vestnik-Moskovskii Universitet Khimia*, 1973, **14**, 13.

21. D.L. Tabb, J.J. Sevcik and J.L. Koenig, *Journal of Polymer Science, Polymer Physics Edition*, 1975, **13**, 815.

22. P.J. Miller, J.F. Jackson and R.S. Porter, *Journal of Polymer Science, Polymer Physics Edition*, 1973, **11**, 2001.

23. M.G. Chan and D.L. Allara, *Polymer Engineering Science*, 1974, **14**, 12.

24. J.F. Heacock, *Journal of Applied Polymer Science*, 1963, **7**, 2319.

25. H. Kveder and G. Ungar, *Nafta (Zagreb)*, 1973, **24**, 86.

26. J.H. Adams and J.E. Goodrich, *Journal of Polymer Science A-1*, 1970, **8**, 1269.

27. R.P. Wood and W.O. Statton, *Journal of Polymer Science, Polymer Physics Edition*, 1974, **12**, 1575.

28. M. Goldstein, M.E. Seeley, H.A. Willis and V.J.I. Zichy, *Polymer*, 1973, **14**, 530.

29. N. Grassie and N.A. Weir, *Journal of Applied Polymer Science*, 1965, **9**, 963.

30. N. Grassie and N.A. Weir, *Journal of Applied Polymer Science*, 1965, **9**, 975.

31. R.G. Schole, J. Bednarezyk and T. Yamanichi, *Analytical Chemistry*, 1966, **38**, 332.

32. J.N. Shaw and M.C. Marshall, *Journal of Polymer Science A-1*, 1968, **6**, 449.

33. *Ageing and Stabilisation of Polymers* (translated from Russian), Ed., M.B. Neiman, Consultants Bureau, New York, NY, USA, 1965, Chapter 4.

34. J. Paulik and G. Paulik, *GIT Labor Fachzeitschrift*, 1972, **16**, 1043.

35. C.S. Meyer, *Industrial and Engineering Chemistry*, 1952, **44**, 1095.

36. P.J. Carlsson, Y. Kato and D.N. Wiles, *Macromolecules*, 1968, **1**, 459.

37. G. Ahlblad, D. Forsstrom, B. Stenberg, B. Terselius, T. Reitberger and L.G. Svensson, *Polymer Degradation and Stability*, 1997, **55**, 287.

38. H. Kaczmarek, A. Kaminska, M. Swiatek and J.F. Rabek, *Angewandte Makromolekulare Chemie*, 1998, **261/262**, 109.

39. S.I. Ohnishi, S.I. Sugimoto and I. Nitta, *Journal of Polymer Science A-1*, 1963, **1**, 625.

40. R. Soliovey and A. Yager, *Journal of Polymer Science A*, 1964, **2**, 219.

41. Y. Hajimoto, N. Tamura and S. Okamoto, *Journal of Polymer Science A*, 1965, **3**, 255.

42. I. Buchi, *Journal of Polymer Science A*, 1965, **3**, 2685.

43. Y. Sakar and H. Iwasaki, *Journal of Polymer Science A-1*, 1969, **7**, 1749.

44. Y. Hama and K. Shinohara, *Journal of Polymer Science A-2*, 1970, **8**, 651.

45. G. Gallet, S. Carroccio, P. Rizzarelli and S. Karlsson, *Polymer*, 2002, **43**, 1081.

46. D. Forsstram, T. Reitberger and B. Terselius, *Polymer Degradation and Stability*, 2000, **67**, 255.

12 Examination of Photopolymers

12.1 Differential Photocalorimetry

12.1.1 Theory

In recent years a significant effort has been undertaken to evaluate the chemical and technological parameters associated with photopolymerisation. Many analytical techniques have been employed to monitor the cure reaction. Commonly used methods are rheology, chromatography, infrared spectroscopy, titration of functionalities, and mechanical property measurements. These methods have a high degree of sensitivity in either the initial or the final stage of polymerisation. They cannot be used, however, for *in situ* analysis of the curing process.

Sepe [1] has reviewed the application of differential photocalorimetry (DPC) to the analysis of photopolymers. DPC measures the heat of reaction absorbed on release by a material in photoinitiated reactions occurring when a material is exposed to ultraviolet (UV)/visible light in a temperature controlled environment or is used to measure the cure rate and degree of cure in photocurable polymers. The technique uses a dual-sample differential scanning calorimeter (DSC) to measure the heat of reaction of one or two samples as they are exposed to UV/visible light usually employing a high-range mercury arc lamp with a maximum intensity in the 200–400 nm range.

Since polymerisation processes are mainly exothermic reactions in which each additional chain formation step generates a defined amount of heat, the reaction process can be monitored directly and continuously. Combining DSC with an irradiation unit enables photoinitiated reactions to be studied. Cure data and extent of polymerisation are determined by chemical parameters such as photoinitiation, resin components, and additives and by technical parameters such as radiation source, temperature, and environmental conditions. With photocalorimetry all these parameters can be investigated individually and coordinated for optimum results.

The technique provides measurement of physical properties before, during, and after exposure to radiation. Such measurements provide qualitative and quantitative information about the performance of light-sensitive materials used as coatings, films, adhesives, inks, and photoinitiators.

DSC is a fast, sensitive, screening technique for the selection of photopolymer constituents and optimisation of formulations. Resins can be tested for degree of cure and curing characteristics: time–heat flow curves for each photoinitiator. The relative concentration of the photoinitiator or other active components of a resin can be optimised by preparing photospeed curves, i.e., time–percentage conversion plots for a range of concentrations of photoinitiator.

DPC measures reaction rate and the effects of processing conditions on photospeed providing direct information for determining optimum processing conditions including exposure time, temperature, wavelength, intensity, and atmospheric environment, e.g., time–percentage conversion plots at different wavelengths or time–heat flow plots under oxygen and nitrogen.

DPC provides a means of characterising polymers supplementing information obtainable by traditional thermal methods, e.g., the physical properties of the photopolymer can be measured before and after exposure to light. The combined physical and light-sensitive properties provide information on rate of cure, as well as the effect of cure on the physical properties of the material, e.g., time–heat flow plots with different pigments, stabilisers, antioxidants, and plasticisers.

12.1.2 Instrumentation

The instrumentation is discussed in Appendix 1.

The TA Instruments DPC (shown schematically in **Figure 12.1**) exposes a sample to a precisely controlled beam of high-intensity UV light. The sample, which can be either liquid or solid, is contained in an open DSC pan placed inside the DSC 2910 cell. This DSC cell provides precise temperature control and the high sensitivity required for the quantitative measurement of the subtle heats of reaction associated with thermal and photoinitiated reactions.

The light source is operator-selectable, with a choice of lamps and wavelengths. The standard high-pressure mercury arc lamp provides high-intensity UV light, while the optional pressurised mercury–xenon and pressurised xenon arc lamps provide a greater proportion of visible light (longer wavelengths).

The light beam is first passed through a series of achroic focusing lenses, which provide constant beam focus for all wavelengths. The focused beam then strikes an infrared-absorbing mirror which reflects the UV and visible wavelengths while absorbing the infrared radiation. Heat build-up in the mirror is dissipated by a metallic heat sink. A photofeedback sensor mounted near the heat sink monitors and controls light intensity.

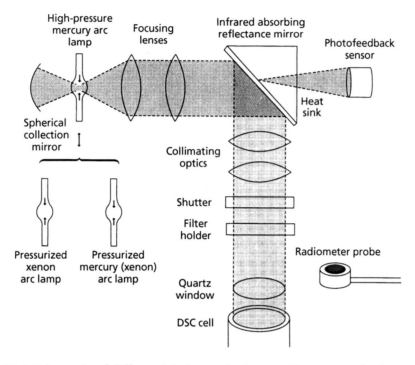

Figure 12.1 Schematic of differential photocalorimeter. (*Source: Author's own files*)

Light reflected from the infrared-absorbing mirror passes through specially designed optics which ensure uniform intensity across the entire DSC cell area. The computer-controlled shutter ensures precise sample exposure. The wavelength of the light beam is determined primarily by lamp selection, and can be refined by use of an operator-selectable band-pass filter.

DuPont also supply a model 930 differential photocalorimeter in their 9900 Thermal Analysis system and Perkin Elmer supply the DSC-7 instrument.

12.1.3 Applications

12.1.3.1 Photocure Rates

The use of radiation-curable coatings and adhesives is growing rapidly.

UV-curable polymer systems typically contain monomer or oligomer (unreacted polymer), a light-sensitive photoinitiator, and other additives for colour and mechanical properties.

409

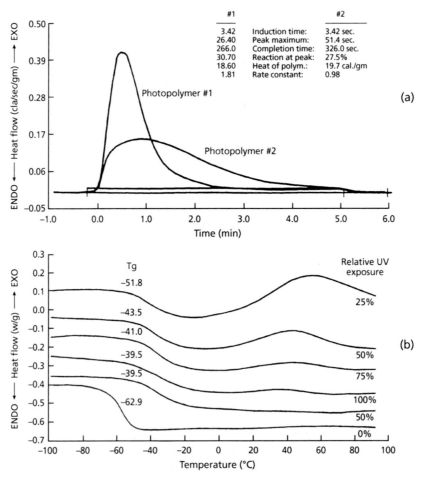

Figure 12.2 Differential photocalorimetry of photopolymers (a) No. 1 rapid cure polymer, No. 2 slow cure polymer; (b) Effect of exposure on glass transition of photopolymers. (*Source: Author's own files*)

Concentrations of these components determine viscosity, cure rate, adhesion, flexibility, and abrasion resistance. These properties are also directly affected by the degree of cure. Therefore, to ensure a consistent product, it is necessary to monitor the response of the photopolymer both before and after processing.

Figure 12.2(a) compares the response of two different photopolymers. Sample 1 cures very rapidly, while Sample 2 takes more than 50 seconds of exposure to reach its maximum rate. Using the data analysis software, the operator can generate a plot of percentage cure *versus* time and quickly determine if the material is within specification.

410

The one component that significantly affects the cure rate of photocurable polymers is the photoinitiator. Decreasing the concentration of a photoinitiator from 0.5 to 0.3% can more than double the induction time for the reaction. Since the photoinitiation is frequently the most expensive of the photopolymer ingredients, DPC offers a fast, reliable method for optimising and verifying the concentration of the photoinitiator.

12.1.3.2 Degree of Cure

Traditionally, DSC has been used with most thermosetting polymers to measure the degree of cure. Two techniques have been developed to make the measurement. The first is to measure the glass transition temperature (T_g) of the material (Chapters 6 and 13). As the cure proceeds, chemical crosslinking increases and the T_g moves towards the maximum value. The second method for determining the degree of cure of thermosets uses the residual heat of reaction of the sample as it is heated through the curing temperature in the DSC. There are advantages and disadvantages for each method. Therefore, since both values can be easily obtained from a single DSC experiment, it is best to measure both.

Photopolymers can be analysed in a way very similar to thermosets. The only difference is that photopolymers use light to initiate chemical crosslinking, while thermosets use heat. **Figure 12.2(b)** shows the effect of varying the exposure time on a photopolymer used for electronic applications. With no exposure (0%), the T_g is −62.9 °C. As the relative amount of UV exposure is increased from 0% to the recommended amount (100%), the T_g is seen to increase to −39.5 °C. Even at five times the normal exposure, no further increase in T_g is seen, indicating that the standard exposure provides nearly complete cure.

Another phenomenon shown in **Figure 12.2(b)** is the effect of heat on samples that have been only partially exposed. As these materials are heated, an exothermic peak is seen between 20 and 100 °C. This peak is caused by additional curing taking place and indicates that once the sample is exposed and cure begins, cure can be continued using either light or heat. Since DPCcan control both of these variables, either sequentially or simultaneously, it promises to be a powerful tool for quality control and research on photopolymer systems.

12.1.3.3 Examples

The following application examples from the field of photoinitiated radical and cationic polymerisation have been selected to demonstrate the multiple use of photocalorimetry. All samples were analysed in open aluminium pans, isothermally at 30 °C. Because of the fact that photopolymerisations via radicals are inhibited by oxygen the sample holder was purged with an inert gas.

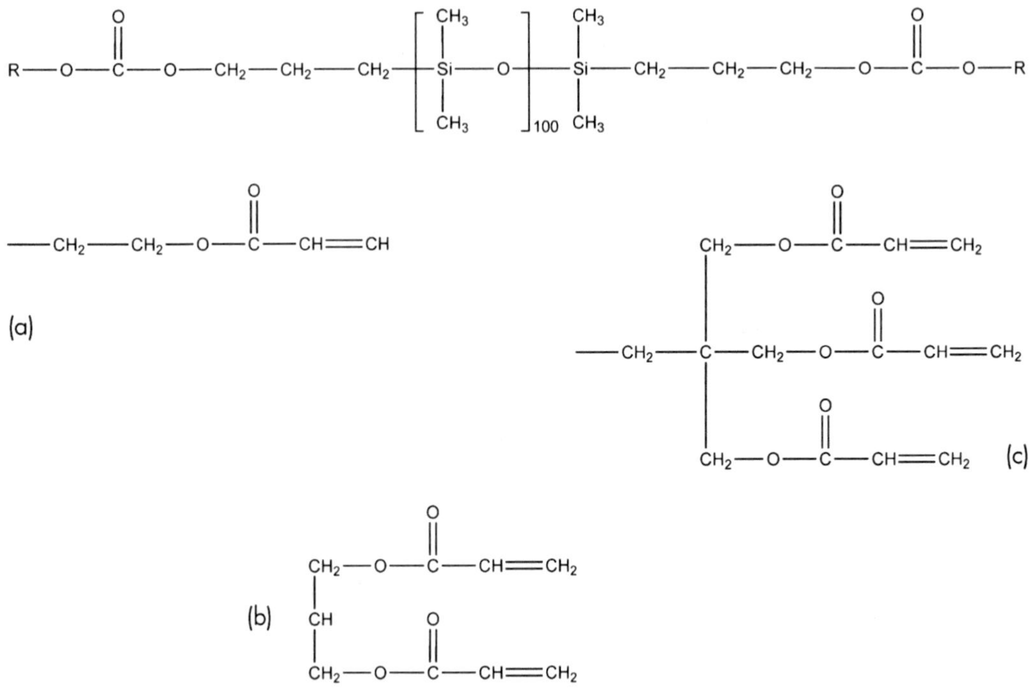

Figure 12.3 (a) is monoacrylate, (b) is diacrylate and (c) is triacrylate, represented by R end groups in polydimethyl siloxanes. (*Reprinted from W. Rogler, H. Markert, B. Stapp and F. Zaff, Polymer Preprints, 1988, 29, 528, with permission from ACS [3]*)

Dependence of reactivity upon functionalisation. During photopolymerisation the functional groups frequently will not react completely, in particular, when polyfunctional monomers are to be crosslinked [2]. Silicone acrylates have been used [3] to investigate the extent to which an accumulation of acrylate functions will influence cure rate and the residual unsaturation that remains after UV exposure. Polydimethylsiloxanes functionalised with mono-, di-, and triacrylates at their chain ends were irradiated in the DSC under identical conditions in the presence of a photoinitiator (**Figure 12.3**).

Figure 12.4 shows that the α,ω-difunctional silicone acrylate is more reactive than the mono- and trifunctional acrylate. The heat of reaction evolved in each case permits the calculation of the conversion of the acrylate groups by means of the standard heat of acrylate polymerisation. A comparison with the initial acrylate contents determined by titration clearly demonstrates that the acrylate functions of the trifunctional siloxane react incompletely.

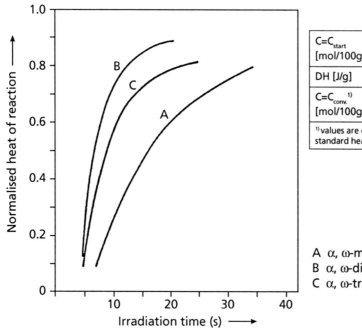

	A	B	C
C=C$_{start}$ [mol/100g]	0.025	0.034	0.050
DH [J/g]	−16.1	−25.1	−29.0
C=C$_{conv.}$ [1] [mol/100g]	0.021	0.032	0.037

[1] values are calculated with 77.9 kJ/mol for the standard heat of acrylate-polymerisation

A α, ω-monoacrylate
B α, ω-diacrylate
C α, ω-triacrylate

Figure 12.4 Dependence of the curing rate of silicone acrylates upon functionalisation. (*Source: Author's own files*)

Influence of wavelength. In a study of the response of a UV-curable composition to different irradiation wavelengths, 2-hydroxy-2-methyl-1-phenylpropan-1-one (Darocur 1173), at a concentration of 3% in an acrylate resin, was used as photoinitiator. The UV spectrum of the photoinitiator showed an absorption maximum at 320 nm, indicating that this should be the best wavelength for irradiation. The photocalorimetric measurements with monochromatic irradiation of the samples at a constant intensity indicate that the highest polymerisation rate occurs at an excitation wavelength of 340 nm.

Influence of photoinitiator concentration. As a basis, a hexanediol-diacrylate (HDDA) was used to which was added 1% and 5% benzyldimethylketal (BMDK) as well as 1% and 5% benzyldiethylketal (BDEK) as photoinitiators. The samples were irradiated in a thickness of 80 μm with a mercury arc lamp at a UV intensity of 1.0 mW/cm². It is apparent from the DSC curves obtained that the HDDA/BDMK mixtures do not show any significant dependence on the photoinitiator concentrations. This is different to the HDDA/BDEK mixtures, where reaction time and reaction rate are distinctly influenced by the photoinitiator concentration.

Influence of humidity. During cationic epoxide polymerisation small amounts of water may operate both as a chain-transfer agent and as a termination agent, depending mostly on the polarity of the surrounding medium [4]. The influence on humidity was investigated for a photoreactive epoxide resin which had been kept for seven days at 25 °C and 0, 50, or 90% relative humidity. The resin was polymerised isothermally in the photocalorimeter under UV irradiation and then, to determine the remaining reaction enthalpy, was heated to 200 °C with a linear temperature programme. The highest polymerisation speed was achieved on the sample that had been kept at 25 °C/90%. The reaction heat released during the isothermal and dynamic scan rises with increasing humidity from 509 to 602 J/g.

DPC has in recent years been applied to studies on a range of photopolymers including DF 2000 photopolymer [5], cinnamoylphenyl methacrylate–glycidyl methacrylate copolymer [6], multiethylene glycol dimethacrylate [7], Ebecryl 270 (aliphatic urethane diacrylate) 1,6-hexanediol diacrylate (Darocur 1173; 2-hydroxy-2-methyl-phenylpropan-1-one) [8], epoxy acrylates [9], epoxy vinyl ether formulations [10], polyacrylates, maleimides, and vinyl ethers [11], hydroxylated poly(imides) [12], and polystyrene–poly(*n*-butylacrylate) copolymers [13].

12.2 Dynamic Mechanical Analysis

The theory and instrumentation of this technique are discussed in Section 10.1.

Bair and co-workers [14] used dynamic mechanical analysis to measure photoshrinkage in photopolymers during cure. The system uses a dynamic mechanical analyser (DMA) modified for optical access. While such a modification reduces the thermal control over the experiment, polymer samples can be exposed to light while the DMA probe is in contact with the sample. In the dynamic mode, using a probe lip with an oscillating frequency, the entire photoinduced polymerisation reaction can be followed with real-time monitoring of shrinkage, sample viscosity, or modulus. Alternatively, using the thermal mechanical analyser (TMA) mode for samples enclosed between glass plates, the probe rests on the top glass plate during the photoreaction and provides an accurate measure of the sample thickness perpendicular to the glass plates. Since dimensional changes are monitored in real time, the timescales for the photopolymerisation shrinkage can be determined directly.

DMA has also been used in curing studies on 2,2-bis(4-(2-hydroxy-3 methacryloxy propoxy)phenyl) propane and triethylene glycol dimethacrylate [15], and bisphenol A epoxy diacrylate [16].

414

12.3 Infrared and Ultraviolet Spectroscopy

Odeberg and co-workers [13] demonstrated grafting reactions in polystyrene–poly(*n*-butylacrylate) copolymers using Fourier transform infrared spectroscopy and proton neutron magnetic resonance spectroscopy. DMA and TMA are used to study crosslinking.

Hrdlovic [17] has reviewed the application of Fourier transform ultraviolet and fluorescence techniques to a study of photochemical reactions and photophysical processes in polymers.

12.4 Gas Chromatography-Based Methods

Subramanian [18] has used gas chromatography–mass spectrometry to study the photodecomposition products such as benzaldehyde, α-hydroxyacetophenone, phenyl glycol, α-methoxyacetophenone, and α-benzyloxyacetophenone produced upon UV irradiation of polystyrene peroxide. Diaz and co-workers [19] carried out similar studies on polybenzyl methacrylates and polybenzyl acrylates.

References

1. M.P. Sepe, *Thermal Analysis of Polymers*, Rapra Review Report No. 95, Rapra Technology Ltd., Shrewsbury, UK, 1997, Vol. 8, No. 11.

2. J.E. Moore in *UV Curing, Science and Technology*, Ed., S.P. Pappas, Technology Marketing Corporation, Stamford, CT, USA, 1978.

3. W. Rogler, H. Markert, B. Stapp and F. Zapf, *Polymer Preprints*, 1988, **29**, 528.

4. N.C. Billingham in *Encyclopaedia of Polymer Science and Engineering*, Ed., J.J. Kroschwitz, Wiley Interscience Publications, New York, NY, USA, 1988, Volume 2, p.789.

5. A.C. Lin, S.R. Liang, J.Y. Jeng, Y.C. Yeh, W.S. Wong and C.T. Ho, *Plastics, Rubbers and Composites*, 2002, **31**, 177.

6. K. Subramanian, S. Nanjundan and A.V.R. Reddy, *Journal of Macromolecular Science*, 2000, **A37**, 1211.

7. K.S. Anseth in the *Proceedings of the ACS, Polymeric Materials Science and Engineering Conference*, Orlando, FL, USA, Fall 1996, **75**, 202.

8. C.S.B. Ruiz, L.D.B. Machado, J.A. Vanin and J.E. Volponi, *Journal of Thermal Analysis and Calorimetry*, 2002, **67**, 335.

9. Q. Wang, P. Zhu and Z. Li, *Polymer Preprints*, 2001, **42**, 2, 216.

10. J.D. Cho, E-O. Kim, H-K. Kim and J-W. Hong, *Polymer Testing*, 2002, **21**, 781.

11. N. Pietschmann, *Macromolecular Symposia*, 2002, **187**, 225.

12. V.Y. Voitekunas, L.G. Komarova, M, Abodie, A.L. Rusanov and M.P. Prigozhina, *Polymer Science, Series A*, 2002, **44**, 463.

13. J. Odeberg, J. Rassing, J.E. Jonsson and B. Wesslen, *Journal of Applied Polymer Science*, 1998, 70, 897.

14. H.E. Bair, M.L. Schilling, V.L. Colvin, A. Hale and N.J. Levinos in *Proceedings of the ACS Polymeric Materials Science and Engineering Conference*, Dallas, TX, USA, Spring 1998, 78, 230.

15. H. Lu, L.G. Lovell and C.N. Bowman, *Polymer Preprints*, 2001, **42**, 2, 763.

16. W.P. Yang, C. Wise, J. Wijaya, A. Gaeta and G. Swei in *Proceedings of RadTech '96, North America*, Nashville, TN, USA, Spring 1996, Volume 2, p.675.

17. P. Hrdlovic, *Polymer News*, 2001, **26**, 161.

18. K. Subramanian, *European Polymer Journal*, 2002, **38**, 1167.

19. F.R. Diaz, J. Moreno, L.H. Tagle, G.A. East and D. Radic, *Synthetic Metals*, 1999, **100**, 187.

13 Glass Transition and Other Transitions

13.1 Glass Transition

The glass transition temperature (T_g) is defined as the temperature at which a material loses its glasslike, more rigid properties and becomes rubbery and more flexible in nature. Practical definitions of T_g differ considerably between different methods, therefore, specification of T_g requires an indication of the method used.

Amorphous polymers when heated above T_g pass from the hard to the soft state. During this process, relaxation of any internal stress occurs. At the T_g many physical properties change abruptly, including Young's and shear moduli, specific heat, coefficient of expansion, and dielectric constant. For hard polymeric materials this temperature corresponds to the highest working temperature; for elastomers, it represents the lowest working temperature. Several methods exist for determining T_g. These include differential thermal analysis (DTA), differential scanning calorimetry (DSC), thermomechanical analysis (TMA), dilatometry, dynamic mechanical analysis (DMA), and nuclear magnetic resonance (NMR) spectroscopic methods. Each method requires interpretation to determine T_g. For this reason exact agreement is frequently not obtained between results obtained by different methods.

13.2 Differential Scanning Calorimetry

13.2.1 Theory

As discussed previously, T_g is shown by a change in the expansion coefficient and the heat capacity as a sample material is heated or cooled through this transition region.

Since DSC measures heat capacity directly, rapidly, and accurately, it is an ideal technique for the determination of T_g. The calorimeter accepts polymers in any form (powder, pellet, or fibre), and only a few milligrams of sample are required. Samples are placed in a standard aluminium sample pan, crimped by a crimping press to ensure good thermal contact, placed in the sample holder, and scanned at an appropriate rate over the temperature range of interest.

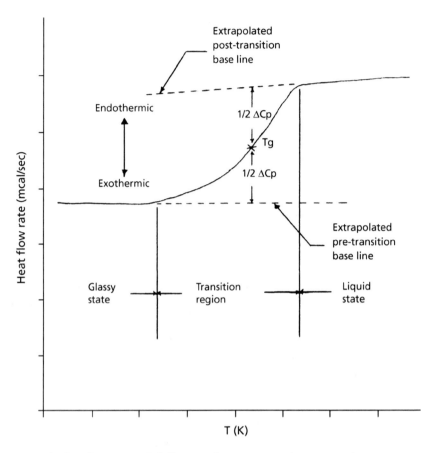

Figure 13.1 Idealised output of differential scanning calorimetry determination of T_g.
(*Source: Author's own files*)

Figure 13.1 shows an idealised output from the DSC. T_g is taken as the mid-point in the thermogram as measured from the extensions of the pre- and post-transition baselines - that is, when the heat capacity change assumes half the value of this change upon going through the transition. This choice is somewhat arbitrary. Other workers have suggested alternative techniques such as taking T_g at the first evidence of the displacement of the thermogram from the pre-transition baseline. The first of these methods is the most reliable and reproducible.

The observed increase in heat capacity is due to the onset of extensive molecular motion, increasing the degrees of freedom of the polymer and, consequently, its heat capacity. This does not infer that molecular motion does not exist in a polymer below its T_g - merely that molecular motion is severely restricted.

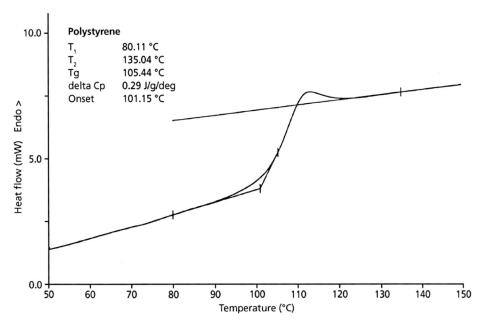

Figure 13.2 Glass transition temperature. Because the DSC 7 directly measures changes in the specific heat of material, the parameters associated with the glass transition of amorphous materials are easily made. As shown in this sample of polystyrene, the temperature where these changes occur, as well as the magnitude of the specific heat changes at the glass transition, are calculated using the TAS 7 TG program.
(*Source: Author's own files*)

Isothermal crystallisation studies by DSC provide a sensitive technique for measuring molecular weight or structural differences between very similar materials (heat flow (mW) *versus* time plots). At the T_g an endothermic shift occurs in the DSC curve corresponding to the increase in specific heat (heat flow (mW) *versus* temperature). Determination of the T_g from the DSC curve is based on the following three auxiliary lines: extrapolated baseline before the transition, inflection tangent through the greatest slope during transition, and the baseline extrapolated after the transition. **Figure 13.2** shows this measurement carried out on a polystyrene (PS) sample using a Perkin Elmer DSC-7 instrument illustrating the temperatures at which the T_g occurs as well as the magnitude of the specific heat changes at the T_g. Frequently, an endothermic peak occurs simultaneously with the change in specific heat the first time a substance is heated above the T_g. This event is caused by a relaxation phenomenon and normally appears only when the T_g is exceeded for the first time. To rule out this effect, a second measurement is carried out immediately after the substance has cooled.

13.2.2 Instrumentation

This is discussed in Section 9.4.4 and Appendix 1.

13.2.3 Applications

Excellent baseline stability helps give the Perkin Elmer DSC 2910 the sensitivity to detect very small T_g. Polypropylene (PP), for example, has been difficult to characterise by DSC because its T_g is small and its heat capacity changes greatly with increasing temperature. The T_g of PP is clearly observed with the Perkin Elmer DSC 2910 instrument. Baseline stability and sensitivity also make it possible for the 2910 to detect the T_g of highly filled or highly crystalline polymers.

DSC has been applied to the determination of T_g in a variety of other polymers including polyimines [1], polyurethanes (PU) [2], Novolac resins [3], polyisoprene, polybutadiene, polychloroprene, nitrile rubber, ethylene–propylene–diene terpolymer, and butyl rubber [4], and bisphenol-A epoxy diacrylate–trimethylolpropane triacrylate [1516].

13.3 Thermomechanical Analysis

13.3.1 Theory

TMA is the measurement of dimensional changes (such as expansion, contraction, flexure, extension, and volumetric expansion and contraction) in a material by movement of a probe in contact with the sample in order to determine temperature-related mechanical behaviour in the temperature range –180 to 800 °C as the sample is heated, cooled (temperature plot), or held at a constant temperature (time plot). It also measures linear or volumetric changes in the dimensions of a sample as a function of time and force.

Plots of sample temperature *versus* dimensional (or volume) changes enable the T_g to be obtained. The T_g is obtained from measurement of sudden changes in the slope of the expansion curve. However, the technique has much wider ranging applications than this in the measurement of the thermal (Chapter 16) and mechanical (Chapter 18) properties of polymers.

13.3.2 Instrumentation

Instrumentation is discussed in Appendix 1.

Perkin Elmer supplies the TMA-7 thermomechanical analyser. In this analyser a quartz probe closely monitors dimensional changes in the sample under study. The position of this probe is continuously monitored by a high-sensitivity linear variable displacement transducer. The transducer itself is temperature controlled to provide excellent stability and reproducibility. The probe mechanism is controlled through a closed loop electromagnetic design circuit. This design allows precise probe control, computer-controlled application of force to the sample, and constant sample loading throughout the experiment. These features provide exceptional temperature control over the range from –170 to +1000 °C. Other features of the TMA-7 include: multiple probe types for multiple modes of operation, computer control for unattended operation, sample load selection through the computer keyboard, automatic zero load calculation, one-touch probe control and position, simultaneous independent instrument operation, precise temperature control, and heating and cooling rates of 0.1 to 100 °C/min.

A series of quartz probes are available that allow the TMA-7 to be used in a variety of different operating modes: expansion, compression, flexure, extension, and dilatometer.

TA Instruments supply the TMA 2940 thermomechanical analyser.

13.3.3 Applications

Johnston [6, 7] studied the effects of sequence distribution on the T_g of alkyl methacrylate–vinyl chloride and α-methylstyrene–acrylonitrile copolymers by DSC, DTA, and TMA.

Figure 13.3 shows an application of a TMA to the characterisation of a composite material, i.e., an epoxy printed circuit board material. The T_g is readily determined from this curve.

When an elastomer was subjected to a penetration load of 0.03 N and a temperature range of –150 to 200 °C, the material showed a slight expansion below the T_g, before allowing penetration at –17.85 °C, resulting in a very marked T_g. The expansion that takes place after the T_g shows that the material is sturdy enough to resist further penetration, even in its rubbery state.

Wohltjen and Dessy [8, 9] have described a surface acoustic wave (SAW) device for performing TMA measurement of T_g on polyethylene terephthalate (PET), bisphenol A polycarbonate, polysulfone, polycarbonate (PC), and polymethyl methacrylate (PMMA). There are several factors that distinguish the SAW device as a useful monitor of polymer T_g. The device is very sensitive and this permits very small samples to be used. Sample preparation and mounting are simple and rapid. The device is quite rugged and possesses a small thermal mass which permits fairly rapid temperature changes to be made.

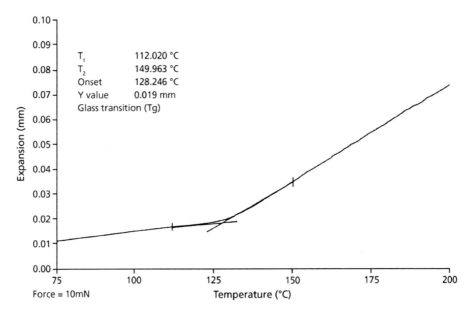

Figure 13.3 Thermomechanical analysis of epoxy resins. Measurements of glass transition temperature. (*Source: Author's own files*)

TMA has been applied to the determination of T_g in neoprene, styrene–butadiene, polyisoprene, polybutadiene, polychloroprene, nitrile, ethylene–propylene–diene, and butyl rubbers [4].

13.4 Dynamic Mechanical Analysis

For the theory and instrumentation of this technique see Section 10.1 and Appendix 1.

13.4.1 Applications

DMA has been applied to the determination of T_g in neoprene, styrene–butadiene, polyisoprene, polybutadiene, polychloroprene, nitrile ethylene–propylene–diene, and butyl rubbers [4], polybutadiene [10], glass-filled low-density polyethylene [11], PU, PMMA, polyimides, acrylonitrile–butadiene–styrene terpolymer, PET, and Nylon 6 [2], bisphenol-A epoxy diacrylate, *N*-vinyl pyrrolidone copolymer, and trimetholpropane triacrylate [5].

13.5 Differential Thermal Analysis and Thermogravimetric Analysis

For the theory and instrumentation of this technology see Section 9.4 and Appendix 1.

Thermogravimetric analysis in conjunction with DSC has been used to measure T_g in phenol bark resins [3]. DTA has been used to measure T_g in bis(trichlorophenolate) di(pyridine) nickel(II) and bis(tribromophenolate) di(pyridine) nickel(II) complexes [12], PS [13], and various other polymers [14].

Combinations of both techniques have been used to obtain T_g values for polymethacrylates containing the *p*-nitrobenzene group [15] and polycarbonate composites [16].

13.6 Nuclear Magnetic Resonance Spectroscopy

High-field ^{13}C Fourier transform NMR spectra have indicated –40 °C as the upper limit for the T_g of linear polyethylene (PE) [17]. Pulsed NMR measurements of ultrahigh-molecular-weight linear PE indicated a second transition of longer relaxation time appearing at the temperature characteristic of the gamma relaxation. The gamma relaxation meets most of the criteria for assignment of T_g [18].

13.7 Dielectric Thermal Analysis

Dielectric thermal analysis (DETA) has been used in measurement of T_g of neoprene, styrene-butadiene, polyisoprene, polybutadiene, polychloroprene, nitrile, ethylene-propylene-diene and butyl rubbers [4].

13.8 Other Transitions (alpha, beta, and gamma)

Four techniques have been used to detect transitions other than T_g in polymers: DTA, DMA, DETA, and infrared spectroscopy.

13.8.1 Differential Thermal Analysis

For a discussion of the theory and instrumentation see Sections 9.2 and 9.3 and Appendix 1.

13.8.1.1 Applications

A linear high-pressure PE blend, upon heating, undergoes three phase changes from its high-pressure form: the 115 °C peak was associated with the high-pressure PE, whereas the 134 °C peak was shown to be proportional to the linear content of the system. Clampitt [19] also applied DTA to a study of the 124 °C peak which he describes as the co-crystal peak. His results appear to indicate that there are two classes of co-crystals in linear high-pressure PE blends with the linear component being responsible for the division of the blends into two groups. The property of the linear component that is responsible for the division is related to the crystallite size of the pure linear crystal.

13.8.2 Dynamic Mechanical Analysis

For a discussion of the theory and instrumentation of this technique see Sections 10.1.1 and 10.1.2 and Appendix 1.

13.8.2.1 Applications

In the dynamic mechanical loss spectra of polymers, the transitions that are related to different molecular motions within polymers are called alpha, beta, and gamma transitions.

The alpha transition involves long segments of the polymer chain where the movement causes other chain segments to move out of the way. These 'cooperative main-chain motions' become increasingly prevalent at the T_g and can be used to define the T_g of a material.

The beta transition involves chain segments that are shorter than those in alpha transitions. For that reason, they occur below the T_g of the material.

The motion producing the gamma transition involves short-chain segments. In many polymeric systems, the gamma transition is caused by the 'crankshaft rotation' of the methylene ($-CH_2-$) groups on a long polymer chain. Since the gamma transition involves short molecular segments, it occurs below the alpha and beta transitions.

DMA is more sensitive to material transitions than traditional thermal analysis techniques (e.g., DSC, TMA). Detection of major transitions such as the T_g, for example, by DMA, is easier in highly filled or reinforced materials because the material modulus changes by several orders of magnitude at T_g, while the material heat capacity (the basis for DSC detection) and expansion coefficient (the basis for TMA detection) change less significantly. Moreover, the detection of weak secondary transitions is possible only by DMA.

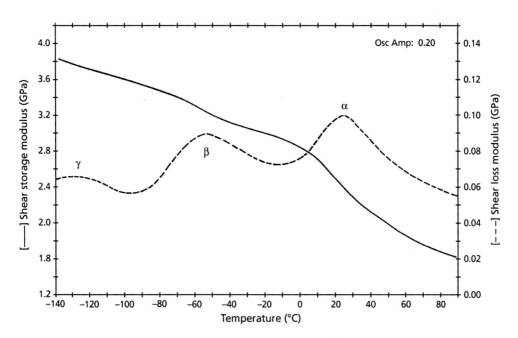

FIgure 13.4 Shear storage modulus – temperature plot for semi-crystalline 6,6 Nylon demonstrating alpha, beta and gamma transitions. (*Source: Author's own files*)

DMA is sensitive enough to detect even weak secondary transitions such as alpha and beta transitions in the resin matrix of a highly filled composite. In fact, all properties measured by this technique generate strong, well-defined signals that are not clouded by background noise or other interferences.

Plots of shear storage modulus (*g*) *versus* temperature in **Figure 13.4** demonstrate the three transitions occurring in a semi-crystalline Nylon 6,6 composite. These results were obtained with a DuPont 983 DTA. The results in **Figure 13.4** provide information about the chemical structure of the resin matrix. The alpha transition at 25 °C is the T_g of the material and is associated with a significant decrease in modulus. The beta transition is due to the local main chain motions of the amide groups found in the polymer backbone structure. The gamma transition is thought to be related to the motion of short sequences of methylene groups and possibly to some molecular motion contributions from the amide groups.

Shear storage modulus (*g*) is a quantitative measure of the stiffness or rigidity of a material defined as:

$$g = \frac{\text{Applied shear}}{\text{Applied strain (}\varepsilon\text{)}}$$

425

Static modulus (E) for homogenous isotropic elastic substances is a quantitative measure of the elasticity of a material defined as:

$$E = \frac{\text{Applied stress } (\sigma)}{\text{Applied strain } (\varepsilon)}$$

The comparative loss modulus curves from Nylon 6,6 after exposure to different humidity levels show three distinct damping peaks (transitions). These are designated as alpha, beta, and gamma in order of decreasing temperature. The alpha transition is the primary T_g of the amorphous phase and is attributed to long-chain segmental motion within the main polymer chain. The beta transition, occurring between –60 and –40 °C, is attributed to local motion within the amide segments of the polyamide. Finally, the gamma transition, occurring between –120 and –110 °C, is attributed to local segmental motion of the (–CH$_2$–) methylene groups between amide functions in the amorphous regions. The primary effect of moisture is in changing the position and intensity of the damping transitions observed for each of these specific relaxations. The most pronounced influence of absorbed moisture is on the position of the alpha or T_g. A shift of 60 to 70 °C is observed by increasing the moisture content from the 'dry as-moulded' case to that of Nylon conditioned at 100% humidity. This temperature shift is primarily the result of water breaking intramolecular hydrogen bonds, thus allowing greater chain mobility and resulting in a large decrease in the T_g. The lowering of the beta transition temperature is due to water molecules weakening the interaction between the chains, shifting the beta transition to lower temperatures. The gamma transition is influenced by water–polymer interaction - increase in water content increases the fraction of hindered segments with a corresponding decrease in gamma intensity.

Figure 13.5 compares the loss moduli (damping) of linear (high-density) and branched (low-density) polyethylene. Three transitions are seen: alpha at 80 °C, beta at –5 °C, and gamma at –110 °C. Each of these transitions corresponds to specific molecular motions which have significance in terms of structure–property relationships:

1. The alpha-transition is associated with crystalline relaxations occurring below the melting point of PE.

2. The beta-transition is due to motion of the amorphous region side chains or branches from the main polymer backbone. The intensity of the beta transition varies with the degree of branching.

3. The gamma transition is a result of crankshaft rotation of short methylene main chain segments and can influence the low-temperature impact stability of PE.

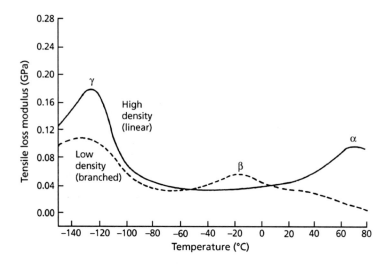

Figure 13.5 DMA comparison of low and high density polyethylene.
(*Source: Author's own files*)

End-use properties such as toughness, acoustic and mechanical damping, and impact resistance are affected by the type, size, and temperature of transitions. Transitions provide the mechanism to convert kinetic energy to heat.

Kwak and co-workers [20] used DMA to determine T_g in reinforcing ethylene–propylene rubber.

13.8.3 Dielectric Thermal Analysis

13.8.3.1 Theory

DETA, which measures a material's response to an applied alternating voltage signal, provides an excellent means of characterising thermoplastics. DETA measures two fundamental electrical characteristics of a material – capacitance and conductance – as a function of temperature, time, and frequency. The capacitive nature of a material reflects its ability to store an electrical charge and this property dominates the electrical response at temperatures below the T_g. The conductive nature is the ability to transfer electrical charge and generally dominates the electrical response at temperatures above the T_g or melting temperature (T_m). While these electrical properties are important in themselves, they acquire more significance when they are correlated to changes in the molecular state of the material. The actual properties monitored using dielectric analysis are e' (permittivity), which is a measure of the degree of alignment of the molecular dipoles to the applied electrical field, and e'' (loss factor), which represents the energy required to align the dipoles or to move trace ions.

The ultra sensitivity of this technique makes it possible to detect transitions that are not seen by other techniques. Its ability to measure bulk or surface properties of materials in solid, paste, or liquid form makes DETA versatile and very useful.

The high sensitivity of DETA makes it ideal for characterising molecular relaxations, which are key predictors of the end-use performance of many polymer products. With its exceptionally wide frequency range, covering eight decades, it can easily resolve different relaxations.

Modern dielectric constant analysers such as the DuPont model DEA2970 can easily separate multiple transitions. These multi-frequency curves show that the alpha and beta transitions become increasingly distinct as the test frequency is decreased - they start to become discernible at about 1 kHz, and are well defined at 100 Hz.

The alpha transition, which involves motion in long segments of the main polymer chain, is related to the T_g. The beta transition involves rotation of short-chain ester side groups and therefore occurs below the T_g. The frequency dependency of the beta T_g can be used to calculate the activation energy for the molecular motion, which provides important information for characterising the structure and predicting the performance of polymeric materials. In a dielectric analysis experiment, the calculated activation energy for the beta transitions in PMMA was 17.7 kcal/mol. This correlates well with the values calculated from DMA and creep experiments.

13.8.2.2 Instrumentation

For a discussion of the instrumentation see Appendix 1.

13.8.3.3 Applications

Figures 13.6(a) and (b) show the alpha and beta transitions obtained in dielectric thermal analysis of PET. The alpha transition (T_g) is affected by the large-scale micro-Brownian motion in the amorphous (non-crystalline) phase. The Brownian motion is observed as a peak in the 1 Hz loss factor curve at about 90 °C. The beta transition is considered to be a result of main-chain motion involving the ester groups.

428

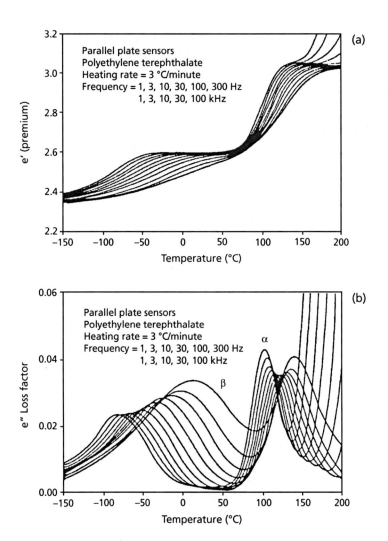

Figure 13.6 Resolution of multiple transitions in amorphous polyethylene terephthalate by dielectric thermal analysis. (a) e′ curves defining polyethylene terephthalate transitions; (b) e″ curves showing frequency dependency of alpha beta transitions. (*Source: Author's own files*)

13.8.4 Thermomechanical Analysis

13.8.4.1 Applications

Savitskii and Gorshkova [21] have applied TMA to the determination of T_g for solutions of polyamide 6 in phenol and formic acid, polyoxadiazoles in sulfuric acid, and polyheteroarylenes in dimethyl acetamide.

13.8.5 Infrared Spectroscopy

13.8.5.1 Applications

While thermal methods predominate for the determination of polymer transitions, infrared spectroscopy has been used to provide information on temperature transitions.

Varob'yev and Vettegren [22] used infrared spectroscopy to determine temperature transitions in PC. Thermal transitions in PC were determined from the concentration variation of the residual solvent or plasticiser.

Structural transitions and relaxation phenomena of PC have been followed by plotting the absorbances at 8.13 μm (stretching vibration of C–O–C groups) and 10.64 μm against temperature [23].

Peppas [24] showed that polycaprolactam (polyamide 6) crystallises in different crystalline forms, which show different characteristic infrared absorbance bands. Based on this he prepared phase diagrams. He evaluated the absorbance coefficients of the bands from the dependence of their intensity on temperature and specific volume of different polyamide 6 samples and described a method for the quantitative determination of the amounts of alpha and gamma modifications and of the amorphous phase. On the basis of these data the densities of the samples were calculated and shown to agree with the values determined experimentally. The specific volume of the amorphous phase was $v = 0.917$ cm^3/g, and was independent of any alpha or gamma modifications the samples may have contained.

References

1. F.R. Diaz, J. Moreno, L.H. Tagle, G.A. East and D. Radic, *Synthetic Metals*, 1999, **100**, 187.

2. R.G. Ferrillo and P.J. Achorn, *Journal of Applied Polymer Science*, 1997, **64**, 191.

3. M.H. Alma and S.S. Kelley, *Polymer Degradation and Stability*, 2000, **68**, 413.

4. A.K. Sircar, M.L. Galaska, S. Rodrigues and R.P. Chartoff in *Proceedings of the 150th ACS Rubber Division Meeting*, Louisville, KY, USA, Fall 1996, Paper No.35, p.65.

5. W.P. Yang, C. Wise, J. Wijaya, A. Gaeta and G. Swei in *Proceedings of RadTech '96, North America*, Nashville, TN, USA, Spring 1996, Volume 2, p.676.

6. N.W. Johnston, *Journal of Macromolecular Science-Chemistry*, 1973, **A7**, 531.

7. N.W. Johnston, *Makromolecules*, 1973, **6**, 453.

8. H. Wohltjen and R. Dessy, *Analytical Chemistry*, 1979, **51**, 1465.

9. H. Wohltjen and R. Dessy, *Analytical Chemistry*, 1979, **51**, 1458.

10. M. Kluppel, R.H. Schuster and J. Schaper in *Proceedings of the 151st ACS Rubber Division Meeting*, Anaheim, CA, USA, Spring 1997, Paper No.54.

11. J.Z. Liang, P.K.Y. Li and S.C. Tjong, *Journal of Thermoplastic Composite Materials*, 2000, **13**, 12.

12. L. Molu and D. Kisakurek, *Journal of Applied Polymer Science*, 2002, **86**, 2232.

13. B. Wunderlich and D.M. Bodily, *Journal of Applied Polymer Science: Part C, Polymer Symposia*, 1964, **6**, 137.

14. L. Rieger, *Polymer Testing*, 2001, **20**, 199.

15. H-Q. Zhang, W-Q. Huang, C-X. Li and B-L. He, *European Polymer Journal*, 1998, **34**, 1521.

16. S.C. Tjong and W. Jiang, *Journal of Applied Polymer Science*, 1999, **73**, 2247.

17. C.L. Beatty and Y.F. Froix, *Polymer Preprints*, 1975, **16**, 628.

18. H.B. Smith, A.J. Manuel and I.M. Ward, *Polymer*, 1975, **16**, 57.

19. B.H. Clampitt, *Journal of Polymer Science: Part A: General Papers*, 1965, **3**, 671.

20. G.H. Kwak, K. Inoue, Y. Tominga, S. Asai and M. Sumita, *Journal of Applied Polymer Science*, 2001, **82**, 3058.

21. A.V. Savitskii and I.A. Gorshkova, *Polymer Science Series A*, 1997, **39**, 356.

22. V.M. Varob'yev and V.I. Vettegren, *Polymer Science USSR*, 1975, **17**, 520.

23. G.G. Andronikashivili, S.A. Samsoniya, M.G. Zhamierashvili and M. Sooleshch, *Akademii Nauk Gruz SSR*, 1977, 85, 73. *Chemical Abstracts*, 1977, **86**, 190545C.

24. N.A. Peppas, *Journal of Applied Polymer Science*, 1976, **20**, 1715.

14 Crystallinity

14.1 Theory

Crystallinity is a state of molecular structure referring to a long-range periodic geometric pattern of atomic spacings. In semi-crystalline polymers, such as polyethylene (PE), the degree of crystallinity (% crystallinity) influences the degree of stiffness, hardness, and heat resistance.

In semi-crystalline polymers, some of the macromolecules are arranged in crystalline regions, known as crystallites, while the matrix is amorphous. The greater the concentration of these crystallites, i.e., the greater the crystallinity, the more rigid the polymer. Morphology denotes the internal structure of a material (separate polymer phases, crystalline regions, amorphous orientation, and so on). Amorphous is a term generally used to describe polymers totally lacking in long-range spatial order (crystallinity). It is also used to denote non-crystalline regions within partially crystalline polymers.

An understanding of the degree of crystallinity for a polymer is important since crystallinity affects physical properties such as storage modulus, permeability, density, and melting point. While most of these manifestations of crystallinity can be measured, a direct measure of the degree of crystallinity provides a fundamental property from which these other physical properties can be predicted.

14.2 Differential Scanning Calorimetry

14.2.1 Theory

Differential scanning calorimetry (DSC) is a technique that measures heat flow into or out of a material as a function of time or temperature. Polymer crystallinity can only be determined by DSC by quantifying the heat associated with the melting (fusion) of the polymer. The heat is reported as % crystallinity by ratioing against the heat of fusion for 100% crystalline samples of the same material or more commonly by ratioing against a polymer of know crystallinity to obtain relative values. This technique is further discussed in Section 9.4.4.

14.2.2 Instrumentation

The Perkin Elmer Pyris Diamond calorimeter has been particularly recommended for crystallinity studies [1].

14.2.3 Applications

Kong and Hay [2] have reviewed the measurement of the degree of crystallinity of polymers by DSC. The inherent problem with all DSC measurements is that the degree of crystallinity changes because crystallisation, partial melting, annealing, recrystallisation, and complete melting occur during the heating of the test sample to the melting point. A first law procedure is suggested in which the sample is heated between two set temperatures, $T1$ and $T2$. $T1$ is taken to be ambient or just above the glass transition temperature (T_g) because it is selected by the requirement that the degree of crystallinity of the sample should not change with either temperature or time. $T2$ is taken to be just above the observed last trace of crystallinity. Integrating the observed specific heat difference between the sample and the completely amorphous material across these temperature ranges determines the residual enthalpy of fusion at $T1$. Kong and Hay [2] discussed the problems with the procedure. The initial fractional crystallinities of metallocene PE and polyethylene terephthalate (PET) were measured using the first law method and were found to be consistent with values determined by density at ambient temperature.

Other polymers for which DSC-based studies of crystallinity have been reported include PET [1, 3], ethylene–propylene copolymers [4], PE gels [5], polyazomettine esters [6], polyvinyl alcohol (PVOH) microfibrils [7], polylactides [8], polycyclohexadiene [9], PET–polycarbonate (PC) copolymer [10], PE-like polyesters [11], cyclohexylene dimethylene terephthalate–PET copolymers [12], poly(trimethylene) 2,6-naphthalate [13], and isotactic polypropylene (iPP) [14].

DSC has been used to study the crystal structure and thermal stability of PVOH modified at low levels by various reagents and by grafting with other vinyl monomers.

The development of crystal modifications of poly-1-butene has been followed using DSC [15].

A method based on DSC measurement of heat of crystallisation has been reported for the determination of the molecular weight of polytetrafluoroethylene [16].

DSC has been used to study crystallisation kinetics of many polymeric systems including amorphous cellulose [17], PE and chlorinated polyethylenes [18–20], aliphatic polyesters [21], Nylon 8 [22], and Nylon 6,6 and 6,10 [23].

Certain thermoplastics, e.g., PET, can be frozen in the amorphous state below the T_g by quench-cooling of the melt. The occurrence of different crystal forms (polymorphism of polyamides (PA), as an example) and melting gaps caused by tempering can be read from the DSC curve. If recrystallisation is studied, crystal growth rate and supercooling can be measured.

Isothermal crystallisation studies by DSC provide a sensitive technique for measuring the molecular weight or structural differences between very similar materials.

DSC and low-angle X-ray scattering have been used to prepare phase diagrams of polyethylene oxide – preferential solvent of polystyrene (PS) [24].

DSC (heat flow *versus* temperature curve) is particularly well suited to the detection and study of liquid crystals. Studies of transition temperatures and enthalpies, range of stability, and states of reversion from one form to another are all possible.

Illers and co-workers [25] have shown that DSC can be used to prepare phase diagrams and illustrated it by work on a binary blend of PS and tetramethyl bisphenol-A–PC. DSC curves obtained for different blends of the same two polymers, ranging in concentration from 100% PS (lower curve) to 100% tetramethyl bisphenol-A–polycarbonate (TMBPA-PC), are shown in **Figure 14.1(a)**. **Figure 14.1(b)** shows a set of curves for 1:1 mixtures of PS (M_w = 320,000, T_g = 109 °C) and TMBPA-PC (M_w = 41,000, T_g = 200 °C) which had been previously heated and chilled up to the temperatures indicated on the curves, then chilled in ice water. The dotted line indicates the phase diagram.

Illers and co-workers [25] showed that a phase diagram of a binary polymer blend can be derived from the T_g of the demixed phases under the following conditions:

1. The T_g values of the pure components are sufficiently different from one another.

2. The T_g values of the one-phase homogeneous mixtures can be determined and vary monotonically with composition.

3. The equilibrium state of the mixture is attained after sufficiently long annealing at temperature T.

4. The equilibrium state at the temperature T can be frozen by quenching.

They showed that between T_g and bimodal, the one-phase mixtures exist in a hindered thermodynamic state, because the crystallisation of the TMBPA-PC component is a very slow process which requires very long annealing. With increasing amounts of PS in the mixture the crystallisation half time of TMBPA-PC is reduced.

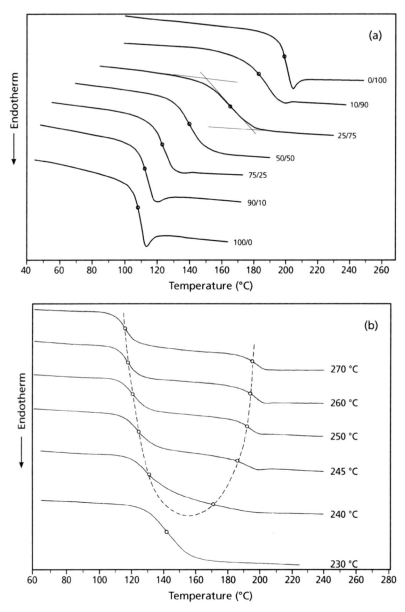

Figure 14.1 Differential scanning calorimetry of binary blends of polystyrene and tetramethyl bisphenol-A-polycarbonate. (a) DSC of various blends between 100% polystyrene (lower curve) and 100% tetramethyl bisphenol-A-polycarbonate (upper curve). Heating rate 20 °C/min. (b) 1:1 mixture of polystyrene:tetramethyl bisphenol-A-polycarbonate, preheated to stated temperature (230-270 °C) then cooled in ice prior to DSC run (*Reprinted from K.H. Illers, W. Hickman and J. Hambrecht, Journal of Colloid and Polymer Science, 1984, 262, 557, with permission from Springer [25]*)

Crystallinity can be calculated from a DSC curve by dividing the measured heat of fusion by the heat of fusion of 100% crystalline material. The crystalline melting point (T_m), which is a characteristic property, is used for quality control and for the identification of semi-crystalline polymers.

The crystallites are destroyed upon melting and reform upon cooling. Their type and quantity depends on the sample's thermal history. Crystallinity values have been determined [26] for poly(*p*-biphenyl acrylate) and poly(*p*-cyclohexylphenyl acrylate) from both heat of fusion and heat capacity measurements by DSC. DSC has also been used to study the degree of crystallinity of Nylon 6 [27, 28] and crosslinked PVOH hydrogels submitted to a dehydration and annealing process [29].

Figure 14.2 shows a melting endotherm for a sample of PE during the initial 'as recovered' heating. The percentage crystallinity was calculated based on 290 J/g for a 100% crystalline material. Typical results are quoted in **Table 14.1**.

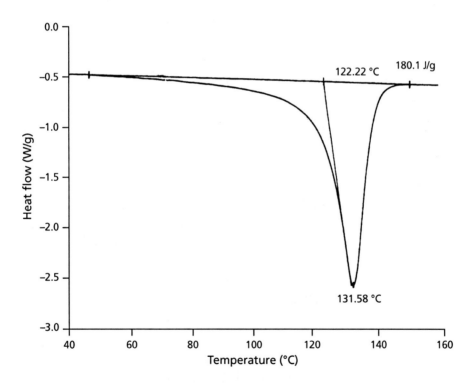

Figure 14.2 Differential scanning calorimetry of polyethylene melt.
(*Source: Author's own files*)

Table 14.1 Determination of crystallinity and other thermal properties of polyethylene film							
Polymer	Polymer as received, heated				DSC of polyethylene polymer given initial thermal treatment (heat at 10 °C/min to 180 °C) then run in DSC analyser			
	a	b	c	d	a	b	c	d
1	121.9	132.9	195.9	67.6	119.9	132.7	187.2	64.6
2	121.3	132.6	194.5	67.1	119.5	132.5	187.7	64.7
3	122.3	131.6	180.1	62.1	119.2	132.6	188.1	64.9
a - melt on-set temperature; b - melt peak temperature; c - enthalpy (Jg); d - %crystallinity *Source: Author's own files*								

14.3 Differential Thermal Analysis

14.3.1 Theory

The theory and instrumentation of this technique are discussed in Sections 9.2.2 and 9.4.3 and Appendix 1.

14.3.2 Applications

Differential thermal analysis has been used for the measurement of crystallinity of random and block ethylene–methacrylate copolymers [30] and of polybutene [31], PET, 1,4-cyclohexanedimethyl terephthalate, and polypropylene (PP) [32, 33] and in the examination of morphologically different structures of PE ionomers [34].

14.4 X-ray Powder Diffraction

14.4.1 Applications

Turley [35] has published a collection of X-ray diffraction patterns of polymers. X-ray diffraction methods have been applied to crystallinity and crystal structure studies of a wide range of polymers including cellulose [36, 37], Nylon 6,6 [38],

Nylon 6 [39, 40], Nylon 8 [39], ethylene–propylene copolymer [41], poly-4-methylpentene [42], polyacrylonitrile [43], polychloroprene [44], PS [45], PVOH [7, 46], polyaniline [47], polycyclohexadiene [9], polylactide [8], polyazomethine [6], poly-α-methylstyrene [48], PET [10], PE-like polyesters [11], and rigid polyphenylene dendrimers [49].

14.5 Wide-Angle X-ray Scattering/Diffraction

14.5.1 Applications

Huang and co-workers [50] used wide-angle X-ray scattering (WAXS) and Fourier transform infrared (FT-IR) spectroscopy to study crystalline transition behaviour under different crystalline conditions for PA with long alkane segments. PA studied included PA 12,20 and PA 10,20, and their transition behaviour was seen to depend strongly on crystal preparation method. Brill transition behaviour (two crystalline transition temperatures) was only displayed by solution-cast samples and after samples had been annealed, and not by dilute solution-grown lamellar crystals. Local melting of the long alkane segments at the Brill transition temperature, rather than at the higher polymer melting temperature, was indicated from infrared studies.

Chaari and co-workers [51] evaluated the development of crystallinity during the drawing of PET samples at different draw rates and at temperatures slightly above the T_g using WAXS methods. It was shown that at low draw rates, crystallinity developed almost exclusively after drawing. At medium draw rates, the whole crystallisation process occurred during drawing, and at high rates crystallisation started during drawing but continued after deformation had stopped. Development of crystallinity was related to the mechanical stress applied to the polymer during drawing, and the strain hardening behaviour was easily observed to coincide with the onset of crystallisation.

Li and co-workers [52] studied the temperature dependence of the crystalline structure and morphology of PA 12 using simultaneous small- and WAXS. The α'-phase was the stable, high-temperature crystalline phase, a transition to the hexagonal gamma-phase occurring on cooling. A discontinuous change in density occurred at the transition. Two crystal reflections were observed at wide angles and low temperatures, attributed to a monoclinic or an orthorhombic structure evolving from the hexagonal gamma-form by anisotropic thermal expansion. From the evolution of the crystalline thickness during cooling and heating, it was proposed that partial surface crystallisation and melting occurred. The transition from gamma- to α'-phase with increasing temperature was attributed to a 'one-dimensional melting' type of rupture between hydrogen bonded sheets. Fast quenching from the melt yielded the γ'-phase, which transformed to either the gamma- or the α'-polymorph on annealing by a melting or recrystallisation process.

Wide-angle X-ray diffraction (WAXD) scattering has been used widely in studies of the characterisation of crystal structure of polymers: (poly(ethylene-2,6-naphthalate) [53], atactic polypropylene [14], poly(ω-pentadecalactone) [54], and PP homopolymer [55]).

Creep studies. These have been carried out on ultrahigh-strength PE fibre [56].

Effect on orientation in amorphous and crystalline regions. Cho and co-workers [57] studied this in the case of poly(glycolide-*co*-ε-caprolactone).

Melting behaviour studies. These have been carried out on PET and PET–cyclohexylenedimethyl terephthalate copolymers [12] and poly(trimethylene-2,6-naphthalate) [13].

Pore size and strain size studies. Carotenuto measured the average grain size of metal polymer nanocomposites [58] and Zamfirova and co-workers [59] carried out mean pore size measurements on poly(heptamethylene-*p*,*p*-tribenzoate) using X-ray diffraction and positron annihilation lifetime spectroscopy (PALS).

14.6 Small Angle X-ray Diffraction Scattering and Positron Annihilation Lifetime Spectroscopy

14.6.1 Theory

Chu [60] has discussed the use of laser light, X-rays, and neutrons with corresponding wavelengths of the order of thousands of angstroms, one angstrom, and several angstroms to probe the structure and dynamics of macromolecules ranging in length scale from angstroms to micrometres and in time periods from picoseconds to tens of seconds. Synchrotron X-rays with a brilliance of the order of 10^{10} of a conventional X-ray source have small beam divergence which can be utilised to design an instrument for simultaneous time-resolved small-angle X-ray scattering (SAXS) and WAXD. The well-defined X-ray beam can also be used for X-ray reflectivity measurements and for the development of X-ray photon correlation experiments, extending the accessible *q*-range of dynamic light scattering (DLS) by orders of magnitude with *q* being the magnitude of the momentum transfer vector. Microbeam X-rays can be developed for X-ray fluorescence correlation, which is complementary to the light and (coherent) X-ray correlation techniques.

The crystalline state of a polymeric material is the most efficient manner in which molecules can be packed into a dense state. However, polymers rarely exist as single crystals - defects and heterogeneities are almost always present. Such heterogeneities are inherent to the imperfect packing nature of a glassy material. It appears that these heterogeneities are very significant in determining various physical and mechanical properties of the glassy material. Thus, it would be useful quantitatively to describe the glassy state of a polymer and hopefully correlate the mechanical behaviour. To date, there exist very few experimental techniques that can accomplish such a description of non-crystalline solids. Hristov and co-workers [61] have discussed the correlations between SAXS and PALS measurements of glassy polymers. These workers examined the glassy structure of several typical polymers. Specifically, strong correlations between the two techniques are sought. On their own, each technique has fundamental shortcomings. It is possible that some of these shortcomings can be alleviated by relating PALS and SAXS in a complementary manner. Through this, an attempt is made to achieve a secondary goal of clarifying exactly what PALS can measure.

14.6.2 Applications

Ho and co-workers [62] used SAXS, among other techniques, to examine microphase separation and crystallisation in a melt-mixed blend of syndiotactic polystyrene (sPS) with a block copolymer of styrene and ethylene propylene.

Spherical micro-domains of sPS were observed in the PS–PE–PP matrix and crystallisation in these areas was observed as it grew to overcome spatial confinement and the immiscibility barrier and crossover surrounding PE–PP domains interconnecting with other sPSmicro-domains. The crystallisation rate increased with increasing sPS content, probably due to decreasing distances between sPS micro-domains, and resulting in a crystalline domain morphology. As mentioned previously, this combination of techniques has been used in pore size measurements on poly(heptamethylene-*p,p*-tribenzoate) [59].

14.7 Static and Dynamic Light Scattering

14.7.1 Applications

Einaga and Fujisawa [63] carried out DLS measurements on benzene solutions of PS in order to characterise the viscoelastic properties. Static light scattering measurements of chloroform and tetrahydrofuran (THF) solutions have been used to characterise copolymers of polymethyl methacrylate and bendazole dyes [64]. Cametti and co-workers [65] have carried out both static and DLS measurements on THF, toluene, and chloroform solutions of polyphenylacetylene. Each solvent produces a different effect on the pristine structure of polyphenylacetylene.

14.8 Infrared Spectroscopy

14.8.1 Applications

Simak [66] has carried out FT-IR measurements of the specific volume of the crystalline phase (V_c; cm³/g) and the specific volume of the amorphous phase (V_a; cm³/g) of PET. Absorptions occurred at 632, 440, and 380 cm⁻¹ originating from C–O–C groups attached to the aromatic ring found in PET. The 632 cm⁻¹ band was used for the evaluation of V_c, and it is shown in **Figure 14.3(a)** that these results lead to a specific volume of V_c = 0.66 cm³/g. The 440 and 380 cm⁻¹ bands were used for the evaluation of V_a, and it is seen in **Figure 14.3(b)** that the results lead to a specific volume of V_a = 0.747 cm³/g.

Other workers [67-87] have discussed the application of infrared spectroscopy to the crystallinity of PET. Infrared band assignments in the 4000–400 cm⁻¹ region have been made for crystalline poly(ethylene glycol dimethyl ether) [88]. Janssen and Yannas [89] measured the infrared dichroism of PC at different strain levels. Below 0.6% strain the dichroism is negligible while above this level the dichroism increases linearly. The infrared spectra of polyesters in the C–O–R stretching region have been discussed [90]. Infrared and Raman bond assignments were made by Holland-Moritz and co-workers [91-93] for linear aliphatic polyesters including the influence of the ester and CH₂ groups.

Elliot and Kennedy [94] measured the infrared spectra of the three crystalline forms of cationically synthesised poly-3-methyl-1-butene. The alpha, beta, and gamma crystalline phases of cationically synthesised poly-3-methyl-1-butene were characterised. McRae and co-workers [95] followed changes in the relative concentrations of methylene groups in crystalline regions in gauche conformations and in tie chains in amorphous regions of cold drawn high-density polyethylene (HDPE) by unpolarised infrared spectroscopy. Luongo [96] studied crystalline orientation effects in the 13.69–13.88 µm doublet of *trans*-crystalline PE. The effect of defects on the infrared spectrum of PE crystals was established by comparison with PE in the extended chain form. Mocheria and Bell [97] used infrared dynamic mechanical and molecular weight measurements in studies of morphology of uniaxially oriented PET. These have also been used in studies of chain folding in annealed polymers [88]. The Raman spectra of iPP at –268 to +250 °C show bands characteristic of the unit cell [98, 99]. Relaxations between crystallinity and chain orientation have been observed in the Raman spectrum of oriented PP [100, 61]. FT-IR spectroscopy has been used in crystallinity transition studies on Nylon 12, 20 and 10, 20 [50].

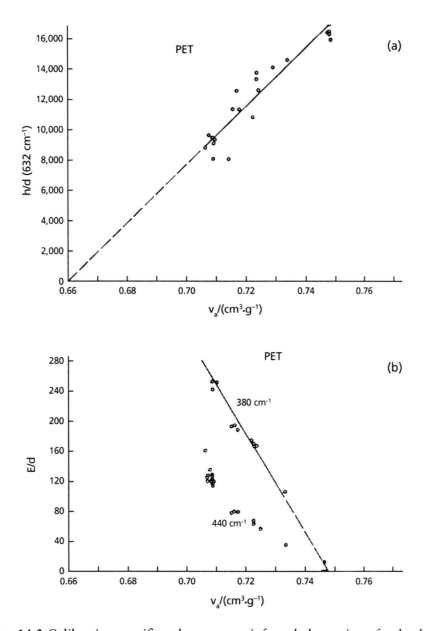

Figure 14.3 Calibration specific volume *versus* infrared absorption of polyethylene terephthalate. (a) Crystalline bands at 632 cm^{-1}, $V_c = 0.66$ cm^3/g; Amorphous bands at 440 and 380 cm^{-1}; $V_a = 0.7470$ cm^3/g

(*Reprinted from P. Simak, Die Makromolekulare Chemie-Makromolekular Symposia, 1986, 5, 61, with permission from Wiley-VCH [66]*)

14.9 Nuclear Magnetic Resonance

14.9.1 Applications

Nuclear magnetic resonance (NMR) line width studies of crystalline polymers are based on the work of Wilson and Pake [102]. This method was, however, unsuccessful due to the rather arbitrary decomposition procedures used, which yielded a crystalline fraction that was not in agreement with crystallinity results obtained by the X-ray method. To overcome this difficulty Bergmann [103–105] decomposed the spectrum into three components and this resulted in an excellent agreement between NMR and X-ray crystallinities. Unfortunately, with this method it was not possible to prove the existence of the two amorphous components of the polymer examined. Also, the two amorphous mobilities could not be predicted theoretically. Bergmann [106] succeeded eventually, as discussed next, in improving the separation procedure by finding more suitable line widths for the crystalline and amorphous components of the polymer. In this procedure a new method was evolved for the determination of the crystalline component and of the amorphous component based on a distribution of correlation times, instead of the two discrete correlation times as used in earlier work [103–105].

Bergmann [106] compared the crystallinities obtained for polymers by line width NMR measurements with those obtained by X-ray crystallography (**Table 14.2**). The NMR crystallinity (W_c NMR) was determined at the specified temperature and X-ray crystallinity (W_c X-ray) at room temperature. The spin-locking time (d) for the measurement of the crystalline component ranged from 10 to 150 milliseconds. For most polymers the results in **Table 14.2** show an agreement of crystallinity within the limits of error, i.e., ±2% for W_c X-ray. The error with regard to W_c NMR mainly depends on the value of alpha. The greater error in conjunction with the broader distribution is explained by the greater difficulty in distinguishing the immobile amorphous from the crystalline protons in these cases.

Axelson and Mandekkern [107] showed that the broad line proton NMR spectrum of drawn HDPE has three component lines: a broad component assigned to crystalline regions, an intermediate component assigned to high molecular weight molecules interconnecting the crystalline regions, and an isotropic narrow component assigned to mobile low molecular weight material and the ends of molecules reflected from the crystalline regions.

Ohta and co-workers [56] studied structural changes leading to creep rupture occurring during creep formation of ultrahigh-strength PE fibres using ^{13}C NMR and WAXD.

Polymer	Density g/cm³	T, K	Crystallinity W_c (%)		d, ms	$\omega_p\,\tau_{cm}$	α
			NMR	x-ray			
HDPE	0.963	310	74±1	75	100	0.7	1.1
LDPE*	0.918	300	47±1	43	30	0.4	1.9
PP	0.918	310	79±1	78	50	0.3	1.0
PP	0.899	310	60±2	57	50	0.3	1.3
POM	1.424	310	72±1	67	150	0.5	0.0
PA 6,10	1.098	370	44±3	52	30	0.3	1.8
PETP	1.410	400	51±3	60	10	3	3.1

Table 14.2 Comparison of crystallinities determined by NMR and x-rays for seven polymers

*Reference crystallinity determined from density measurement
HDPE = high density polyethylene; LDPE = low density polyethylene;
PP = polypropylene; PA6,10 = Nylon 6,10;
PETP = polyethylene terephthalate
Reprinted from K. Bergmann, Polymer Bulletin, 1981, 5, 355, with permission from Springer [106]

References

1.   L. Li, G. Curran and G. Garrabe, *Revue Générale des Caoutchoucs et Plastiques*, 2002, **78**, 28.

2.   Y. Kong and J.N. Hay, *Polymer*, 2002, **43**, 3873.

3.   D.P.R. Kint, A.M. de Ilarduya and S. Munoz-Guerra, *Polymer Degradation and Stability*, 2003, **79**, 353.

4.   X. Colom, J. Canavate, J.J. Sunol, P. Pages, J. Saurina and F. Carrasco, *Journal of Applied Polymer Science*, 2003, **87**, 1685.

5.   P.M. Pakhomov, S. Khizhnyak, E. Ruhl, V. Egorov and A. Tshmel, *European Polymer Journal*, 2003, **39**, 1019.

6.   U. Shukla, K.V. Rao and A.K. Rakshit, *Journal of Applied Polymer Science*, 2003, **88**, 153.

7.   H.D. Ghim, J.P. Kim, I.C. Kwo, C.J. Lee, J. Lee, S.S. Kim, S.M. Lee, W.S. Yoon and W.S. Lyoo, *Journal of Applied Polymer Science*, 2003, **87**, 1519.

8. M. Takasaki, H. Ito and T. Kikutano, *Journal of Macromolecular Science B*, 2003, B42, 57.

9. R.P. Quirk, F. You, L. Zhu and S.Z.D. Cheng, *Macromolecular Chemistry and Physics*, 2003, **204**, 755.

10. P. Marchese, A. Celli, M. Fiorini and M. Gabaldi, *European Polymer Journal*, 2003, **39**, 1081.

11. M.G. Menges, K. Schmidt-Rohr and J. Penelle, *Polymer Preprints*, 2002, **43**, 2, 472.

12. C.A. Avrila-Orta, F.J. Medellín-Rodrígues, Z.G. Wang, D. Navarro-Rodríguez, B.S. Hsiao and F. Yeh, *Polymer*, 2003, **44**, 1527.

13. Y.G. Jeong, W.H. Jo and S.C. Lee, *Polymer*, 2003, **44**, 3259.

14. W. Di, P. Liu and X. Wang, *Journal of Applied Polymer Science*, 2003, **88**, 2791.

15. V.A. Era and T. Jauhiainen, *Angewandte Makromolekulare Chemie*, 1975, **43**, 157.

16. T. Suwa, M. Takehisa and S. Machi, *Journal of Applied Polymer Science*, 1973, **17**, 3253.

17. M. Kimura, T. Hatakeyama and J. Nakano, *Journal of Applied Polymer Science*, 1974, **18**, 3069.

18. V. Erä and H. Venäläinen, *Journal of Polymer Science: Polymer Symposia*, 1973, No. 42, 879.

19. H. Heyns and S. Heyer in *Thermal Analysis*, Volume 3, Ed., H.G. Wiedermann, Birkhaeuser, Basel, Switzerland, 1972, p.341.

20. U. Johnsen and G. Spilgies, *Kolloid Zeitschrift und Zeitschrift für Polymere*, 1972, **250**, 1174.

21. M. Gilbert and F. Hybart, *Polymer*, 1974, **15**, 407.

22. G. Ceccorulli and F. Manescalchi, *Makromolekulare Chemie*, 1973, **168**, 303.

23. A. Savolainen, *Journal of Polymer Science: Polymer Symposia*, 1973, **42**, 885.

24. K.H. Illers, *European Polymer Journal*, 1974, **10**, 991.

25. K.H. Illers, W. Hickman and J. Hambrecht, *Journal of Colloid and Polymer Science*, 1984, **262**, 557.

26. V. Frosini, P.L. Hagagnini and B.A. Newman, *Journal of Polymer Science, Physics Edition*, 1974, **12**, 23.

27. G. Coppola, R. Filippini and B. Pallesi, *Polymer*, 1975, **16**, 546.

28. K. Kamide and A. Imanaka, *Kobunshi Ronbunshu*, 1975, **32**, 537.

29. N.A. Peppas, *Journal of Applied Polymer Science,* 1976, **20**, 1715.

30. T. Tanaka and K. Hatada, *Journal of Polymer Science*, 1973, **11**, 2057.

31. C. Geacintov, R.S. Schotland and R.B. Miles, *Journal of Polymer Science: Part C*, 1964, **6**, 197.

32. R.F. Schwenker and R.K. Zuccarello, *Journal of Polymer Science: Part C*, 1964, **6**, 1.

33. H.J. Donald, E.S. Humes and L.W. White, *Journal of Polymer Science*, 1964, C-6, 93.

34. C.L. Marx and S.L. Cooper, *Makromolekulare Chemie*, 1973, **168**, 339.

35. J.W. Turley, *Journal of Polymer Science A*, 1965, **3**, 2400.

36. K.J. Heritage, J. Mann and L. Roldan Gonzalez, *Journal of Polymer Science A*, 1963, **1**, 671.

37. K.C. Ellis and J.O. Warwicker, *Journal of Polymer Science A*, 1963, **1**, 1185.

38. R. Johnson, *Journal of Polymer Science A*, 1963, **1**, 715.

39. D.C. Vogelsang, *Journal of Polymer Science A*, 1963, **1**, 1055.

40. M. Hirami, *Journal of Polymer Science A*, 1966, **4**, 967.

41. J.F. Jackson, *Journal of Polymer Science A*, 1963, **1**, 2119.

42. A. Mitt, *Journal of Polymer Science A*, 1963, **1**, 2219.

43. R. Chang, *Journal of Polymer Science A*, 1963, **1**, 2765.

44. W.R. Frigbaum and F.J. Doe, *Journal of Polymer Science A*, 1964, **2**, 4391.

45. A. Dutschke, C. Diegelmann and P. Lobmann, *Journal of Materials Chemistry*, 2003, **13**, 1058.

46. H.M. Zidan, *Journal of Applied Polymer Science*, 2003, **88**, 516.

47. C. He, Y. Tan and Y. Li, *Journal of Applied Polymer Science*, 2003, **87**, 1537.

48. V. Petraccone, O. Tarallo and V. Califano, *Macromolecules*, 2003, **36**, 685.

49. D. Grebel-Koehler, D. Liu, S. De Feyter, V. Enkelmann, T. Weil, C. Engels, K. Samyn and F.C. De Schryver, *Macromolecules*. 2003, **36**, 578.

50. Y. Huang, W. Li and D. Yan, *European Polymer Journal*, 2003, **39**, 1133.

51. F. Chaari, M. Chaouche and J. Doucet, *Polymer*, 2003, **44**, 473.

52. L. Li, M.H.K. Koch and W.H. de Jen, *Macromolecules*, 2003, **36**, 1626.

53. M.C. Garcia Guttierrez, D.R. Rueda, F.J. Balta Calleja, N. Stribeck and R.K. Bayer, *Polymer*, 2003, **44**, 451.

54. M. Gazzano, V. Malta, M.L. Focarete and M. Scandola, *Journal of Polymer Science, Polymer Physics Edition*, 2003, **41**, 1009.

55. P.N. Vashi, A.K. Kulshreshtha and K.P. Dhake, *Journal of Applied Polymer Science*, 2003, **87**, 1190.

56. Y. Ohta, A. Kaji, H. Sugiyama and H. Yasuda, *Journal of Applied Polymer Science*, 2001, **81**, 312.

57. D.K. Cho, J.W. Park, S.H. Kim, Y.H. Kim and S.S. Im, *Polymer Degradation and Stability*, 2003, **80**, 223.

58. G. Carotenuto, *Polymer News*, 2002, **27**, 82.

59. G. Zamfirova, M. Krasteva, M. Misheva, M. Mihaylova, E. Perez and J.M. Perena, *Polymer International*, 2003, **52**, 46.

60. B. Chu in *Proceedings of ACS Polymeric Materials Science and Engineering Conference*, Dallas, TX, USA, Spring 1998, Volume 78, p.108.

61. H.A. Hristov, C.L. Soles, B.A. Bolan, D.W. Gidley and A. Yee in *Proceedings of ACS Polymeric Materials Science and Engineering Conference*, San Francisco, CA, USA, Spring 1997, Volume 76, p.431.

62. R-M. Ho, C.C. Chang, T-M. Chung, Y-W. Chiang and J-Y. Wu, *Polymer*, 2003, **44**, 1459.

63. Y. Einaga and T. Fujisawa, *Polymer*, 2002, **43**, 5105.

64. F.S. Rodembusch, N.P. Da Silveira, D. Samios, L.F. Campo and V. Stefani, *Molecular Crystals and Liquid Crystals, Section A*, 2002, **374**, 367.

65. C. Cametti, P. Codastefano, R. D'Amato, A. Furlani and M.V. Russo, *Synthetic Metals*, 2000, **114**, 173.

66. P. Simak, *Die Makromolekulare Chemie - Makromolecular Symposia*, 1986, **5**, 61.

67. W.H. Cobbs and R.L. Burton, *Journal of Polymer Science*, 1953, **10**, 275.

68. A. Thomson and D.W. Woods, *Nature (London)*, 1955, **176**, 78.

69. R. Miller and H. Willis, *Journal of Polymer Science*, 1956, **19**, 485.

70. J. Novak, V. Suskov and L. Zosin, *Beitrage Faserforsch und Textiltechnik*, 1969, **6**, 513.

71. K. Edelmann and H. Wyden, *Kautschuk und Gummi Kunststoffe*, 1972, **25**, 353.

72. W. Statton, J. Koenig and M. Hannon, *Journal of Applied Physics*, 1970, **41**, 4290.

73. K.H. Illers, *Colloid Polymer Science*, 1980, **258**, 11.

74. J. Pitha and R.N. Jones, private communication, 1968.

75. R. Danz and J. Deshant and C. Rushcer, *Faserforschung und Textiltechnik*, 1970, **21**, 503.

76. A. Miyake, *Journal of Polymer Science*, 1959, **38**, 497.

77. T.R. Manley and D. Williams, *Journal of Polymer Science, Part C*, 1969, **22**, 1009.

78. F. Boerio, S.K. Bahl and G. McGraw, *Journal of Polymer Science*, 1976, **14**, 1029.

79. R. Mehta and I.P. Bell, *Journal of Polymer Science, Physics Edition*, 1973, **11**, 1793.

80. A. Baumgarter, S. Blasenbrey and W. Pechhold, *Kolloid Zeitschrift und Zeitschrift für Polymere*, 1972, **250**, 1026.

81. F. Schonherr, *Faserforschung und Textiltechnik*, 1970, **21**, 246.

82. G. Schmidt, *Journal of Polymer Science, Part A*, 1963, **1**, 1271.

83. W. Astbury and C.I. Brown, *Nature (London)*, 1946, **158**, 871.

84. R. Daubeny, C.W. Bunn and C.I. Brown, *Proceedings of the Royal Society London A*, 1954, **226**, 531.

85. H. Zahn and R. Krzikalla, *Die Makromolekulare Chemie*, 1957, **23**, 1.

86. Y. Kinoshita, R. Nakamura, Y. Kitano and T. Ashida, *Polymer Preprints*, 1979, **20**, 454.

87. H.G. Kilian, H. Halboth and E. Jenkel, *Kolloid Zeitschrift und Zeitschrift für Polymere*, 1960, **172**, 166.

88. H. Matsura, T. Miyazawa and K. Machida, *Spectrochimica Acta A*, 1973, **29**, 771.

89. J.F. Jansson and I.V. Yannas in *Proceedings of the 7th International Congress on Rheology*, Göteborg, Sweden, 1976, p.274.

90. A.L. Smolyanski and G.V. Gusakova, *Trudy Yologod Moloch Inst*, 1970, **60**, 105.

91. K. Holland-Moritz, *Kolloid Zeitschrift und Zeitschrift für Polymere*, 1973, **251**, 906.

92. K. Holland-Moritz and D.O. Hummel, *Journal of Molecular Structure*, 1973, **19**, 289.

93. K. Holland-Moritz and D.O. Hummel, *Quad Ric Science*, 1973, **84**, 158.

94. J.J. Elliot and J.P. Kennedy, *Journal of Polymer Science: Polymer Chemistry Edition*, 1973, **11**, 2993.

95. M.A. McRae, W.F. Maddams and J.E. Preedy, *Journal of Material Science*, 1976, **11**, 2036.

96. J.P. Luongo, *Journal of Polymer Science, Polymer Chemistry Edition*, 1974, **12**, 1203.

97. K.K. Mocheria and J.P. Bell, *Journal of Polymer Science, Polymer Physics Edition*, 1973, **11**, 1779.

98. J.V. Derby and R.W. Freedman, *American Laboratory*, 1974, May, 94.

99. G.V. Fraser, P.J. Hendra, D.S. Watson, M.J. Gall, H.A. Willis and M.E.A. Cudby, *Spectrochimica Acta A*, 1973, **29**, 1525.

100. R.T. Bailey, A.J. Hyde and J.J. Kim, *Advances in Raman Spectroscopy*, 1972, **1**, 296.

101. R.T. Bailey, A.J. Hyde and J.J. Kim, *Spectrochimica Acta A*, 1974, 30, 91.

102. C.W. Wilson and G.E. Pake, *Journal of Polymer Science*, 1853, **10**, 503.

103. K. Bergmann and K. Nawotki, *Kolloid Zeitschrift und Zeitschrift für Polymere*, 1967, **219**, 132.

104. K. Bergmann, *Kolloid Zeitschrift und Zeitschrift für Polymere*, 1973, **251**, 962.

105. K. Bergmann, *Journal of Polymer Science, Polymer Physics Edition*, 1978, **16**, 1611.

106. K. Bergmann, *Polymer Bulletin*, 1981, 5, 355.

107. D.E. Axelson and L. Mandekkern, *Journal of Polymer Science, Polymer Physics Edition*, 1978, **16**, 1135.

15 Viscoelastic and Rheological Properties

15.1 Dynamic Mechanical Analysis

15.1.1 Theory

Dynamic mechanical analysis (DMA) provides putative information on the viscoelastic properties – modulus and damping – of materials. Viscoelasticity is the characteristic behaviour of most materials in which a combination of elastic properties (stress proportional to strain rate) are observed. A DMA simultaneously measures both elastic properties (modulus) and viscous properties (damping) of a material.

DMA measures changes in mechanical behaviour such as modulus and damping as a function of temperature, time, frequency, stress, or combinations of these parameters.

The technique also measures the modulus (stiffness) and damping (energy dissipation) properties of materials as they are deformed under periodic stress. Such measurements provide quantitative and qualitative information about the performance of the materials. The technique can be used to evaluate elastomers, viscous thermoset liquids, composite coatings, and adhesives, and materials that exhibit time, frequency, and temperature effects or mechanical properties because of their viscoelastic behaviour.

Some of the viscoelastic and rheological properties of polymers that can be measured by DMA are [1, 2]:

- Modulus and strength (elastic properties)
- Viscosity (stress–strain rate)
- Damping characteristics
- Low- and high-temperature behaviour (stress–strain)
- Viscoelastic behaviour
- Compliance
- Stress relaxation and stress relaxation modulus
- Creep

- Gelation
- Projection of material behaviour
- Polymer lifetime prediction

Basically this technique involves the measurement of the mechanical response of a polymer as it is deformed under periodic stress and is used to characterise the viscoelastic and rheological properties of polymers.

DMA is the measurement of the mechanical response of a material as it is deformed under periodic stress. Material properties of primary interest include modulus (E'), loss modulus (E''), tan δ (E''/E'), compliance, viscosity, stress relaxation, and creep. These properties, expressed in quantitative units of measurement, characterise the viscoelastic performance of a material.

DMA provides material scientists and engineers with the information necessary to predict the performance of a material over a wide range of conditions. Test variables include temperature, time, stress, strain, and deformation frequency. Because of the rapid growth in the use of engineering plastics and the need to monitor their performance and consistency, dynamic thermal analysis has become the fastest growing thermal analysis technique.

Recent advances in the materials sciences have been both the cause and effect of advances in materials characterisation technology. The technique of DMA is one of the most important developments because it allows materials scientists and engineers to predict accurately the performance of a material under a wide range of conditions.

The theory of DMA, which measures the mechanical response of a material as it is deformed under periodic stress, has been understood for many years. But because of the complexity of measurement mechanics and the mathematics required to translate theory into application, DMA did not become a practical tool until the late 1970s when DuPont developed a device for reproducibly subjecting a sample to appropriate mechanical and environmental conditions. The addition of computer hardware and software capabilities several years later made DMA a viable tool for the industrial scientist because it greatly reduced analysis times and the labour intensity of the technique.

DMA provides information on the viscoelastic properties – modulus and damping – of materials. Viscoelasticity is the characteristic behaviour of most materials in which a combination of elastic properties (stress proportional to strain) and viscous properties (stress proportional to strain rate) are observed. The DMA simultaneously measures both elastic properties (modulus) and viscous properties (damping) of a material. Such data are particularly useful because of the growing trend towards the use of polymeric materials as replacements for metals in structural applications.

454

Whereas elastic modulus at a single temperature (as measured by a static-mode instrument) is indicative of the properties and quality of metals, this is not necessarily true of polymeric materials in which molecular relaxations dramatically influence the temperature and time (reciprocal frequency) dependences of material properties.

DMA is the preferred tool for evaluating the effects of the following variables on the mechanical properties of materials:

Temperature. Temperature scanning can determine whether product performance will remain within specifications at end-use temperatures.

Frequency (1/time) deformation. Frequency sweeps can be used to study the frequency dependence of material properties.

Load. Creep and stress relaxation measurements can determine the long-term mechanical properties of materials and their ability to withstand loading and deformation influences.

The DuPont 983 DMA is a versatile laboratory instrument for characterising the viscoelastic and rheological behaviour of materials. This system is designed to operate in any of four modes: fixed frequency, resonant frequency, stress relaxation, or creep. It is discussed in detail next as an example of the type of equipment now available.

System components are the 983 DMA module, an optional liquid nitrogen cooling accessory, a thermal analyser/temperature controller, data analysis software (model 9000 computer controller or DuPont Thermal-Analyst 2100), and a plotter.

The 9900 controls the test temperature and collects, stores, and analyses the resulting data. Data analysis software for the 983/9900 DMA system provides hardcopy reports of all measured and calculated viscoelastic properties, including time/temperature superpositioning of the data for the creation of master curves. A separate calibration program automates the instrument calibration routine and ensures highly accurate, reproducible results. In addition to measuring specific viscoelastic properties of interest, the four modes provide a more complete characterisation of materials, including their structural and end-use performance properties. The system can also be used to simulate and optimise processing conditions, such as those used with thermoset resins.

The 983 DMA can evaluate a wide variety of materials, ranging from the very soft, such as elastomers and supported viscous liquids, to the very hard, such as reinforced composites, ceramics, and metals. The clamping system can accommodate a range of sample geometries, including rectangles, rods, films, and supported liquids. The 983 instrument produces quantitative information on the viscoelastic and rheological properties of a material by measuring the mechanical response of a sample as it is deformed under periodic stress.

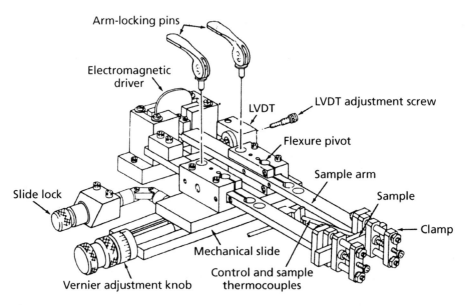

Figure 15.1 DMA electromechanical system. LVDT: Linear variable differential transformer
(*Source: Author's own files*)

Figure 15.1 shows the mechanical components of the 983 DMA. The clamping mechanism for holding samples in a vertical configuration consists of two parallel arms, each with its own flexure point, an electromagnetic driver to apply stress to the sample, a linear variable differential transformer for measuring sample strain, and a thermocouple for monitoring sample temperature. A sample is clamped between the arms and the system is enclosed in a radiant heater and Dewar flask to provide precise temperature control.

Easily interchanged clamp faces are designed to accommodate various geometries: smooth and serrated faces for rectangular shapes, and notched faces to hold cylinders or tubing. A horizontal clamping system with smooth clamp faces is available to hold soft samples and supported liquids.

The sample is clamped between the ends of two parallel arms, which are mounted on low-force flexure pivots allowing motion in only the horizontal plane. The distance between the arms is adjustable by means of a precision mechanical slide to accommodate a wide range of sample lengths. An electromagnetic motor attached to one arm drives the arm/sample system to a strain amplitude selected by the operator. As the original sample system is displaced, the sample undergoes a flexural deformation. A linear variable differential transformer mounted on the driven arm measures the sample's response (strain and frequency) to the applied stress, and provides feedback control to the motor.

The sample is positioned in a temperature-control chamber which contains a radiant heater and a coolant distribution system. The radiant heater provides precise and accurate control of sample temperature. The coolant distribution system uses cold nitrogen gas for smooth, controlled sub-ambient operation and for quench cooling at the start or end of a run. Both the radiant heater and the coolant accessory are controlled automatically by the 983 system to ensure reproducible temperature programming. An adjustable thermocouple, mounted close to the sample, provides precise feedback information to the temperature controller, as well as a readout of sample temperature.

15.1.1.1 Fixed Frequency Mode

This mode is used for accurate determination of the frequency dependence of materials and prediction of end-use product performance. In the fixed frequency mode applied stress (i.e., a force per unit area that tends to deform the body, usually expressed in Pa (N/m)) forces the sample to undergo sinusoidal oscillation at a frequency and amplitude (strain), i.e., the deformation from a specified reference state, measured as the ratio of the deformation to the total value of the dimension in which the strain occurs. Strain is non-dimensional, but is frequently expressed in reference values (such as %strain) selected by the operator. Energy dissipation in the sample causes the sample strain to be out of phase with the applied stress (**Figure 15.2(a)**). In other words, since the sample is viscoelastic, the maximum strain does not occur at the same instant as maximum stress. This phase shift or lag, defined as phase angle (δ), is measured and used with known sample geometry and driver energy to calculate the viscoelastic properties of the sample.

The 983 analyser can be programmed to measure a sample's viscoelastic characteristics at up to 57 frequencies during a single test. In such multiplexing experiments, an isothermal step method is used to hold the sample temperature constant while the frequencies are scanned. The sample is allowed to reach mechanical and thermal equilibrium at each frequency before the data are collected. After all the frequencies have been scanned, the sample is automatically stepped to the next temperature and the frequency scan is repeated. Multiplexing provides a more complete rheological assessment than is possible with a single frequency.

15.1.1.2 Resonant Frequency Mode

This mode is used for detection of subtle transitions, which are essential for understanding the molecular behaviour of materials and structure–property relationships. Allowing a sample to oscillate at its natural resonance provides higher damping sensitivity than at a fixed frequency, making possible detection of subtle transitions in the material. Because of its high sensitivity the resonance mode is particularly useful in analysing polymer blends and filled polymers, such as reinforced plastics and composites.

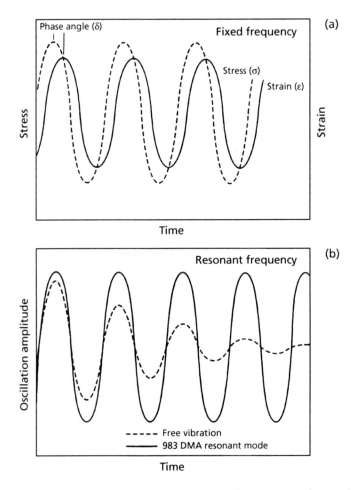

Figure 15.2 Modes of operation of DuPont 983 dynamic mechanical analyser.
(*Source: Author's own files*)

In the resonance mode, the 983 operates on the mechanical principle of forced resonant vibratory motion at a fixed amplitude (strain), which is selected by the operator. The arms and sample are displaced by the electromagnetic driver, subjecting the sample to a fixed deformation and setting the system into resonant oscillation.

In free vibration, a sample will oscillate at its resonant frequency with a decreasing amplitude of oscillation (dashed line in **Figure 15.2(b)**). The resonance mode of the 983 analyser differs from the free vibration mode in that the electromagnetic driver puts energy into the system to maintain a fixed amplitude, as illustrated in **Figure 15.2(b)**. The make

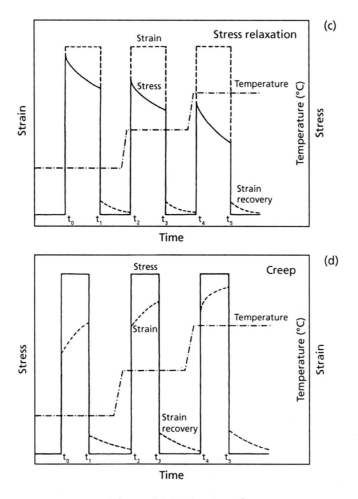

Figure 15.2 *Continued*

up energy, oscillation frequency, and sample geometry are used by the DMA software to calculate the desired viscoelastic properties.

15.1.1.3 Stress Relaxation Mode

Stress relaxation is defined as a long-term property measured by deforming a sample at a constant displacement (strain) and monitoring decay over a period of time. Stress is a very significant variable that can dramatically influence the properties of plastics and composites.

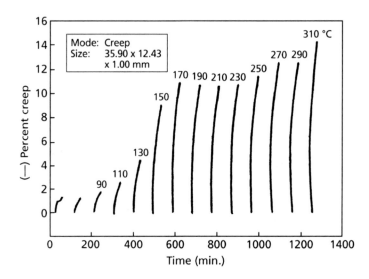

Figure 15.3 Stress relaxation modulus of polycarbonate. (*Source: Author's own files*)

Induced stress is always a factor in structural applications and can result from processing conditions, thermal history, phase transitions, surface degradation, and variations in the expansion coefficient of components in a composite. Since the modulus of a material is not only temperature dependent but time dependent as well, the stress relaxation behaviour of polymers and composites is of great importance to the structural engineer. Stress relaxation is also important to the polymer chemist developing new engineering plastics because relaxation times and moduli are affected by polymer structures and transition temperatures. Therefore, it is essential that polymeric materials that will be subjected to loading stress be characterised for stress relaxation and creep behaviour.

Figure 15.3 shows the stress relaxation curves obtained for polycarbonate (PC). The large drop in modulus of the material above 130 °C is caused by the increased mobility of the polymer molecules as the glass transition temperature (T_g) is approached.

The stress relaxation mode is used for measurement of ultralow-frequency relaxation in polymers and composite structures, a primary indicator of the long-term performance of materials. This mode of operation provides definitive information for prediction of the long-term performance of materials by measuring the stress decay of a sample as a function of time and temperature at an operator-selected displacement (strain).

Figure 15.2(c) shows a schematic representation of the process for measuring stress relaxation in a viscoelastic material. Using an isothermal step method, the sample is allowed to equilibrate at each temperature in an unstressed state. The sample is then stressed to the selected strain. The amount of driver energy (stress) required to maintain that displacement is recorded as a function of time for a period selected by the operator. After the measurements are recorded, the driver stress is removed and the sample is allowed to recover in an unstressed state. This sample recovery (strain) can be recorded as a function of time for a period selected by the operator. When the measurements at one temperature are complete, the temperature is increased and the experiment is repeated.

15.1.1.4 Creep Mode

Creep is defined as a long-term property measured by deforming a sample at a constant stress and monitoring the flow (strain) over a period of time. Viscoelastic materials flow or deform when subjected to loading (stress). In a creep experiment, a constant stress is applied and the resulting deformation is measured as a function of temperature and time. Just as stress relaxation is an important property to structural engineers and polymer scientists, so is creep behaviour.

Elastic modulus is a quantitative measure of the stiffness or rigidity of a material. For example, for homogeneous isotropic substances in tension, the strain (ε) is related to the applied stress (σ) by the equation $E = \sigma/\varepsilon$, where E is defined as the elastic modulus. A similar definition of shear modulus (g) applies when the strain is shear.

For a polyether ether ketone laminate below the T_g the laminate exhibits less than 2% creep, while above the T_g creep increases to greater than 10%.

The thermal curve in **Figure 15.4** shows the flexural storage modulus and loss properties of a rigid, pultruded oriented-fibre-reinforced vinyl ester composite. Since the flexural modulus and T_g increase dramatically with post-curing, the test can be used to evaluate the degree of cure, as well as to identify the high-temperature mechanical integrity of the composite.

The creep mode is used for measurement of flow at constant stress to determine the load-bearing stability of materials, a key to prediction of product performance. The creep mode is used to measure sample creep (strain) as a function of time and temperature at a selected stress. Using the isothermal step method, the sample is allowed to equilibrate at each temperature in a relaxed state. After equilibration, the sample is subjected to a constant stress, as illustrated in **Figure 15.2(d)**. The resulting sample deformation (strain) is recorded as a function of time for a period selected by the operator. After the first set of measurements is made, the driver stress is removed and the sample is allowed to recover in an unstressed state. Sample recovery (strain) can be recorded as a function of time for any

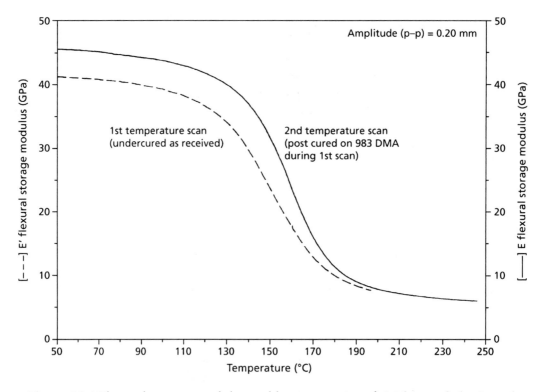

Figure 15.4 Flexural storage modulus and loss properties of rigid extruded oriented fibre reinforced vinyl ester composites. (*Source: Author's own files*)

desired period. When the measurement at one temperature is complete, the temperature is changed and the measurement is repeated.

Creep studies are particularly valuable for determining the long-term flow properties of a material and its ability to withstand loading and deformation influences.

15.1.2 Instrumentation

For further details see Section 10.1.2 and Appendix 1.

Seiko Instruments supplies the Exstar 6000 Series, DMM 6100 [3]. The instrument can determine characteristics such as T_g, damping intensity, heat resistance, and creep and stress relaxation of various materials, allowing the user to obtain complete characterisation of a processed material. The instrument can also be used to evaluate the compatibility, anisotropy, vibration absorbency, molecular weight, degree of crystallinity, and degree of orientation of polymeric and elastomeric materials.

Recently introduced instruments from TA Instruments include the DMA Q800 analyser [4, 5]. The instrument can be used for the testing of mechanical properties of a broad range of viscoelastic materials at temperatures ranging from –150 °C to 600 °C. The DMA Q800 is claimed to provide unmatched performance in stress–strain control and measurement. It uses a proprietary non-contact linear motor to provide precise stress control, and optical encoder technology for unmatched sensitivity and resolution in strain deflection.

The Mettler-Toledo Inc., SDTA 861 DMA analyser [6] is capable of providing DMA measurements over wide frequency ranges (from 1 mHz to 1 kHz) and large dynamic stiffness ranges. A schematic of the instrument is illustrated and examples are presented of test data for various rubbers obtained using this equipment.

Other suppliers include Rheometric Scientific, Thermocahn, and Seiko Instruments [7].

15.1.3 Applications

A number of applications are discussed here. Modern DMA allow for multiple mechanical tests to be performed with a single instrument. The different modes of the analyser permit a complete transformation of the analyser for measurement of the effects of temperature, frequency, stress, strain, and change of volume in a material. The additional ability of holding either stress or strain constant during a mechanical test increases the sensitivity and versatility of these types of mechanical analyser. This capability permits quantitative measurements of viscoelastic material behaviour to be observed while scanning frequency, scanning temperature, scanning stress, monitoring long-term creep behaviour, or holding force constant thermomechanical analysis (TMA) all with a single instrument. Previously, these analyses required the use of multiple instruments to measure all of these mechanical properties of materials.

Flexibility is particularly important in mechanical analysis for testing many different sizes and geometries of polymer samples over a wide temperature and modulus range.

Modern instruments can handle many different sample types such as bars, films, fibres, coatings, pellets, rods, and cylinders. Different sample types are handled using the six measuring systems available: three-point bending, parallel plate, dual cantilever, single cantilever, fibre tension, and film extension.

15.1.3.1 Storage Modulus

One of the values obtained in a DMA is the storage modulus, which, roughly speaking, quantifies the flexural or tensile strength of a material. **Figure 15.5** shows the modulus of some common materials analysed over three decades of frequency using the frequency scan mode. In this mode, the temperature is held constant and the frequency at which the sample is oscillated is scanned from low to high, or from high to low, frequencies. Polymer melts at the low-frequency end display a large frequency dependence at low frequencies.

In polymeric materials exhibiting viscoelastic behaviour, the modulus and viscosity are dependent upon the frequency of the DMA measurement, and this frequency dependence is quite different for materials with different degrees of molecular branching, crosslinking, or molecular weight distribution. The use of DMA in the frequency scan mode of operation is particularly important in material analysis since these types of molecular differences are very difficult to distinguish using any other thermal analysis techniques. For example, a frequency scan can show clear differences between polyethylene (PE) samples having a difference in average molecular weight of only a few percent.

The effect of frequency on materials and the corresponding rate dependence of materials on it can supply additional information about the viscoelastic characteristics of polymers.

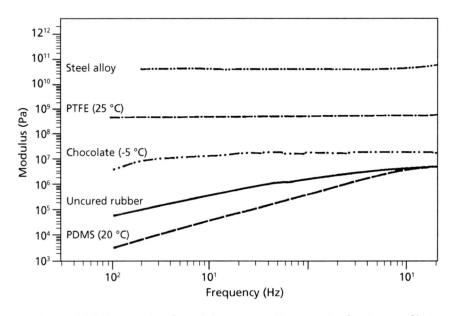

Figure 15.5 Examples of modulus range. (*Source: Author's own files*)

When frequency is being scanned, small differences in molecular weight and the distribution of molecular weight can be detected by shifts in the viscosity and modulus curves.

15.1.3.2 Stress–Strain Relationships

Choosing the best conditions of dynamic stress and strain is often difficult when creating mechanical analysis methods for a series of samples. Since sample behaviour will often change when the stress or strain imposed on a sample is increased or decreased, a curve of stress *versus* strain is very valuable. The Newtonian or linear region for a material can be measured with the stress scan mode of the DMA. This is the region in which quantitative data can be obtained.

The stress scan (dynamic stress (10^4 Pa) *versus* strain (%)) of an unvulcanised rubber material depicted the linear region of this material to be within 0.08–0.27% strain and 2,000–4,200 Pa stress. Without this capability, this linear region of stress and strain can only be approximated.

15.1.3.3 Frequency Dependence of Modulation and Elasticity

As shown in **Figure 15.6**, the frequency dependence of modulus and viscosity can be measured at different temperatures through the glass transition T_g to predict long-term behaviour of polymers. This information can be used to calculate how the polymer will perform over long durations at various temperatures and mechanical stress.

15.1.3.4 Elastomer Low Temperature Properties

Flexibility is an important end-use property for elastomers. The T_g, which is the point at which an elastomer (on cooling) goes from a flexible more rubber-like form to a more rigid inflexible form, is a critical parameter in determining the elastomer's suitability for specific applications. Plots of flexural storage modulus G (Pa) *versus* specimen temperature are very useful in evaluating the stiffness and flexibility of polymeric materials.

15.1.3.5 Effect of Increasing Stress on Polymers

A stress scan (i.e., dynamic stress (Pa) *versus* strain (%) plots) will show the effect of increasing stress on a polymer. There is usually an initial region where the strain is proportional to stress. Then, with increasing strain there can be deviations from linearity due to various molecular effects. Calculations can determine proportional limits, yield modulus, draw strength, and ultimate modulus.

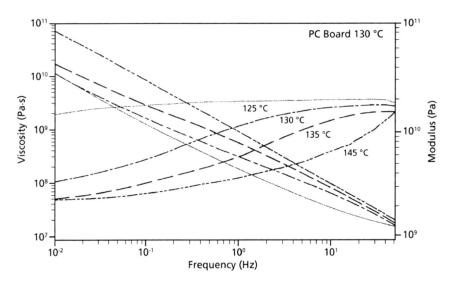

Figure 15.6 Viscoelastic behaviour through T_g of epoxy PC board.
(*Source: Author's own files*)

15.1.3.6 Projection of Material Behaviour Using Superpositioning

The effects of time and temperature on polymers can be predicted using time–temperature superpositioning principles. These superposition principles are based upon the premise that the processes involved in molecular relaxation or rearrangements occur at greater rates at higher temperatures. The time over which these processes occur can be reduced by conducting the measurement at elevated temperatures and transposing the data to lower temperatures. Thus, viscoelastic changes that occur relatively quickly at higher temperatures can be made to appear as if they occurred at longer times simply by shifting the data with respect to time.

Viscoelastic data can be collected by performing static measurements under isothermal conditions (e.g., creep or stress relaxation), or by performing frequency multiplexing experiments where a material is analysed at a series of frequencies. By selecting a reference curve and then shifting the other data with respect to time, a 'master curve' can be generated. A master curve is of great value since it covers times or frequencies outside the range easily accessible by experiment.

Figure 15.7(a) shows the DMA frequency multiplexing results for an epoxy/fibreglass laminate. The plot shows the flexural storage modulus (E') and the loss modulus (E'') as a function of temperature at the various analysis frequencies. The loss modulus peak

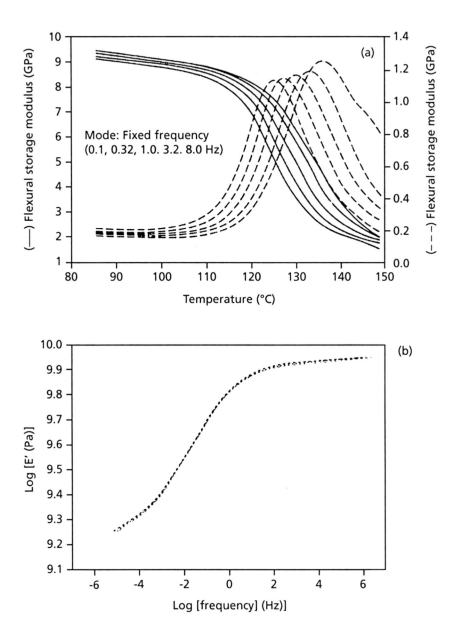

Figure 15.7 Dynamic mechanical analysis. (a) Frequency multiplexing results for an epoxy/fibreglass laminate; (b) Master curve generated from fixed frequency multiplexing data for epoxy fibreglass laminate. (*Source: Author's own files*)

temperatures show that the T_g moves to higher temperatures as the analysis frequency increases. Through suitable calculations, the storage modulus data can be used to generate a master curve (**Figure 15.7(b)**). The reference temperature of this is 125 °C, and shows the effects of frequency on the modulus of the laminate at this temperature. At very low frequencies (or long times) the material exhibits a low modulus and would behave similarly to a rubber. At high frequencies (or short times) the laminate behaves like an elastic solid and has a high modulus. This master curve demonstrates that data collected over only two decades of frequency can be transformed to cover more than nine decades.

Examples of practical uses for superpositioning are:

- Gaskets: to measure flow (creep and stress relaxation) effects, which reduce seal integrity over time.

- Force-fit of snap parts: to measure stress relaxation effects, which can lead to joint failure.

- Structural beams: to measure modulus drop with time, which leads to increasing beam deflection under load over time.

- Bolted plates: to measure creep of the polymer, which reduces the stress applied by the fastener.

- Hoses: to measure creep of the polymer, which can lead to premature rupture of the hose.

- Acoustics: to aid in the selection of materials that exhibit high damping properties in specific frequency ranges.

- Elastomeric mounts: to assess the long-term creep resistance of mounts used for vibration damping with engines, missiles, and other heavy equipment.

- Structural parts: to measure high deformation frequencies, which cause shifting of molecular transitions to high temperatures and can result in impact failure or microcracking.

15.1.3.7 Predictions of Polymer Lifetime

A practical and accurate method for predicting the useful service life of polymers has long been sought. The need has become increasingly critical with the development of new materials and demanding applications, particularly those in which engineering plastics and composites are substituted for metals.

The proliferation of materials gives scientists and engineers new design freedom. It also presents a considerable challenge. Before the best material for an application can be selected, the required performance properties (such as rigidity, strength, impact resistance, and creep) and the environment in which the product will operate must be defined. Then, the desired life expectancy for the product must be determined. Only then can the material selection process begin.

Traditional evaluation procedures are generally laborious, time consuming, and expensive because they require fabrication of prototype parts and testing under actual end-use or simulated service conditions. These processes are more empirical than analytical, making the results of questionable value. The processes are generally impractical because they require months or years to produce results.

Sichina [8] has discussed the application of the DuPont 983 DMA to the prediction of polymer lifetimes and long-term performance, e.g., creep in gaskets, stress relaxation in snap-fit parts, modulus decay in composite structural beams, creep in bolted plates, and heat deformation frequencies in structural parts. The ability of this system to generate master curves makes prediction of product performance fast and easy, and facilitates the correlation of DMA data with evaluations performed by traditional time- and labour-intensive methods.

Figure 15.8 shows the creep data obtained for PC in the temperature range 130–155 °C. The logarithm of creep compliance (S) is shown as a function of the logarithm of decay time. One of the curves is selected as the reference (in this case, T_o = 145 °C), then the other curves are shifted along the log time axis and superimposed upon the reference curve. The final master curve based on creep data is shown in **Figure 15.9**. The curve shows that at small time intervals the material exhibits relatively low compliance (or high modulus). At longer times, viscous flow occurs and the material exhibits a high compliance (or modulus). Thus this master curve clearly demonstrates the effects of time on the mechanical properties of PC.

15.1.3.8 Other Applications

DMA is now being used very extensively in the testing of viscoelastic and rheological properties of polymers [9–14]. Some recent examples include polymer damping peaks, storage modulus and loss modulus in polypropylene (PP)–wood composites [15], graft PP–styrene–maleic anhydride polymer [16], storage modulus of basalt fibre-reinforced PP [17], viscoelastic properties of PE and PP foams [18], study of phase transitions in PE [19], relaxation modulus, etc., of metallocene-catalysed low-density polyethylenes [20], stress–strain curves of PP, PC, and 30% glass-fibre-reinforced polyamide 6,6 (PA)

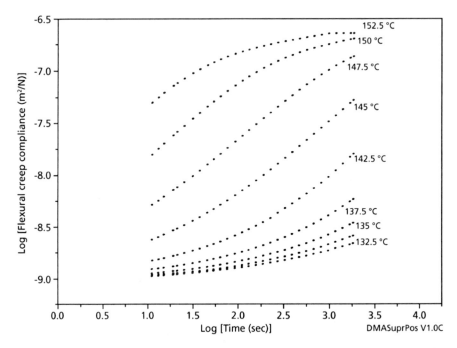

Figure 15.8 Creep data obtained on polycarbonate in the temperature range of 130 °C to 155 °C. (*Reprinted from W.J. Sichina, International Laboratory, 1988, 81, 4, 36 [8]*)

and polybutylene terephthalate [21], isothermal dynamic mechanical properties at different frequencies of PA 12 and polyvinylidene fluoride [22], thermal denaturation of collagen in biopolymers [23], viscoelastic properties of ethylene–propylene–diene elastomers [24], storage modulus curves of ethylene–propylene–diene–polyvinyl chloride and monomethyl methacrylate-g-ethylene–propylene–diene–polyvinyl chloride blends [25], viscoelastic [26] and dynamic properties [27] of natural rubber, rheological properties of styrene–butadiene–styrene (SBS) polymer [28], stress–strain curves of styrene–butadiene block copolymers [29], gel and vitrification transitions during the cure of epoxy resin–amine systems [30], molecular motions in beta relaxations of diglycidyethene–bisphenol A epoxy resin networks [31], relaxations in styrene–butadiene block copolymer with doped polyaniline [32], storage modulus of polyvinylidene fluoride clay nanocomposites [33], dynamic modulus properties of polyvinyl alcohol (PVOH) [34], rheological study of acrylate-terminated polyesters with styrene [35], modulus and viscosity of thermosetting resins [36], storage modulus and loss tangent of polymeric glycol monomethacrylates and methacrylate–urethane systems [37], viscoelastic properties of polyester-based clear coat resins [38], viscoelastic properties of polyvinyl esters and unsaturated polyester resins [39], viscoelastic and mechanical properties of

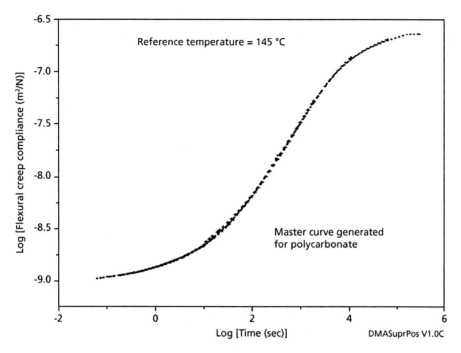

Figure 15.9 Final master curve based on creep data for polycarbonate. (*Reprinted from W.J. Sichina, International Laboratory, 1988, 81, 4, 36 [8]*)

polyimide–clay nanocomposites [40], elastic and viscoelastic properties of polyether sulfone–styrene acrylonitrile and polystyrene (PS)–polymethyl methacrylate (PMMA) blends and PMMA, PS, and polytetrafluoroethylene homopolymers [41], thermal properties of poly(l-lactide)–polyoxyethylene and poly(l-lactide–poly(ε-caprolactone) copolymers [42], tensile modulus and relaxation behaviour of poly-*p*-phenylene [43], dynamic and thermal properties of benzooxazine-based phenolic resins [44], viscoelastic properties of cyanate ester resin–carbon fibre composites [45], storage modulus, loss modulus, and loss tangent variations of water-based adhesives [46, 47], viscoelastic behaviour of polycyanate epoxy resins [48], shear strength modulus and loss factor of cold seal adhesives [49], viscoelastic properties of styrenic block copolymer [50], transition temperatures of cured acrylate-terminated unsaturated copolyesters [51], mechanical properties of polyethylene terephthalate [52], dynamic viscosity of polyvinyl chloride–ethylene–vinyl acetate–carbon monoxide terpolymer [53], cure rheo-kinetics of a phenol–urethane composition [54], rheological characterisation of long-chain branched metallocene-catalysed ethylene homopolymers [55], and additive compatibility with polyvinyl chloride [56].

Other applications of DMA are discussed in Sections 10.1 (resin cure studies), 12.2 (photopolymers), 13.3 and 13.7 (phase transition studies), and 18.1 (mechanical properties).

15.2 Thermomechanical Analysis

Some of the viscoelastic or rheological properties that can be measured using this technique include viscosity, modulus tensile compliance, creep-stress relaxation, gel time and gel temperature, tensile compliance, and stress–strain properties.

The principles of the technique are discussed in Section 13.2.1. The instrumentation is discussed in Section 13.2.2 and Appendix 1.

15.2.1 Applications

Gel time measurement. Gel time is a critical processing time, since after the gel point the material is no longer able to flow and is therefore unprocessable. Traditional methods for the measurement of gel time are of questionable reliability because they tend to be very operator dependent. For example, in the electronics (printed circuit board) industry, a prepreg is evaluated by grinding it to a fine powder, passing the powder through a sieve to remove the glass reinforcement particles, placing the filtered resin powder on a preheated plate, and then stirring with a glass rod until the resin adheres to the stirring rod. The time from initial softening until adhesion occurs as monitored by a stopwatch is considered to be the gel time.

Thermomechanical analysis (TMA) provides a convenient and more reproducible, scientific approach to gel time measurement. Using a specific TMA probe configuration (parallel plate rheometer), TMA-measured dimensional changes can be converted to gel time and viscoelastic values.

Plots of dimensional change (mm) *versus* temperature during gelation shows that as the prepreg is heated from room temperature, a slight expansion occurs. After a certain time (*T*1), the curve starts to drop, indicating softening of the resin. This continues until another time (*T*2) when the curve begins to flatten due to the crosslinking process of gelation.

The gel time of the prepreg is observed as the difference between the time to the onset of gelation minus the time to the onset of softening. The total displacement during cure (in mm) in conjunction with a knowledge of the initial sample thickness can be used to calculate the percent displacement during flow. Such data are useful for optimisation of resin bleed *versus* void formation during laminate processing, as in a 'scaled flow' test.

Fibre stress–strain measurements. Stress–strain measurements are widely used to assess and compare materials. Although conventional physical testing devices can accommodate single-filament fibres, the results are difficult to obtain and accuracy is doubtful since the mass and inertia of the grips is much greater than the tensile strength of the fibres being evaluated. With proper mounting in the fibre probe configuration to ensure elimination of end effects, TMA curves like that shown in **Figure 15.10** for a 25.4 µm diameter PA filament can be obtained. From such a curve, it is possible to determine information about yield stress and Young's modulus in the elastic region.

Riga [57] comments that a TMA has been used as a tool in many diverse projects, such as quality control, verification of standards, failure analysis, and characterisation of polymeric materials.

Workers at the Institute for Dynamic Materials Testing in Ulm [58] have developed a new method for determining viscoelastic characteristics of thin coatings. Elasticity and damping components of the dynamic shear modulus as a function of temperature or time can be determined from the resonance frequency and damping of the natural vibration of a coated aluminium carrier plate.

TMA has also been used in rheological studies carried out on nadimide resin [59], crosslinkable aromatic polyimides [60], and various PA [61]. Other measurements that can be carried out by TMA are discussed in Sections 13.2 (phase transition studies), 16.2.1 and 16.4.1 (thermal properties), and 18.1 (mechanical properties).

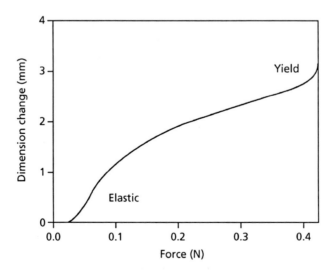

Figure 15.10 Measurement of fibre stress/strain properties by thermomechanical analysis. (*Source: Author's own files*)

15.3 Dielectric Thermal Analysis

15.3.1 Theory

This technique has the ability to provide information on polymer chemistry, rheology, and molecular mobility (see Chapters 10 and 13).

15.3.2 Instrumentation

The instrumentation is discussed in Section 10.2.2 and Appendix 1.

15.3.3 Applications

Flow and cure of an aerospace adhesive. The rheological changes in a material during complex thermal histories can provide valuable information about processing, chemical structure, and end-use performance. Dielectric analysers can characterise dramatic and rapid changes in a polymer's physical state, even into the final stages of cure, which can significantly influence the physical and chemical properties of the finished product.

Figures 15.11(a) and **(b)** describe the flow and cure of a B-staged aerospace adhesive during a complex thermal history, using parallel plate sensors. Specifically, **Figure 15.11(a)** shows how the loss factor (log e'') changes with time and temperature. The increase in log e'' between 25 and 80 °C is due to softening of the adhesive. As the temperature is held isothermally at 80 °C for 30 minutes, log e'' remains constant. Then, between 60 and 75 minutes as heat is applied at a rate of 1 °C/min, temperature-induced fluidity causes the viscosity of the resin to decrease. After 75 minutes, the resin polymerises and develops a three-dimensional network, as indicated by the decrease in log e''.

Figure 15.11(b) shows a plot of log ionic conductivity (reciprocal ohm/cm) against time for the same thermal profile. That portion of the time–temperature schedule in which the log ionic conductivity curves are independent of the test frequency shows that the loss factor measurement (**Figure 15.11(a)**) is dominated by ionic movement (or DC conductivity). Those regions of the curve in which the value for log ionic conductivity displays a frequency dependency indicate that the measurement is strongly influenced by dipole relaxations. The log ionic conductivity curves are frequency-dependent between 0 and 20 minutes because the B-staged resin passes through its T_g region as it is heated. Between 85 and 160 minutes, the ionic conductivity curves are frequency dependent because dipole molecular relaxations dominate the measurement as the resin reacts and develops molecular weight.

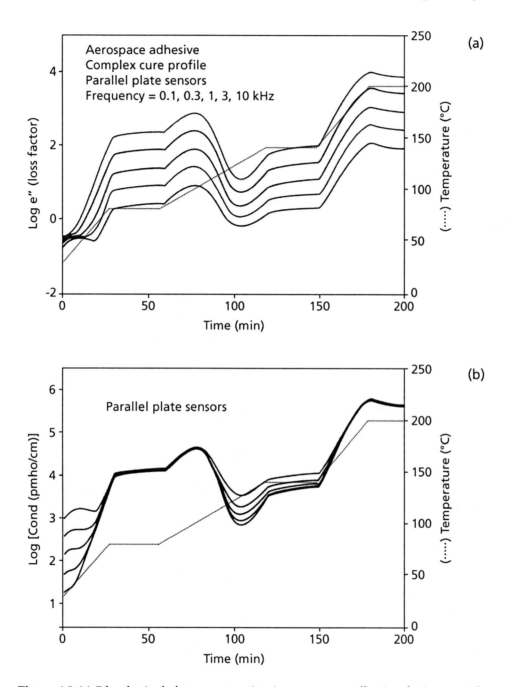

Figure 15.11 Rheological changes occuring in aerospace adhesive during complex thermal history by dielectric thermal analysis. (a) Dielectric and temperature profiles for flow and curve; (b) Ionic conductivity. (*Source: Author's own files*)

By comparing log ionic conductivity profiles with loss factor measurements, the operator can identify the window in the process during which the material is fluid (and therefore workable), develops molecular weight and crosslinks (which is critical to product performance and appearance), and is completely cured (which identifies the proper time to demould the product or remove jigs). The dielectric analyser can also record the dielectric properties of the resin during cooling.

Through off-line product development, the scientist or engineer can use dielectric properties to optimise the processing thermal history of products, and predict material processibility and end-use product performance.

Influence of thermal history on Nylon 6,6. Dielectric analysis is an effective tool for characterising the influence of thermal history on the molecular structure of a material and, in turn, on molecular relaxation.

Figures 15.12(a) and (b) show the results of experiments on a sample of Nylon 6,6. The sample, a commercial film of Nylon, was tested at 1 and 100 Hz with a temperature scan of 5 °C/min from –100 to 150 °C. In the dielectric profile plot for the Nylon film, '1st Temperature Scan' represents the as-received sample material.

After ramping the temperature to 150 °C, the sample was held isothermally at 150 °C for 20 minutes, cooled to –100 °C, and then exposed to a second 5 °C/min heating up to 150 °C. The dielectric profiles of the retested sample are identified as '2nd Temperature Scan'. All of the curves in **Figures 15.12(a) and (b)** were generated in a single experiment, which had been programmed by customising the temperature profile.

As seen in **Figure 15.12(a)**, the α transition (T_g) in Nylon 6,6 is identified by the peak in the dielectric loss factor (e'') curve. The first temperature scan has a higher permittivity (e'), a higher loss factor (e''), and a lower T_g than those recorded during the second temperature scan.

These results characterise the effect of water loss from the Nylon 6,6 during the first scan. T_g increases by approximately 30 °C due to the loss of water. The β transition in **Figure 15.12(b)** is due to local main-chain motions of the amine groups in the polymer's backbone structure. This transition becomes well defined during the second temperature scan at low test frequencies.

The β transition is ill defined in the first temperature scan because molecular motion in the as-received sample is restricted by hydrogen bonding caused by the presence of small amounts of water. Evaporation of the water during the first heating cycle makes the β transition more detectable during the second scan.

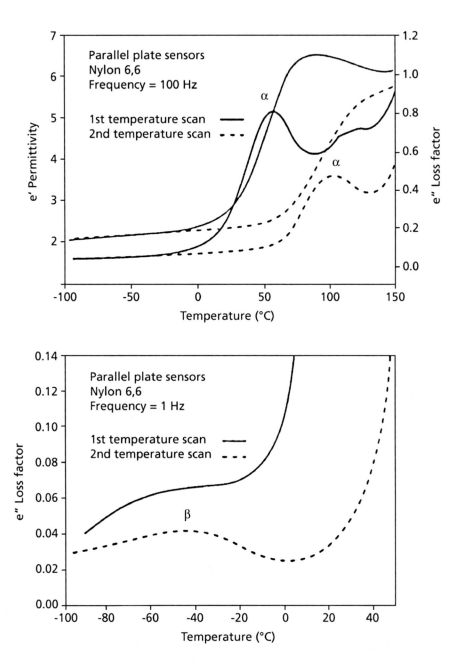

Figure 15.12 Influence of thermal history on molecular structure of Nylon 6,6 (a) showing that heat history causes a shift in transitions in Nylon film, (b) showing that beta transition becomes well defined in the second scan.
(*Source: Author's own files*)

A broad range of test frequencies is important for isolation of different types of molecular relaxations. As demonstrated by this experiment with Nylon, low frequencies are important for characterising the β transition, whereas high frequencies help in the separation of the α transition from other complex phenomena that occur at high temperatures.

The broad capabilities of dielectric thermal analysis make possible the definitive evaluation of many influences on measured dielectric properties and, therefore on the performance of finished products. These influences include crystallinity, cooling rates, annealing temperatures, and post-curing. Experiments to evaluate these factors are possible because of the ability of this technique to measure and control force and plate spacing (sample thickness). These data will help the scientist and engineer correlate end-use performance with such vital factors as chemical structure, molecular relaxations, and materials storage and processing conditions.

Dynamic mechanical thermal analysis and dielectric thermal analysis methods have been used to examine relaxations in blends of SBS block copolymer with doped polyaniline [32]. Doping was carried out using emeraldine base or dodecylbenzenesulfonic acid. Different degrees of interaction between the doped polyaniline and the polybutadiene and PS blocks of the SBS copolymer were observed, the damping curve peak width of the PS being the more affected. Activation energy of relaxations changed little when dodecylbenzenesulfonic acid doping was used, but changed more when emeraldine base was used. In contrast, both dielectric constant and loss factor increased when dodecylbenzenesulfonic acid was used compared to emeraldine base. Dielectric loss factor increased sharply at temperatures higher than the T_g of the PS phase due to interfacial polarisation and contributions from ionic conductivity. In some cases, transitions could only be observed at measurements below 1000 Hz.

15.4 Further Viscoelastic Behaviour Studies

Various workers have discussed aspects other than those mentioned above in studies of the viscoelastic properties of polymers. These include PVOH [62], hydroxy-terminated polybutadiene [63], styrene–butadiene and neoprene-type blends [64], and polyamidoimides [65]. Other aspects of viscoelasticity that have been studied include relaxation phenomena in PP [66] and methylmethacrylate–N-methyl glutarimide copolymers [67], shear flow of high-density polyethylene [68], T_g of PMMA and its copolymers with N-substituted maleimide [69] and ethylene–vinyl acetate copolymers [70], and creep behaviour of poly(p-phenylene terephthalate) [71] and PE [72].

15.5 Further Rheology Studies

Aspects of viscosity, elasticity, and morphology have been discussed in general terms by various workers [73–76]. Rheological studies specific to particular polymers include dynamic rheological measurements and capillary rheometry of rubbers [77], capillary rheometry of PP [78], degradation of PP [79], torsion rheometry of PE [80], viscosity effects in blends of PC with styrene–acrylonitrile and acrylonitrile–butadiene–styrene [81], peel adhesion of rubber-based adhesives [82], and the effect of composition of melamine–formaldehyde resins on rheological properties [83].

References

1. C. Treny and B. Duperray in *Proceedings of the ANTEC '97 conference*, Toronto, Canada, Spring 1997, Volume III, p.2805.

2. P.C. Haschke, *Machine Design*, 2001, **73**, 61.

3. Exstar 6000 Series, Dynamic Mechanical Spectrometer, Seiko Instruments Co.

4. *Sampe Journal*, 2002, **38**, 3, 91.

5. H. Wang, D.G. Thompson, J.R. Schoonover, S.R. Ambushon and R.A. Palmer, *Macromolecules*, 2001, **34**, 7084.

6. J. Foreman, J.E.K. Schawe and M. Wagner in *Proceedings of the 162nd ACS Rubber Division Conference*, Pittsburgh, PA, USA, Fall 2002, Paper No.13.

7. F. Gouin, *Plastiques et Elastomères Magazine*, 2001, **53**, 17.

8. W.J. Sichina, *International Laboratory*, 1988, **181**, 36.

9. M. Buhk, W. Schlesing and W. Bosch, *Macromolecular Symposia*, 2002, **187**, 873.

10. M. Sepe, *Injection Molding*, 1999, **7**, 52.

11. J. Otaigbe, H.S. Kim, J. Xiao and T. Tacke in *Proceedings of ANTEC '99 Conference*, New York, NY, USA, 1999, Volume III, p.3936.

12. M.A. Rodríguez-Perez, S. Rodríguez-Llorente and J.A. De Saja, *Polymer Engineering and Science*, 1997, **37**, 959.

13. P.S. Gill in *Proceedings of the ACS Polymeric Materials Science and Engineering Conference*, Orlando, FL, USA, Fall 1996, Volume 75, p.162.

14. J. Gearing in *Handbook of Polymer Testing*, Ed., R. Brown, Marcel Dekker Inc., New York, NY, USA, 1999, p.501-3.

15. J. Son, D.J. Gardner, S. O'Neill and C. Metaxas, *Journal of Applied Polymer Science*, 2003, **89**, 1638.

16. D. Jia, Y. Luo, Y. Li, H. Lu, W. Fu and W.L. Cheung, *Journal of Applied Polymer Science*, 2000, **78**, 2482.

17. M. Botov, H. Betchev, D. Bikiaris and D. Panayiotou, *Journal of Applied Polymer Science*, 1999, **74**, 523.

18. M.A. Rodríguez-Perez, S. Rodríguez-Llorente and J.A. De Saja, *Polymer Engineering and Science*, 1997, **37**, 959.

19. L.T. Pick and E. Harkin-Jones, *Polymer Engineering and Science*, 2003, **43**, 905.

20. E. Kontou and G. Spathis, *Journal of Applied Polymer Science*, 2003, **88**, 1942.

21. O. Schroder and E. Schmachtenberg in *Proceedings of the ANTEC 2001 Conference*, Dallas, TX, USA, 2001, Paper No.405.

22. N.L.A. McFerran, C.G. Armstrong, T. McNally, G.M. McNally and W.R. Murphy in *Proceedings of 60th SPE Annual Technical Conference, ANTEC 2002*, San Francisco, CA, USA, 2002, Session M33, Paper No.166.

23. B. Twombly, B. Cassel and A.T. Miller in *Proceedings of the ANTEC '98 Conference*, Atlanta, GA, USA, 1998, Volume II, p.2177.

24. H.S. Liu, C.P. Richard, J.L. Mead and R.G. Stacer in *Proceedings of the ANTEC 2000 Conference*, Orlando, FL, USA, 2000, Paper No.570.

25. D. Singh, V.P. Malhotra and J.L. Vats, *Journal of Applied Polymer Science*, 1999, **71**, 1959.

26. P.K. Patra, P.K. Das, M.S. Banerjee and A.R. Tripathy, *Rubber India*, 2000, **52**, 9.

27. V. Noparatonakailis, *Journal of Rubber Research*, 2000, **3**, 95.

28. X. Lu and U. Isacsson, *Journal of Testing and Evaluation*, 1997, **25**, 383.

29. R.A. Palmer, P. Chen and M.H. Gilson in *Proceedings of the ACS Polymeric Materials Science and Engineering Conference*, Orlando, FL, USA, Fall 1996, Volume 75, p.43.

30. J. Lange, N. Altmann, C.T. Kelly and P.J. Halley, *Polymer*, 2000, **41**, 5949.

31. L. Heux, J.J. Halary, F. Laupretre and L. Monnerie, *Polymer*, 1997, 38, 767.

32. M.E. Leyva, B.G. Soares and D. Khastgir, *Polymer*, 2002, **43**, 7505.

33. L. Priya and J.P. Jog, *Journal of Polymer Science, Polymer Physics Edition*, 2003, **41**, 31.

34. H-H. Wang, T-W. Shyr and M-S. Hu, *Journal of Applied Polymer Science*, 1999, **73**, 2219.

35. A. Zlatanic, B. Dunjic and J. Djonlagic, *Macromolecular Chemistry and Physics*, 1999, **200**, 2048.

36. M. Sepe, *Injection Molding*, 1999, **7**, 52.

37. M. Kurcok and P. Szewczyk, *Polimery*, 2002, **47**, 637.

38. M. Korhonen, P. Starck, B. Loefgren, P. Mikkila and O. Hormi, *Journal of Coatings Technology*, 2003, **75**, 67.

39. A. Valea, M.L. Gonzalez and I. Mondragon, *Journal of Applied Polymer Science*, 1999, **71**, 21.

40. M.O. Abdalla, D. Dean and S. Campbell, *Polymer*, 2002, **43**, 5887.

41 F. Oulevey, N.A. Burnham, G. Gremaud, A.J. Kulik, H.M. Pollock, A. Hammiche, M. Reading, M. Song and D.J. Houston, *Polymer*, 2000, **41**, 3087.

42. G. Maglio, A. Migliozzi and R. Palumbo, *Polymer*, 2003, **44**, 369.

43. D. Dean, M. Husband and M. Trimmer, *Journal of Polymer Science, Polymer Physics Edition*, 1998, **36**, 2971.

44. M.T. Huang and H. Ishida, *Polymer and Polymer Composites*, 1999, **7**, 233.

45. K. Chung and J.C. Seferis, *Polymer Degradation and Stability*, 2001, **71**, 425.

46. D.W. Bamborough in *Proceedings of Technomic Conference on Pressure Sensitive Adhesive Technology*, 2001, Paper No. 10.

48. P. Ponteins, B. Medda, P. Demont, D. Chatain and C. Lacabanne, *Polymer Engineering and Science*, 1997, **37**, 1598.

49. R.G. Polance and J.P. Baetzold in *Journal of the Adhesive and Sealant Council, Conference Proceedings*, Chicago, IL, USA, 1998, p.189.

47. D.W. Bamborough in *Proceedings of Technomic Pressure Sensitive Adhesives Conference*, Basel, Switzerland, 1993, Sections 111-2, 111-17. **Check in library**

50. *Proceedings of Technomic Conference on Pressure Sensitive Adhesive Technology*, Ed., D.W. Bamborough, 2001, Paper No.9. **Check in library**

51. J. Djonlagic, A. Zlatanic and B. Dunjic, *Macromolecular Chemistry and Physics*, 1998, **199**, 2029.

52. P.T. Mather and A. Romo-Uribe, *Polymer Engineering and Science*, 1998, **38**, 1174.

53. A. Zarraga, J.J. Pena, M.E. Munoz and A. Sanamaria, *Journal of Polymer Science, Polymer Physics Edition*, 2000, **38**, 469.

54. I.Y. Gorbunova, M.L. Kerber, I.N. Balashov, S.I. Kazakov and A.Y. Malkin, *Polymer Science, Series A*, 2001, **43**, 826.

55. C. Gabriel, E. Kokko, B. Löfgren, J. Seppala and H. Münstedt, *Polymer*, 2002, **43**, 6383.

56. R. Bacaloglu, B. Hegranes and M. Risch, *Journal of Vinyl and Additive Technology*, 1997, **3**, 112.

57. A. Riga in *Proceedings of the Thermal and Mechanical Analysis of Plastics in Industry and Research Conference*, Newark, DE, USA, 1999, p.105.

58. Institute for Dynamic Materials Testing, *Gummi Fasern Kunststoffe*, 2001, No. 2, 85. **check in Library**

59. J.P. Habas, J. Peyrelasse and M.F. Grenier-Loustalot, *High Performance Polymers*, 1996, **8**, 515.

60. J.A. Mikroyannidis, *Journal of Polymer Science, Polymer Chemistry Edition*, 1997, **35**, 1303.

61. T.A. Godfrey, *Journal of Textile Institute - Part 1: Fibre Science and Textile Technology*, 2002, **92**, 16.

62. J-S. Park, J-W. Park and E. Duckenstein, *Journal of Applied Polymer Science*, 2001, **82**, 1816.

63. J.L. de la Fuente, M. Fernandez-Garcia and M.L. Cerrada, *Journal of Applied Polymer Science*, 2003, **88**, 1705.

64. H.H. Yang, L. Nikiel, H. Gerspacher and C.P. O'Farrell, *Rubber and Plastics News*, 2000, **30**, 12.

65. I.I. Perepechko, V.A. Danilov, T.S. Shcherbakova and L.I. Chudinova, *Plasticheskie Massy (USSR)*, 1996, No. 2, 25.

66. V.P. Privalko, B.B. Dolgoshey, E.H. Privalko, V. Shumsky, A. Lisouskii, M. Rodensky and M.S. Eisen, *Journal of Macromolecular Science B*, 2002, **B-41**, 539.

67. L. Teze, J.L. Halary, L. Monniere and L. Canora, *Polymer*, 1999, **40**, 971.

68. M.S. Song, C. Zhang, G.X. Hu and X.Y. Li, *Polymer Testing*, 2002, **21**, 823.

69. P. Tordjeman, L. Teze, J.L. Halary and L. Monniere, *Polymer Engineering and Science*, 1997, **37**, 1621.

70. A. Arsac, C. Carrot and J. Guillet, *Journal of Applied Polymer Science*, 1999, **74**, 2625.

71. J.J.M. Baltussen and M.G. Northold, *Polymer*, 2001, **42**, 3835.

72. G. Pilz and R.W. Long in *Proceedings of the IOM Plastic Pipes XI Conference*, Munich, Germany, 2001, Paper No.93, p.903.

73. P. Maltese, *Materie Plastiche ed Elastomeri*, 1995, **4**, 225.

74. Rheometric Scientific Inc., *China Plastic and Rubber Journal*, 2002, No. **10-11**, 68.

75. M. Ezrin in *Proceedings of the 60th Annual Technical Conference ANTEC 2002*, San Francisco, CA, USA, 2002, Session M40, Paper No.202.

76. J.L. Leblanc, *Revue Générale des Caoutchoucs et Plastiques*, 1995, **745**, 41.

77. F. Declerq in *Proceedings of the Technical Rubber Goods: Part of our Everyday Life Conference*, Puchov, Slovakia, 1996, p.173.

78. M. Fujiyama, Y. Kitajima and H. Inata, *Journal of Applied Polymer Science*, 2002, **84**, 2128.

79. B. Vergnes and F. Berzin, *Macromolecular Symposia*, 2000, **158**, 77.

80. H. Kwang, D. Rana, K. Cho, J. Rhee, T. Woo, B.H. Lee and S. Choe, *Polymer Engineering and Science*, 2000, **40**, 1672.

81. J. Bouton, Rheo SA, *Plastiques et Elastomères* Magazine, 2000, **52**, 34.

82. A.B. Kummer in *Proceedings of the Pressure Sensitive Tape Council 23rd Annual Technical Seminar: Pressure Sensitive Tapes for the New Millenium*, New Orleans, LA, USA, 2000, p.25.

83. M. Doyle and J-A.E. Manson in *Proceedings of the ANTEC 2000 Conference*, Orlando, FL, USA, 2000, Paper No.169.

16 Thermal Properties

16.1 Linear Coefficient of Expansion

The curve in **Figure 16.1** illustrates a method for determining the expansion coefficient of a polymer.

16.1.1 Dilatometric Method

Details of this method are given in **Table 16.1**.

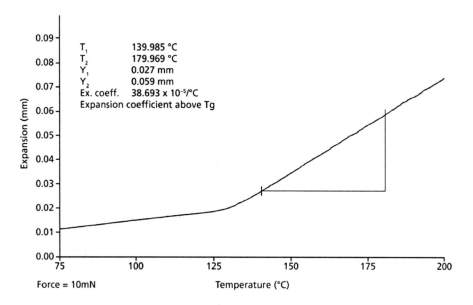

Figure 16.1 Theromechanical analysis of epoxy printed board material. Measurement of expansion coefficient. (Source: *Author's own files*)

Table 16.1 Thermal properties of polymers				
Method	ATS FAAR Apparatus Code No.	Measurement units	Test suitable for meeting following standards	Notes
Temperature of deflection under load or heat deflection or distribution temperature (Martens Method)		°C	ASTM D648-04 [01]	
Softening point Vicat	10.01004 10.01009 10.01003 10.01010 10.01041 10.01042	°C	ASTM D648-04 [01] ASTM D1525 [2] DIN EN ISO 306 [3] DIN EN ISO 75-1 [4] DIN EN ISO 75-2 [5] ISO 75-1 [6] ISO 75-2 [7] ISO 75-3 [8] ISO 306 [9] UNI EN ISO 75-1 [10] UNI EN ISO 75-2 [11] UNI EN ISO 75-3 [12] UNI EN ISO 306 [13]	
Melt flow index (or rate)	1002013 10.02017		ASTM D1238 [14] ASTM D2116 [15] ASTM D3364 [16] UNI EN ISO 1133 [17]	
Low temperature embrittlement (or brittleness temperature)	10.12000 10.12006	°C	ASTM D746 DIN ISO	
Melting point (Fischer Johns)	070173C	°C	ASTM	Also measured from differential thermal analysis peak

			Table 16.1 Continued...		
Method	ATS FAAR Apparatus Code No.	Measurement units	Test suitable for meeting following standards	Notes
Thermal conductivity		w/mk	ASTM C177 [19] DIN 52612 [20]	
Linear expansion coefficient	10.46000	mm/mm cm/cm per °C in/in °F	ASTM D696-44 [21] UNI 5284 [22]	
Specific heat		kJ Kg^{-1}	ASTM DIN	Can also be measured by differential scanning calorimetry
Minimum filming temperature	10.35000 10.35001 10.35002	°C	ASTM DIN	
Ageing in air	10.69000		ASTM	
Source: Author's own files				

16.2 Melting Temperature

The melting temperature (T_m) is defined as the temperature at which crystalline regions in a polymer melt.

16.2.1 Thermal Methods

In semi-crystalline polymers, some of the macromolecules are arranged in crystalline regions, known as crystallites, while the matrix is amorphous. The greater the concentration of crystallites, i.e., the greater the crystallinity, the more rigid is the polymer, i.e., the higher the T_m value.

Figure 16.2 shows a sample temperature–heat flow differential scanning calorimetric (DSC) curve obtained for high-density polyethylene (HDPE) using a Perkin Elmer DSC-7 instrument illustrating the measurement of the T_m and the heat of melting in a single run.

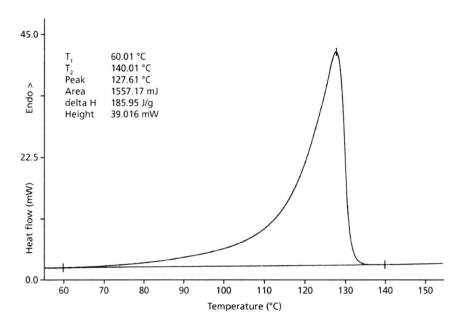

Figure 16.2 DSC run on high density polyethylene (Perkin Elmer DSC-7 instrument) showing measurement of T_m and heat of melting in a single run.
(Source: *Author's own files*)

The true melting points of crystalline polymers can be determined by plotting the DSC melting peak temperatures as a function of the square root of heating rate and linear extrapolations to zero heating rate.

Differential thermal analysis has been used to study the effect of side-chain length in polymers on melting point, and the effect of heating rate of polymers on their melting point. DSC has been used to evaluate multiple peaks in polystyrene (PS) [23–25].

DSC measurements of the energy during melting of aqueous polymer solutions and gels yield heats of mixing and sorption [26]. The technique has also been used to study the heat changes occurring in a polymer as it is cooled (plots of temperature *versus* heat flow (mW)).

Thermomechanical analysis (TMA) has been used for softening measurements of polymers and the measurement of the amount of probe penetration into a polymer at particular applied forces as a function of temperature. This technique allows evaluations of T_m and the evaluation of dimensional properties over the temperature range of use or under actual accelerated conditioning cycles (plots of temperature *versus* compression (mm)).

488

16.2.2 Fisher-Johns Apparatus

Details of this standard ASTM method are given in **Table 16.1.**

16.3 Softening Point (Vicat)

The Vicat method measures the temperature at which an arbitrary determination or a specified needle penetration (Vicat) occurs when polymers are subjected to an arbitrarily standardised set of testing conditions. Details are given in **Table 16.1.**

This property can also be measured by TMA (plots of temperature *versus* compression (mm)).

Tang and co-workers [27] found that the Vicat softening temperature of acrylonitrile–butadiene–styrene–calcium carbonate composites increased with addition of the filler indicating the beneficial effect of the filler on the heat resistance of the terpolymer.

A ring and ball apparatus has been used to determine a suitable temperature range, i.e., softening temperature, for the extrusion of low-density polyethylene (LDPE) and HDPE/PP blends [28].

16.4 Heat Deflection/Distortion Temperature

16.4.1 Thermomechanical Analysis

A flexure analysis accessory is available on TMA that allows the determination of the deflection (distortion) temperatures of polymers at selected temperatures and sample loading forces (plots of temperature *versus* flexure (mm)).

16.4.2 Martens Method

Details of the ASTM D648-04 [1] and the DIN 53458 (withdrawn) methods for determining the temperature of deflection under load are given in **Table 16.1.**

16.4.3 Vicat Softening Point Apparatus

Nam and co-workers [29] used this method to measure heat distortion temperatures of polyphenylene sulfide/acrylonitrile–butadiene–styrene blends.

16.4.4 Dynamic Mechanical Analysis

Nam and co-workers [29] used this method to measure heat distortion temperatures of polyphenylene sulfide/acrylonitrile–butadiene–styrene blends.

16.5 Brittleness Temperature (Low-Temperature Embrittlement)

Details are given in **Table 16.1**.

16.6 Minimum Filming Temperature

Details are given in **Table 16.1**.

16.7 Delamination Temperature

TMA has been applied to the measurement of this characteristic.

16.8 Melt Flow Index

This property is measured by a melt flow index extrusion plastomer, which measures the weight of polymer extruded through a standard orifice in 10 minutes (see Table 16.1). Alternatively, a multi-functional extrusion plastomer that measures melt flow rate resistance to thermal degradation is available.

Melt flow studies have been reported on PP [30], isotactic butadiene [31], glass fibre-filled PP [32], and LDPE [33].

16.9 Heat of Volatilisation

The heats of volatilisation of polymers have been determined. Values obtained for polymethyl methacrylate (PMMA) agreed well with calculated values [34]. Chemically crosslinked and oriented LDPE have been investigated using differential photocalorimetry and DSC [35, 36].

16.10 Thermal Conductivity

Details of the apparatus for performing this measurement to ASTM C177 [19] and DIN 52612 [20] standards are given in **Table 16.1**.

Yu and co-workers have reported results of thermal conductivity measurements on PS–aluminium nitride composites [37].

Dos Santos and Gregorio [38] measured the thermal conductivity of Nylon 6,6, PP, PMMA, rigid polyvinyl chloride (PVC), and polyurethane (PU) foam.

16.11 Specific Heat

Details are given in **Table 16.1** for carrying out measurements to ASTM and DIN standards.

16.11.1 Transient Plane Source Technique

This technique has been used to measure a specific heat of fibre-reinforced phenol–formaldehyde resins [39].

16.11.2 Hot Wire Parallel Technique

This technique has been used to measure the specific heat of Nylon 6,6, PP, PMMA, rigid PVC, and PU foam [38].

16.12 Thermal Diffusivity

Thermal diffusivity measurements have been reported for fibre-reinforced phenol–formaldehyde resins [39], Nylon 6,6, PP, PMMA, rigid PVC, and PU foam [38, 40].

16.13 Ageing in Air

Details of ageing tests according to ASTM are given in **Table 16.1**.

References

1.   ASTM D648-04, *Standard Test Method for Deflection Temperature of Plastics Under Flexural Load in the Edgewise Position*, 2004.

2. ASTM D1525, *Standard Test Method for Vicat Softening Temperature of Plastics*, 2000.

3. DIN EN ISO 306, *Plastics - Thermoplastic Materials - Determination of VICAT Softening Temperature (VST)*, 2004.

4. DIN EN ISO 75-1, *Plastics - Determination of Temperature of Deflection Under Load - Part 1: General Test Method*, 2005.

5. DIN EN ISO 75-2, *Plastics - Determination of Temperature of Deflection Under Load - Part 2: Plastics and Ebonite*, 2004.

6. ISO 75-1, *Plastics - Determination of Temperature of Deflection Under Load - Part 1: General Test Method*, 2004.

7. ISO-75-2, *Plastics - Determination of Temperature of Deflection Under Load - Part 2: Plastics and Ebonite*, 2004.

8. ISO 75-3, *Plastics - Determination of Temperature of Deflection Under Load - Part 3: High-Strength Thermosetting Laminates and Long-Fibre-Reinforced Plastics*, 2004.

9. ISO 306, *Plastics - Thermoplastic materials - Determination of Vicat Softening Temperature (VST)*, 2004.

10. UNI EN ISO 75-1, *Plastics - Determination of Temperature of Deflection Under Load - Part 1: General Test Method*, 2004.

11. UNI EN ISO 75-2, *Plastics - Determination of Temperature of Deflection Under Load - Part 2: Plastics, Ebonite and Long-Fibre-Reinforced Composites*, 2004.

12. UNI EN ISO 75-3, *Plastics - Determination of Temperature of Deflection Under Load - Part 3: High-Strength Thermosetting Laminates*, 2004.

13. UNI 5642, superseded by UNI EN ISO 306, *Plastics - Thermoplastic Materials - Determination of VICAT Softening Temperature (VST)*, 1998.

14. ASTM D1238-04c, *Standard Test Method for Melt Flow Rates of Thermoplastics by Extrusion Plastometer*, 2004.

15. ASTM D2116, *Specification for FEP-Fluorocarbon Molding and Extrusion Materials*, 2003.

16. ASTM D3364-99, *Test Method for Flow Rates for Poly (Vinyl Chloride) with Molecular Structural Implications*, 2004.

17. UNI EN ISO 1133, *Plastics - Determination of the Melt Mass-Flow Rate (MFR) and the Melt Volume-Flow Rate (MVR) of Thermoplastics*, 2001.

18. ASTM D746, *Test Method for Brittleness Temperature of Plastics and Elastomers by Impact*, 2004.

19. ASTM C177, *Standard Test Method for Steady-State Heat Flux Measurements and Thermal Transmission Properties by Means of the Guarded-Hot-Plate Apparatus*, 2004.

20. DIN 52612, *Testing of Thermal Insulating Materials; Determination of Thermal Conductivity by Means of the Guarded Hot Plate Apparatus*, 1984.

21. ASTM D696, *Test Method for Coefficient of Linear Thermal Expansion of Plastics Between -30 Degrees C and 30 Degrees C with a Vitreous Silica Dilatometer*, 2003.

22. UNI 5284, *Compassi Per Scuole. Ricambi Portamina*, 1994.

23. P.J. Lemstra, A.J. Schouten and G. Challa, *Journal of Polymer Science, Physics Edition*, 1972, **10**, 2301.

24. P.J. Lemstra, A.J. Schouten and G. Challa, *Journal of Polymer Science, Physics Edition*, 1974, **12**, 1565.

25. A. Lety and C. Noel, *Journal de Chimie Physique et de Physico-Chemie Biologique*, 1972, **69**, 875.

26. E. Ahad, *Journal of Applied Polymer Science*, 1974, **18**, 1587.

27. C.Y. Tang, L.C. Chon, J.Z. Liang, K.W.E. Cheng and T.L. Wong, *Journal of Reinforced Plastics and Composites*, 2002, **21**, 1337.

28. E.B.A.V. Pacheco and E.B. Mano, *Polimeros: Ciencia e Tecnologia*, 1996, **6**, 22.

29. J.D. Nam, J. Kim, S. Lee and C. Park, *Journal of Applied Polymer Science*, 2003, **87**, 661.

30. A. Gordillo, O.O. Santana, A.B. Martinez, M.L. Maspoch, M. Sanchez-Soto and J.I. Velasco, *Revista de Plásticos Modernos*, 2002, **83**, 395.

31. M. Le Bras, S. Bourbigot, C. Siat and R. Delobel in *Fire Retardancy of Polymers*, Eds., M. Le Bras, S. Bourbigot, G. Camino and R. Delobel, Royal Society of Chemistry, Cambridge, UK, 1998, Paper 54F, p.266.

32. C.W. De Maio and S. Baushke, *Machine Design*, 2002, **74**, 58.

33. N. Belhaneche-Bensemra and M.A. Chabou in *Proceedings of ISFR 2002*, Ostend, Belgium, 2002, Paper No.A01.

34. W.J. Frederick, Jr., and C.C. Mentzer, *Journal of Applied Polymer Science*, 1975, **19**, 1799.

35. J.L. Haberfield and J.A. Reffner, *Society of Plastics Engineers - Technical Paper*, 1975, **21**, 858.

36. D.P. Pope, *Journal of Polymer Science, Polymer Physics Edition*, 1976, **14**, 811.

37. S. Yu, P. Hing and X. Hu, Composites, Part A: *Applied Science and Manufacturing*, 2002, **33-A**, 289.

38. W.N. Dos Santos and R. Gregorio, *Journal of Applied Polymer Science*, 2002, **85**, 1779.

39. R. Agarwal, N.S. Saxena, K.B. Sharma, S. Thomas and M.S. Sreekala, *Journal of Applied Polymer Science*, 2003, **89**, 1708.

40. *Rubber Asia*, 2003, **17**, 93.

17 Flammability Testing

17.1 Combustion Testing and Rating of Plastics

17.1.1 Introduction

The progress of a fire is often divided into three phases consisting of:

1. ignition and early development,

2. total involvement, and

3. fire recession and extinguishing.

The temperature profile of such processes is depicted schematically in **Figure 17.1**.

The fire characteristics of a material are characterised by ease of ignition, contribution to flame spread, and heat contribution, as well as other factors generally associated with fires, including smoke density and toxicity and corrosiveness of the combustion by-products. Fire behaviour, however, cannot be considered a material property because it is markedly affected by both material and environmental factors. These include the distribution of material in the room, material geometry and other physical factors, temperature history, thermal conductivity, intensity and type of ignition source, exposure time to the ignition source, integration of the material, and ventilation effects [1-3]. The varying nature of the fire risk situation and the influence of materials and environmental factors make it very difficult to establish tests and ratings criteria that would be generally applicable. That is, no doubt, one of the main reasons that many different tests have been developed in research and industry and by regulatory agencies. Many of these, in part material-specific, procedures serve well as quality control or developmental guideposts. The results obtained from within the regime of a test standard are, at best, of limited use in drawing conclusions concerning the performance in real fire situations [4, 5]. Hence we must ensure that the environmental conditions adequately simulate the fire risk situation of concern.

The risk assessment must differentiate between the production, storage, and end-use applications. Different risks in production are often due to the amounts of combustible material. The end-use application, transportation, construction, furniture, or furnishings

495

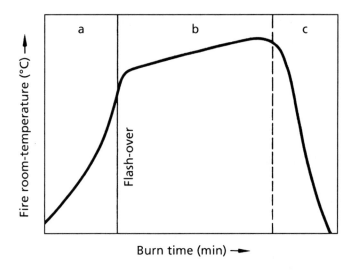

Figure 17.1 Temperature/time fire profile: (a) Fire development phase; (b) Fully developed fire; (c) Receding fire. (*Source: Author's own files*)

all have a major impact in determining risk. Duplication of natural ignition sources such as cigarettes, sparking contacts, and overheated wiring are a first step towards proper applications and risk-oriented testing.

Small laboratory tests, e.g., ASTM D1692 [5], oxygen index test (OIT) [6], Setchkin test [8], or similar tests under certain circumstances provide a means for production quality control. Their results can only be applied towards the characterisation of fire risk assessment if it can be demonstrated that the tests adequately simulate the actual fire situation.

The small laboratory test previously designated ASTM D1692 [5] was abandoned without replacement as a consequence of the 1978 Federal Trade Commission action [7]. As part of the UL test procedure, Subject 94 (**Table 17.1**), this test was used to determine classes of fire retardants [6]. The laboratory procedure consists of exposing a test specimen to the flame from a Bunsen burner fitted with a wing tip flame spreader. The rate and extent of burning were measured. The OIT [6] measures the oxygen/nitrogen concentration needed to support combustion of a test specimen. The specimen is positioned vertically and ignited at the top. Results from such small laboratory tests are, by definition, not suitable for drawing conclusions in real fire situations. Enriched oxygen environments are found in the space industry. In such a cases, the OIT could be used to study and draw conclusions about the ignition phase from ignition sources of low intensity.

Table 17.1 Instrumentation available from ATS FAAR for testing flammability properties of polymers			
Method	ATS FAAR Apparatus Code No.	Test suitable for meeting following Standards	Notes
Flammability by glow wire	10.05300	DIN 695 [8] VDE 0471 [9]	
Rate of burning of rigid specimens exposed to ignition flame of 45°	10.05000	ASTM D635 [10]	
Oxygen index	10.04070 10.04050	DIN ASTM BS CEI	
Ignition properties of plastics	10.04020	ASTM	
Incandescence resistance of rigid plastics in horizontal position	10.04000	DIN ASTM CEI UNI	
Smoke density combustion of materials	10.05470	ASTM	
Resistance to combustion of materials used inside automotive vehicles	10.05600	DIN FED Fiat ISO	
Fire resistance of building materials	10.05200	ISO	
Flame resistance of vertical specimens	10.05050	DIN CSE UNI	Thermogravimetric analysis can be used to measure this property
Fire reaction of specimens exposed to radiant heat	10.05020	CSE ISO	
Fire reaction of upholstered furniture	10.05150	CSE	

Table 17.1 *Continued*			
Method	ATS FAAR Apparatus Code No.	Test suitable for meeting following Standards	Notes
Flammability test	10.05454	UL 1581 [11]	
Flammability test	10.5700	UL 94 [12] UL 60950 [13]	
Burning rate and flame resistance of rigid insulating materials	10.05400	D635-T [10] UL 94 [12]	
Response of plastics to ignition by small flame	10.05500	ASTM UL	
Source: Author's own files			

The Setchkin test, according to ASTM D1926 [14], is used to determine the temperature of solids at which ignitable decomposition products are formed. This is analogous to determining the flashpoint temperature of liquids [15, 16]. The term flash ignition temperatures (FIT) refers to ignition points determined using a small gas pilot flame. Self-ignition temperature (SIT) refers to the temperature where the gases ignite through contact with the hot oven walls.

The application of combustion testing of plastics and rubbers in different applications is discussed in further detail next.

17.1.2 Mining Applications

Mining applications require that specific test criteria be met because of the extraordinary difficulties in rescue and fire extinguishing efforts underground. Highly expandable and combustible blowing agents may not be used if they increase the fire hazard underground. In case of an accidental fire, the blowing agents used should not propagate the fire. The tunnel test facility as per DIN 22118 [17] is used, for example, to evaluate the contribution to fire spread of conveyor belts with textile inserts, such as would be used in anthracite mines. The test samples measure 1200 mm × 90 mm and are placed horizontally into the laboratory test tunnel. The test sample is exposed from underneath to the flame of a special propane burner for a period of 15 minutes. The burner is positioned 170 mm from the front edge of the sample. A maximum burn extent and afterburn are specified in order to meet the minimum test criteria. The basis for the standardisation of this test procedure was a series of full-scale tests carried out by the Tremonia Mining Organisation in Dortmund.

17.1.3 Electrical Applications

In Germany performance requirements for electric applications are regulated through VDE guidelines [18, 19]. Localised smouldering or ignition may result from overhead wires, sparking contacts, or other failure of electrical equipment. Underwriter Laboratories, Inc., has developed universally applicable fire protection test procedures. According to UL-94 [12, 20] (**Table 17.1**) horizontally (or, if applicable, vertically) oriented test samples are exposed to a gas flame source. Test criteria consist of the rate of burning, the burn distance, or the continued burning of test samples themselves and/or drippings from melted materials after removal of the flame source.

Tests have been reported for polyamide 6,6 formulations containing melamine polyphosphate as flame retardant [21], ethylene–vinyl acetate copolymer containing ammonium polyphosphate as flame retardant [22], and elastomeric materials [23].

A horizontal test procedure using a 30 second flame exposure is used for determining the HB classification. This test is analogous to ASTM D635 [10] (**Table 17.1**) and the discontinued ASTM D1692. Bar-like samples measuring 127 mm × 12.7 mm × 6.4 mm are used in both the vertical and the horizontal test procedures. The vertical test procedure also requires that the test specimens be conditioned for seven days at 70 °C [12, 20]. The sample is ignited at the lower end. The afterburn time of the sample or dripped material is used for classifying samples as V0, V1, and V2.

The ASTM E162 test procedure [24] normally used for construction applications is used to test larger-sized electrical boxes or housings. The requirements for such are outlined in UL 94 [12]. In this test, the contribution of the sample to surface flame spread is measured under conditions where the sample is exposed to a specified radiant heat flux. Another application-specific procedure is the test for hot wires or contacts [25] according to DIN EN ISO 60695 [25] (**Table 17.1**) in which the sample is exposed to a glowing wire at temperatures between 550 and 960 °C.

Potential ignition sources through failure of overheated wires or termination screws are simulated in this procedure. The performance of insulation is determined by contact with hot, glowing wires. The test procedure specifies stepwise increasing temperatures in the range 450–906 °C when wire temperatures at the point of failure are determined.

Test criteria include ignition time, dripping, and subsequent burning. The insulation test sample is moulded on a movable fixture and pressed against the hot wire loop with a force of 1 N. The penetration distance is restricted. The exact temperature of the wire loop is measured with a miniature thermocouple.

The failure of electrical appliances or fixtures can initiate smouldering which could lead to localised fires, such as are simulated in, for example, DIN EN 60695-2-2 [26], through, for example, flame exposure from a Bunsen burner. Depending on the application (TV cabinets, cable, etc.), other fire sources are used.

A widely used test procedure is the glowing rod test. Here a rod-shaped test sample is pressed tip-to-tip against a glowing rod heated to 960 °C with a force of 1 N. Test criteria specify burning rate and extent.

Flame retardancy for electrical and electronic appliances and wiring has been discussed by various workers including Dawson and Landry [27], Canaud and co-workers [28]. Porro investigated the electrical properties of filled and unfilled aliphatic polyketones [29]. Flame-retardant thermoplastic elastomers have been discussed by De Maio and Baushke [23].

17.1.4 Transportation Applications

The Federal Motor Vehicle Safety Standard (FMVSS) [30] serves as a worldwide basis for specification in outfitting automotive interiors. This test provides for a limiting burning rate of '10 cm per minute' and falls under the jurisdiction of the Federal Highway Administration (FHA) of the United States. It has been incorporated worldwide into industrial specifications, national standards, as well as international ISO standards [31]. The test sample measures 356 mm × 100 mm × d mm (d is the sample thickness) and is supported horizontally. The free end is subjected to a specific flame exposure. Actual application-ready materials are to be tested [35]. These include composites, such as laminates prepared using adhesives, flame-lamination, and so on. The decisive test criterion is the burning rate. To ensure that the parts that might be individually exported satisfy the requirements of the American statutes, the automotive industry has, in part, adopted more stringent internal requirements such as limitation of burn distance.

The vertically or horizontally arranged test samples are subjected to a wide specified flame for three minutes in the vertical and two minutes in the horizontal configurations. The extent of surface destruction is used to determine the ratings ranging from 'B1, leichtbrennbar' (light burning) to 'B4, nichtbrennbar' (non burning). The dripping characteristics as well as the visually determined smoke characteristics are divided into four additional classifications. The finished component characteristics are divided into four additional classifications. The finished composites used for padding or cushioning applications are tested. The European Organisation of Railroad Authorities is considering a French proposal for fire safety determination and testing of seat upholstery. The origin of the procedure comes from model tests designed to simulate actual applications. The fire source is a 100 g paper cushion measuring 36 cm × 27 cm. The cushion is conditioned for four hours at 70 °C. Performance criteria require that the test sample be not completely consumed, that the burn time be less than ten minutes, and that the combination of flaming and dripping does not occur.

The regulations of the US Federal Aviation Administration (FAA) are applicable worldwide to the aircraft transportation industry. The test requirements are specified by Federal Air Regulations - FAR 25.853 [33, 34]. Test samples are oriented vertically, horizontally, or at a 45° angle and exposed to a specified Bunsen burner ignition source. Test criteria consist of burn distance, occurrence of dripping behaviour, and continued burning after removal of the ignition source. Internal industry specifications [35] also limit, among other factors, smoke density and toxicity of combustion products. The determination of toxicity is made from an analytical perspective. The concentrations and compositions of the combustion products are determined from gas samples obtained during smoke density measurements in a National Bureau of Standards (NBS) chamber. Concentrations of sulfur dioxide, carbon monoxide, carbon dioxide, hydrogen cyanide, hydrogen chloride, and so on, are determined.

The requirements of the Seeberufsgenossenschaft apply to ocean liners operating under the German flag. While many aspects are regulated through the Inter Governmental Maritime Consulting Organization (IMCO), the setting of test and performance criteria for combustible materials is left to national agencies. For ships with German registration, a limited use of combustible products is permitted. These must, however, conform to the construction class B1 in accordance with DIN 4102 [36]. In addition, the judgement of relative toxicity must be favourable. Through animal experiments, the relative toxicity of the combustion products of the test material may be no worse than that of the combustion products of wood or cork. In order to minimise the potential for fire spread, these construction elements must additionally be covered by steel with a minimum thickness of 1 mm. Analogous safety requirements, for example those covering combustible insulation, are specified as part of the Merchant Shipping Notice M592 in the UK [37].

17.1.5 Furniture and Furnishing Applications

No general regulations apply to furniture or furnishings. German requirements for theatre curtains, decorations, and so forth, used in public meeting places or department stores are generally not well defined. Based on detailed fire statistics [38-40] the UK has prepared a law using test and performance criteria of BS 5852 [41]. As in the USA [42], the primary consideration is the reduction of fire risk in homes from cigarette ignition sources. In those cases where the cigarette ignition and/or the simulated match ignition conditions are not successfully passed, the furniture must be labelled to that effect and identified. All furniture is required to pass this cigarette ignition test.

The test procedures are based on numerous full-scale model fire tests that were done worldwide. Aside from governmentally required testing [43, 44], industry carried out extensive testing and trials [45–48] to ensure statistically correct test and performance criteria.

The US additionally imposes material-specific requirements. Examples are the California State statutes and regulations of the New York Port Authority [49]. Among other requirements for cushioning materials are the construction-specific tests such as ASTM E162 [24] and ASTM D2843 [50]. If the cushioning material fails, upholstery combinations may be substituted.

The British Property Service Agency (PSA) responsible for furniture procurement requires application-oriented tests with correspondingly staggered ignition sources from cigarettes to a wooden crib [51].

17.1.6 Construction Material Applications

The construction area is forced to comply with the most encompassing set of regulations. In Germany the requirements for fire prevention and protection are contained in or regulated through zoning regulations, building permits, and guidelines, as well as construction-specific codes such as DIN 4102 [36]. Analogous regulations are promulgated through, for example, the building regulations in the UK, building codes in the USA, or the guidelines of the fire police in Switzerland. According to DIN 4102 [36], combustible building materials are categorised into 'B1: schwerentflammbar (heavy burning), B2: normal entflammbar (normal burning) or B3: leichtenflammbar (light burning)', using both a small burner test and a large chimney test procedure.

The small burner test consists of a vertically oriented specimen which is exposed on either edge or side to a specific ignition flame for 15 seconds. To obtain a B2 classification, the flame front may not have reached a previously marked line at 150 mm within a 20 second time interval inclusive of the 15 second flame exposure time. The test for possible B1 performance uses four vertically arranged test samples, 1000 mm × 190 mm. The samples, in this in a case chimney arrangement, are exposed to a flame source at their lower edge using a gas ring burner for periods of ten minutes. Test criteria consist of undamaged sample distances and combustion gas temperatures. Dripping behaviour, if any, is noted even though it is not part of the test classification. However, such an observation may be required to obtain the necessary building permits.

Enhanced tests and performance criteria have been developed for classification of composites into B1, B2, and B3 categories. Floor coverings are evaluated for B1 performance, using both a critical flux test and a modified small burner test procedure. To obtain a B2 rating using this modified test requires an increased time of 30 seconds. The critical radiant flux test measures the contribution to flame spread of a horizontally oriented test sample. The sample is exposed to a radiant energy flux of varying intensity. The flames may not propagate past the 0.45 W/cm^2 exposure limit. This so-called 'critical limit' was determined using wood parquet flooring.

Recently, single burning item test for class B, C, and D materials of intermediate combustibility have been introduced [52] and equipment is now available for carrying out six new test methods for European-wide classification of construction products.

It has proved difficult to determine any correlation between the test result of the various national procedures [53]. Because of this, the experts of TC92 of the International Standards Organisation (ISO) have undertaken the development of a test procedure to characterise independently ignitability, flame spread, rates of heat release, and other fire-related parameters [54-56]. Worldwide efforts continue to correlate laboratory tests to real-life fires [57]. Examples of such programmes are the corner test programme carried out by Factory Mutual and the corrugated metal tool deck [58] trials [59, 60] carried out by TNO. The corner test has been used to determine the fire behaviour of rigid foam materials when exposed to severe wood crib fire.

Construction elements and special constructions. The fire resistance of constructions that form an enclosed space is determined by the phenomenon of flashover, i.e., the extension of the fire from the room of origin to adjacent spaces. Containment of the fire is through a sufficiently high fire resistance performance of the components forming the enclosed space, such as walls, ceilings, and doors. For proper classification, it is necessary to ensure that construction details, such as holes for cables, water, or other piping, as well as joint details, do not result in weak spots allowing fire penetration. To assess properly fire safety of construction components, their performance is determined relative to thermal loading of the panels using a standard time–temperature profile designated (ETK) in DIN 4102 [36]. Such a profile has been adopted both naturally and as ISO Standard 834 [61]. Construction elements are assembled into the wall or ceiling test furnace in the same form as they would be used. The elements are then subjected to the standard time–temperature exposure for the time interval corresponding to the rating desired. Acceptance requires that fire breakthrough does not occur, that structural integrity be preserved, and that temperature gradients remain within specified limits. In Germany speciality constructions such as parapets have reduced fire performance requirements because of the lower fire risk associated with such applications. The reduced test conditions essentially consist of a modified (flattened) standardised curve according to DIN 4102, Section 3 [36].

For single case evaluations of the fire performance of parapet elements, application-like model tests can be conducted using a test device currently used for testing facades [62]. Roofing is evaluated for fire spread caused by external fire sources and radiant heat. According to DIN 4102, Section 7 [36] the fire source consists of 600 g of wood shavings. The test is carried out at roof inclinations of 15 and 30°. Neither burn-through nor unacceptable fire spread may occur in order to pass the performance criteria. The corresponding testing in the Netherlands is regulated under Standard NEN 6065 and 6066 [63, 64].

For proper risk assessment of the fire performance of metal roof constructions, model fire test studies were done [65]. The trials carried out by TNO indicated that the fire performance classification of the insulation was the main factor influencing fire spread along the upper roof system to adjacent sections.

17.1.7 Other Fire-Related Factors

In any fire, liquid, solid, and gaseous combustion products are formed. For the category of 'nicht brennbare' (non burning) building materials (Category A2), judgements are based on 'expert opinion'. The smoke density measurements are made using an XP-2 cabinet [66] and the DIN 'pyrolysis apparatus' [67].

The test sample in the XP-2 chamber (ASTM D2843 [50]) is supported on a wire mesh and ignited. The build-up of combustion products results in attenuation of a horizontal light beam. A strip-like test sample is decomposed in air in the combustion tube as per DIN 53436 [67]. Again, a photoelectric system is used to measure the dependency on test temperature.

The aircraft industry uses, on a worldwide basis, the smoke density apparatus developed by the NBS. A vertically oriented test sample is subjected to a radiant heat flux of 2.5 W/cm^2. The test is run both with and without a pilot ignition flame.

The attenuation of the light beam of a vertically oriented measuring system is determined. A promising development is the ISO smoke box [56], developed by ISO TC92. One of the benefits of this method is the ability to also test composite materials. The smoke development is measured in the presence of a radiant energy flux in the range 0–5 W/cm^2 with piloted ignition.

The combustion tube apparatus according to DIN 53436 [67] also serves to generate the combustion products for animal experiments. Toxicologists are of the opinion that neither the determined chemical composition of the combustion products nor the analytically determined concentrations are sufficient to determine acute toxicity [68, 69]. Possible additive, synergistic, or antagonistic effects can only be determined through animal experiments. However, such studies and assessments can only be made using combustion products with a composition representative of that developed in an actual fire situation. The combustion tube apparatus consists of a quartz tube, heated by a movable electric ring furnace. The ring furnace is moved uniformly and in the opposite direction to the air stream. The thermal load on the test specimen is determined by test temperatures in the range 100–600 °C. The reason for the counter-current air stream is to guarantee uniform thermal conditions for the test sample over the test interval.

The corrosiveness of combustion products has been considered in a variety of ways [70-72] in relation to actual fire damage as well as large-scale experiments. Electric insulation studies have been carried out [70-72] using the combustion tube apparatus to provide the data necessary for standards work. DIN EN 50267 [73-76], Section 813, provides information concerning test and performance criteria for the determination of the corrosiveness of combustion products from fires of electric origin [73-76].

Flame retardancy in various polymers has been discussed both in general terms [77] and for particular types of polymers including polyvinyl chloride (PVC) [78], polystyrene foam [79], unsaturated polyester resins [80, 81], polyisocyanurate–polyurethane (PU) water-blown foams [82], ethylene–vinyl acetate copolymers [83], and plastic and rubber cables [84].

17.2 Instrumentation

Some instrumentation available from ATS FAAR for flammability testing of polymers is listed in **Table 17.1**. Other suppliers are listed in Appendix 1.

17.3 Examination of Combustible Polymer Products

When polymers are burnt or smoulder in air, the combustion products are extremely complex, often consisting of several hundred compounds. Because of the toxic or unknown nature of these products, it is important to know their composition in some detail. This information is also essential for mechanistic and modelling studies of the smoke formation process, which can lead to the design of less hazardous polymers in the future.

A number of analytical methods [85, 86] involving pyrolysis of polymers have been reported in the literature. Michal and co-workers [87] developed a method using direct gas chromatography (GC)–mass spectrometry (MS) for their study of the combustion of polyethylene (PE) and polypropylene (PP). Morikawa [88] used GC to determine polycyclic aromatic hydrocarbons in combustion of polymers. Liao and Browner [89] described a method for the determination of polycyclic aromatic hydrocarbons. Many other workers have studied soot and smoke formation and mechanisms in the combustion of polymers. Generally in these studies, relatively simple and specific methods were used, which were appropriate for the intended tasks. However, these methods are not suitable for complete analysis of the very complex smoke particulates resulting from combustion of many polymers. Most methods have been developed either for volatile compounds of low molecular weight or for polycyclic aromatic hydrocarbons. Joseph and Browner [90] developed a method that can be used to analyse all classes of compounds produced by the combustion of polymers and applied this method to rigid PU foams prepared by the polymerisation of diphenylmethane diisocyanate and glycols:

$$R(NCO)_2 + R^1(OH)_2 \rightarrow OCNHRNCOOR^1O$$

The method is directed towards the particulates produced and excludes the volatile compounds of very low molecular weight.

Smoke particulates from the smouldering of PU foam in air were collected on glass fibre filters and extracted with chloroform. The concentrated extract was subjected to acid and base extractions. The acid compounds were converted to methyl esters and analysed by a GC-MS data system. The basic and phenolic compounds were analysed using the same system without derivatisation. The neutrals were separated into different classes by high-performance liquid chromatography on a bonded amine column. Different fractions were collected and each fraction was analysed by GC-MS with data collection.

Some nitrogen-containing compounds were identified in the neutral fractions. There were five-membered nitrogen-containing ring compounds, including indole, isoxazole, indazole, and carbazoles which do not show basic properties, and consequently do not react with 1 M hydrochloric acid in the basic extraction step. A number of phthalate esters were also present in different fractions.

Derby and Freedman [91] have described apparatus for the determination of the products (benzene, toluene, carbon dioxide, and hydrogen chloride) produced upon combustion of PVC at 800 °C.

Zhu and co-workers [84] have studied the thermal degradation of a new flame-retardant phosphate methacrylate polymer. Degradation was monitored by *in situ* Fourier transform infrared spectroscopy, X-ray photoelectron spectroscopy, X-ray diffraction, and Raman measurements. The carbon structure of the final char after burning was examined.

17.4 Oxygen Consumption Cone Calorimetry

One of the most important parameters that can be used to characterise an unwanted fire is the rate of heat release. It provides an indication of the size of the fire, the rate of fire growth, and consequently the release of smoke and toxic gases, the time available for escape or suppression, the types of suppressive action that are likely to be effective, and other attributes that define the fire hazard. Methods based on the oxygen consumption principle are now available to measure the rate of heat release reliably and accurately. The principle depends upon the fact that the heats of combustion of organic materials per unit of oxygen consumed are approximately the same. This is because the processes in the combustion of all these products involve breaking of C–C and C–H bonds (which release approximately the same amount of energy) with the formation of carbon dioxide and water.

The cone calorimeter is the most generally accepted and powerful instrument in this field. Originally developed by Vytenis Babrauskas of the Center for Fire Research at NIST (formerly the National Bureau of Standards) it is now available as a complete instrument embodying specific safety and design features while retaining the design set out in the proposals. The design enables the following to be determined and computed automatically:

- Rate of heat release

- Rate of heat release per unit area

- Mass loss rates

- Time to ignition

- Effective heat of combustion

This instrument has been used for the evaluation of building, furniture, transport, aerospace, wood, and electrical materials and composites. Polymers to which this technique has been applied include PP and PP-composites [92], ethylene–vinyl acetate copolymers [93], polyamide 6,6 [21], silicate–siloxane composites [94], epoxy composites and glass–phenolic composites [95], polyolefins [96], conveyor belt materials [97], and roofing materials [98].

Thermal Sciences Ltd supply a cone calorimeter produced by Stanton Redcroft.

17.5 Laser Pyrolysis–Time-of-Flight Mass Spectrometry

This is a technique being pioneered by Price and co-workers [99, 100]. Two systems have been investigated [100] to model different aspects of flame-retarded polymer behaviour in a fire. One system uses a continuous laser to model radiative heat at a level similar to that from a burning item in a room fire and the other uses a pulsed laser to model conditions immediately behind the flame front. It was found that [99] the laser pyrolysis of aluminium oxide trihydrate-retarded polymethyl methacrylate (PMMA) produces a large amount of water and carbon dioxide in the volatiles. Also, the amount of the monomer evolved is reduced significantly compared to that obtained from pure PMMA. The implication of these results is that in a real fire situation aluminium oxide trihydrate influences PMMA pyrolysis in such a manner as to bring about a reduction in the evolved 'fuel' while at the same time adding non-combustible gases to the flame region. Thus the PMMA is flame retarded.

17.6 Pyrolysis–Gas Chromatography–Mass Spectrometry

This technique has been applied to studies of thermal decomposition and combustion behaviour of polyhydroxyamide and its bromine, fluorine, phosphate, and methoxy derivatives [101], poly(3,3′-dihydroxybiphenylisophthalimide) and its halogenated phosphinite and phosphate derivatives [102], and rigid PU foams [103]. Sato and co-workers [104] studied the thermal degradation of a polyester-based material flame retarded with antimony trioxide–brominated polycarbonate by means of various temperature-programmed analytical pyrolysis techniques such as temperature-programmed pyrolysis–MS, temperature-programmed atomic emission detection, and temperature-programmed pyrolysis–gas chromatography. During the degradation of the flame-retarded polyester, brominated phenols were first observed to evolve at temperatures slightly lower than those for the flammable product evolution from the substrate polyester polymer, followed by the evolution of hydrogen bromide over the whole range of degradation temperatures for the substrate polymer. These degradation processes were closely related to the synergistic effects of antimony trioxide on the decomposition of brominated polycarbonate in the flame-retardant system to promote the thermal degradation of brominated polycarbonate at lower temperatures than those for pure brominated polycarbonate. Furthermore, the evolution of the flame poisoning antimony tribromide formed through the reaction between brominated polycarbonate and antimony trioxide could also be monitored directly by temperature-programmed pyrolysis techniques.

17.7 Thermogravimetric Analysis

Wang and co-workers [105] studied the thermal degradation of flame-retarded PE–magnesium hydroxide–PE-*co*-PP/elastomer composites using thermogravimetric analysis.

508

References

1. W. Becker in *Brandverhalten von Kunststoffen*, Ed., J. Troitzsch, Carl Hanser Verlag, Munich, Germany, 1982.

2. G. Schreyer, *Konstruieren mit Kunststoffen*, Carl Hanser Verlag, Munich, Germany 1972.

3. F.H. Prager, *Kunststoffe im Bau*, 1978, **13**, 2, 45.

4   DIN 75200, *Determination of Burning Behaviour of Interior Materials in Motor Vehicles*, 1980. (In German)

5. ASTM D1692, *Standard Method of Test for Flammability of Plastic Sheeting and Cellular Plastics*.

6. ASTM D2863, *Test Method for Measuring the Minimum Oxygen Concentration to Support Candle-Like Combustion of Plastics (Oxygen Index)*, 2000.

7. *Proposed Trade Regulation Rules Concerning Disclosure of Cellular Plastics Products: 39 Federal Regulation 28, 292, August 1974; 40 Federal Regulation 30, 842*, Federal Trade Commission (FTC), July 1975.

8. DIN 695, *Grade 2 Chain Slings With Hook or Ring Type Terminal Fittings*, 1986.

9. VDE 0471, *Fire Hazard Testing*, 2000.

10. ASTM D635, *Test Method for Rate of Burning and/or Extent and Time of Burning of Plastics in a Horizontal Position*, 2003.

11. UL 1581, *Reference Standard for Electrical Wires, Cables, and Flexible Cords*, 2003.

12. UL 94, *Tests For Flammability of Plastic Materials for Parts in Devices And Appliances*, 2003.

13. UL 60950, *Safety of Information Technology Equipment*, 2002.

14. ASTM D1926, *Test Methods for Carboxyl Content of Cellulose*, 2000.

15. DIN 51755, *Testing of Mineral Oils and Other Combustible Liquids; Determination of Flash Point by the Closed Tester According to Abel-Pensky*, 1974. (In German)

16. DIN EN ISO 1523, *Determination of Flash Point - Closed Cup Equilibrium Method*, 2002. (In German)

17. DIN 22118, *Conveyor Belts with Textile Plies for use in Coal Mining; Fire Testing*, 1991.

18. DIN VDE 0472-814, *Testing of Cables, Wires and Flexible Cords; Continuance of Isolation Effect Under Fire Conditions, 1991.* (In German)

19. VDE 0860, *Audio, Video and Similar Electronic Apparatus - Safety Requirements, 2003.* (In German)

20. UL 1416, *Overcurrent and Overtemperature Protectors for Radio- and Television-Type Appliances*, 2004.

21. F. Dabrowski, M. Le Bras, R. Delobel, D. Le Maguer, P. Barddlet and J. Aymani in *Proceedings of an Interscience Conference: Flame Retardants 2002*, London, UK, 2002, Paper 15, p.127.

22. M. Le Bras, S. Bourbigot, C. Siat and R. Delobet in *Fire Retardancy of Polymers: The Use of Intumescence*, Eds., M. Le Bras, G. Camino, S. Bourbigot and R. Delobet, Royal Society of Chemistry, Cambridge, UK, 1998, p.266.

23. C.W. De Maio and S. Baushke, *Machine Design*, 2002, **74**, 58.

24. ASTM E162, *Test Method for Surface Flammability of Materials Using a Radiant Heat Energy Source*, 2002.

25. DIN EN 60695-5-1, *Fire Hazard Testing - Part 5-1: Corrosion Damage Effects of Fire Effluent - General Guidance*, 2003.

26. DIN EN 60695-2-2, *Fire Hazard Testing - Part 2: Test Methods; Section 2: Needle-Flame Test*, 1996.

27. R.B. Dawson and S.D. Landry in *Proceedings of GPEC 2003: Plastics Impact on the Environment*, Detroit, MI, USA, 2003, SPE 2003, p.191.

28. C. Canaud, L.L.Y. Visconte and R.C.R. Nunes, *Macromolecular Materials and Engineering*, 2001, **286**, 377.

29. P. Porro, *Materie Plastiche ed Elastomeri*, 1998, **63**, 732.

30. *Federal Motor Vehicle Safety Standard (FMVSS) 302, Test Procedure and Specimen Preparation*, Docket 3-3, Notice 6: Federal Regulation 38, No. 95, May 1973.

31. ISO 3795, *Road Vehicles and Tractors and Machinery for Agriculture and Forestry - Determination of Burning Behaviour of Interior Materials,* **1989.**

32. *Federal Motor Vehicle Safety Standard (FMVSS) 302, Proposed Covered Component,* Notice 7: Federal Regulation 38, No.133, July 1973.

33. *Federal Air Regulation (FAR), 25, 853*; Federal Regulation 34, 135, August 1968.

34. *Federal Air Regulation (FAR), 25, 853*; Federal Regulation 37, 37, February 1979.

35. H.H. Cantow, M. Kowalski and C. Krozer, *Angewandte Chemie International Edition in English,* 1972, **11**, 334.

36. DIN 4102-7, *Fire Behaviour of Building Materials and Building Components - Part 7: Roofing; Definitions, Requirements and Testing,* **1998.** (In German)

37. *Merchant Shipping Notice M 592,* Board of Trade, Marine Division, UK, 1970.

38. S.E. Chandler, *The Incidence of Residential Fires in London – The Effect of Housing and Other Social Factors,* BRE Information Paper, IP 20/79, 1979.

39. S.E. Chandler, *Some Trends in Furniture Fires in Domestic Premises,* BRE-CP, 66/76.

40. S.E. Chandler and R. Baldwin, *Fire and Materials,* 1976, **1**, 76.

41. BS5852-1990, *Methods of Test for Assessment of the Ignitability of Upholstered Seating by Smouldering and Flaming Ignition Sources,* **1994.** This supersedes the one you cited

42. W.G. Berl and B.M. Halpan, *Fire Journal,* 1979, 105.

43. K.N. Palmer, W. Taylor and K.T. Paul, *Fire Hazards of Plastics in Furniture and Furnishing: Characteristics of the Burning,* BRE-CP 3/75, HMSO, London, UK, 1985.

44. K.N. Palmer, W. Taylor and K.T. Paul, *Fire Hazards of Plastics in Furniture and Furnishing: Fully Furnished Room,* BRE-CP 21/76, HMSO, London, UK, 1976.

45. W.J. Wilson in *Proceedings of the SPI 4th Annual Combustion Symposium,* London, UK, 1975, p.62

46. F.H. Prager in *Proceedings of the 5th International Conference Brandschutseminar,* Karlsruhe, Germany, 1976.

47. R.P. Marchant, *The Ignitability of Upholstery by Smokers' Materials*, FIRA, Stevenage, UK, 1977.

48. *State of California Technical Information Bulletins 116 and 117.*

49. *Specifications Governing the Flammability of Upholstery Materials and Plastics in Furniture*, The New York Authority, NY, USA, 1977.

50. ASTM D2843-99, *Test Method for Density of Smoke from the Burning or Decomposition of Plastics*, **2004.**

51. *DoE/PSA, Fire Retardant Specifications, 551*, Department of the Environment, London, UK.

52. Fire Testing Technology, *Plastics and Rubber Weekly*, 1998, **1730**, 9.

53. H.W. Emmons, *Fire Research Abstracts and Review*, 1968, **10**, 133.

54. ISO TR 3814, *Test for Measuring 'Reaction-to-Fire' of Building Materials - Their Development and Application*, 1989.

55. ISO 5658, *Reaction to Fire Tests - Spread of Flame Test*, 1996.

56. ISO 5659, *Plastics - Smoke Generation*, 1996.

57. ISO TR 3814, *Test for Measuring 'Reaction-To-Fire' of Building Materials - their Development and Application*, **1989.**

58. *A Fire Study of Rigid Cellular Plastic Materials for Insulated Wall and Roof/ Ceiling Constructions*, Factory Mutual Research Corporation, Project 2000.

59. H. Zorgman in *Proceedings of the 5th International Brandshutzseminar*, Karlsruhe, Germany, 1976.

60. F.H. Prager and H. Zorgman, *Kunststoffe*, 1979, **2**, 14.

61. ISO 834-1, *Fire Resistance Tests, Elements of Building Construction – Part 1 General Requirements*, 1999.

62. W. Klöker, H. Niesel, F.H. Prager, H.W. Schiffer, O. Bökenkamp and H.G. Klingelhofer, *Kunststoffe*, 1977, **67**, 438.

63. NEN 6065 *Determination of the Contribution to Fire Propagation of Building Products*, 1997.

64. NEN 6066, *Determination of the Smoke Production During Fire of Building Products*, 1997.

65. D. Brein and P.G. Seeger, *Kunststoffe im Bau*, 1979,14, **2**, 51.

66. ASTM D2843-99, *Test Method for Density of Smoke from the Burning or Decomposition of Plastics*, 2004.

67. DIN 53436, *Producing Thermal Decomposition Products from Materials in an air Stream and their Toxicological Testing*, 1981.

68. ISO TS 19706, *Guidelines for Assessing the Fire Threat to People*, 2004.

69. G. Kimmierle, *Journal of Combustion Toxicology*, 1974, 1.

70. F.W. Locher and S. Sprung, *Beton*, 1970, **2**, 63.

71. F.W. Locher and S. Sprung, *Beton*, 1970, **3**, 99.

72. C. Hammer and K. Fischer, *Beton*, 1971, **9**, 20.

73. DIN EN 50267-1, *Common Test Methods for Cables Under Fire Conditions - Tests on Gases Evolved During Combustion of Materials from Cables - Part 1: Apparatus, 1999.*

74. DIN EN 50267-2-1, *Common Test Methods for Cables Under Fire Conditions - Tests on Gases Evolved During Combustion of Material From Cables - Part 2-1: Procedures - Determination of The Amount of Halogen Acid Gas, 1999.*

75 DIN EN 50267-2-2, *Common Test Methods for Cables Under Fire Conditions - Tests on Gases Evolved During Combustion of Material From Cables - Part 2-2: Procedures - Determination of Degree of Acidity of Gases for Materials by Measuring pH and Conductivity, 1999.*

76 DIN EN 50267-2-3, *Common Test Methods for Cables Under Fire Conditions - Tests on Gases Evolved During Combustion of Material From Cables - Part 2-3: Procedures - Determination of Degree of Acidity of Gases for Cables by Determination of the Weighted Average of pH and Conductivity, 1999.*

77. F.J. del Portillo, *Plast '21*, 1996, October, 55, 139.

78. *Modern Plastics International*, 1998, **28**, 106.

79. *Plastics in Building Construction*, 1997, **21**, 2.

80. *Reinforced Plastics*, 1999, **43**, 18.

81. V.S. Patel, R.G. Patel and M.P. Patel in *Handbook of Polymer Blends and Composites*, Eds., A.K. Kulshreshtha and C. Vasile, Rapra Technology Ltd., Shrewsbury, UK, 2002, Volume 1, p.333.

82. M. Modesti and A. Loernzetti, *European Polymer Journal*, 2003, **39**, 263.

83. G. Beyer in *Proceedings of Flame Retardants 2002 Conference*, London, UK, 2002, Paper No.22, p.209.

84. S. Zhu and W. Shi, *Polymer Degradation and Stability*, 2003, **80**, 217.

85. F.D. Hileman, K.J. Vorhees, L.H. Wojiek, M.M. Birky, P.W. Ryan and I.N. Eihorn, *Journal of Polymer Science*, 1975, **13**, 571.

86. E.A. Boettrer, G. Ball and B.Weiss, *Journal of Polymer Science*, 1969, **13**, 377.

87. J. Michal, J. Mitera and S. Tardon, *Fire and Materials*, 1976, **1**, 160.

88. J. Morikawa, *Journal of Combustion Technology*, 1978, **3**, 349.

89. J.C. Liao and R.F. Browner, *Analytical Chemistry*, 1978, **50**, 1683.

90. K.T. Joseph and R.F. Browner, *Analytical Chemistry*, 1980, **52**, 1083.

91. J.V. Derby and R.W. Freedman, *American Laboratory*, 1974, **6**.

92. S-B. Kwak and J-D. Nam, *Polymer Engineering and Science*, 2002, 42, **1674.**

93. M. Le Bras, S. Bourbigot and B. Revel, *Journal of Materials Science*, 1999, **34**, 5777.

94. J.E. Connell, E. Metcalfe and M.J.K. Thomas, *Polymer International*, 2000, **49**, 1092.

95. F.Y. Hshieh and H.D. Beeson, *Fire and Materials*, 1997, **21**, 41.

96. S. Bourbigot and M. Le Bras, in *Fire Retardancy of Polymers: The Use of Intumescence*, Eds., M. Le Bras, G. Camino, S. Bourbigot and R. Delobel, Royal Society of Chemistry, Cambridge, UK, 1998, 222.

97. J. Wachowicz, *Fire and Materials*, 1998, **22**, 213.

98. P. Eigenmann, *Materie Plastiche ed Elastomeri*, 1995, **4**, 194.

99. D. Price, F. Gao, G.J. Milnes, B. Eling, C.I. Lindsay and P.T. McGrail, *Polymer Degradation and Stability*, 1999, **64**, 403.

100. D. Price, G.J. Milnes and F. Gao, *Polymer Degradation and Stability*, 1996, **54**, 235.

101. H. Zhang, P.R. Westmoreland and R.J. Farris in *Proceedings of ACS Polymeric Materials Science and Engineering Meeting*, Chicago, IL, USA, 2001, Volume 85, p.463.

102. H. Zhang, R.J. Farris and P.R. Westmoreland, *Macromolecules*, 2003, **36**, 3944.

103. T. Zhong, M.M. Maroto-Valer, J.M. Andresen, J.W. Miller, M.L. Listemann, W.R. Furlan and D. Morita, *Polymer Preprints*, 2001, **42**, 7, 687.

104. H. Sato, K. Kondo, S. Tsuge, H. Ohtani and N. Sato, *Polymer Degradation and Stability*, 1998, **62**, 41.

105. Z. Wang, K. Hu, Y. Hu and Z. Gui, *Polymer International*, 2003, **52**, 1016.

18 Mechanical, Electrical, and Optical Properties

18.1 Mechanical Properties of Polymers

In the past, results of standard tests such as tensile strength, Izod impact strength, and softening point have been given major emphasis in plastic technical literature. More recently, however, with the increasing use of plastics in more critical applications, there has been a growing awareness of the need to supplement such information with data collated from tests more closely simulating operational conditions.

For many applications the choice of material used depends upon a balance of stiffness, toughness, processability, and price. For any particular application, a compromise between these features will usually be necessary. For example, it is generally true that within a given family of grades of a particular polymer, the rigidity increases as the impact strength decreases. Again, processing requirements may well place an upper or a lower limit on the molecular weight (MW) of the polymer that can be used, and this will frequently influence the mechanical properties quite markedly. Single test values of a given property are a less reliable guide to operational behaviour with plastics than with metals, since for plastics the great mass of empirical experience and tradition of effective design which has been built up over centuries in the field of metals is not yet available. Furthermore, no single value can be placed on either the stiffness or the toughness of a plastic as:

- Stiffness will vary with time, stress, and temperature.

- Toughness is influenced by the design and size of the component, the design of the mould, processing conditions, and the temperature of use.

- Both stiffness and toughness can be affected by environmental effects such as thermal and oxidative ageing, ultraviolet (UV) ageing, and chemical attack (including the special case of environmental stress corrosion).

In addition, a change in a specific polymer parameter may affect both processability and basic physical properties and both of these can interact in governing the behaviour of a fabricated article. Comprehensive experimental data are therefore necessary to understand effectively the behaviour of plastic materials and to give a realistic and reliable guide to material and grade selection.

Since in many applications' plastics are replacing traditional materials, there is often a natural tendency to apply to plastics, tests similar to those that have been found suitable for gauging the performance of the traditional material. Dangers can obviously arise if plastics are selected on the basis of these tests without clearly recognising that the correlation between laboratory and field performance values may well be quite different for the two classes of materials.

It is therefore important to realise that standard tests are not devised to give direct prediction of end-use performance. In general, the reverse is the case, in that if a particular grade of a particular polymer is found to perform satisfactorily in a given end-use, it can then be characterised with reasonable accuracy by standard tests, and the latter tests then used to ensure the maintenance of the required end-use quality.

18.1.1 Load-Bearing Characteristics of Polymers

An increasing number of applications are being developed for thermoplastics in which a fabricated article is subjected to a prolonged continuous stress. Typical examples are pipes, crates, cold water tanks, and engine cooling fans. Under such conditions of constant stress, materials exhibit, to varying extents, a continuous deformation with increasing time. This phenomenon is termed 'creep'.

A wide variety of materials will, under appropriate conditions of stress and temperature, exhibit a characteristic type of creep behaviour as shown in **Figure 18.1**.

The general form of this creep curve can be described as follows. Upon application of the load an instantaneous elastic deformation occurs (O–A). This is then followed by an increase in deformation with time represented by the portion of the curve A–D. This is the generally accepted classic creep curve, and is usually considered to be divided into three parts:

1. A–B: the primary stage in which creep rate decreases linearly with time.

2. B–C: the secondary stage where the change in dimensions with time is constant, i.e., a constant creep rate.

3. C–D: the tertiary stage where the creep rate increases again until rupture occurs.

With thermoplastics the secondary stage is often only a point of inflection, and the final tertiary stage is usually accompanied by crazing or cracking of the specimen, or, at higher stresses, by the onset of necking (i.e., a marked local reduction in cross-sectional area). In assessing the practical suitability of plastics, we are interested in the earlier portions of the curve, prior to the onset of the tertiary stage, as well as in the rupture behaviour itself.

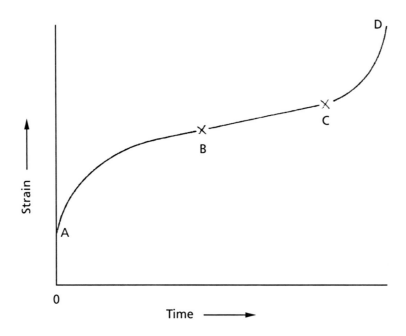

Figure 18.1 Creep behaviour of polyethylene. (*Source: Author's own files*)

In order to minimise as far as possible the influence of processing variables, studies have been carried out using tensile creep tests on carefully prepared compression moulded specimens. It must be realised that, with injection moulded articles, the creep properties will also be subject to variation with the amount and direction of residual flow orientation, while, with crystalline polymers such as the polyolefins, the creep effects will also be influenced by variations in density caused by a combination of flow orientation, compressive packing, and cooling effects. Stresses will generally be complex and will often involve compressive and flexural components. However, articles should normally be designed to limit the strains occurring to quite low levels, where a reasonable correlation can be expected between tensile, compressive, and flexural creep data.

18.1.1.1 Basic Creep Data

The first step in building up a comprehensive picture of creep behaviour is generally to obtain creep curves (elongation *versus* time) at a series of stress levels, each at a series of test temperatures. It is common practice to plot strain on a linear scale against time on a logarithmic scale. Each curve should preferably cover several decades on the logarithmic time scale so that any subsequent extrapolations will have a firm foundation.

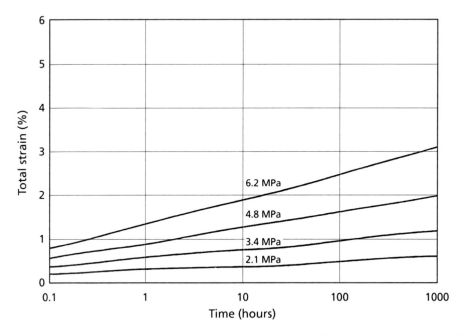

Figure 18.2 Creep of high-density polyethylene at 23 °C. (*Source: Author's own files*)

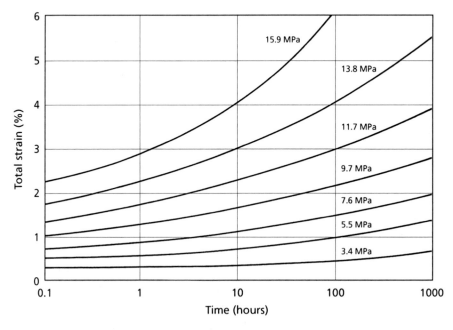

Figure 18.3 Creep of polypropylene homopolymer at 23 °C. (*Source: Author's own files*)

As the values of the stress, the temperature, and the time of loading vary with differing applications, the type of information extracted from these basic creep data will also vary with the application. There are several alternative methods, some of which are discussed later, of expressing this derived information but whichever form is used, whether tabular or graphical, it is unlikely to cover all eventualities. It is thus considered of most value to give the actual creep curves and to discuss by means of examples the methods by which particular data can be extracted from these basic creep curves.

Typical creep curves for polyethylene (PE) and polypropylene (PP) at 23 °C are shown in **Figures 18.2 and 18.3**. Lin and co-workers [1] have discussed creep phenomena accuracy in reinforced polyamide (PA) and polycarbonate (PC) composites. The phenomenon of the increasing dynamic creep property and the temperature under tension–tension fatigue loading are compared between semi-crystalline and amorphous composites.

18.1.1.2 Isochronous Stress–Strain Curves

Where it is necessary to compare several different materials, basic creep curves alone are not completely satisfactory. This is particularly so where the stress levels used are not the same for each material. If the stress endurance time relevant to a particular application can be agreed, a much simpler comparison of materials for a specific application can be made by means of isochronous stress–strain curves.

As an example, let us consider a milk crate in which we can assume that the moulding will not be continuously loaded for times greater than 100 hours. From the basic creep curves, the strain values at the 100 hour point for various stresses can easily be determined. For each material, a stress–strain curve can now be drawn corresponding to the selected loading time of 100 hours. Now let us suppose that for this application it is further stipulated that after 100 hours' continuous loading the strain shall not exceed 1%. The stress that can be sustained without violating the strain-endurance stipulation can then be conveniently read off for each material from the isochronous stress–strain curves, thus indicating with reasonable confidence the stress that can be used for design purposes for each of the materials considered. Examples of these interpolations are demonstrated in **Figure 18.4**.

Hay and co-workers [2] developed an approximate method for the theoretical treatment of pressure and viscous heating effects on the flow of a power-law fluid through a slit die. The flow was assumed to remain one-dimensional and the accuracy of this approximation was checked via finite element simulations of the complete momentum and energy equations. For pressures typically achieved in the laboratory, it was seen that the one-dimensional approximation compared well with the simulations. The model therefore offered a method

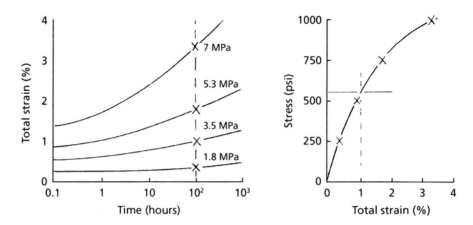

Figure 18.4 Isochronous stress/strain curve of polyethylene. (*Source: Author's own files*)

of including pressure and viscous heating effects in the analysis of experiments and was used to rationalise experimentally pressure profiles obtained for the flow of polymer melts through a slit die. Data for the flow of a low-density polyethylene (LDPE) and a polystyrene (PS) melt in a slit die showed that these two effects were significant under normal laboratory conditions. The shear stress–strain rate curves would thus be affected to the point of being inaccurate at high shear rates. In addition, it was found that the typical technique to correct for a pressure-dependent viscosity was also inaccurate, being affected by the viscous heating and heat transfer from the melt to the die.

18.1.1.3 Stress–Time Curves

With thermoplastic materials, above a certain stress the stress–strain curve shows a marked departure from linearity, i.e., above such values further increases in stress lead to disproportionately greater increases in elongation. By the study of the isochronous stress–strain curves of a material it is possible to decide upon a certain strain that should not be exceeded in a given application. The stress required to produce this critical strain will naturally vary with the time of application of the load, so that the longer the time of application, the lower will be the permissible stress.

From isochronous stress–strain curves relating to endurance times of, say, 1, 10, 100, and 1000 hours, and so on, the magnitude of the stress to give the critical strain at each duration of loading can be easily deduced. This procedure can be repeated for different selected levels of strain. In general, the more critical the application and the longer the time factor involved, the lower will be the maximum permissible strain. With the polyolefin family of materials, the upper limit of critical strain will always be governed by the onset of brittle-type rupture, e.g., hair cracking at elongations that are low compared with those expected from the short-term ductility of the material.

With thermoplastics of intermediate modulus (e.g., high-density polyethylene (HDPE) and PP) the stress–strain curves depart slightly from linearity at quite low strains. Thus, if accurate results are to be obtained from standard formulae, it is sometimes necessary to limit the critical strain and the corresponding design stress to low values. One commonly selected criterion with high-modulus thermoplastics (e.g., polyacetal) is to base the design stress and the design modulus on that point on the stress–strain curve at which the secant modulus falls to 85% of the initial tangent modulus. This procedure is illustrated in **Figure 18.5**.

It would be wrong, however, in many applications to limit the critical strain to a low value if the ductility of the material is to be fully utilised. Extreme examples are those such as pipes where rupture behaviour is the controlling factor.

The above type of data can be conveniently summarised by the presentation of stress–time curves corresponding to different levels of permissible strain. Such curves enable the time dependency of different materials to be conveniently compared. Where possible, rupture data should also be included on such plots so that at longer times an adequate safety margin over the rupture strength is always maintained.

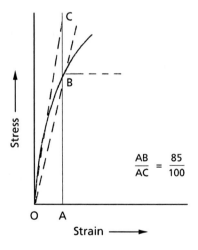

Figure 18.5 Stress-time curve. (*Source: Author's own files*)

As with normal creep curves the time scale will normally be logarithmic. For extrapolation purposes, however, a linear relationship can be obtained by also using a logarithmic scale for the stress axis. In fact, if rupture data are to be included, a double logarithmic plot is preferred.

18.1.1.4 Stress–Temperature Curves

The methods discussed so far have been concerned with the presentation of combined creep data obtained at a single temperature. Thus, a further method is required to indicate the influence of temperature.

One convenient method is to combine the information available from either isochronous stress–strain curves or stress–time curves obtained on the same materials at different temperatures. For example, suppose the performance criterion for a particular application is that the total strain should not exceed 2% in 1000 hours. Using the 1000 hour isochronous stress–strain curve for each temperature, and erecting an ordinate at the 2% point on the strain axis, the individual working stresses for each temperature can be obtained. Alternatively, by erecting an ordinate at the 1000 hour point on the stress–time curve for 2% strain for each temperature investigated, the individual working stresses can be similarly obtained. From these interpolated results the stress–temperature curve can be drawn.

By similar procedures, stress–temperature plots for other specific 'time-permissible strain' combinations can be obtained.

18.1.1.5 Extrapolation Techniques

Normal creep curves in which percentage total strain (or total strain) on a linear scale is plotted against time on a logarithmic scale show an increasing curvature with time, particularly at higher stress levels, Due to this curvature it is not possible to extrapolate curves to longer times with certainty. If, however, a double logarithmic plot is used for a series of stresses at a given temperature, a family of parallel straight lines is obtained. These can then be extrapolated quite easily. With PE and PP the slope of such curves remains constant or tends to decrease at very long times. Furthermore, an increase in temperature also leads to a slight reduction in the slope of the family of curves for a given material. A straightforward linear extrapolation of the log (strain)/log (time) creep curves may thus involve an additional safety factor.

As has already been mentioned, stress–time curves on a double logarithmic scale are also linear and can be easily extrapolated. In both cases, however, over-extrapolation should be avoided (i.e., one decade and preferably not more than two).

18.1.1.6 Basic Parameters

Within a given family of materials of the same basic polymer a further condensation of data can be achieved if the principal polymer parameters influencing creep can be separated. For example, with PE, density and melt index would be obvious possibilities.

If such factors can be found then all the results can be combined in suitable trend curves and the results for any intermediate grade interpolated. If good correlations are obtained the reliability in the individual results is automatically increased and one can feel confident of predicting the creep behaviour of any material within a particular family.

18.1.1.7 Recovery in Stress Phenomena

In some applications, the load, instead of being constantly applied, will only be applied intermittently for limited periods. In the period between consecutive loadings the part will thus have an opportunity to recover. The extent to which this occurs will depend upon the ratio of the loaded time to the unloaded time. The recovery will be approximately exponential, and thus to achieve virtually complete recovery, the time without load must be substantially longer than the time under load. The residual strain for a fixed ratio of loaded to unloaded time will also depend upon the magnitude of the strain experienced at the time of removal of the load.

With intermittent loading for relatively long time periods, higher permissible stresses will be possible than with continuous loading. With repeated loading over short time periods, however, fatigue effects could considerably reduce the value of the permissible stress for a long total endurance.

18.1.1.8 Stress Relaxation

In many applications (e.g., those involving an interference fit, such as pipe couplings, closures, and so on) we are concerned more with the decrease in stress with time at constant strain (stress relaxation), rather than with the increase in strain with time under constant stress (creep).

To a first approximation, such data can be obtained from the basic family of creep curves by sectioning through them at the relevant strain value parallel to the time axis. That is to say, similar procedures are used as for obtaining stress–time curves.

18.1.1.9 Rupture Data

In applications in which strains up to the order of a few percent do not interfere with the serviceability of the component, the controlling factor as far as permissible stress is concerned will probably be the rupture behaviour of the material. It must be realised that the rupture curve forms the limiting envelope of any family of creep curves. Care must therefore be exercised in extrapolating creep data to ensure that for long-term applications a suitable safety factor is always allowed over the corresponding extrapolated rupture stress. In general, with thermoplastics an increase in MW leads to improved rupture resistance, which is not necessarily reflected by an improvement in creep resistance, and in fact may be associated with a decrease in creep resistance. An increase in MW will also generally be associated with a decrease in processability which will tend to give higher residual stresses in fabricated articles which can in turn influence rupture properties. In designing injection moulded articles particular care should always be taken to ensure that the principal stresses experienced by the article in service are, wherever possible, parallel to the direction of flow of material in the mould. Avoidance of stress concentration effects, by appropriate design, will of course tend to minimise the interference of rupture phenomena, and allow greater freedom to design on the basis of creep modulus.

It is known that certain environments reduce rupture performance with particular materials, e.g., PE in contact with detergents, and due allowance must be made for this, where necessary, by the incorporation of an additional safety factor to the rupture stress. In general the resistance to rupture will be greater when the stress is applied in compression than when in tension.

As differing stress levels have been used for HDPE and LDPE, for the reasons previously discussed, an easier comparison between grades can be made by the use of isochronous stress–strain curves. These are shown in **Figures 18.6–18.9** for times of 1, 10, 100, and 1000 hours. The data have been further combined in **Figure 18.10** which demonstrates the marked but systematic influence of density on the creep behaviour of PE. The curves in **Figure 18.10** are relevant to a total strain of 1% but similar plots for other permissible strains can easily be derived from the isochronous stress–strain curves. The linear relationship between creep and density for PE at room temperature, irrespective of melt index over the range investigated (i.e., 0.2 to 5.5), has enabled the stress–time curves of **Figure 18.11** to be interpolated for the complete range of PE and for a range of PP (**Figure 18.12**). In this case the data have been based on a permissible strain of 2% but again data for other permissible strains can be similarly interpolated from the creep curves as previously explained. The effect of temperature on the creep properties of PE is summarised in **Figure 18.13**.

It should be again emphasised that for PE the principal factor controlling creep is density but the principal factor influencing rupture behaviour is melt index. Thus, for a given density, for long-term applications, the higher the melt index the lower will be the maximum permissible strain. This will be particularly important at elevated temperatures.

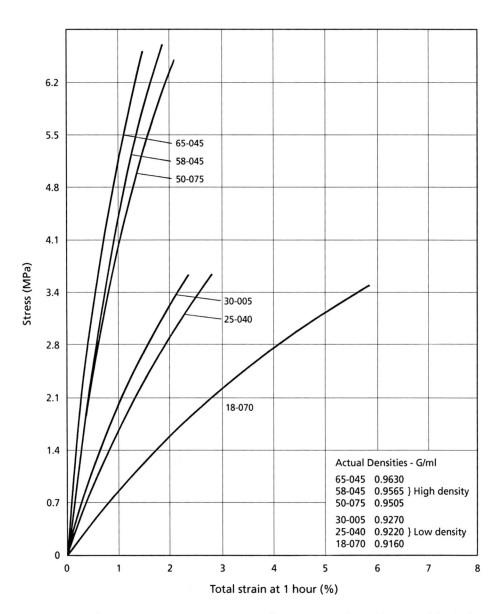

Figure 18.6 Isochronous stress-strain curves of various grades of low- and high-density polyethylene at 23 °C (1 hour data). (*Source: Author's own files*)

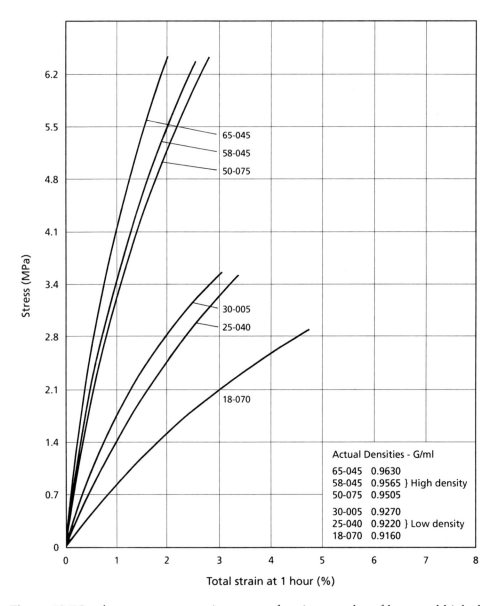

Figure 18.7 Isochronous stress-strain curves of various grades of low- and high-density polyethylene at 23 °C (10 hour data). (*Source: Author's own files*)

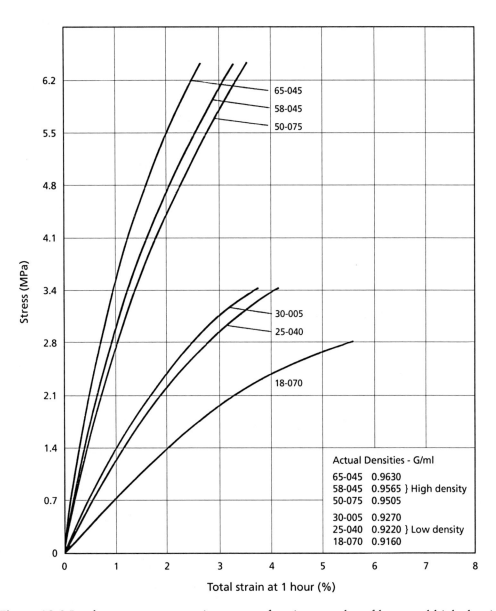

Figure 18.8 Isochronous stress-strain curves of various grades of low- and high-density polyethylene at 23 °C (100 hour data). (*Source: Author's own files*)

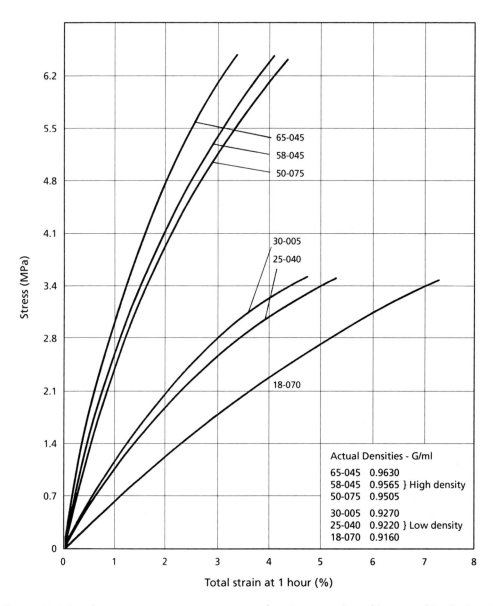

Figure 18.9 Isochronous stress-strain curves of various grades of low- and high-density polyethylene at 23 °C (1000 hour data). (*Source: Author's own files*)

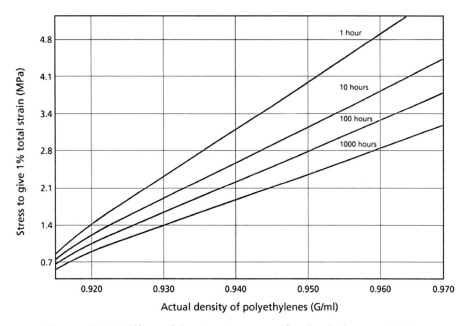

Figure 18.10 Effect of density on creep of polyethylene at 23 °C.
(*Source: Author's own files*)

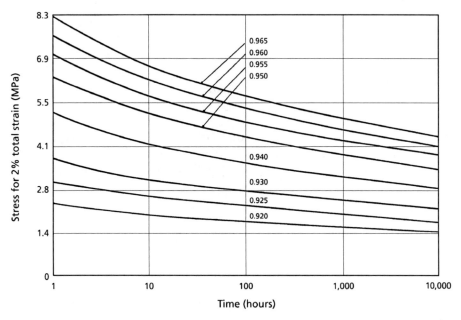

Figure 18.11 Interpolated stress-time curves of polyethylene of various densities
(2% total strain). (*Source: Author's own files*)

531

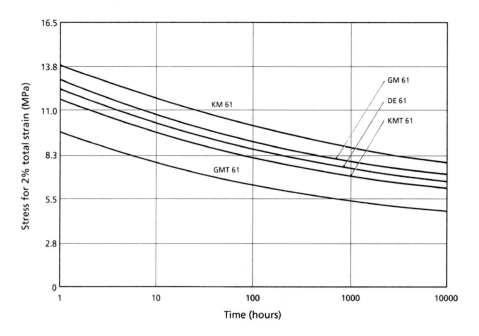

Figure 18.12 Interpolated stress-time curves of various grades of polypropylene at 23 °C (2% total strain). KMT 61 and GMT 61 = polypropylene copolymers. KM 61, GM 61 and DE 61 = polypropylene homopolymer. (*Source: Author's own files*)

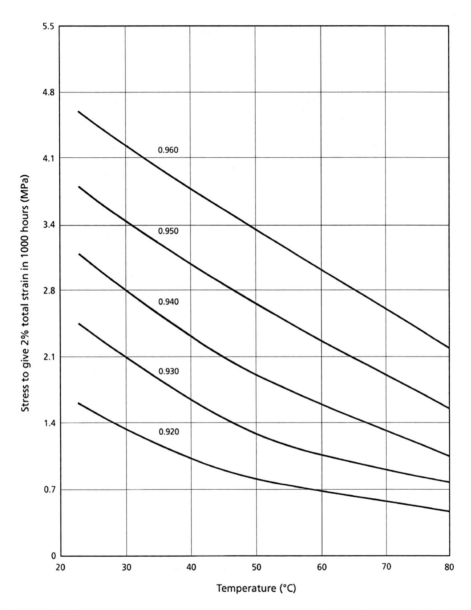

Figure 18.13 Interpolated stress-temperature curves of polyethylenes of different densities (2% total strain at 1000 hours). (*Source: Author's own files*)

The effect of temperature on the creep of PP is also conveniently summarised in **Figures 18.14 and 18.15**. Although these curves refer specifically to 0.5 and 2% total strain in 1000 hours, similar interpolations have shown that to a first approximation the fall off in stress with temperature is similar for other strain–time combinations.

In contrast to PE, an excellent correlation has been found between the creep and melt index of PP. An increase in melt index is associated with an increase in creep resistance. This has been found for both homopolymers and copolymers and is illustrated in **Figure 18.16**. For similar melt indices the inferior creep resistance of the copolymers is also indicated, this difference being more marked for grades of lower melt index.

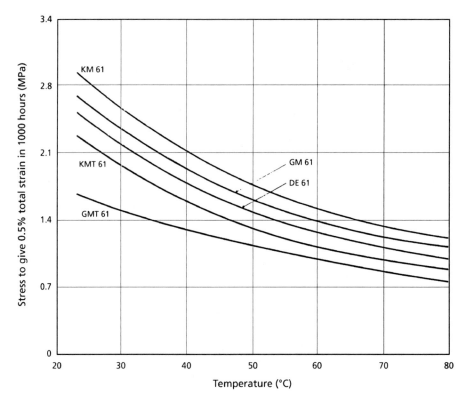

Figure 18.14 Interpolated stress-temperature curves for various grades of polypropylene (0.5% total strain at 1000 hours). KMT 61 and GMT 61 = polypropylene copolymers, KM 61, GM 61 and DE 61 = polypropylene copolymers. (*Source: Author's own files*)

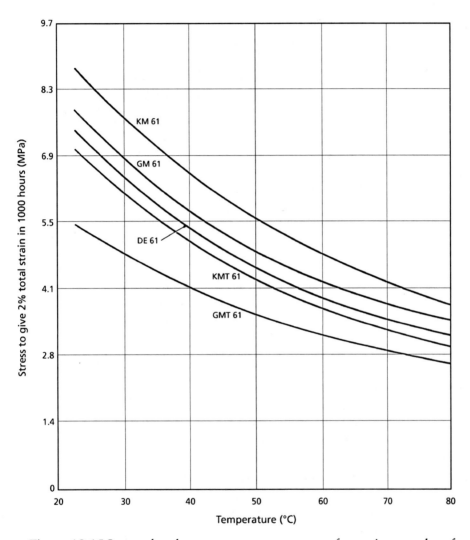

Figure 18.15 Interpolated stress-temperature curves for various grades of polypropylene (2% total strain at 1000 hours). KMT 61 and GMT 61 = polypropylene copolymers. KM 61, GM 61 and DE 61 = polypropylene homopolymers. (*Source: Author's own files*)

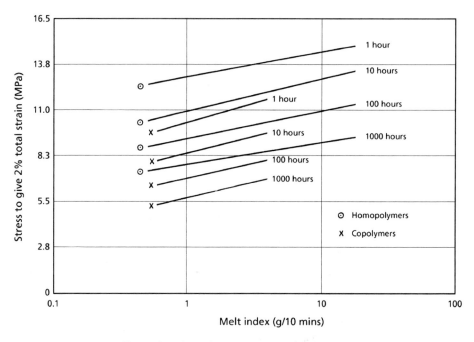

Figure 18.16 Effect of melt index on creep of polypropylenes at 23 °C.
(*Source: Author's own files*)

18.1.1.10 Long-Term Strain–Time Data

To illustrate the linearity of double logarithmic plots of percentage total strain against time, **Figure 18.17** shows a typical family of curves at a range of stress levels for a HDPE and **Figure 18.18** shows that for a PP copolymer. From the uniform pattern of behaviour, the linear extrapolation of the curves at lower stresses to longer times appears well justified. As emphasised previously, with the higher melt index materials at higher temperatures, over-extrapolation is dangerous because of the possible onset of rupture at comparatively low strains.

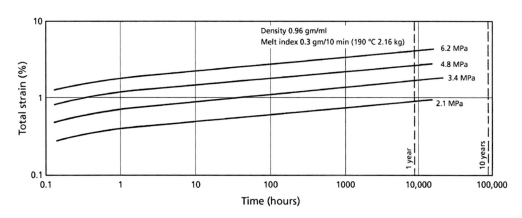

Figure 18.17 Long-term creep of high-density polyethylene at 23 °C.
(*Source: Author's own files*)

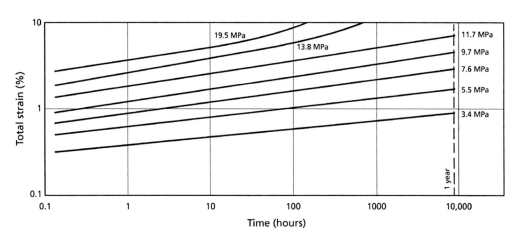

Figure 18.18 Long-term creep of high-density polypropylene copolymer at 23 °C.
(*Source: Author's own files*)

Figures 18.19, 18.20, 18.21 and **18.22** show creep curves, isochronous stress–strain curves, and stress–time curves, for a typical acrylonitrile–butadiene–styrene (ABS) terpolymer and a typical unplasticised polyvinyl chloride (PVC).

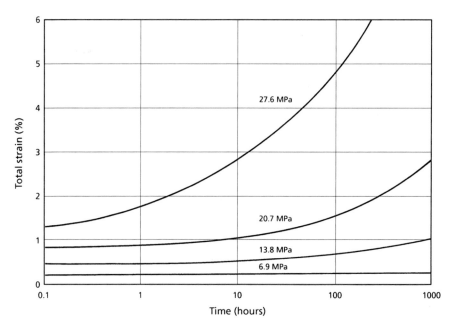

Figure 18.19 Creep of typical acrylonitrile butadiene styrene terpolymer (normal impact grade) at 23 °C. (*Source: Author's own files*)

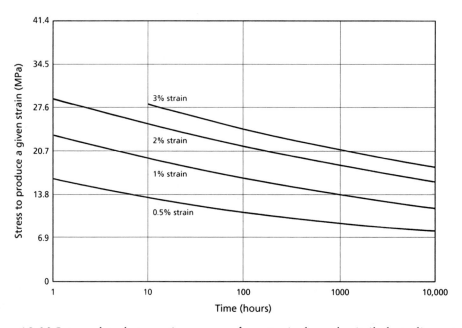

Figure 18.20 Interpolated stress-time curves for a typical acrylonitrile butadiene styrene terpolymer (normal impact grade) at 23 °C. (*Source: Author's own files*)

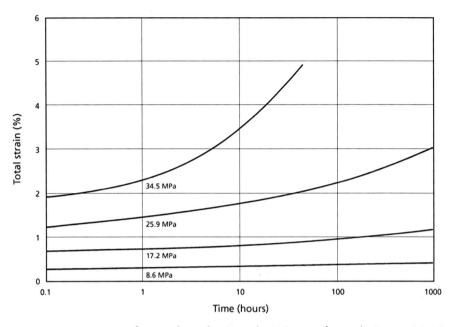

Figure 18.21 Creep of typical unplasticised PVC pipe formulation at 23 °C. (*Source: Author's own files*)

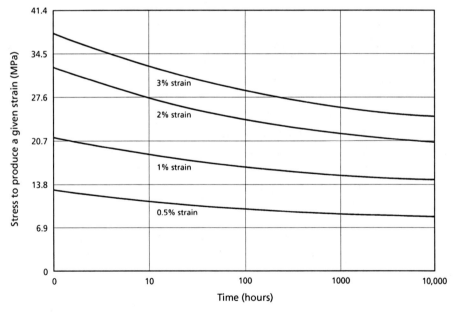

Figure 18.22 Interpolated stress-time curves for a typical unplasticised PVC pipe formulation at 23 °C. (*Source: Author's own files*)

18.1.2 Impact Strength Characteristics of Polymers

Of the many properties of a plastic material that influence its choice for a particular article or application, the ability to resist the inevitable sharp blows and drops met in day-to-day use is often one of the most important. The prime object of impact testing should be to give a reliable guide to practical impact performance. However, the performance requirements and the design and size of articles can vary considerably. The method of fabrication can also vary, and as all these factors can influence impact performance, a reasonably wide range of data are required if most of these eventualities are to be covered and materials and grades compared sensibly.

18.1.2.1 Izod Test

Of all the types of mechanical test data normally presented for grade characterisation and quality control purposes, the standard notched Izod test is probably the one most vigorously criticised, on the grounds that it may give quite misleading indications regarding impact behaviour in service. In most cases, however, the criticism should properly be levelled at too wide an interpretation of the results, particularly the tendency to overlook the highly important influence of the differences between the standard and operational conditions in terms of dimensions, notch effects, and processing strains. For example, in the Izod test, the specimens are of comparatively thick section (e.g., 6.25 cm × 1.25 cm × 1.25 cm or 6.25 cm × 1.25 cm × 0.625 cm) and are consequently substantially unoriented. Even with injection moulded specimens, where some orientation does exist, the specimens are usually tested in their strongest direction (with the crack propagating at a right angle to the orientation direction) which generally leads to high values. Furthermore, the specimens are notched with the specific intention of locating the point of fracture and ensuring that as wide a range of materials as possible fracture in a brittle manner in the test. It is thus reasonably safe to infer that if a material breaks in a ductile manner at the high impact energies involved in the Izod test, it will, in nearly all circumstances involving thick sections, behave in a ductile manner in practice. Because of the severe nature of the test, it would be wrong, however, to assume that because a material gives a brittle type fracture in this test it will necessarily fail in a brittle manner at low impact energies in service.

It should also be remembered that in cases where high residual orientation is present, additional weakening can result from the easier propagation of cracks in the direction parallel to the orientation.

A low Izod value does give a warning that care should be exercised in design to avoid sharp corners or similar points of stress concentration. It does not by itself, however, necessarily indicate a high notch sensitivity.

18.1.2.2 Falling Weight Impact Test

The normal day-to-day abuse experienced by an article is more closely simulated by the falling weight impact type of test (see BS 2782) [3]. It is also much easier to vary the type and thickness of specimens in this test. Furthermore any directional weaknesses existing in the plane of the specimen are easily detected.

One would normally expect the impact energy required for failure to increase with specimen thickness, but the rate of increase will not be the same for all materials. In addition to this, the effect of the fabricating process cannot be over-emphasised. It has already been mentioned that residual orientation can lead to a marked weakening and to a susceptibility to cracking parallel to the orientation direction. Such residual orientation is extremely likely to be brought about in the injection moulding process, where a hot plastic melt is forced at a high shear rate into a narrow, relatively cold, cavity. Naturally, the more viscous the melt and the longer and narrower the flow path, the greater the degree of residual orientation expected. Within a given family of grades the melt viscosity, besides being influenced by temperature and shear rates, will also be affected by polymer parameters such as MW and molecular weight distribution. Thus it cannot be assumed that the magnitude of this drop in impact strength will be the same for differing materials or for differing grades of the same material. For a satisfactory comparison it is, therefore, essential to know the variation in impact strength with thickness of both unoriented compression moulded samples and injection moulded samples moulded over a range of conditions.

In some investigations a flat, circular, centre-gated, 15 cm diameter dish mould, as shown in **Figure 18.23**, has been used. The dish is mounted in the falling weight impact tester in its inverted position and located so that the striker hits the base at a distance of 1.875 cm from the centre, as the area near the sprue is one of the recognised weak spots of an injection moulding. By means of inserts in the mould the thickness of the base of the moulding can be varied. For each material a family of curves is obtained, for variation of impact strength with thickness of compression moulded sheets moulded under standard conditions, and of injection moulded dishes produced at a series of temperatures. An estimate of the weakening effects of residual orientation and of the tolerance that can be allowed on injection moulding conditions for satisfactory impact performance is thus obtained.

A further factor that can adversely affect impact performance is often encountered in injection moulding. Whether a single gate or multiple gates are used, it is usually inevitable, due to a non-uniform flow pattern during the mould filling operation, that separate flow fronts meet to form a weld line. The strength at the weld line will vary with the material, the injection moulding conditions, and the length and thickness of the flow path, and can be considerably weaker than adjacent parts of the moulding. Weld effects have been studied by means of a flat 37.5 cm × 3.125 cm ruler mould as shown in **Figure 18.24** which can be gated at either one or both ends. The thickness of the specimens can again be varied by means of inserts. In this way a series of comparisons can be made between

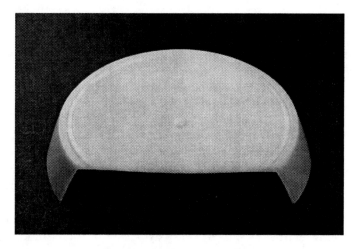

Figure 18.23 Injection moulded dish used in falling weight strength measurement. (*Source: Author's own files*)

Figure 18.24 Injection moulded ruler used in study of weld effects (length 37.5 cm). (*Source: Author's own files*)

the falling weight impact strength of welded and non-welded specimens at the same distance from the gate for varying path length to cavity thickness ratios and differing moulding conditions.

- Importance of the Type of Failure in the Falling Weight Test

Besides the average level at which failures occur in the falling weight impact test, the type of failure can also be of importance in judging the relative impact performance of a material in practice. There are generally three types of failure:

1. A *ductile* or *tough* failure in which the material yields and flows at the point of impact, producing a hemispherical depression, which at sufficiently high impact energies eventually tears through the complete thickness.

2. A *brittle* failure in which the specimen shatters or cracks through its complete thickness with no visible signs of any yielding having taken place prior to the initiation of the fracture. (With specimens possessing a high degree of residual orientation a single crack parallel to the orientation direction is obtained.)

3. An intermediate or '*bructile*' failure in which some yielding or cold flow of the specimen occurs at the point of impact prior to the initiation and propagation of a brittle-type crack (or cracks) through the complete thickness of the specimen.

These different types of failure are shown in **Figure 18.25**.

An idea of the impact energy associated with these types of failure can be ascertained by consideration of a typical load–deformation curve. This in general will be of the form shown in **Figure 18.26**. From the definitions given previously a brittle-type failure will occur on the initial, essentially linear part of the curve, prior to the yield point. A

Figure 18.25 Examples of types of failure occurring in the falling weight impact strength test. (*Source: Author's own files*)

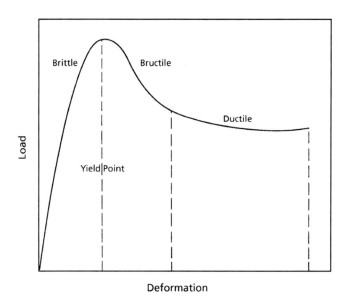

Figure 18.26 Typical load deformation curve. (*Source: Author's own files*)

'bructile' failure will occur on that part of the curve near to but following the yield point. A ductile failure will be associated with high elongations on the final part of the curve which follows the yield point. With the last type of failure, the deformation at failure will be reasonably constant for a striker of given diameter. As the area beneath the curve up to failure gives a measure of the impact energy to cause failure, the conclusion can be drawn that, in general, for a given thickness, the tough-type failure is associated with a high and reasonably constant impact energy. The 'bructile' failure is also associated with a relatively high but more variable impact energy. Brittle failures are generally associated with a low impact level.

If in a given test, failures are all of the tough type, a reliable and reproducible measure of the impact strength of the sample can be obtained. If, however, differing types of failures are observed a much wider variation in impact values can be obtained from repeat tests.

However, the difference in energy level required to produce tough- and brittle-type failure varies, for differing materials. As in practice the lowest level at which a failure is likely to occur is as important as the estimated level at which 50% of the samples are likely to fail, it is necessary to have a measure of this variability. The F_{50} or 50% failure level is the value normally reported in a falling weight impact test. This can be obtained by repeating the falling weight test using the 'Probit' rather than the 'Staircase' procedure. In the former, sets of samples are tested at each of a series of increasing energy levels

544

between the limits at which, none or all the specimens in a given set fail. If percentage failure is now plotted against impact energy on probability paper, the slope of the curve gives an estimate of the variability and the likelihood of occasional samples failing at low impact levels.

A final important factor that requires consideration for many applications is the influence of temperature, as it is well known that some thermoplastics undergo a tough–brittle transition, with a correspondingly marked drop in impact performance, just below ambient temperature. Changes in notch sensitivity with temperature have been ascertained by the use of notched bar tests, as previously described, and changes in impact strength with temperature by both the falling weight test and the standard Izod test. Comparatively thick compression and/or injection moulded specimens have been used in these investigations to ensure that the starting value of the impact strength at room temperature is high and any change in this value is easily detected.

18.1.2.3 Notch Sensitivity

The sharpness of a notch is of course well known to have a strong influence on the impact performance of notched specimens. The extent of this effect is found to vary considerably from polymer to polymer. To investigate this point, impact values obtained with the BS standard notch of 0.10 cm tip radius have been compared with those obtained with other tip radii, namely 0.025 (the ASTM standard), 0.050, and 0.20 cm. The results for notch sensitivity are conveniently expressed either by plotting the impact value as a ratio of the BS value against the notch tip radius or alternatively plotting the actual Izod value on a logarithmic scale against notch tip radius on a linear scale. Obviously, the steeper the slope of the curve, the greater the notch sensitivity of the material and the more essential it is to remove sharp radii and points of stress concentration in design if optimum performance is to be obtained.

If the foregoing limitations are remembered, the results of notched bar tests give useful information, but for realistic comparisons of materials additional impact data are required.

Reference was made previously as to how a low Izod value gives a warning regarding the care that should be exercised in design to remove points of stress concentration if optimum impact behaviour is to be obtained with a component in practice. It has also been mentioned that a more reliable measure of notch sensitivity can be obtained by carrying out tests using a series of notch radii rather than a single standard notch. **Figure 18.27** shows the relative notch sensitivities of some polyolefins and a few other well-known thermoplastics. It can be seen that polyolefins are of intermediate

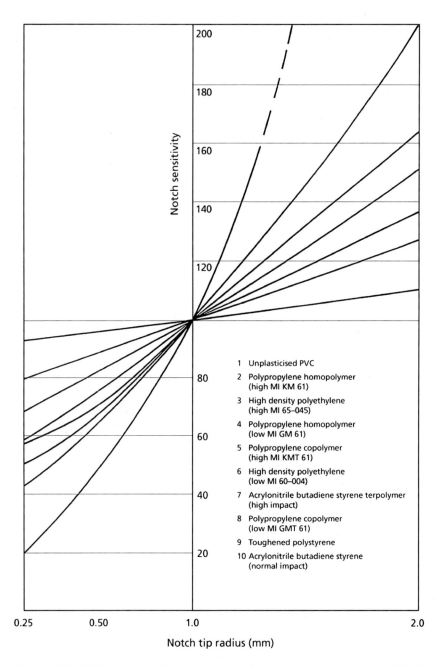

Figure 18.27 Typical notch sensitivites of some thermoplastics at 23 °C
(*Source: Author's own files*)

notch sensitivity, this being more marked with the higher melt index grades. Among the polyolefins, the low melt index PP copolymer designated GMT 61 has the lowest notch sensitivity and the higher melt index PE and PP homopolymers (e.g., 65-045 and KM 61) the highest.

Figure 18.28 shows the variation in standard Izod impact strength (0.10 cm notch tip radius) with temperature for the same range of materials. By far the most temperature sensitive is the low melt index PP copolymer, and it is interesting to compare the sharp fall off in notched impact strength with this material between 23 and 0 °C with the steady values obtained for falling weight impact strength for the same temperature range (**Figure 18.29**). The removal of points of stress concentration is thus the most vital factor governing the impact performance of this grade at temperatures of 0 °C and below. **Figure 18.30(a)** shows in more detail the notched impact behaviour of PP at room temperature. A semi-logarithmic plot has been used so that both the overall level and the notch sensitivity can be determined. The marked influence of melt index and the superiority of the copolymers over the homopolymers are clearly demonstrated.

Figures 18.30(b) and (c) show similar data for temperatures of 0 and –20 °C, respectively. With the exception of the low melt index PP copolymer, the notch sensitivities are similar to those at room temperature but the marked drop in overall level brought about by a decrease in temperature is again clearly demonstrated.

Figures 18.31 and 18.32 show the combined data for room temperature and –20 °C for HDPE, and for ABS terpolymer and toughened PS, respectively. Of particular note is the very small effect of temperature on both the impact level and notch sensitivity of HDPE, although the effect of MW (or melt index) is again clearly demonstrated. This indicates the suitability of HDPE for low-temperature applications.

One important practical example of the controlling influence of notch sensitivity in governing impact behaviour is that of embossed surfaces. **Table 18.1** gives the results of falling weight impact tests in which a series of different textured finishes has been investigated for a range of PP. The injection moulded disc specimens possessed one textured and one plain, smooth surface. When the embossed surfaces were struck high impact values were obtained. However, when the smooth surface was struck, so that the textured surface was placed in tension, very low impact values were obtained, the magnitude of the drop depending upon the depth and sharpness of the embossing. The sharper the pattern the lower the value obtained. Fortunately in practice it will generally be the textured surface that is struck but where the reverse applies considerable care should be exercised in the choice of the finish.

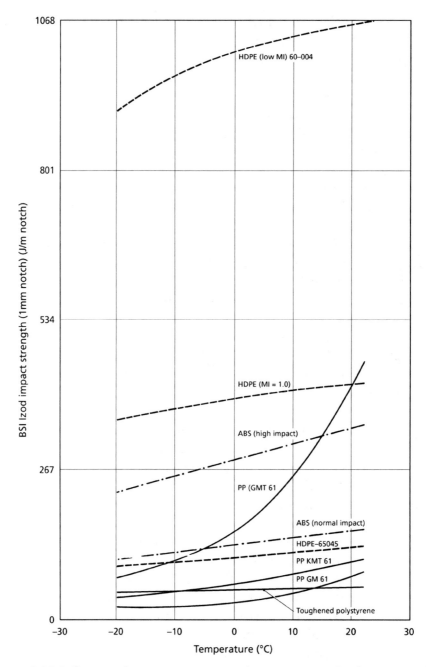

Figure 18.28 Influence of temperature on Izod impact strength of various polymers. HDPE = high-density polyethylene low melt index and high melt index. ABS = acrylonitrile-butadiene-styrene terpolymer, GMT 61 and KMT 61 = polypropylene copolymers, GM 61 = polypropylene homopolymer. (*Source: Author's own files*)

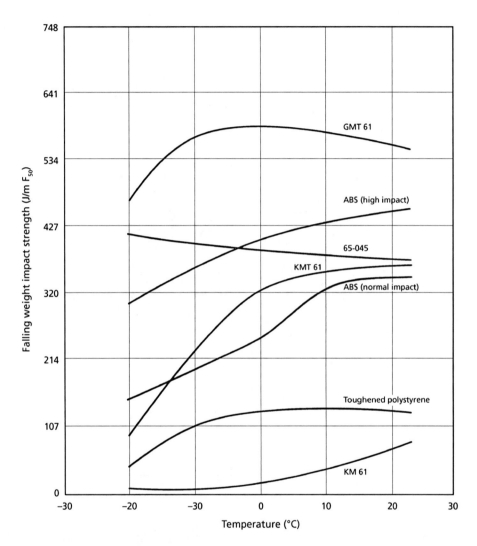

Figure 18.29 Effect of temperature on falling weight impact strength of 0.060 inch compression moulded specimens of various polymers. GMT 61 and KMT 61 = polypropylene copolymer. ABS = acrylonitrile-butadiene-styrene terpolymer, KM 61 = polypropylene homopolymer. (*Source: Author's own files*)

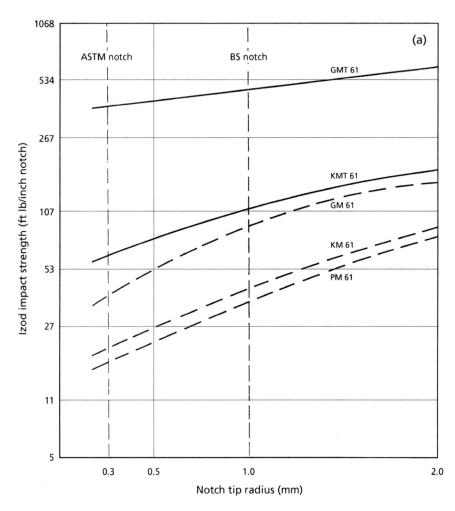

Figure 18.30 Variation of Izod impact strengths of various grades of polypropylenes with notch tip radius (a) at 23 °C, (b) at 0 °C and (c) at –20 °C. KMT 61 and GMT 61 = polypropylene copolymer, KM 61, PM 61 and GM 61 = polypropylene homopolymer. (*Source: Author's own files*)

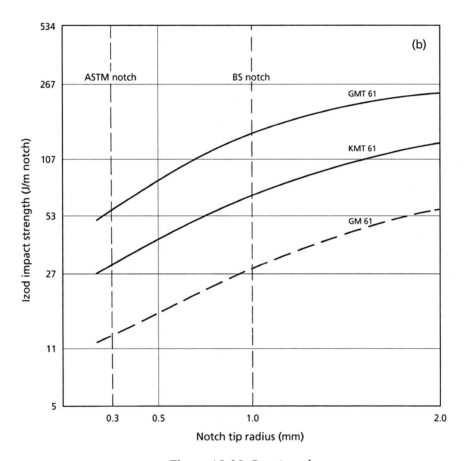

Figure 18.30 *Continued*

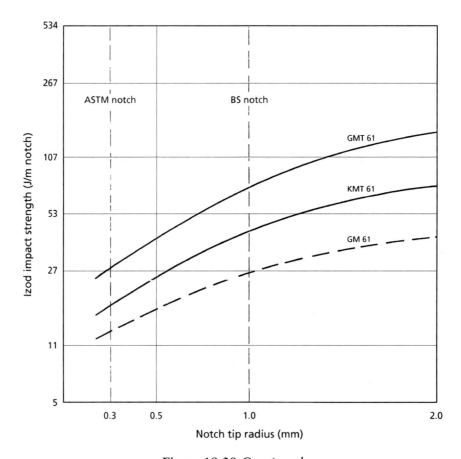

Figure 18.30 *Continued*

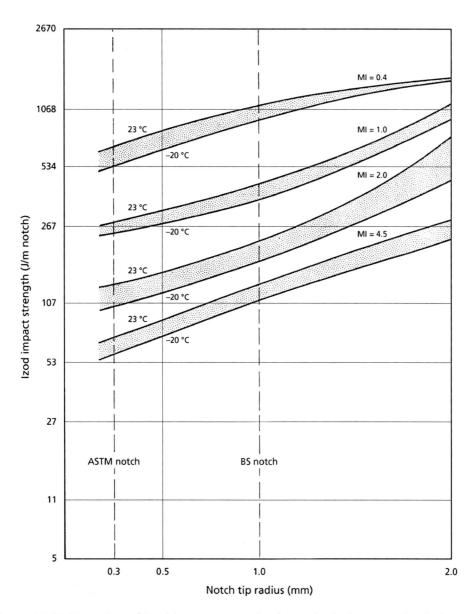

Figure 18.31 Variation of Izod impact strength of 0.96 high-density polyethylenes with notch tip radius at 23 °C and –20 °C. (*Source: Author's own files*)

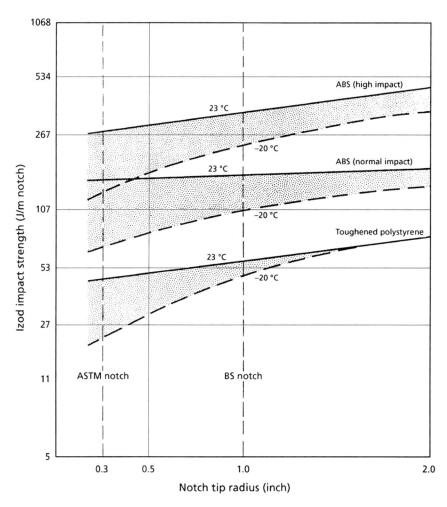

Figure 18.32 Variation of Izod impact strength of ABS terpolymer and toughened PS with notch tip radius and temperature. (*Source: Author's own files*)

Table 18.1 Effect of embossing on falling weight impact strength at 23 °C

Polypropylene grade	Surface towards striker	Falling weight impact strength at 23 °C - J (F_{50})					Specimen thickness (mm)
		Surface 1 Smooth grain (Mould Decoration Incorporated)	Surface 2 Rough finish (Martins)	Surface 3 Rough finish (Dornbush)	Surface 4 Smooth grain (Mould Decoration Incorporated)	Surface 5 Medium finish (Dornbush)	
PM61	Smooth	0.4	0.4	0.4	0.5	0.4	2.7
	Embossed	2.2	4.5	1.9	3.8	2.0	2.7
GM61	Smooth	0.5	0.4	0.4	0.5	0.4	2.8
	Embossed	11.7	12.5	1.9	21.7	10.7	2.8
KMT61	Smooth	1.5	0.8	0.9	4.6	1.9	2.7
	Embossed	24.4	20.3	13.6	>24.4	>24.4	2.7

Source: Author's own files

18.1.2.4 Falling Weight Impact Tests: Further Discussion

Figure 18.33 gives the variation in falling weight impact strength with thickness, at 23 °C, of a range of materials in the form of substantially stress-free compression moulded samples. However, in practice, fabricated articles are seldom stress-free, this being particularly true with injection moulded articles. Any residual orientation can result in a marked reduction in impact strength and this is well illustrated in **Figures 18.34–18.40** covering high and low melt index PP (**Figures 18.34 and 18.35**), PP copolymer (**Figure 18.36**), HDPE (**Figure 18.37**), normal-impact and high-impact ABS terpolymers (**Figures 18.38 and 18.39**), and toughened PS (i.e., styrene–butadiene copolymer, **Figure 18.40**).

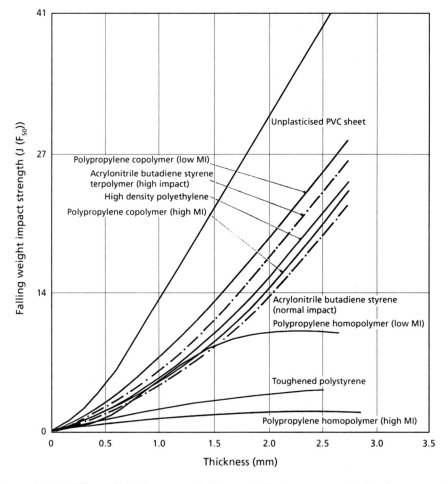

FIgure 18.33 Effect of thickness on falling weight impact strength of some typical thermoplastics at 23 °C (compression moulded samples). (*Source: Author's own files*)

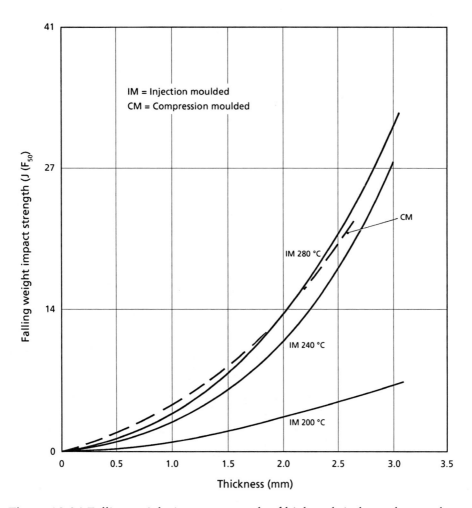

Figure 18.34 Falling weight impact strength of high melt index polypropylene copolymer at 23 °C. (*Source: Author's own files*)

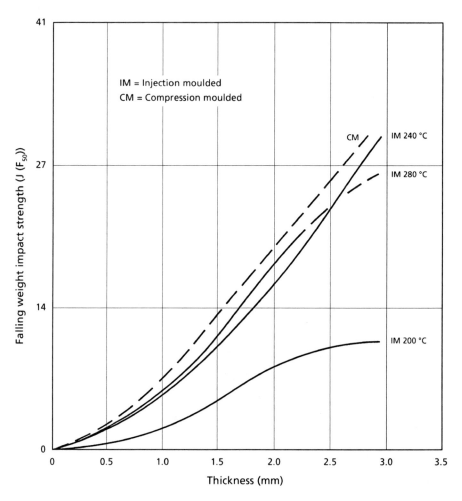

Figure 18.35 Falling weight impact strength of low melt index polypropylene copolymer at 23 °C. (*Source: Author's own files*)

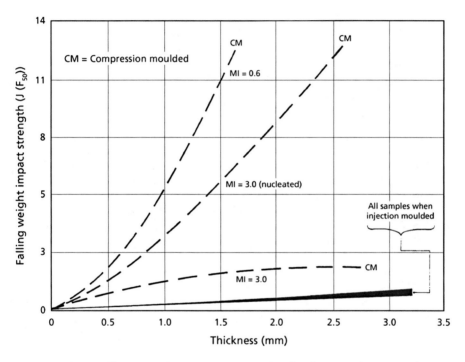

Figure 18.36 Falling weight impact strength of polypropylene copolymers (comparison of moulded and injection moulded specimens at 23 °C). (*Source: Author's own files*)

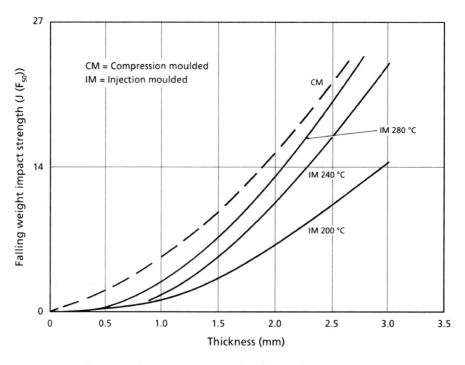

Figure 18.37 Falling weight impact strength of high density polyethylene at 23 °C).
(*Source: Author's own files*)

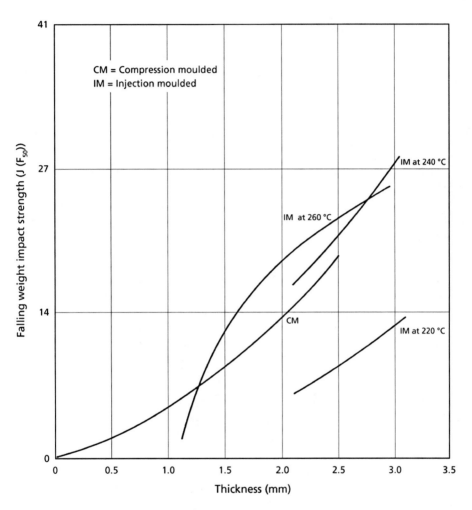

Figure 18.38 Falling weight impact strength of ABS terpolymer (normal impact grade) at 23 °C. (*Source: Author's own files*)

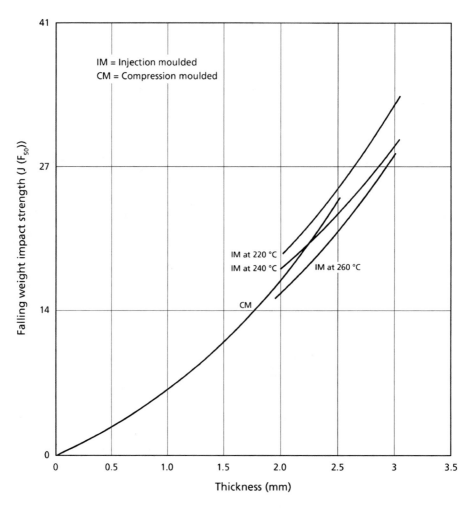

Figure 18.39 Falling weight impact strength of ABS terpolymer (high impact grade) at 23 °C. (*Source: Author's own files*)

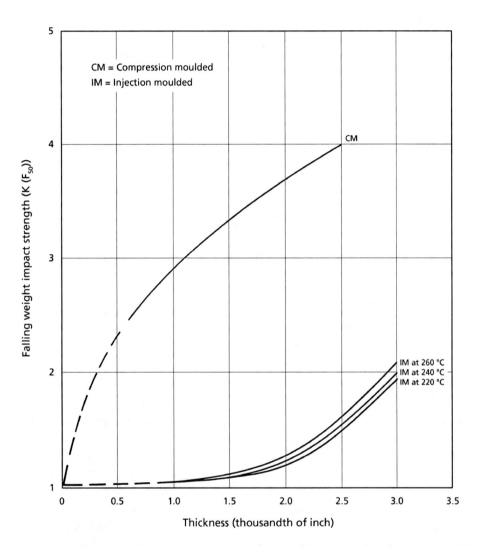

Figure 18.40 Falling weight impact strength of typical toughened polystyrene at 23 °C. (*Source: Author's own files*)

The change in impact strength with injection moulding temperature gives an indication of the process adaptability of each material. With the high-impact ABS terpolymers there is little difference in the level of falling weight impact strength over the range of injection moulding temperatures. With all the materials except the PP homopolymers and the toughened PS, the impact strength of injection moulded samples produced at the higher temperatures is similar to that of the virtually non-oriented compression moulded samples. With the toughened PS the impact strength of injection moulded samples is markedly lower than that of compression moulded samples, although little effect of injection moulding temperature has been observed.

The marked drop in impact strength that can result from weld lines is demonstrated in **Figure 18.41** for a PP copolymer. At the lower thicknesses, where the flow path to cavity thickness ratios are correspondingly higher, the impact strength measured at the weld is very much lower than that measured with a similar non-welded specimen. The positioning of gates to minimise weld effects is thus of vital importance in obtaining optimum behaviour in an injection moulded article of a given design.

Figures 18.29 and 18.42 show the effect of temperature on the falling weight impact strength for compression moulded and injection moulded specimens, respectively, of a range of materials. The reduction in impact strength with decrease in temperature is particularly marked with the injection moulded samples of the two otherwise higher impact resistant types, PP copolymers and ABS terpolymers. The superior impact strength at low temperatures of injection moulded HDPE is clearly illustrated.

It has been previously emphasised that with regard to end-use performance, the minimum level at which failures are likely to occur is of equal importance to the usually quoted 50% failure level. This is particularly true where different types of failure are obtained in the falling weight impact test. Such variations in the type of failure frequently occur as the temperature is reduced below ambient and the 'Probit' method is then the preferred procedure for evaluating impact strength behaviour. The additional information that can be obtained by this method is demonstrated in **Figures 18.43–18.45**. For a low melt index HDPE (**Figure 18.43**) the Probit is almost horizontal at both ambient and –20 °C. For a typical high-impact ABS terpolymer (**Figure 18.44**) the slopes of the Probit increase as the temperature is decreased. With the PP copolymers (**Figure 18.45**) the slopes of the curves are relatively steep compared to HDPE, with the exception of those instances where the impact strength is very low. The probability of failure occurring at low temperatures and at low impact levels is thus higher with PP copolymers and ABS than with HDPE. In fact the results indicate that the impact strength of the latter material at –20 °C is slightly superior to that at 23 °C.

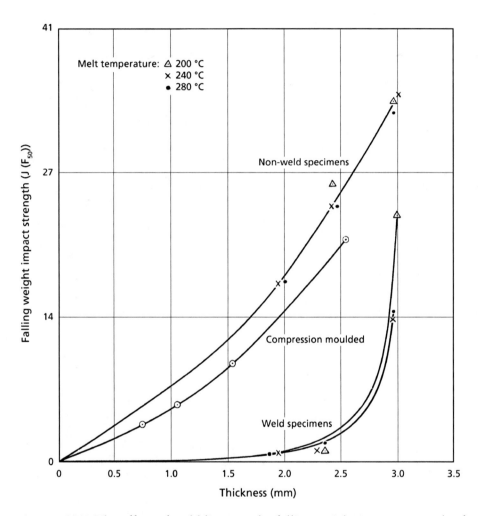

Figure 18.41 The effect of weld lines on the falling weight impact strength of polypropylene copolymer at 23 °C. (*Source: Author's own files*)

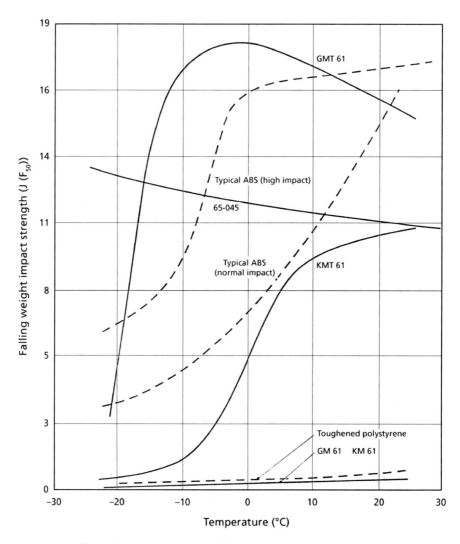

Figure 18.42 Effect of temperature on falling weight impact strength of 0.080 inch thick injection moulded specimens of various polymers moulded at 240 °C. KMT 61 and GMT 61 = polypropylene copolymers, 65-045 = high-density polyethylene, GM 61 and KM 61 = polypropylene homopolymer, ABS = acrylonitrile-butadiene-styrene terpolymer. (*Source: Author's own files*)

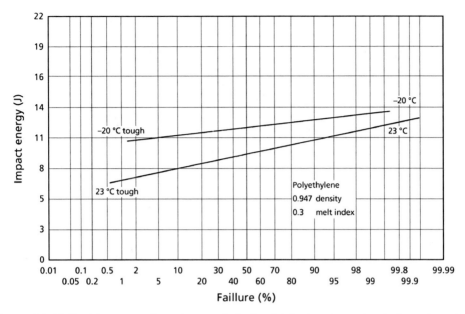

Figure 18.43 Probit curves for a low melt index high-density polyethylene (nominal thickness 2.0 mm, injection moulded at 270 °C). (*Source: Author's own files*)

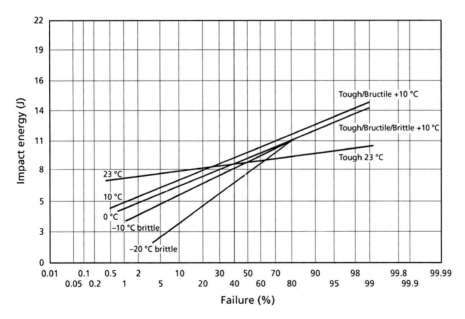

Figure 18.44 Probit curves for a typical high impact acrylonitrile-butadiene-styrene terpolymer (nominal thickness 2.0 mm, injection moulded at 240 °C). (*Source: Author's own files*)

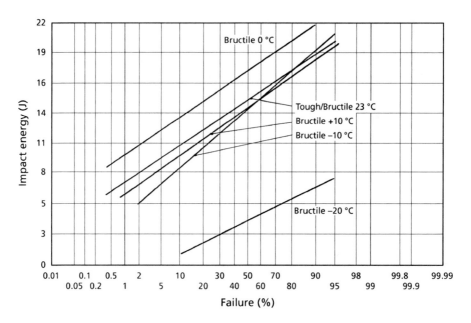

Figure 18.45 Probit curves for a polypropylene copolymer (nominal thickness 2.0 mm, injection moulded at 240 °C). (*Source: Author's own files*)

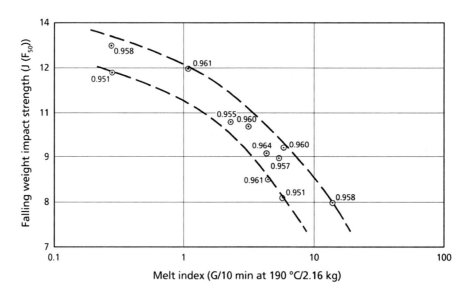

Figure 18.46 Effect of melt index on falling weight impact strength of various high-density polyethylenes at 23 °C (1.5 mm compression moulded samples). (*Source: Author's own files*)

18.1.2.5 Effect of Molecular Parameters

The major influence of MW on impact behaviour of polyolefins, both with regard to notched Izod and falling weight impact tests, is emphasised. With HDPE, MW is the prime factor controlling impact behaviour and density a secondary one. This is clearly demonstrated in **Figures 18.46 and 18.47,** which show results for commercial HDPE. It should be realised, however, that it is generally thought that for any given density there is a critical melt index which increases as the density decreases. Above this a marked fall-off in impact strength occurs. Furthermore, within a given family of grades, a decrease in melt index is usually accompanied by a slight reduction in density.

A comparison of the impact data with the creep data also confirms, particularly with PP copolymers, that an increase in impact strength is accompanied by a decrease in creep resistance.

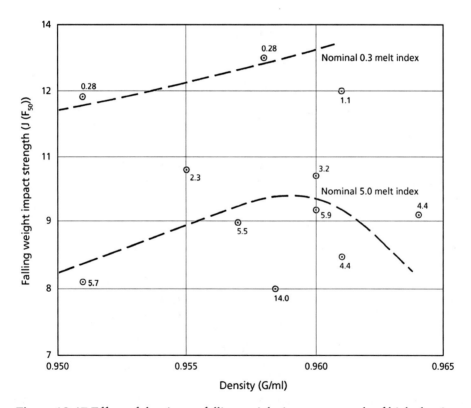

Figure 18.47 Effect of density on falling weight impact strength of high-density polyethylenes in various melt indices at 23 °C (1.5 mm compression moulded samples). (*Source: Author's own files*)

569

18.1.3 Measurement of Mechanical Properties in Polymers

18.1.3.1 Electronic Dynamometer Testing of Tensile Properties

ATS FAAR supplies the Series TC 200 computer-governed dynamometer which can perform tensile, compression, and flexural tests on a variety of materials. All details of the tests are managed including computing the final results and presenting them in either alphanumerical or graphical form. Moreover, the use of a PC for the management of the dynamometers gives the extra advantage of storing – at will – the results of the tests in an electronic data file which can be utilised to obtain organised print-outs, historical series, go-no-go checks, and the like. The full automation of all operations, computations included, increases the precision, reproducibility, and speed of analysis due to the elimination of otherwise unavoidable human errors.

Tests performable by this dynamometer include (see **Table 18.2**):

- Tensile strength (tensile modulus) tests with or without preloading, according to specifications ASTM D638-03 [4], DIN EN ISO 527-1 [5] and DIN EN ISO 572-2 [6].

- Compression and compressive strength tests according to specifications ASTM D695-02a [7] and DIN EN ISO 179-1 [8].

- Flexural and flexural strength tests according to specifications ASTM D790-03 [9], ASTM D732 [10], and DIN EN ISO 178 [11].

The instrument can also perform cycles between preset limits, creep and relaxation studies, peeling experiments, and measurement of friction coefficient and puncture resistance of films.

18.1.3.2 Impact Resistance (Izod–Charpy) and Notched Impact Resistance Strength

(1) *Pendulum Izod–Charpy instrument.* An instrument available from ATS FAAR (**Table 18.2**) is designed for testing against the following specifications: ASTM D256 [13], ISO 179-1 [14], ISO 180 [15], DIN EN ISO 178 [11], BS 2782 Method [3], UNI EN ISO 180 [16] and UNI EN ISO 8256 [17]. The energies available with this instrument are 2–25 J.

Janik and Krolikowski [53] investigated the effect of Charpy notched impact strength and other mechanical properties such as tensile strength, flexural strength, elastic modulus, flexural modulus, melt viscosity, and fracture on the mechanical and rheological properties of PE–polyethylene terephthalate (PET) blends.

Table 18.2 Mechanical properties of polymers			
Method	ATS FAAR Apparatus Code No.	Measurement units	Test suitable for meeting following standards
Dyanometer Test			
Tensile impact strength	16.10050	kJ/m^2	DIN EN ISO 8256 [12]
Tensile strength (tensile modulus)	16.00121 16.00122 16.00126 16.00127	N/mm^2 psi kg/cm^2	ASTM D638-03 [4] DIN EN ISO 527-1 [5] DIN EN ISO 527-2 [6]
Compression strength and compressive stress	16.00121 16.00122 16.00126 16.00127	N/mm^2 psi N/mm^2	ASTM D695-02a [7] DIN EN ISO 604 [29]
Flexural strength (dyanometer)	16.00121 16.00122 16.00126 16.0127	N/mm^2 psi kg/cm^2	ASTM D790-03 [9] ASTM D732 [10] DIN EN ISO 178 [11]
Izod Chapy Impact Test			
Pendulum impact strength (Izod and Charpy) and notched impact strength	16.10000 16.10500 16.10600	ft lb/in kJ/m^2 N m/m	ASTM D256-56 [13] DIN EN ISO 179 [14] ISO 179-1 [8] ISO 180 [15] BS2782 [3] UNI EN ISO 180 [16] UNI EN ISO 8256 [17]
Falling Dart Impact Test			
Impact resistance (free falling dart)	16.20003 16.20004 16.20001 16.20002		ASTM D5420 [18] ASTM D5628 [19] DIN EN ISO 6603 [20]
Guided Universal Impact Test			
Impact resistance at low temperature (cables)	10.65050		

		Table 18.2 Continued ...	
Method	ATS FAAR Apparatus Code No.	Measurement units	Test suitable for meeting following standards
Torsion Tests			
Modulus elasticity versus temperature (torsion test)	10.22000 10.22001		ASTM D1043 [21] DIN 53477 [22] ISO 458 [23] BS 2782 [3]
Bend Tests			
Modulus of elasticity or flexural modulus (bend test)	10.29000	kN/mm^2 psi Kg/cm^2	ASTM D1896 [24] ASTM D3419 [25] ASTM D3641 [26] ASTM D4703 [27] ASTM D5227 [28] DIN EN ISO 178 [11] DIN EN ISO 527-1 [5] DIN EN ISO 527-2 [6] DIN EN ISO 604 [29]
Cold bend test (cables)	10.65053		
Flexing Tests			
Resistance to dynamic fatigue on bending (De Mattia apparatus)	10.23000 10.23001 10.23003		ASTM D430 [30] ASTM D813 [31]
Compressive set to measure permanent deformation	10.65000 (stress) 10.07000 (strain)		ASTM D395 [32] ISO 815 [33] UNI 6121 [35]
Shear strength		psi kg/cm^2	ASTM D732 [36]
Deformation under load		%	
Elongation in tension		%	ASTM D638-03 [4]
Abrasion resistance	10.55000		DIN

Method	ATS FAAR Apparatus Code No.	Measurement units	Test suitable for meeting following standards
Mold shrinkage a) injection b) compression and post shrinkage		%	DIN 53464 [37]
Hardness Meters			
Hardness Ball indentation Rockwell IRHD	10.48000 11.03000 10.25130	N/mm^2	DIN EN ISO 2039-1 [38] ASTM D785 [39] UNI EN ISO 2039-2 [40] ASTM D1415 [41] ISO 48 [42] UNI 7318 [43] UNI 7319 [49]
Shore	11.03100 11.03101 11.3102 11.3110 11.3120		ASTM D2240 [45] DIN 53505 [46]
Fiat	10.25140 10.25150		Fiat 50430/02
Bulk density		g/l	DIN EN ISO 60 [47] DIN EN ISO 61 [48] DIN ISO 171 [49]
Density (specific volume) gradient column	10.06000 10.14000	g/cm^3	ASTM D792-00 [50] DIN EN ISO 1183-1 [51] DIN EN ISO 1183-2 [52] (See also MIC apparatus)

Table 18.2 Continued...

Source: Author's own files

(2) *Free falling dart.* ATS FAAR supplies the Fall-o-Scope Universal apparatus for impact resistance tests with free or guided fall taps (**Table 18.2**). The apparatus can be set to obtain the rate of energy absorption during the impact. The apparatus is capable of performing tests to different specifications with computerised operation in the temperature range 70–200 °C.

ATS FAAR also supplies guided universal impact resistance testers and a low-temperature impact resistance tester for electrical cable (**Table 18.2**). The application of dynamic mechanical analysis (DMA) to the determination of impact resistance is discussed in Section 15.1.

(3) *Gardner impact test.* Lavach [54] has discussed the factors that affect results obtained by the Gardner impact test. This test is used by plastics producers to approximate the mean failure energy for many plastics. The test is inexpensive and easy to operate and the test equipment can be placed close to the manufacturing equipment, permitting fast and nearly on-line determination of an article's impact resistance. The test is useful for finding the mean failure energy for brittle thermoplastics like acrylic and high-impact polystyrene, with standard deviations between 8 and 10%.

18.1.3.3 Apparent Modulus of Elasticity (Torsion Test)

AST FAAR supplies a torsion test instrument for the measurement of the apparent modulus of elasticity *versus* temperature of plastics and elastomers (**Table 18.2**), according to the following specifications: ASTM D1043 [21], DIN 53477 [22], ISO 498 [55], and BS 2782 Method 156 [3].

This apparatus consists of an assembly of interdependent units. It is compact and very easy to operate and features an accurate torque applications system with a broad angular measurement range.

The temperature is electronically controlled and the fully automatic system of presetting heating periods and torque is electronic. The apparatus is equipped with a set of weights allowing a wide range of torques to be applied to specimens having different dimensions. The distance between the clamps can be varied in the range 20–100 mm.

It is thus possible to measure accurately the apparent modulus of elasticity of specimens obtained from a wide range of materials (the modulus is, in this case, defined as apparent as it is obtained by measuring the angular torsion of the specimens under test).

In its regular version the apparatus can perform tests in the temperature range –100 to +100 °C; a special version, however, allows one to run tests between –100 and +300 °C. The torque application system is assembled on ball bearings of very reduced dimensions and utilises the same 'pivot' contrivance used in watch making. As a consequence all radial frictions are reduced to zero and the main pulley can be rotated even when applying torques smaller than 0.005 N/m. The initial zero point can be adjusted so as to allow tests to be performed even on specimens having planar deformation between +5° and –5°. The cooling is obtained either by solid carbon dioxide or by liquid nitrogen so as to achieve the maximum operational flexibility.

ATS FAAR also supplies a hand test apparatus for the measurement of the modulus of elasticity (flexural modulus) according to specifications ASTM D1869 [24], D3419 [25], D3641 [26], D4703 [27], D5227 [28] and DIN EN ISO 178 [11], DIN EN ISO 527-1 [5], DIN EN ISO 527-2 [6] and DIN EN ISO 604 [29] (**Table 18.2**). This company also supplies a flexing machine (De Mattia) for the measurement of resistance to dynamic fatigue according to ASTM D430 [30] and ASTM D813 [31]. These test methods are utilised to test the resistance to dynamic fatigue of rubber-like materials when subjected to repeated bending. The tests simulate the stresses – either in tension or in compression of inflexion or in a combination of the three modes of load application – to which the materials will be subjected when in actual use.

The failure of the tested specimens is indicated by cracking of the surface or – as prescribed by ASTM D813 [31] specifications – by the dimensional increase of a nick made on the specimen before starting the test. In the case of composite materials the failure can show up as separation of the different layers.

The apparatus can test up to 12 specimens and is equipped with a presettable cycle counter with presentation of the cycle number on a luminous digital display. Optionally the apparatus can also perform tests at temperatures higher or lower than room temperature.

18.1.3.4 Compressive Set (Permanent Deformation)

The equipment described in the following is supplied by ATS FAAR (**Table 18.2**).

Tests can be carried out under constant load Method A according to ASTM D359-03 [32]. The test is used to determine the capability of rubber compounds, and elastomers in general, to maintain their elastic properties after prolonged action of compression stresses. According to Method A of ASTM D395 [32], this stress is applied by a constant load.

The tester incorporates a spring with calibration curve and is accurately calibrated to have a slope of 70 ± 3.5 kN/m at 1.8 kN force. Spring deflection is indicated by a vernier system with an accuracy of 0.05 mm applied directly to the moving plate.

The tester may be placed in an oven or in a cryogenic chamber to perform tests in a temperature range of +70 to –30 °C as indicated by the ASTM specifications.

Tests can be carried out under constant deflection Method B according to ASTM D395 [32], ISO 815 [33], and UNI 6121 [35]. This method applies a constant compression stress to specimens having a diameter of 29 ± 0.5 mm or 13 ± 0.2 mm with a thickness of 12.5 ± 0.5 mm or 6 ± 0.2 mm. After the introduction of the specimens the apparatus is placed in a conditioning chamber for a preset time at a preset temperature. At the end of the heating period the final thickness of the specimen is measured.

18.1.3.5 Shear Strength

Apparatus for measurement of this property according to ASTM D732 [10] is available from ATS FAAR (**Table 18.2**).

18.1.3.6 Deformation Under Load

The use of light scattering and X-ray beams to measure polymer deformation has been reported by Lee and co-workers [56] and Riekel and co-workers [57]. See also **Table 18.2**.

18.1.3.7 Elongation in Tension

Apparatus for the measurement of this property according to ASTM D638-03 [4] is available from ATS FAAR (**Table 18.2**).

18.1.3.8 Abrasion Resistance

Apparatus for the measurement of this property according to the DIN specification is available from ATS FAAR (**Table 18.2**).

18.1.3.9 Mould Shrinkage

Apparatus for the measurement of this property according to DIN 53464 [37] is available from ATS FAAR (**Table 18.2**). Bertacchi and co-workers [58] have reported on computer-simulated mould shrinkage studies on talc-filled PP, glass-reinforced PA, and a PC/ABS blend.

18.1.3.10 Hardness

Various types of hardness meters are available from ATS FAAR (**Table 18.2**) including IRHD, Shore, Rockwell, and Fiat for testing against ASTM, ISO, UNI, DIN, and FIAT specifications. Martin Instrument Company is also suppliers of Shore hardness meters. Hardness measurements have recently been reported on methacrylate dental resins [59], polymeric coatings based on polyester resin, polyurethane (PU) and acrylic acid [60], polyacrylamide [61], and aged PE [62].

18.1.3.11 Density

ATS FAAR supplies equipment for the determination of density according to ASTM and DIN specifications (**Table 18.2**). Martin Instrument Company supplies one- and three-column density gradient instruments for the measurement of density.

18.1.3.12 Miniature Material Tester

The Minimat miniature materials tester supplied by PL Polymer Laboratories is capable of carrying out measurements in the following fields:

- Fracture mechanics
- Stress–strain analysis
- Stress relaxation studies
- Changes in crystallinity on cold drawing of fibres
- Failure mechanisms in fibre composites
- Plastic deformation studies in polarised light
- Stress optical analysis for stress concentrations around moulding defects
- Stress/creep relaxation
- Adhesion studies

18.1.3.13 Thermomechanical Analysis

This technique has been used to measure the following mechanical properties of polymers in addition to those discussed in Sections 13.2 (transitions [63]), 15.2 (viscoelastic and rheological properties [64]), 16.1 (coefficient of expansion), and 16.4 (heat deflection temperature):

- Linear or volumetric changes in samples
- Dimensions as a function of time, temperature, and force
- Temperature extension plots
- Temperature expansion plots
- Temperature compression plots
- Thermal relaxation studies

- Effect of temperature of physical, mechanical and dimensional properties of polymers, e.g., thermal stress analysis (see next)

- Rigidity studies

- Thermal stress analysis

Thermomechanical analysis is an ideal technique for analysing fibres since the measured parameters – dimension change, temperature, and stress – are all major variables that affect fibre processing. **Figure 18.48** shows the thermal stress analysis curves for a polyolefin fibre as received and after cold drawing. In this experiment, the fibres are subjected to initial strain (1% of initial length) and the force required to maintain that fibre length is monitored. Obviously, as the fibre tries to shrink, more force must be exerted to maintain a constant length. The result is a direct measurement of the fibre's shrink force. Shrink force reflects the orientation frozen into the fibre during processing, which is primarily related to the amorphous portions of the fibre. Techniques that track fibre crystallinity, therefore, are not as sensitive a measure of processing conditions as thermomechanical analysis. In this case, the onset of the shrink force peak indicates the draw temperature, while the magnitude of the peak is related to the fibre's draw ratio. It has been shown that the area

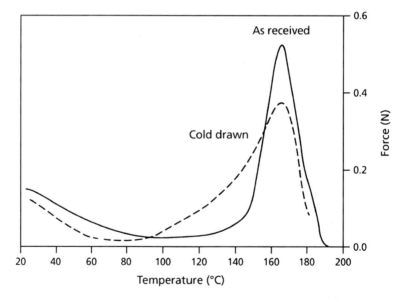

Figure 18.48 Application of thermomechanical analysis to thermal stress analysis of polyolefin films (as received and cold drawn) (*Source: TA Instruments, New Castle, DE, USA*)

under the shrink force curve (from onset to maxima) can be correlated to properties such as elongation at break and knot strength. Other portions of the thermomechanical analysis thermal stress plot can yield additional information. For example, the initial decreasing slope is related to the fibre's expansion properties, and the appearance of secondary force peaks can be used to determine values such as heat set temperature in Nylon.

18.1.3.14 DMA

This technique has found the following applications in addition to those discussed in Sections 10.1 (resin cure studies on phenol urethane compositions) [65], 12.2 (photopolymer studies [66-68]), and 13.3 (phase transitions in PE) [66], Chapter 15 (viscoelastic and rheological properties), and Section 16.4 (heat deflection temperatures): epoxy resin–amine system [67], cured acrylate-terminated unsaturated copolymers [68], PE and PP foam [69], ethylene–propylene–diene terpolymers [70], natural rubbers [71, 72], polyester-based clear coat resins [73], polyvinyl esters and unsaturated polyester resins [74], polyimide–clay nanocomposites [75], polyether sulfone–styrene–acrylonitrile, PS–polymethyl methacrylate (PMMA) blends and PS–polytetrafluoroethylene PMMA copolymers [76], cyanate ester resin–carbon fibre composites [77], polycyanate epoxy resins [78], and styrenic copolymers [79].

Some particular mechanical property measurements carried out by DMA and discussed in this text include:

- Loss modulus in PP–wood composites [80] and wood-based adhesives [81, 82].

- Tensile modulus of poly-*p*-phenylene [83], relaxation modulus in LDPE [84], diglycidyl ether bisphenol A epoxy resins [85], and styrene–butadiene block copolymers with doped polyaniline [86].

- Dynamic viscosity of PVC–ethylene–vinyl acetate–carbon monoxide terpolymers [87].

- Storage modulus of PP–wood composites [80], basalt fibre-reinforced PP [88], ethylene–propylene–diene terpolymer [89], polyvinyl fluoride–clay nanocomposites [90], thermosetting resins [91], and water-based adhesives [81, 82].

- Stress–strain curves of PE, PC, PA, and polybutylene terephthalate [92], and styrene–butadiene block copolymers [93].

- Shear strength modulus curves of cold seal adhesives [94].

- Polymer damping peak measurement of PP–wood composites [80].

Some examples of these applications are discussed next.

Prediction of polymer impact resistance. Impact resistance is critical in end uses of many commercial plastic products. Measurement of impact properties, however, requires lengthy sample preparation and is often irreproducible. For example, the ASTM drop weight (falling dart) impact (DWI) method D5420 [18] and D5628 [19] (**Table 18.2**) requires an impact measurement on as many as 30 samples. Each sample must be prepared by high-quality injection moulding, followed by temperature conditioning at –29 °C for 24 hours.

Low-temperature loss peak measured by DMA correlates with the ASTM DWI values. Since DMA values are more precise than DWI measurements, as few as four determinations can be used to rank impact resistance. Sample preparation for DMA is not as lengthy and does not require sophisticated processing equipment since surface effects are not critical and smaller samples are used. **Figure 18.49** shows the comparative DMA profiles for a series of impact-modified PP. The intensity of the damping peak at –110 °C correlates well with the DWI values at –29 °C. Using these kinds of data, a suitable calibration curve can be developed so that future formulations can be rapidly screened by DMA.

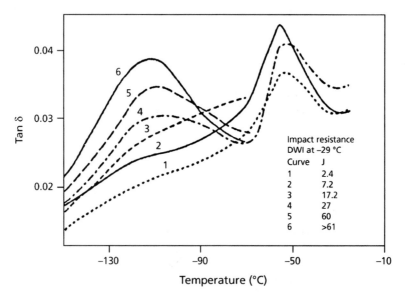

Figure 18.49 Prediction of polymer impact resistance. Comparative dynamic mechanical analysis profiles for a series of impact modified polypropylenes. (*Source: TA Instruments, New Castle, DE, USA*)

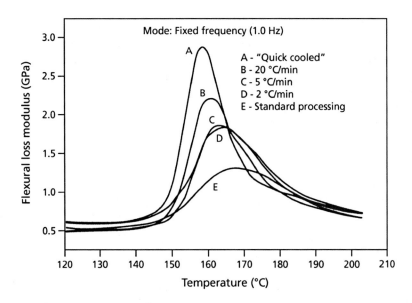

Figure 18.50 Effect of polymer processing on loss modulus by dynamic mechanical analysis. Effect of cooling rate of polyether ether ketone (PEEK) composite loss modulus at T_g. (*Source: TA Instruments, New Castle, DE, USA*)

- *Effect of processing on loss modulus.* Polyether ether ketone (PEEK) is an important matrix material for thermoplastic composite applications. The properties of PEEK laminate depend primarily on this morphology developed during processing. Cooling rate, time in the melt, and sub-melt annealing are all critical processing variables that can be simulated, and the effects of changes immediately evaluated by DMA. **Figure 18.50** illustrates, for example, shows the effect of cooling rate on loss modulus results. The peak maximum temperature, reflective of the glass transition (T_g) of the matrix, increases significantly as the cooling rate is decreased. The magnitude of the loss peak decreases and the peak width increases significantly as the cooling rate is decreased. These changes in the loss modulus T_g peak can be explained in terms of increased free volume as well as decreased level of crystallinity as the cooling rate is increased. A very fast cooling rate results in greater free volume in the amorphous phase and imparts greater mobility to the polymeric backbone. A decrease in the matrix crystalline content also results in greater mobility when passing through the T_g.

- *Material selection for elevated temperature applications.* The task of evaluating new materials and projecting their performance for specific applications is a challenging one for engineers and designers. Often materials are supplied with short-term test information such as deflection temperature under load (DTUL) (ASTM D693-03a

[95], **Table 18.2**) which is used to project long-term, high-temperature performance. However, because of factors such as polymer structure, filler loading and type, oxidative stability, part geometry, and moulded-in stresses, the actual maximum long-term use temperatures may be as much as 150 °C below or above the DTUL. DMA continuously monitors material modulus with temperature and, hence, provides a better indication of long-term elevated temperature performance.

Figure 18.51 illustrates the DMA modulus curves for three resins with nearly identical DTUL: a PET, a polyethersulfone (PES), and an epoxy. The PET begins to lose modulus rapidly at 60 °C as the material enters the glass transition. The amorphous component in the polymer achieves an increased degree of freedom and at the end of the T_g the modulus of the material has declined by about 50% from room temperature values. Because of its crystalline component, the material then exhibits a region of relative stability. The modulus again drops rapidly as the crystalline structure approaches the melting point. Because the T_g in a semi-crystalline thermoplastic typically occurs 150–200 °C below the melting point, the actual modulus of a resin of this type at the DTUL is only 10–30% of the room temperature value. The DTUL of highly filled systems based on these resins is more closely related to the melting point than to the significant structural changes associated with T_g.

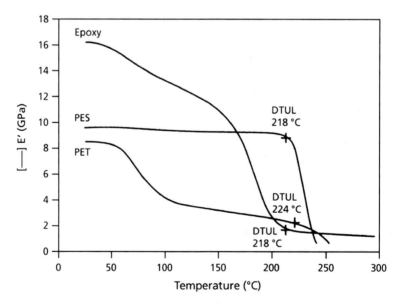

Figure 18.51 Comparison of DTUL and DMA results.
(*Source: TA Instruments, New Castle, DE, USA*)

PES is a high-performance amorphous resin. Amorphous materials exhibit higher T_g than their semi-crystalline counterparts, and maintain a high percentage of their room temperature properties up to that point. However, with the onset of the T_g the loss in properties is sudden and complete, even for highly reinforced grades. The DTUL of these systems are closely associated with the T_g, but almost always fall on the steep-sloped part of the modulus curve. Thus the DTUL occurs in a region of great structural instability and the actual maximum temperature for reliable performance under load is 15–30 °C below the DTUL.

The epoxy is a crosslinked system with a well-defined T_g. The temperature dependency of the modulus in such materials is related to the crosslink density. The relationship of the DTUL modulus to the room temperature modulus in this case is similar to that observed for the PET. However, in this case the crosslinked system provides an extended region of stability well beyond T_g and the DTUL. Thus, while both thermoplastic systems are no longer solid above 250 °C, the epoxy has structural integrity and virtually the same modulus at 300 °C as it has at 250 °C. It is therefore still serviceable for short-term excursions above the DTUL and may prove useful for extended periods under reduced loads providing that it possesses good thermal and oxidative stability.

DMA offers an enhanced means of evaluating the performance of polymeric systems at elevated temperatures. It provides a complete profile of modulus *versus* temperature as well as a measurement of mechanical damping. Operating in the creep mode and coupled with the careful use of time–temperature superpositioning, projections can be made regarding the long-term time-dependent behaviour under constant load. This provides a much more realistic evaluation of the short- and long-term capabilities of a resin system than the values for DTUL offered in the data sheets.

18.1.3.15 Coefficient of Friction

Coefficient of friction measurements have been reported for nitrile rubber [96].

18.1.3.16 Gas Permeability

Cerrada [97] has examined the gas barrier properties of food-grade ethylene–vinyl alcohol copolymers used in film and bottle manufacturing as a function of the ethylene and vinyl alcohol content. Oxygen transmission rates have been reported for polyolefin nanocomposites [98].

18.1.3.17 Organic Liquid Transmission Tests

Lagaron and Powell [99] made measurements of the permeation rates of methanol, toluene, and a 50:50 toluene/isooctane fuel (fuel C) and fuel C with 10% ethanol or 15% methanol through ethylene–vinyl alcohol films. Large-scale methanol- and water-induced plasticisation was suggested by large solvent uptakes and confirmed by the observation of considerable decreases in the T_g. The permeability of fuel C increased significantly when methanol was present, while the high barrier properties of ethylene–vinyl alcohol to hydrocarbons was maintained for fuel C containing ethanol.

18.1.3.18 Water Vapour Transmission

Lagaron and Powell [99] used NMR imaging technologies to characterise super water absorbency polyacrylate polymers swollen with water.

18.1.3.19 Ageing Tests

This subject has been reviewed [100-102] and details of ageing/weathering tests have been discussed [62, 100-102]. In particular, the Xenotest 250 T apparatus [62] and SAE J 1885 [102, 103] are mentioned.

18.1.4 Properties of Polymer Film and Pipe

Some common tests on polymer films and pipes and the relevant standard tests are summarised in **Table 18.3**. Equipment for all these tests is available from ATS FAAR (Appendix 1).

Additional tests for polymer films that have been mentioned in the literature are the tear strength of ethylene–vinyl acetate plasticiser [108], coefficient of friction of polyolefin plasticiser blown film [109], and oxygen absorption in food packaging films [110].

Additional tests on polymers mentioned in the literature include hydrostatic pressure tests on polyolefin [111] and PA [112] pipes, ageing and service life tests on polyolefin [113] and composite off-shore oil pipes, and general discussions on the mechanical properties of pipes [114, 115].

Table 18.3 Tests on films and pipes

	DIN	ASTM	ISO	BS	CEI	UNI EN ISO	IEC	VDE	Others	Code No. Suppler ATS FAAR
Plastic films										
Tear resistance of films Elmendorf pendulum unit		✓							TAPPI scan APPITA AFNOR ATICELCA	12.00001
Impact resistance on films (falling dart)		D1709 [104]	✓			7765-1 [105]				16.20005 16.20007
Static and dynamic friction coefficient on films		✓								10.21003 10.21000
Gas transmission through films	✓	✓		✓						10.58000
Water vapour transmission through films		✓		✓						10.08000
Plastic pipes										
Impact resistance (falling dart)		D2944 [106]				7448 [107]				16.20077 16.20079
Resistance to internal pressure	✓	✓	✓			✓				10.24000 10.24010

Source: Author's own files

585

18.1.5 Polymer Powders

Methods for the testing of polymer powders have been reviewed by Chambers [116] and Klaren [117] and are listed in **Table 18.4**. With the exception of particle size distribution (Chapter 19) and thermal analysis measurements (by differential scanning calorimetry and differential thermal analysis) none of these tests requires commercially produced equipment, i.e., they are simple laboratory tests. Full ASTM test specifications for polymer powders are described in ASTM D3451 [118]. Brief comments on the various tests on powders included in **Table 18.4** are given below.

18.1.5.1 Storage Stability

The ability of a powder to remain in a usable condition in storage is of considerable practical importance. Thermoplastic powders can be considered to have indefinite shelf lives although blocking may occur with PVC powders containing appreciable quantities of plasticiser. Thermosetting powders are chemically reactive and hence changes will take place over a period of time depending on the nature of the system and the storage temperature. In addition, powders used for electronic applications are of small particle size and hence surface effects will be more important.

The deterioration of a powder is the result of either physical or chemical processes or a combination of the two. For the former, changes in particle size distribution or the onset for caking and blocking can occur, whereas a reduction in reactivity or an increase in melt viscosity are evidence of chemical change.

Test methods for storage stability are described in DIN ISO 8130-10 [133] and DIN ISO 8130-8 [134] and in ASTM D3451 [119] where the conditions relating to temperature and the duration of the test are laid down. These tests quantify the degree of change and therefore are of value for comparison of different powders. However, a simple coatings test whereby any change in the general appearance or the ease of application is noted would seem to be of more practical use to the end user.

18.1.5.2 Chemical Stability

To determine whether pre-polymerisation has taken place after storage at elevated temperature, the gelation time can be determined. The powder from the physical stability test is used for this. From practical experience it is known that for an epoxy powder a reduction in gelation time of 10–15% at 40 °C and 25% at 50 °C as compared with freshly prepared powder may be considered acceptable.

Table 18.4 Test methods for polymer powders	
Measurement	Standard test method(s)
Polymer Powders	
Particle size distribution	ASTM D1921 [119] BS ISO 13319 [120]
Storage stability	ASTM D609 [121] DIN ISO 8130 [122]
Chemical and Physical Stability	
Ultraviolet and outdoor stability Reactivity	ASTM D822-01 [123] ASTM G155 [124] ASTM D3451 [118] DIN ISO 8130 [123]
Melt viscosity	ASTM D3451 [118]
Volatiles (Tesson stoving)	ASTM D3451 [118] DIN ISO 8130 [123]
True density	DIN ISO 8130 [123] ASTM D792 [125]
Bulk density	BS 2782 [3] BS 2701 [126] ASTM D1895 [127]
Powder flow	ASTM D1895 [128]
Test for Cure	
Electrical properties Thermal analysis	by DSC and DTA
Tests on Fused Powder Coatings	
Defects	
Glass Whiteness Film thickness Hardness	ASTM 523-89 [128] DIN 53157 [129]
Adhesion	
Flow deformation	ASTM D2794-93 [130]
Bend test	ASTM D522-93 [131]
Chemical Resistance	
Corrosion resistance	ASTM B117-64 [132]
Glass Transition Melting Cure	
Source: Author's own files	

18.1.5.3 Ultraviolet and Outdoor Resistance

Although the physical properties of a powder coating may be studied in order to predict its probable behaviour in practice, there are occasions when UV and/or outdoor resistance are required. There are two main procedures: artificial weathering and natural weathering.

18.1.5.4 Artificial Weathering

The Atlas Xenon Arc Weather-O-Meter is used according to ASTM D822-01 [123]/G155 [124]. In this cabinet the coated panel is exposed to a cycle of 102 minutes light, followed by 18 minutes light together with water spray.

18.1.5.5 Natural Weathering

The area most frequently used to evaluate coatings for their resistance to natural weathering is the weathering site in Miami, Florida, USA. There is another field in the desert of Arizona, 40 miles north of Phoenix, which greatly differs from Florida in atmospheric conditions. Because of the elevation (2000 feet above sea level) and the clear, dry atmosphere (38% relative humidity as the annual average) the percentage of UV in the solar radiation is higher in Arizona than in the humid climate at sea level in Florida. (Arizona has 4,000 sun hours per year compared to Great Britain with 1,000–1,400 hours.) The yearly average climatological data of Arizona and Florida are as follows:

- Arizona: hours of sunlight: 3.993; cm of rain: 16; relative humidity: 38%; average temperature: 21 °C

- Florida: hours of sunlight: 2.721; cm of rain: 148; relative humidity: 58%; average temperature: 24.5 °C

18.1.5.6 Reactivity

Thermosetting powders vary in their degree of reactivity and some means of quantifying this property is required. The basic method described in DIN ISO 8130-10 [133] and ASTM D3451 [118] is to melt the powder sample on a heating block at a temperature in the range 160–200 °C and stir the molten material until gelation occurs. The time to achieve gelation is noted.

There is undoubtedly some degree of subjectivity in this test and slight differences in test method could have a significant effect on the final result. Hence, although repeatability is reasonable, there is considerable doubt as to how reproducible the method is from one laboratory to another. Thus, at present, the gelation test should be treated only as a useful laboratory tool until sufficient evidence is available to justify inclusion in a national standard.

18.1.5.7 Melt Viscosity

The fluidity of a material during the stoving process determines to a large degree the smoothness of the final coating and the degree of edge coverage achieved. It may be argued that the final appearance is the sole criterion for excellence but some means of quantitative measurement of melt viscosity is required for research work and is useful for quality control. A method is described in ASTM D3451 [118] which employs a cone and plate viscometer. A simpler method is to use an inclined plane whereby a standard pellet of the compressed powder is allowed to flow down a heated glass plate set at a suitable angle. The total length of flow after a fixed time, usually ten minutes, is related to the melt viscosity.

18.1.5.8 Loss on Stoving

Although stoving losses of coating powders are small when compared to conventional finishes, some emission does occur particularly with PU powders. Basically there are two methods of determination:

1. Spraying onto a test plate and measuring mass loss after stoving.

2. Placing powder on a tared dish and determining loss mass after heating in an oven.

The first approach is favoured by the French (NFT 30-502) [135] whereas DIN ISO 8130-10 [133] and ASTM D3451 [118] describe dish methods. The dish method has the advantage that it follows normal laboratory practice. The temperature of the test and its duration can conveniently be the recommended cure schedule.

18.1.5.9 True Density

For powders of specific gravity above unity a pyknometer with water plus wetting agent may be used. DIN ISO 8130-7 [136] describes an air pyknometer which caters for any powder.

18.1.5.10 Bulk Density

Bulk density refers to the mass per unit volume of the powder including air trapped between the particles. Thus the measurement enables the calculation to be made of appropriate capacities for containers, hoppers, and so on. It is essential to ensure that the sample is not compacted during the test and this is normally achieved by pouring the powder through a funnel into a measuring container. Methods are described in BS 2782 [3] and ASTM D1895 [129]. A Rees–Hugill powder density flask is also described in BS 2701 [126].

18.1.5.11 Powder Flow

The flow characteristics of a powder bear a complex relationship to particle size distribution, shape, and structure. It is of particular significance in connection with the ability of a powder to pass easily along feed lines to give the necessary consistency of applications when using electrostatic spraying equipment. The flow may be measured by determining the time for a prescribed volume of powder to flow through a standard funnel as described in ASTM D1895 [127].

This method has poor reproducibility in the case of thermosetting powders and this has led to the development of an empirical technique by TNO in the Netherlands which is described in the French standard NFT 30-500 [137].

The method consists of placing the powder in a standard fluidised bed with a plugged orifice in the side of the chamber. The height of the powder during and after fluidisation is measured and the powder is then refluidised and the plug removed for 30 seconds allowing powder to flow into a tared container. The powder is weighed and the flow factor calculated as the product of this mass and the height difference. This is then compared to flow factors previously obtained on a wide range of powders, the flow characteristics of which are known. This enables a prediction of flow behaviour to be obtained.

18.1.5.12 Test for Cure

Unlike coatings made from thermoplastic powders which require only a visual inspection to see whether the coating is satisfactory, thermosetting powder coatings require additional examination to determine whether the coating has achieved its full physical properties. Simple mechanical tests may be used, as is the practice with conventional paint finishes, but, particularly for epoxy powder coatings, it is normal to use chemical resistance as a criterion for suitability. A common method is to swab the coating continuously for 30 seconds with a cotton wool pad soaked in methyl isobutyl ketone. Any deterioration of the film with the exception of a minor loss of gloss may be taken as an indication that the coating is insufficiently cured.

18.1.5.13 Electrical Properties

The application of the electrostatic technique to powders has drawn attention to the need for powders to possess the proper electrical characteristics if they are to hold an electrical charge and be successfully processed. This has led to the development of a special apparatus which in addition to measuring resistivity and powder charge also acts as a diagnostic kit for the electrostatic spraying equipment. This apparatus was developed by the Wolfson unit at Southampton University and is manufactured by Industrial Development (Bangor), University College of North Wales.

18.1.5.14 Thermal Analysis

The performance of powder coatings during fusion and subsequent cure may be followed in detail by thermal analytical procedures such as differential thermal analysis and differential scanning calorimetry. These techniques offer an opportunity for the study of phase transitions and cure behaviour under closely controlled conditions. Differential scanning calorimetry is a requirement in the British Gas standard for epoxy powder coating materials for the external coatings of steel pipes where it is used as a guide to ascertain optimum cure conditions.

18.1.6 Physical Testing of Rubbers and Elastomers

18.1.6.1 Mechanical Properties

Suppliers and standard test details of equipment for the testing of rubbers and elastomers are listed in **Table 18.5**. Measured properties include:

- Rheology
- Viscosity
- Processability and rheological behaviour
- Density
- Hardness

Table 18.5 Physical testing of rubbers and elastomers

Properties measured	Instrument	Testing to the following standards	Supplier	Notes
Rheology				
Measurement of processability and rheological behaviour of raw (unvulcanised) elastomers at variable shear rates and temperatures (viscosity and elasticity)	THS Rheometer	BS EN ISO 9001 [39]	Negretti Automation	
Oscillating disc rheometry. Full analysis of vulcanisation curing	Rheocheck 100C Barco Rheomicro Computer package	BS ASTM ISO	Negretti Automation	
Oscillating die rheometer	Oscillating die rheometer	BS ASTM ISO	Gibitre, Italy available from Negretti Automation	
Viscometry				
Viscosity measurement of rubbers and elastomers	Mark III Mooney viscometer	BS ISO ASTM NFT DIN JIN	Negretti Automation	Part No. 8500000

Table 18.5 *Continued ...*

Properties measured	Instrument	Testing to the following standards	Supplier	Notes
	Mark 8 Mooney Viscometer	BS ISO ASTM NFT DIN JIN	Negretti Automation	
Weight-density-volume variation	Balance check and Rapid Direct Reading density check		Gibitre, Italy available from Negretti Automation	
Hardness	IRHD Micro durometer and Durometer Support		Gibitre, Italy available from negretti Automation	
Brittleness point TR test	Low temperature brittleness check	ASTM D1329 [140]	Gibitre, Italy available from Negretti Automation and ATS FAAR	Code Nos. 10.2900 10.29005 10.12010
Ross flexing test	Measurement of cut growth when subject to repeat bend flexing	ASTM 1052 [141]	ATS FAAR	Code No. 10.64002
De Mattia type flexing test	De Mattia type dynamic testing machine	ASTM DIN	Gibitre, Italy available from Negretti Automation	

Table 18.5 Continued ...

Properties measured	Instrument	Testing to the following standards	Supplier	Notes
Mechanical properties Resilience etc.	Yersley mechanical oscillograph	ASTM D945 [142]	ATS FAAR	Code No. 16.65100
Universal tensile testing			Negretti Automation	
Mechanical stability of natural and synthetic latices	-	ASTM	ATS FAAR	Model 10.67000
Abrasion test		DIN	Negretti Automation	
Peel adhesion test		BS AFERA		
Ozone degradation	Ozone Check	-	Gibitre, Italy available from Negretti Automation	

Source: Author's own files

- Brittleness point (low-temperature crystallisation)

- Flexing tests

- Resilience

- Dynamic and static moduli

- Kinetic energy

- Creep and set

- Tensile properties

- Mechanical stability

- Abrasion resistance

- Elastic modulus

- Peel tests

- Ozone resistance

18.1.6.2 Rheological Properties

Suppliers of instrumentation are reviewed in Appendix 1. Negretti Automation offers a range of rheological testing instruments for rubber and elastomers. Probably the top of the market instrument is the THS rheometer which is a unique and versatile instrument offering a complete characterisation of a polymers flow behaviour in the unvulcanised state. Variable shear rates between 0.1 and 100/s are available using a die and rotor principle with a preheating chamber. A typical curve obtained with this instrument is illustrated in **Figure 18.52** which shows the curve obtained in the measurement of rheological behaviour.

A further instrument, the Rheocheck 100C, is a classic oscillating disc rheometer offering a full analysis of the vulcanisation curve with pre-set limits for quality control purposes available from the standard cost-effective computer/printer/plotter package. The instrument can be extended to include a Mooney viscometer.

The Barco Rheomicro computer package enables existing oscillating disc rheometers to be converted to a powerful dynamic testing system.

The Gibitre oscillating die Rheocheck available from Negretti Automation is a moving die rheometer which enables the vulcanisation curve for any compound to be produced rapidly at closely controlled temperatures. It is designed for rapid computer-controlled quality checks in a production environment.

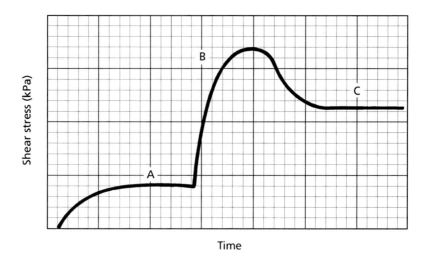

Time

Figure 18.52 Characterisation of polymer flow behaviour with TMS rheometer showing (A) viscosity at very low shear rates, the stress measured is a function of viscosity, (B) transient flow-step changes from low to high shear rates generate a peak stress value which is a product of the thixotropic and structural features of the sample, (C) elasticity – high shear rates produce results that represent elastic behaviour. (*Source: Negretti Automation, Aylesbury, UK*)

18.1.6.3 Viscometry

Depending upon temperature, shear rate, and other flexural factors, the elements of viscosity and elasticity impose a varying influence on processing behaviour. A rheometer measures these in isolation in a single test, allowing prediction of processing behaviour at an early stage. It also allows close control of raw polymers of nominally identical specification from various sources, which often behave very differently when processed.

Negretti Automation supplies the Negretti Mooney viscometer which meets the requirements of all international standards. It is available with a digital control unit and recorder to provide semi-automatic operation.

In this instrument, viscosity of a material is derived by measuring the torque required to rotate a disc squeezed between two samples of the material at a defined pressure. The diagram of the system (**Figure 18.53**) shows how the presence of the sample around the shearing rotor causes a breaking reaction which can be measured as a thrust against the calibrated U-beam. The net displacement of the U-beam, which is proportional to the torque required to keep the rotor turning, is measured directly on a gauge and can also

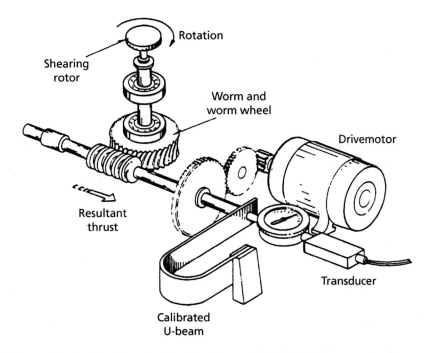

Figure 18.53 Working details of a Negretti Mark 3 Mooney viscometer. (*Source: Negretti Automation, Aylesbury, UK*)

be plotted on a chart recorder. The results are read or plotted directly in Mooney units, which are universally used for viscosity measurement in the rubber industry.

In accordance with the standards, the sample is preheated for a set time to a temperature suitable for the test. For example, viscosity may be measured at 100 °C, whereas scorch tests normally use a higher temperature, typically 125 °C.

In the mechanical layout of the viscometer (**Figure 18.53**) the rotor is driven by the motor through the gear train and worm wheel. The axle carrying the worm can move laterally in the direction of the arrow. When the motor is started with the rotor surrounded by a sample, rotation is resisted by the viscosity of the material. As a result, the worm rotates against the worm wheel and tries to move in the direction of the arrow. This movement is resisted by the U-beam and eventually the rotor is forced to turn.

The deflection of the U-beam is related to the torque required to turn the rotor and thus to the viscosity of the sample. The deflection can be read directly on the gauge which is calibrated in Mooney units. It can also control the linear deflection transducer whose output drives a chart recorder.

Negretti Automation also supplies the Negretti Mark 8 Mooney viscometer which is a computer-controlled instrument offering improved temperature control, repeatability, hard copy of result, statistical memory, and peripheral computer interface.

18.1.6.4 Brittleness Point (Low-Temperature Crystallisation)

Negretti Automation supplies the computer-controlled Gibitre low-temperature check apparatus designed to test the low-temperature properties (−73 °C, liquid carbon dioxide), i.e., crystallisation effects and elastic recovery, of rubbers and elastomers according to ASTM D1329 [140] by means of the TR test and brittleness point. ATS FAAR supplies an instrument for carrying out the same measurements.

- *TR test*. After positioning the test pieces in the relative mechanical system using the special test piece extensions, it is placed in the test chamber. The chamber contains liquid cooled with carbon dioxide (−73 °C) and then gradually heated to 20 °C (2 °C/min). The test assesses the temperature difference between the two return values as a percentage.

Brittleness point. After positioning the test pieces in the relative mechanical system, it is placed in the test chamber. This contains liquid cooled with carbon dioxide (−73 °C) to the required temperature. A ram comes into impact with the test pieces at a speed established by the standards and this appears on the display. The test is repeated several times using new test pieces to establish the brittleness temperature.

18.1.6.5 Flexing Test

The Ross flexing apparatus supplied by ATS FAAR measures, according to ASTM D1052 [141], the cut growth of rubber or elastomer specimens when subjected to repeated blend flexing. The machine allows the pierced flexed area of the specimen to bend freely over a 10 mm diameter rod at an angle of 90° with a frequency of 100 ± 5 or 50 ± 5 cycles/min. The test is considered as ended when the length of the cut, made with a special nicking tool, has increased by 500%.

The test is widely used by all industries manufacturing rubber or elastomer goods subjected, when in use, to repeated bending, such as shoes, tyres, and conveyor belts.

Negretti Automation supplies the Gibitre De Mattia type dynamo tester which meets the requirements of ASTM, BS, and ISO standards. The equipment has been specially designed by Gibitre to meet different requirements and can be used for other tests besides the De Mattia type. Repeated flexing tests can be carried out on De Mattia type test pieces or other types.

The equipment is constructed entirely of stainless steel and can therefore be used in thermostatic stoves and for immersion in different liquids to test the resistance of test pieces under these conditions. The stroke is fixed, but the clamps can be moved to suit requirements. Stress speed can be regulated from 30 to 350 impulses per minute.

8.1.6.6 Deformation

The Yerseley Mechanical Oscillograph supplied by ATS FAAR measures, according to ASTM D945 [142], the mechanical properties of rubber vulcanisations in the small range of deformation that characterises many technical applications. These properties include resilience, dynamic modulus, static modulus, kinetic energy, creep, and set under a given force.

18.1.6.7 Tensile Properties

Negretti Automation supplies universal testing machines with capacities up to 5 kN (500 kg) and 10 kN (1000 kg) complete with computer systems and printers.

18.1.6.8 Mechanical Stability of Natural and Synthetic Lattices

ATS FAAR supplies equipment for carrying out this test to the ASTM specifications.

18.1.6.9 Abrasion Test

Negretti Automation supplies equipment to carry out abrasion tests to the DIN specification.

18.1.6.10 Peel Adhesion Test

Negretti Automation supplies equipment for carrying out 180° peel tests for adhesive tapes and similar products to British Standard specifications.

18.1.6.11 Ozone Resistance Test

The Ozone Check supplied by Negretti Automation determines the resistance to ozone of elastomers used in the tyre industry.

In a test chamber with a capacity of approximately 155 litres, air is circulated in a closed circuit with an adjustable flow so that there are 0.5 to 3 air changes per minute. The air temperature can be set from 20 to 7.5 °C and is controlled with a precision of ±1 °C. The ozone concentration can be set from 0.1 to 5.5 ppm and an electrical signal is emitted, proportional to the concentration being used, by means of a continuous analyser, based on the bubbling of the air–ozone mixture in a potassium iodide buffer solution. This electrical signal acts automatically on the regulator of an UV lamp that generates the ozone.

18.1.6.12 Compression Properties

Orjela and co-workers [143] carried out precise measurement of tyre cord. They described the use of the Pirelli pure moment dynamometer for these measurements.

18.2 Electrical Properties

18.2.1 Volume and Surface Resistivity

Apparatus for carrying out these measurements in the range 0.05 MΩ to 2×10^{16} Ω is discussed in **Table 18.6**. The ATS FAAR Teraohmmeter model M 150 E or equivalent is suitable for carrying out these measurements according to ASTM D1401-02 [144] and DIN IEC 60093 [145] and DIN IEC 600167 [146].

18.2.2 Dielectric and Dissipation Factor

In addition to the classic methods, thermomechanical analysis has been used to carry out this measurement according to DIN 53483 [149] and ASTM D150-98 [148].

18.2.3 Dielectric Strength (Dielectric Rigidity)

Details of equipment for carrying out this measurement according to ASTM D149-97a [150] and similar specifications are given in **Table 18.6**.

18.2.4 Surface Arc Resistance

Details of equipment for carrying out this measurement according to ASTM D495-55T [155] are given in **Table 18.6**.

Table 18.6 Apparatus available from ATS FAAR for determining electrical properties of polymers			
Method	ATS FAAR apparatus	Measurement units	Test suitable for meeting following standards
Volume and surface resistivity	Teraohmeter Model M510E	Ω cm (volume) Ω (surface)	ASTM D257-99 [147] DIN IEC 60093 [145] DIN IEC 60167 [146]
Dielectric constant and dissipation (power) factor	10.33510 10.33500	Hz	ASTM D150-98 [148] DIN 53483-1 [149]
Dielectric strength Dielectric rigidity		kv/un v/mil	ASTM D149-97 [150] DIN 53481-1 [149]
Surface arc resistance		seconds	ASTM D495-99 [151]
Tracking resistance	10.62000	stage	DIN EN 60112 [152] CEI 15-50 [153] UNI 4290 [154]
Behaviour during and after contact with glow rod		stage	
Source: Author's own files			

18.2.5 Tracking Resistance

The apparatus supplied by ATS FAAR (**Table 18.6**) evaluates, during a short period of time, the low-voltage (up to 600 V) track resistance (or comparative tracking index) of insulating materials.

The test is performed by applying two electrodes – with a force of 1 N – to the surface of the specimen. A 0.1% solution of ammonium chloride is dropped at the rate of 1 drop having a volume of 20 ± 2 mm^3 every 30 seconds while a variable-tension 50 Hz AC voltage is applied. The test ends when the short circuit current overpasses 1 amp.

18.3 Optical Properties and Light Stability

Properties such as refractive index, clarity, transparency, and effect of sunlight are determined according to ASTM D542 [156] and are usually measured by in-house procedures developed by polymer manufacturers.

AST FAAR manufactures a colour fastness meter suitable for testing against ASTM, UNI, and Fiat specifications (Appendix 1).

18.3.1 Stress Optical Analysis

One of the applications of the PL (Polymer Laboratories) Minimat Materials Tester is for stress optical analysis in which transmitted polarised light shows stress concentrations developing around defects in polymer mouldings.

18.3.2 Light Stability of Polyolefins

In contrast to ferrous metals, plastic materials do not suffer from the disadvantage of corrosion, but in common with most organic materials they are degraded by sunlight. The chemical basis of such photodegradation consists, in polyolefins, of chain scission accompanied by oxidation. This results in deterioration of the mechanical and electrical properties, and of the surface appearance of the polymer. Resistance to photodegradation varies and the following groups of polymers show decreasing light stability in the order LDPE > HDPE > PP > PS.

It is, of course, well known that opaque pigments can have a valuable screening effect when incorporated into a plastic, and can help to confine the photodegradation to the extreme surface layers, thus protecting the inner layers of the material. The most effective of such pigments is carbon black, although care has to be exercised in the choice of the grade of carbon black and its concentration to ensure maximum protection while avoiding any side effects on other polymer characteristics. In addition to the opaque pigments, a wide range of non-pigmenting UV stabilisers has been developed. These may function in one of two ways: by preferentially absorbing the UV portion of the solar spectrum or by inhibiting the chain reactions initiated by the UV light.

In considering the possible choices of pigment and/or UV stabiliser systems for sunlight protection, it has to be remembered that many PE, and all PP, contain thermal antioxidants, and it is known that interactions can occur between antioxidants, UV stabilisers, and pigments. These interactions may either be synergistic or antagonistic in terms of the weathering life of the polymer.

An enclosed carbon arc has been used for many years for testing the colour fastness of dyes and pigments. More recently it has been used as a convenient laboratory 'standard' source for the accelerated testing of the susceptibility of polymers to photodegradation. Unfortunately, correlations between results from carbon arc exposure and from direct sunlight have been poor. A xenon lamp is at present regarded as a more reliable light source than the carbon arc because its spectral distribution is similar to that of solar radiation and it is now probably more widely used for laboratory purposes. The correlation between xenon lamp results and direct sunlight is also, however, far from perfect, and varies in its reliability from polymer to polymer. For instance, a much better correlation between xenon lamp and sunlight results has been found for PE than for PP.

It is probably true to say that the more data are amassed on the correlation between laboratory weatherometers and direct sunlight, the more is there a tendency to use direct sunlight testing at suitably chosen sites in various parts of the world, and to restrict laboratory accelerated weatherometer work almost entirely to control testing, where a correlation has already been fully established on a given polymer/additive system.

Direct sunlight stability studies normally involve exposures in the UK and, for example, at Curaçao (off the northern coast of Venezuela), and sometimes also at Phoenix, Arizona, which has much drier climatic conditions than Curaçao. The number of sunshine hours per day in tropical areas is at least double that in the UK and the light is of higher intensity. Once reliable basic data have been established, it is possible from the results obtained from exposures in tropical areas to predict the outdoor life of plastic formulations in temperate climates with reasonable accuracy.

Numerous tests are available for following the progress of photodegradation. Tests such as infrared spectroscopy and measurements of change in MW and of mechanical and electrical properties have been employed. Convenient tests have been found to be the measurement of flexural endurance for PE, and MW change and hard brittleness determinations for PP. As photodegradation is more intense at the surface of materials, thin specimens have been used (0.05 cm) so that changes are more rapidly followed, although correlations have been established by means of fall off in impact strength of thicker specimens. Change in MW of PP is conveniently followed by measuring the change in solution viscosity of a 1% solution of the polymer in Decalin at 135 °C, and the critical MW below which a PP sample will exhibit brittleness for 0.05 cm thick samples has been found from tests on a wide range of samples to correlate with a 'limiting viscosity number' (LVN) of 1.9.

As mentioned previously, the most important pigment additive employed for light stability purposes is carbon black. The addition of 2% by weight of a well-dispersed carbon black of fine particle size should extend the outdoor life of most low- and HDPE to as much as 20 years in temperate climates, while even the addition of 0.2% should give 10 years' life in temperate climates and 3 years in tropical climates. PP pigmented with 2% carbon black has shown such resistance to photodegradation that no obvious adverse effects could be detected after two years' exposure under Curaçao conditions.

The assessment of the life under outdoor exposure conditions of materials with other pigment systems is a very complex matter, since the range of types and concentrations of pigments used in normal industrial practice is extremely wide, and it is known that certain additives interact to affect the outdoor weathering life.

The effect on the life of PP in which is incorporated various coloured pigments is judged by comparing polymer life for the unpigmented and the pigmented polymers both alone and in conjunction with a UV stabiliser.

It is natural to expect that the degradation of a polymer resulting from sunlight exposure will be most acute at the surface of the plastic, and this is borne out by all experimental evidence. UV stabilisers, apart from reducing the rate of photodegradation, also restrict the depth to which photodegradation can occur (effectively 0.05 cm for natural PP). These intense surface effects must be clearly borne in mind in considering the effect of outdoor weathering on all mechanical properties involving a flexural type of stress. For example, although PP shows powdering rather than gross surface crazing after sunlight exposure, minute surface cracks are visible under microscopic examination. Under certain circumstances these micro-cracks can act as points of stress concentration similar to the sharp notch used in the Izod impact test. For example, where the unexposed surface is struck and the exposed surface is thereby placed in tension, surface micro-crazing can substantially reduce the impact strength of the component. Conversely, however, where the exposed micro-crazed surface is struck and placed in compression, the reduction in impact strength is considerably less. It is fortunate that the latter generally happens in practice. The rate of loss of impact strength on exposure to sunlight has been found to be approximately independent of thickness for sections in excess of 0.3 cm thick up to say 0.6 cm thick. In other words, while the impact strength of very thin sections is highly influenced by surface degradation effects, this phenomenon becomes of relatively small practical importance for sections greater than 0.3 cm, provided it is the exposed surface that is struck.

In any discussion of the application of standard test piece exposure results to the prediction of the exposure life of an actual component, consideration must, of course, be given to geometric and environmental factors. The most important factors to be taken into account in estimating the outdoor service life of an article are as follows:

• Whether or not the article will be exposed continuously outdoors.

• The position of the article relative to the horizon, and the latitude of exposure.

- Whether the article will be exposed in the open or in partial shade.

- The thickness of the article in relation to impact strength requirements.

- The importance of surface appearance.

18.3.3 Effect of Pigments

Table 18.7 gives the 'outdoor lives' of a pigmented PP homopolymer without UV stabiliser, as estimated from exposure at a UK site, Curaçao, and Arizona. The results from all three locations indicate the marked differences in the degree of protection afforded by the differing pigment formulations and emphasise the care that should be exercised in the choice of colours for outdoor applications.

With few exceptions, the degree of protection afforded by the different pigment formulations falls roughly in the same order at all three locations. Under the conditions used, photodegradation was found to be most severe at Curaçao and least in the UK. The results comparing exposure at Curaçao with that in the UK and in Arizona with those in the UK are shown in **Figures 18.54 and 18.55**.

18.3.4 Effect of Pigments in Combination with a UV Stabiliser

Table 18.8 gives the 'outdoor lives' of PP containing both a UV stabiliser and pigment systems as estimated from exposures under tropical conditions at Curaçao. The relative light stabilities of these formulations compared firstly with the natural polymer without UV stabiliser and secondly with the corresponding pigmented formulations are also given in the table. The latter correlation is also shown in **Figure 18.56**. Although there are obvious exceptions, with the majority of formulations the addition of the UV stabiliser approximately doubles the 'outdoor life'. It is also interesting to note that the addition of a UV stabiliser is most effective with the colour systems that give least protection.

The results of a limited experiment on the effect of pigment concentration are summarised in **Table 18.9**. These indicate that increasing the pigment concentration leads to a better light stability in non-UV-stabilised formulations but that little is to be gained by increasing the pigment concentration.

All samples showed a significant loss of surface gloss but it was found possible to restore some of this gloss by rubbing the exposed surface with a cloth. Surface powdering or visual crazing was not apparent, although microscopic examinations showed numerous shallow cracks after the longest exposure.

Table 18.7 Environmental factors affecting the lives of non-UV stabilised polypropylene pigmented with colours					
	Direct sunlight			Under glass	
Colours	Outdoor life (sunshine hours)			Outdoor life (sunshine hours)	
	Carrington, UK	Curaçao	Arizona	Curaçao	Arizona
Natural Control (1)	590	135	50	380	880
Woodtone Tan	>2700	>3500	>3000	>>2500	>>4000
Scarlet	2970	1620	2540	–	3660
Vermilion	2550	1340	1800	1980	3550
Crimson	2870	1740	2660	–	>4000
Brown	>>2700	3500	>3000	–	>>4000
Maroon	1900	920	1360	870	1710
Carmine Red	>2700	2450	2600	–	2920
Mushroom	2210	1200	–	–	–
Orange	2700	1450	1580	–	3800
Red-Orange	1170	375	940	670	1480
Flame Red	3100	2400	2840	–	3800
Pink	2200	1100	–	–	–
Lilac	1740	820	1640	980	2000
Ultra Blue	1400	540	800	710	1360
Natural Control (2)	710	140	440	350	920
Sandalwood	2150	1200	–	–	–
Peach Bloom	2090	1360	2000	1950	2400
Rose	2870	1050	2400	1780	2900
Coral	2200	800	2000	2060	2200
Antique Gold	1970	1100	–	–	–
Golden Rod	2150	1300	–	–	–
Bright Yellow	1320	600	1060	640	1960
Soft Yellow	2630	1300	2420	2100	>4000

	Direct sunlight			Under glass	
	Outdoor life (sunshine hours)			Outdoor life (sunshine hours)	
Colours	Carrington, UK	Curaçao	Arizona	Curaçao	Arizona
Primrose Yellow	2480	1400	2080	–	3500
Cream	2270	1625	–	–	–
Ivory	2100	1255	–	–	–
Emerald Green	3100	870	1600	810	2600
Forest Green	>2700	2650	3220	–	>4000
Spring Green	2100	970	1860	970	2450
Bright Green	2700	1300	–	–	–
Lime Green	1690	570	1580	1000	2000
Sprout Green	2440	1020	–	–	–
Fern Green	2530	1480	2300	2200	>4000
Sage Green	2200	1200	–	–	–
Navy Blue	2650	1430	1720	1420	2860
Purple	1070	370	–	–	–
Orchid	1580	730	1160	840	1740
Turquoise	2130	1160	–	–	–
Sapphire Blue	2200	950	1260	830	1820
Glacier Blue	1790	1100	–	–	–
Cadet Blue	2300	1000	–	–	–
Aqua Blue	1950	1050	–	–	–
Sea Blue	1740	750	1080	600	1750
Dusk Grey	2470	1500	–	–	–
Charcoal	>2700	3500	3220	–	>4000
Granite Grey	>2700	2050	–	–	–
Pearl Grey	2650	1200	1910	–	2200

Table 18.7 Continued …

Source: Author's own files

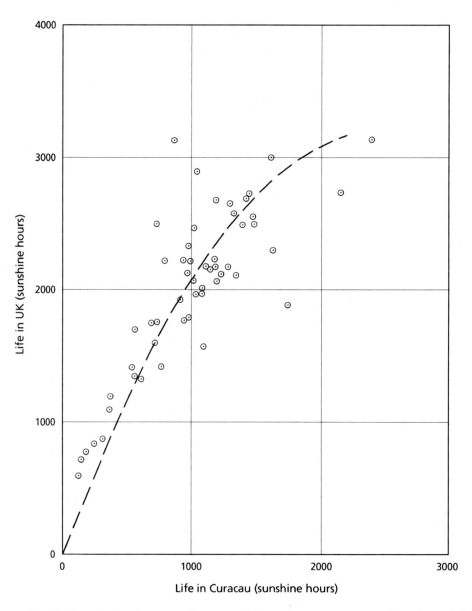

Figure 18.54 Correlation between direct sunlight exposures in UK and at Curaçao for pigmented polypropylene without UV stabiliser. (*Source: Author's own files*)

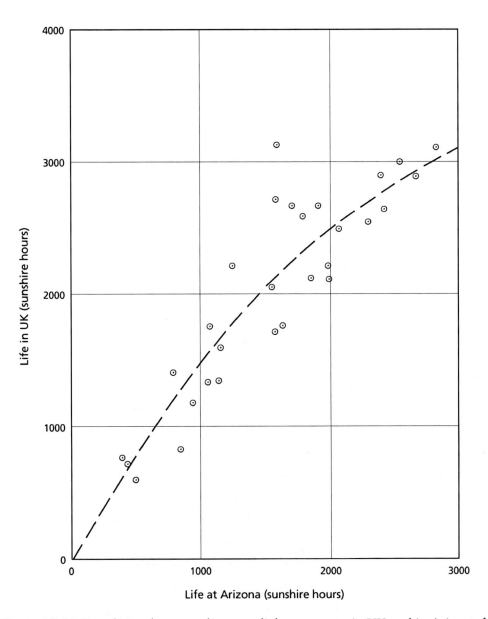

Figure 18.55 Correlation between direct sunlight exposures in UK and in Arizona for pigmented polypropylene without UV stabiliser. (*Source: Author's own files*)

Table 18.8 Outdoor lives at Curaçao of UV stabilised polypropylene with colour pigments			
Colours + 0.5% UV stabiliser	Outdoor life (sunshine hours)	Screening power relative to natural non-UV stabilised polymer	Screening power relative to non-UV stabilised pigment formulation
Natural Control (1)	1900	14	–
Woodtone Tan	>>4000	>>30	–
Scarlet	3000	22	1.85
Vermilion	2630	20	1.96
Crimson	2200	16	1.26
Brown	4000	30	1.14
Maroon	1880	14	2.04
Carmine Red	3400	25	1.39
Mushroom	2400	18	2.00
Orange	3050	23	2.10
Red-Orange	**2800	21	7.47
Flame Red	3500	26	1.46
Pink	1950	14	1.77
Lilac	1850	14	2.26
Ultra Blue	*1500	11	2.78
Natural Control (2)	2050	15	–
Sandalwood	2500	18	2.08
Peach Bloom	2300	16	1.69
Rose	**3620	26	3.43
Coral	2750	20	3.43
Antique Gold	2400	17	2.18
Golden Rod	2250	16	1.73
Bright Yellow	2100	15	3.50
Soft Yellow	2220	16	1.71
Primrose Yellow	2350	17	1.68
Cream	2400	17	1.48

Colours + 0.5% UV stabiliser	Outdoor life (sunshine hours)	Screening power relative to natural non-UV stabilised polymer	Screening power relative to non-UV stabilised pigment formulation
	Table 18.8 Continued ...		
Ivory	1950	14	1.55
Emerald Green	*1800	13	2.07
Forest Green	3250	23	1.23
Spring Green	2040	15	2.10
Bright Green	2270	16	1.74
Lime Green	*1730	12	3.04
Sprout Green	*1720	12	1.69
Fern Green	2570	18	1.74
Sage Green	*1560	11	1.30
Navy Blue	2900	21	2.03
Purple	*1500	11	4.05
Orchid	*1800	13	2.47
Turquoise	2000	14	1.72
Sapphire Blue	*1710	12	1.80
Glacier Blue	*1700	12	1.55
Cadet Blue	2140	15	2.14
Aqua Blue	*1850	13	1.76
Sea Blue	*1900	14	2.53
Dusk Grey	2900	21	1.93
Charcoal	>4000	>30	1.14
Granite Grey	4000	30	1.95
Pearl Grey	2250	16	1.88
* Pigment system antagonistic towards UV stabiliser ** Synergism between pigment system and UV stabiliser			
Source: Author's own files			

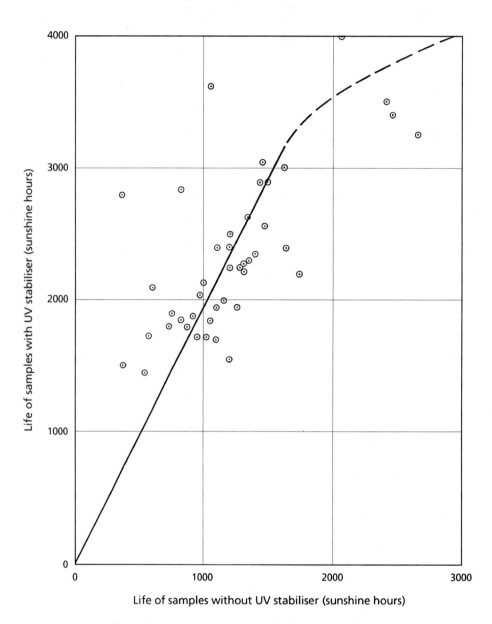

Figure 18.56 Effect of UV stabiliser on the lives of pigmented polypropylene. (*Source: Author's own files*)

Table 18.9 Effect of pigment concentration on life of polypropylene homopolymer (KM61 grade) exposed at Curaçao		Estimated outdoor life (sunshine hours)
Material		
KM61 + 0.5%	Green DC 2366	145
KM61 + 2.0%	Green DC 2366	500
KM61 + 4.0%	Green DC 2366	1120
KM61 + 0.5%	Green DC 2366 + UV stabiliser	1620*
KM61 + 2.0%	Green DC 2366 + UV stabiliser	2160*
KM61 + 4.0%	Green DC 2366 + UV stabiliser	1630*
KM61 + 0.5%	Rutiox CR	690
KM61 + 2.0%	Rutiox CR	1570*
KM61 + 4.0%	Rutiox CR	2000*
KM61 + 0.5%	Rutiox CR + UV stabiliser	1720*
KM61 + 2.0%	Rutiox CR + UV stabiliser	1960*
KM61 + 4.0%	Rutiox CR + UV stabiliser	2000*
Extrapolated values - tests still in progress Source: Author's own files		

18.3.5 Effect of Carbon Black

Excellent protection is afforded by a range of carbon blacks for the weatherability of PP. Although the average particle size of the carbon blacks tested, as measured by electron microscopy, varied from 16 to 38 µm, at a concentration of 2%, none of the PP showed a significant fall-off in MW after 6,000 sunshine hours exposed under tropical conditions at Curaçao. For the small-particle-size carbon black, a reduction in concentration to 0.5% gave similar results (see **Figure 18.57**).

With the higher melt index formulations a slight embrittlement was detected when specimens were flexed by hand after 6,000 hours tropical exposure even though there was no change in MW as determined from viscosity in solution. This is in contrast to the behaviour observed with other pigments where embrittlement was always accompanied by a sharp drop in viscosity in solution. No embrittlement was observed with the low melt index samples.

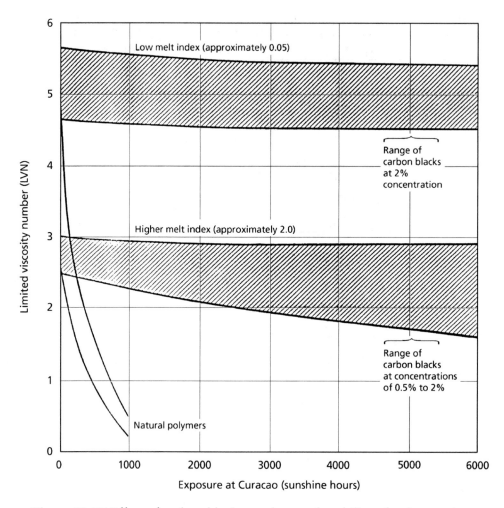

Figure 18.57 Effect of carbon blacks on the weatherability of polypropylene
(*Source: Author's own files*)

18.3.6 Effect of Window Glass

Many applications involve exposure of samples behind a layer of glass instead of direct exposure to sunlight. It is therefore important to know whether there is an improvement in life resulting from the screening action of the glass. In general the effect of common window glass is to increase the life by approximately one and a half times. This rather surprisingly low factor emphasises the need for similar careful choice of pigment formulations for behind-glass applications.

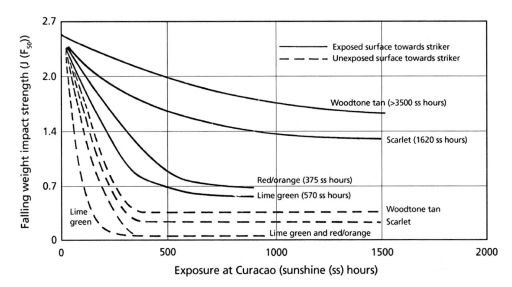

Figure 18.58 Change in falling weight impact strength of pigmented 1.5 mm compression moulded polypropylene (nominal melt index 2.0) on exposure at Curaçao. (*Source: Author's own files*)

18.3.7 Effect of Sunlight on Impact Strength

As previously discussed, the 'outdoor lives' quoted in the tables have been based on a correlation between the rate of fall-off in MW (as determined by viscosity measurements) with hand embrittlement tests on a very large number of samples. In order to confirm the practical significance of this chosen end point (i.e., LVN = 1.9) the effect of exposure on the falling weight impact strength of a range of pigmented formulations has been investigated. **Figure 18.58** shows some typical results obtained and illustrates the differences in impact strength depending upon whether the exposed or the unexposed surface of the specimens is struck. Where the exposed surface is struck, it can be seen that approximately one-third of the original impact strength is retained at the point at which 'exposure life' is quoted, and the general indications are that this remains approximately constant for exposure times considerably in excess of the quoted lives.

18.3.8 Effect of Thickness

Figure 18.59 demonstrates that photodegradation is confined to the surface layers and that the depth of degradation is considerably less for the UV-stabilised than for the natural polymer. The results were obtained by measuring the change in MW of successive samples microtomed from the surface of exposed specimens.

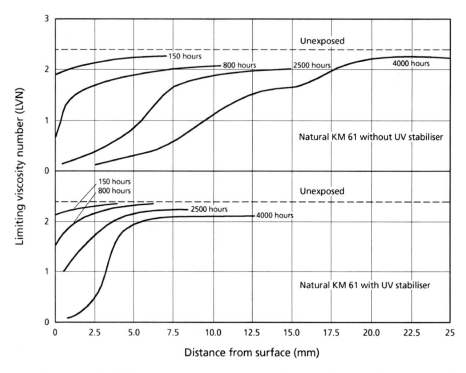

Figure 18.59 Depth of degradation on thick samples of polypropylene homopolymer exposed at Curaçao. (*Source: Author's own files*)

An increase in thickness leads to an improvement in impact strength and the effect of placing the exposed surface in tension is clearly demonstrated. The beneficial effect of the UV stabiliser in limiting the reduction in impact strength is also well illustrated.

18.3.9 Effect of Stress During Exposure

The data presented hitherto have been derived from tests involving the unstressed exposure of specimens. In practice, however, many applications naturally involve plastic components being exposed to sunlight under conditions of stress. Interim results obtained under an applied stress of 6.9 MPa are summarised in **Table 18.10**. The indications are encouraging in that after one and a half years' exposure, rupture has only been observed with the natural materials, while the pigmented samples are still free of the surface crazing marks that were exhibited by the natural samples prior to failure. It would appear that the quoted outdoor lives are realistic even for stressed applications.

Researched colour	Sample thickness (mm)	Number of failures*	Time to rupture (sunshine hours)	Time to onset of crazing (sunshine hours)	Outdoor life in non-stressed state (sunshine hours)
Natural Control (1)	0.6	4	900	750	710
Natural Control (1)	1.5	2	1200	1100	–
Soft Yellow	0.6	None	>1400	>1400	2630
Soft Yellow	1.5	None	>1400	>1400	–
Orchid	0.6	None	>1400	>1400	1580
Orchid	1.5	None	>1400	>1400	–
Charcoal	0.6	None	>1400	>1400	3800
Charcoal	1.5	None	>1400	>1400	–
Natural Control (2)	0.6	4	900-1200	750	590
Natural Control (2)	1.5	None	>1400	1100	–
Ultra Blue	0.6	None	>1400	>1400	1400
Ultra Blue	1.5	None	>1400	>1400	–
Vermilion	0.6	None	>1400	>1400	2550
Vermilion	1.5	None	>1400	>1400	–
Woodtone Tan	0.6	None	>1400	>1400	>3000
Woodtone Tan	1.5	None	>1400	>1400	–

Table 18.10 Stress rupture behaviour of polypropylene exposed to sunlight in the UK

*A total number of 4 samples of each formation are being exposed

**Determined from limiting LVN value of 1.9 as previously discussed

Source: Author's own files

18.3.10 Effect of Molecular Weight

The MW does not have a significant effect on the rate of degradation of PP. However, it should be remembered that PP of higher original MW have better basic impact properties and resistance to stress rupture effects, so it may be concluded that the higher the original MW, the greater the margin of safety before the critical limits for performance are reached due to sunlight degradation.

18.3.11 Effect of Sunlight on the Surface Appearance of Pigmented Samples

The criteria of failure under sunlight exposure previously considered have been related only to mechanical performance, either directly or by correlations separately established between mechanical performance and molecular weight. Changes in surface appearance during exposure may also be important, since photodegradation may result in a loss of gloss and the formation of easily removable powder on the surface of the samples. Observations regarding these aspects of PP behaviour are summarised in **Table 18.11**. Generally, exposures at Curaçao were found to be more severe than those at Arizona, which in turn were, quite naturally, more severe than those in the UK. At Arizona, surface powdering and the loss of gloss both appeared to be delayed by the presence of a UV stabiliser and by exposure behind window glass, whereas at Curaçao these factors only appeared to retard the loss of gloss and not the appearance of surface powdering. In the UK, surface powdering was not observed after 2,700 sunshine hours' exposure, but unlike the effects produced under more intense exposure conditions, surface crazing was in fact noted.

Where outdoor lives of specimens have been prolonged by the use of pigments or pigment/ UV stabiliser combinations, crazing or powdering may occur before the quoted outdoor lives (estimated from MW change) have been reached.

Table 18.11 Effect of sunlight on the surface appearance of pigmented samples			
Exposure site	Type of exposure	Comments on surface powdering and crazing	Comments on loss of gloss
Curaçao	Non-UV stabilised to direct sunlight	No crazing, but distinct surface powdering after 2,400 sunshine hours exposure	Slight loss of gloss observed after 800 sunshine hours exposure
Curaçao	UV stabilised to direct sunlight	No crazing but distinct surface powdering after 2,400 sunshine hours exposure	Slight loss of gloss observed after 1,600 sunshine hours exposure
Curaçao	Non-UV stabilised behind window glass	No crazing, but distinct surface powdering after 2,400 sunshine hours exposure	Slight loss of gloss observed after 1,600 sunshine hours exposure
Arizona	Non-UV stabilised to direct sunlight	No crazing, but slight surface powdering after 3,000 sunshine hours exposure	Slight loss of gloss observed after 2,000 sunshine hours exposure
Arizona	UV stabilised to direct sunlight	No crazing, but very slight surface powdering after 4,000 sunshine hours exposure	Very slight loss of gloss observed after 2,500 sunshine hours exposure
Arizona	Non-UV stabilised behind window glass	No crazing, but very slight surface powdering after 4,000 sunshine hours exposure	Slight loss of gloss observed after 2,500 sunshine hours exposure
UK	Non-UV stabilised to direct sunlight	Very slight crazing, observed after 1,500 sunshine hours exposure, but no surface powdering after 2,700 sunshine hours	Very slight loss of gloss observed after 2,000 sunshine hours exposure
UK	UV stabilised to direct sunlight	Very slight crazing observed after 2,700 sunshine hours exposure, but no surface powdering	Very slight loss of gloss observed after 2,700 sunshine hours exposure
Source: Author's own files			

References

1. S.H. Lin, C.C.M. Ma, N.H. Tai and L.H. Perng in *Proceedings of the Materials Challenge, Diversification and the Future Symposium*, Anaheim, CA, USA, 1995, Volume 40, Book 2, p.1046.

2. G. Hay, M.E. Mackay, P.N. Awati and Y. Park, *Journal of Rheology*, 1999, 43, 1099.

3. BS 2782-0, *Methods of Testing Plastics - Part 0: Introduction*°4.   ASTM D638-03, *Standard Test Method for Tensile Properties of Plastics*, 2003.

5. DIN EN ISO 527-1, *Plastics - Determination of Tensile Properties - Part 1: General Principles*, 1996

6. DIN ES ISO 527-2, *Plastics - Determination of Tensile Properties - Part 2: Test Conditions for Moulding and Extrusion Plastics*, 1996.

7. ASTM D695-02a, *Standard Test Method for Compressive Properties of Rigid Plastics*, 2002.

8. DIN EN ISO 179-1, *Determination of Charpy Impact Properties - Part 1: Non-Instrumented Impact Test*, 2001.

9. ASTM D790-03, *Standard Test Methods for Flexural Properties of Unreinforced and Reinforced Plastics and Electrical Insulating Materials*, 2003.

10. ASTM D732, *Standard Test Method for Shear Strength of Plastics by Punch Tool*, 2002.

11. DIN EN ISO 178, *Plastics - Determination of Flexural Properties*, 2003.

12. DIN EN ISO 8256, *Plastics - Determination of Tensile-Impact Strength*, 2005

13. ASTM D256-56, *Standard Test Methods for Determining the Izod Pendulum Impact Resistance of Plastics*, 2005.

14. ISO 179-1, *Plastics - Determination of Charpy Impact Properties - Part 1: Non-Instrumented Impact Test*, 2001.

15. ISO 180, *Plastics - Determination of Izod Impact Strength*, 2000.

16. UNI EN ISO 180, *Plastics - Determination of Izod Impact Strength*, 2001.

17. UNI EN ISO 8256, *Plastics - Determination of Tensile- Impact Strength*, 2005.

18. ASTM D5420, *Standard Test Method for Impact Resistance of Flat, Rigid Plastic Specimen by Means of a Striker Impacted by a Falling Weight (Gardner Impact)*, 2004.

19. ASTM D5628, 96e1, *Standard Test Method for Impact Resistance of Flat, Rigid Plastic Specimens by Means of a Falling Dart (Tup or Falling Mass)*, 2001.

20. DIN EN ISO 6603, *Plastics - Determination of Puncture Impact Behaviour of Rigid Plastics*, 2000.

21. ASTM D1043, *Standard Test Method for Stiffness Properties of Plastics as a Function of Temperature by Means of a Torsion Test*, 2002.

22. DIN 53477, *Testing of Plastics; Determination of Particle Size Distribution of Moulding Materials by Dry Sieving Analysis*, 1992.

23. ISO 458, *Plastics - Determination of Stiffness in Torsion of Flexible Materials - Part 1: General Method*, 1985.

24. ASTM D1896-99, *Standard Practice for Transfer Molding Test Specimens of Thermosetting Compounds*, 2004.

25. ASTM D3419, *Standard Practice for In-Line Screw-Injection Molding Test Specimens From Thermosetting Compounds*, 2000.

26. ASTM D3641, *Standard Practice for Injection Molding Test Specimens of Thermoplastic Molding and Extrusion Materials*, 2002.

27. ASTM D4703, *Standard Practice for Compression Molding Thermoplastic Materials into Test Specimens, Plaques, or Sheets*, 2003.

28. ASTM D5227-01, *Standard Test Method for Measurement of Hexane Extractable Content of Polyolefins*, 2001.

29. DIN EN ISO 604, *Plastics - Determination of Compressive Properties*, 2003.

30. ASTM D430-95, *Standard Test Methods for Rubber Deterioration-Dynamic Fatigue*, 2000.

31. ASTM D813-95, *Standard Test Method for Stiffness Properties of Plastics as a Function of Temperature by Means of a Torsion Test*, 2000.

32. ASTM D395-03, *Standard Test Method for Stiffness Properties of Plastics as a Function of Temperature by Means of a Torsion Test*, 2003.

33. ISO 815, *Rubber - Vulcanized or Thermoplastic - Determination of Compression Set at Ambient, Elevated or Low Temperatures*, 1993.

34. UNI ISO 815, *Rubber, Vulcanized or Thermoplastic - Determination of Compression Set at Ambient, Elevated or Low Temperatures*, 2001.

35. UNI 6121, *Elastomeri: Prodotti Finiti. Prodotti di Gomma Spugnosa aa Lattice. Definizioni E Prove*, 1967.

36. ASTM D732, *Standard Test Method for Stiffness Properties of Plastics as a Function of Temperature by Means of a Torsion Test*, 2002.

37. DIN 53464, *Testing of Plastics; Determination of Shrinkage Properties of Moulded Materials from Thermosetting Moulding Materials*, 1962.

38. DIN EN ISO 2039-1, *Plastics - Determination Of Hardness - Part 1: Ball Indentation Method*, 2003.

39. ASTM D785, *Standard Test Method for Stiffness Properties of Plastics as a Function of Temperature by Means of a Torsion Test*, 2003.

40. UNI EN ISO 2039-2, *Plastics - Determination of Hardness - Rockwell Hardness*, 2001.

41. ASTM D1415-88, *Standard Test Method for Rubber Property—International Hardness*, 2004

42. ISO 48, *Rubber, Vulcanized or Thermoplastic - Determination of Hardness (Hardness Between 10 IRHD and 100 IRHD)*, 1999.

43. UNI 7318, *Elastomeri: Prove su Vulcanizzati. Determinazione della Durezza in Gladi Internazionali su Provette Normali Mediante Durometro e Microdurometro*, 1974.

44. UNI 7319, *Elastomeri: Prove su Vulcanizzati. Determinazione della Durezza in Gradi Internazionali su Provette di Spessore Diverso dal Normale e su Prodotti Finiti*, 1974.

45. ASTM D2240-04e1, *Standard Test Method for Rubber Property—Durometer Hardness*, 2004.

46. DIN 53505, *Testing of Rubber - Shore A and Shore D Hardness Test*, 2000.

47. DIN EN ISO 60, *Plastics - Determination of Apparent Density of Material that can be Poured from a Specified Funnel*, 2000.

48. DIN EN ISO 61 *Plastics - Determination of Apparent Density of Moulding Material that Cannot be Poured From a Specified Funnel*, 2000.

49. DIN ISO 171, *Plastics - Determination of Bulk Factor of Moulding Materials*, 2000.

50. ASTM D792-00, *Standard Test Methods for Density and Specific Gravity (Relative Density) of Plastics by Displacement*, 2000.

51. DIN EN ISO 1183-1 *Plastics - Methods for Determining the Density of Non-Cellular Plastics - Part 1: Immersion Method, Liquid Pyknometer Method and Titration Method*, 2004.

52. DIN EN ISO 1183-2, *Plastics - Methods for Determining the Density of Non-Cellular Plastics - Part 2: Density Gradient Column Method*, 2004.

53. J. Janik and W. Krolikowski, *Polimery*, 2002, **47**, 250.

54. M. Lavach, *Plastics Engineering*, 1999, **55**, 41.

55. ISO 498, *Natural Rubber Latex Concentrate - Preparation of Dry Films*, 1992.

56. E.C. Lee, M.J. Solomon and S.J. Muller, *Macromolecules*, 1997, **30**, 7313.

57. C. Riekel, M. Burghammer, M.C. Garcia, A. Gourrier and S. Roth, *Polymer Preprints*, 2002, **43**, 1, 215.

58. G. Bertacchi, A. Pipino and G. Boero, *Macplas*, 2001, **26**, 78.

59. C. Pianelli, J. Devaux, S. Bebelman and G. Leloup, *Journal of Biomedical Materials Research (Applied Biomaterials)*, 1999, **48**, 675.

60. W. Shen, S.M. Smith, F.N. Jones, J. Caigui, R.A. Ryntz and M.P. Everson, *Journal of Coatings Technology*, 1997, **69**, 123.

61. P. Hron, J. Slechtova, K. Smetana, B. Dvorankova and P. Lopour, *Biomaterials*, 1997, **18**, 1069.

62. J. Borek and W. Osoba, *Polymer*, 2001, **42**, 2901.

63. A.B. Savitskii and A. Gorschkova, *Polymer Science, Series A*, 1997, **39**, 356.

64. Institute for Dynamic Materials Testing, *Gummi Fasern Kunststoffe*, 2001, **54**, 2, 85.

65. M.T. Huang and H. Ishida, *Polymer and Polymer Composites*, 1999, **7**, 233.

66. L.T. Pick and E. Harkin-Jones, *Polymer Engineering and Science*, 2003, **43**, 905.

67. J. Lange, N. Altmann, C.T. Kelly and P.J. Halley, *Polymer*, 2000, **41**, 5949.

68. J. Djonlagic, A. Zlatanic and B. Dunjic, *Macromolecular Chemistry and Physics*, 1998, **199**, 2029.

69. M.A. Rodríguez-Perez, S. Rodríguez-Llorente and J.A. De Saja, *Polymer Engineering and Science*, 1997, **37**, 959.

70. H.S. Liu, C.P. Richard, J.L. Mead and R.G. Stacer in *the Proceedings of the ANTEC 2000* Conference, Orlando, FL, USA, 2000, Paper No.570.

71. P.K. Patra, P.K. Das, M.S. Banerjee and A.R. Tripathy, *Rubber India*, 2000, **52**, 9.

72. V. Noparatonakailis, *Journal of Rubber Research*, 2000, **3**, 95.

73. M. Korhonen, P. Starck, B. Loefgren, P. Mikkila and O. Hormi, *Journal of Coatings Technology*, 2003, **75**, 67.

74. A. Valea, M.L. Gonzalez and I. Mondragon, *Journal of Applied Polymer Science*, 1999, **71**, 21.

75. M.O. Abdalla, D. Dean and S. Campbell, *Polymer*, 2002, **43**, 5887.

76 F. Oulevey, N.A. Burnham, G. Gremaud, A.J. Kulik, H.M. Pollock, A. Hammiche, M. Reading, M. Song and D.J. Houston, *Polymer*, 2000, **41**, 3087.

77. K. Chung and J.C. Seferis, *Polymer Degradation and Stability*, 2001, **71**, 425.

78. P. Ponteins, B. Medda, P. Demont, D. Chatain and C. Lacabanne, *Polymer Engineering and Science*, 1997, **37**, 1598.

79. D.W. Bamborough in *Proceedings of the Technomic Pressure Sensitive Adhesive Technology Conference*, 2001, Milan, Italy, 2001, Paper No.9.

80. J. Son, D.J. Gardner, S. O'Neill and C. Metaxas, *Journal of Applied Polymer Science*, 2003, **89**, 1638.

81. D.W. Bamborough in *Proceedings of the Technomic Pressure Sensitive Adhesive Technology Condference*, 2001, Milan, Italy, 2001, Paper No.10.

82. D.W. Bamborough in *Proceedings of the Technomic Pressure Sensitive Adhesives Conference*, 1993, Section 111, p.2-17.

83. D. Dean, M. Husband and M. Trimmer, *Journal of Polymer Science, Polymer Physics Edition*, 1998, **36**, 2971.

84. E. Kontou and G. Spathis, *Journal of Applied Polymer Science*, 2003, **88**, 1942.

85. L. Heux, J.J. Halary, F. Laupretre and L. Monnerie, *Polymer*, 1997, **38**, 767.

86. M.E. Leyva, B.G. Soares and D. Khastgir, *Polymer*, 2002, **43**, 7505.

87. A. Zarraga, J.J. Pena, M.E. Munoz and A. Sanamaria, *Journal of Polymer Science, Polymer Physics Edition*, 2000, **38**, 469.

88. M. Botov, H. Betchev, D. Bikiaris and D. Panayiotou, *Journal of Applied Polymer Science*, 1999, **74**, 523.

89. D. Singh, V.P. Malhotra and J.L. Vats, *Journal of Applied Polymer Science*, 1999, **71**, 1959.

90. L. Priya and J.P. Jog, *Journal of Polymer Science, Polymer Physics Edition*, 2003, **41**, 31.

91. M. Sepe, *Injection Molding*, 1999, **7**, 52.

92. O. Schroder and E. Schmachtenberg in *the Proceedings of the ANTEC 2001 Conference*, Dallas, TX, USA, 2001, Paper No.405.

93. R.A. Palmer, P. Chen and M.H. Gilson in *the Proceedings of the ACS Polymeric Materials Science and Engineering Conference*, Orlando, FL, USA, Fall 1996, Volume 75, p.43.

94. R.G. Polance and J.P. Baetzold in *Journal of the Adhesive and Sealant Council, Conference Proceedings*, Chicago, IL, USA, 1998, p.189.

95. ASTM D693-03a, *Standard Specification for Crushed Aggregate for Macadam Pavements*, 2003.

96. G. Evrard, A. Belgrine, F. Carpier, E. Valot and P. Dang, *Revue Générale des Caoutchoucs et Plastiques*, 1999, **777**, 95.

97. M.L. Cerrada, *Revista de Plásticos Modernos*, 2001, **81**, 236.

98. G. Qian, J.W. Cho and T. Lan in *Proceedings of the Polyolefins 2001 Conference*, Houston, TX, USA, 2001, p.553.

99. J.M. Lagaron and A.K. Powell, *Revista de Plásticos Modernos*, 2001, **81**, 244.

100. M. Bertucci, *Macplas*, 2002, **27**, 70.

101. J. Lemaire, *Revue Générale des Caoutchoucs et Plastiques*, 2002, **802**, 26.

102. C. Granel and M. Tran in *Proceedings of the 60th SPE Annual Technical Conference*, ANTEC 2002, San Francisco, CA, USA, 2002, Session M36, Paper No. 175.

103. SAE J 1885, *Coating Powders - Part 8: Assessment of the Storage Stability of Thermosetting Powders*, 2005.

104. ASTM D1709, *Standard Test Methods for Impact Resistance of Plastic Film by the Free-Falling Dart Method*, 2004.

105. UNI EN ISO 7765-1, *Plastics Film and Sheeting - Determination of Impact Resistance by the Free-Falling Dart Method - Part 1: Staircase Methods*, 2005.

106. ASTM D 2944-71, *Standard Test Method of Sampling Processed Peat Materials*, 1998.

107. UNI 7448, *Tubi di Pvc Rigido (Non Plastificato). Metodi di Prova*, 1975.

108. R.M. Patel, P. Saavedra, J. de Groot, C. Hinton and R. Guerra in *Proceedings of the TAPPI 1997: Polymers, Laminations and Coatings Conference*, Toronto, Ontario, Canada, 1997, Book 1, p.239.

109. S.S. Woods and A.V. Pocius in *Proceedings of the ANTEC 2000 Conference*, Orlando, FL, USA, 2000, Paper No.33.

110. M.G. Cernak and W.L. Chiang in *Proceedings of the TAPPI 1997: Polymers, Laminations and Coatings Conference*, Toronto, Ontario, Canada, 1997, Book 2, p.479.

111. M. Ifwarson and K. Aoyama in *Proceedings of the Plastic Pipes X Conference*, Göteborg, Sweden, 1998, p.731.

112. Y. Germain, *Polymer Engineering and Science*, 1998, **38**, 657.

113. R. Vegas, *Revista de Plásticos Modernos*, 2002, **83**, 589.

114. M.A. Gómez, G. Ellis and C. Marco, *Revista de Plásticos Modernos*, 2002, **83**, 582.

115. C. Bonten and E. Schmachtenberg, *Kunststoffe Plast Europe*, 2000, **90**, 38.

116. R.H. Chambers, *Polymers Paint and Colour Journal*, 1979, **169**, 4013, 1109.

117. K.H.J. Klaren, *Journal of the Oil and Colour Chemists' Association*, 1977, **60**, 205.

118. ASTM D3451-01, *Standard Guide for Testing Coating Powders and Powder Coatings*, 2001.

119. ASTM D1921, *Standard Test Methods for Particle Size (Sieve Analysis) of Plastic Materials*, 2001.

120. BS ISO 13319, *Determination of Particle Size Distributions - Electrical Sensing Zone Method*, 2000.

121. ASTM D609, *Standard Practice for Preparation of Cold-Rolled Steel Panels for Testing Paint, Varnish, Conversion Coatings, and Related Coating Products*, 2000.

122. DIN ISO 8130, *Coating Powders*, 2005.

123. ASTM D822-01, *Standard Test Method for Stiffness Properties of Plastics as a Function of Temperature by Means of a Torsion Test*, 2001.

124. ASTM G155-04a, *Standard Practice for Operating Xenon Arc Light Apparatus for Exposure of Non-Metallic Materials*, 2004.

125. ASTM D791, replaced by E308, *Standard Practice for Computing the Colors of Objects by Using the CIE System*, 2001.

126. BS 2701, *Specification for Rees-Hugill Powder Density Flask*, 1965.

127. ASTM D1895-96, *Standard Test Methods for Apparent Density, Bulk Factor, and Pourability of Plastic Materials*, 2003.

128. ASTM 523-89, *Standard Test Method for Specular Gloss*, 1999.

129. DIN 53157, *Paints and Varnishes - Pendulum Damping Test*, 2000.

130. ASTM D2794-93, *Standard Test Method for Resistance of Organic Coatings to the Effects of Rapid Deformation (Impact)*, 2004.

131. ASTM D522-93a, *Standard Test Methods for Mandrel Bend Test of Attached Organic Coatings*, 2001.

132. ASTM B117-64, *Standard Practice for Operating Salt Spray (Fog) Apparatus*, 2003.

133. DIN ISO 8130-10, *Coating Powders - Part 10: Determination of Deposition Efficiency*, 2005.

134. DIN ISO 8130-8, *Coating Powders - Part 8: Assessment of the Storage Stability of Thermosetting Powders*, 2005.

135. NFT 30-502, *Peintures en Poudre Thermodurcissables - Determination de la Temperature d'agglomeration*, 1977.

136. DIN ISO 8130-7, *Coating Powders - Part 7: Determination of Loss of Mass on Stoving*, 2005.

137. NFT 30-500, *Peintures en Poudre Thermodurcissables - Determination de l'aptitude a la Projection*, 1976.

138. BS EN ISO 9001, *Quality Management Systems – Requirements*, 2000.

139. ISO 9001, *Quality Management Systems – Requirements*, 2001.

140. ASTM D1329, *Standard Test Method for Evaluating Rubber Property— Retraction at Lower Temperatures (TR Test)*, 2002.

141. ASTM D1052, *Standard Test Method for Measuring Rubber Deterioration-Cut Growth Using Ross Flexing Apparatus*, 2005.

142. ASTM D945-92e1, *Standard Test Methods for Rubber Properties in Compression or Shear (Mechanical Oscillograph)*, 2001.

143. G. Orjela, G. Riva and F. Fiorentini, *Tire Science and Technology*, 1998, **26**, 208.

144. ASTM D1401-02, *Standard Test Method for Water Separability of Petroleum Oils and Synthetic Fluids*, 2002.

145. DIN IEC 60093, *Methods of Test for Insulating Materials for Electrical Purposes; Volume Resistivity and Surface Resistivity of Solid Electrical Insulating Materials*, 1993.

146. DIN IEC 60167, *Methods of Test for Insulating Materials for Electrical Purposes; Insulation Resistance of Solid Materials*, 1993.

147. ASTM D257-99, *Standard Test Methods for DC Resistance or Conductance of Insulating Materials*, 1999.

148. ASTM D150-98, *Standard Test Methods for AC Loss Characteristics and Permittivity (Dielectric Constant) of Solid Electrical Insulation*, 2004.

149. DIN 53483-1, *Testing of Insulating Materials; Determination of Dielectric Properties; Definitions, General Information*, 1969.

150. ASTM D149-97a, *Standard Test Method for Dielectric Breakdown Voltage and Dielectric Strength of Solid Electrical Insulating Materials at Commercial Power Frequencies*, 2004.

151. ASTM D495-99, *Standard Test Method for High-Voltage, Low-Current, Dry Arc Resistance of Solid Electrical Insulation*, 2004.

152. DIN IEC 60112, *Method for the Determination of the Proof and the Comparative Tracking Indices of Solid Insulating Materials*, 2004.

153. CEI 15-50, *Guida per la Determinazione delle Proprieta' di Resistenza alla Sollecitazione Termica dei Materiali Isolanti Elettrici - Parte 2: Scelta dei Criteri di Prova*, 1997.

154. UNI 4290, *Tests on Thermosetting Plastics - Determination of Resistance to Tracking*, 1959.

155. D495-99, *Standard Test Method for High-Voltage, Low-Current, Dry Arc Resistance of Solid Electrical Insulation*, 2004.

156. ASTM D542, *Standard Test Method for Index of Refraction of Transparent Organic Plastics*, 2000.

19 Miscellaneous Physical and Chemical Properties

19.1 Introduction

Miscellaneous properties of polymers for which test standards exist are listed in **Table 19.1** together with details of suppliers of test equipment.

19.2 Particle Size Characteristics of Polymer Powders

Procedures based on several principles are used for the measurement of the particle size distribution of polymer powders.

19.2.1 Methods Based on Electrical Sensing Zone (or Coulter Principle)

Figure 19.1 shows schematically a simple form of the apparatus. Particles, suspended homogeneously at a low concentration in an electrolyte solution, are made to flow through a small aperture (or orifice) in the wall of an electrical insulator, which is commonly called the aperture (or orifice) tube - the aperture creates the sensing zone. In addition, a current path is established between two immersed electrodes, across this aperture, setting a certain base impedance to the electrical detection circuitry. A direct current is generally used. As each particle enters the aperture, it has effectively displaced a volume of electrolyte solution equal to its own immersed volume, and the base impedance is therefore modulated by an amount proportional to the displaced volume of the particle. This results in an electrical pulse of short duration being created by each particle, the height of the pulse being essentially proportional to the volume of the particle. The pulse may be measured, for example as the change in resistance, current, or voltage across the electrodes.

The passage of a number of particles produces a train of pulses that can be observed on an oscilloscope and analysed by counter and pulse height analyser circuits to produce a number against particle volume, or equivalent spherical diameter, distribution. A volume or mass ('weight') against size distribution can also be measured, calculated, or computed; the 'weight' percentages being possible if all of the particles have uniform density or a known density distribution across their size range.

Table 19.1 Test equipment for miscellaneous physical and chemical tests on polymers		
Method	Specification test	Equipment supplier
Water Absorption		
Water absorption of polymers	ASTM D570 [1] DIN 7708 [2]	
Gas Permeability		
Gas permeability of elastomers (water vapour, CO_2 and O_2)	DIN	ATS FAAR Code Nos. 10.08050 and 10.08100
Chemical Resistance		
Chemical resistance	ASTM D543 [3]	
Solvent resistance	ASTM D543 [3]	ATS FAAR Code Nos. 10.47000 and 10.47005
Ozone resistance	ASTM	Gibitre Ozone Cabinet available from Negretti Automation Ltd.
Environmental stress cracking resistance	ASTM	ATS FAAR Code No. 10.36000
Miscellaneous Properties		
Hydrogen chloride content of PVC	CEI	ATS FAAR Code No. 11.00050
Carbon black content of polyethylene	ASTM	ATS FAAR Code No. 11.00001
Carbon black dispersion in rubbers	-	Tangent Electroscanner Negretti Automation Ltd
Particle distribution		See Table 19.2
Source: Author's own files		

Simple models have only one counter and size level circuit (and so are called single-channel models); more complex instruments can obtain number and/or mass (weight) distributions automatically in up to 256 size channels within a few seconds. Counting and sizing rates of up to some 10,000 particles per second are possible, with each pulse height being measured to within 1 or 2%.

Since most particulate materials are irregularly shaped, the volumetric response is invaluable, as volume is the only single measurement that can be made of an irregular particle in order to characterise its size. In biological applications the size response is usually left calibrated in volume units (femtolitres, or 'cubic microns'), but industrially it is conventional to report the equivalent spherical diameter calculated from it. This volumetric method makes no assumptions about particle shape, and indeed is not greatly affected by particle shape except in extreme cases like flaky materials, such as some clays. To count the number of particles in a known volume of suspension, such as for particulate contamination studies or for a blood cell count, the sample volume is accurately metered by means of a calibrated 'manometer'. **Figure 19.1** illustrates the original, simple mercury siphon and metering system.

The shape of the aperture tube can vary according to application - for example a very narrow design, which can be inserted into glass ampoules as small as 1 ml capacity, allows the particle contamination of injectable solutions to be measured.

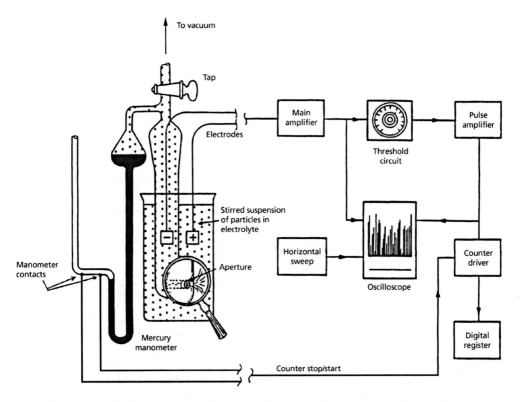

Figure 19.1 Schematic drawing of a simple form of particle size analyser.
(*Source: Author's own files*)

Modern instrument designs have a range of extra features, including embedded microprocessors, and various data reduction handling and presentation methods and the manufacturers should be contacted for details. The volumetric sizing resolution, speed of data collection, statistical accuracy of counting, freedom from any optical response effects, and the reliability and simplicity of calibration make these devices unsurpassed for providing particle counting and size distribution analyses.

Coulter type instrumentation available from Coulter Electronics Ltd., is listed in **Table 19.2** (Multisizer and Model 2M) and Appendix 1.

19.2.2 Laser Particle Size Analysers

Light diffraction is one of the most widely used techniques for measuring the size of particles in the range 0.1–1000 µm. This popularity is partly due to the way precise measurements can be made quickly and easily. It also stems from the flexibility of the technique, particularly the way it can be adapted to measure samples presented in various forms.

The method depends on analysing the diffraction patterns produced when particles of different sizes are exposed to a collimated beam of light. As the patterns are characteristic of the particle size, mathematical analysis can produce an accurate, reproducible picture of the size distribution.

Coulter Electronics Ltd., supplies a range of laser particle size distribution apparatus (LS100 and LS130) (see **Table 19.2** and Appendix 1).

Christianson Scientific Equipment supply the Fritsch Analysette 22 laser particle sizer which operates over the particle size range 0.16–1250 µm. This is a universally applicable instrument for determining particle size distributions of all kinds of solids which can be analysed either in suspension in a measuring cell or dry by feeding through a solid particle feeder. In the Fritsch Analysette 22, laser diffraction apparatus the measured particle size distribution is displayed on the monitor in various forms, either as a frequency distribution or as a summary curve in tabular form, and can be subsequently recorded on a plotter, stored on hard disk, or transferred to a central computer via an interface. The time required for one measurement is approximately two minutes.

Table 19.3 lists some suppliers and working ranges of various particle size distribution methods.

Terray [4] examined laser diffraction as a method for the in-line measurement of particle size, and examples of the application of Insitec laser granulometers developed by Malvern Instruments are described. These include their use for particle size measurement in the cryogenic grinding of plastics.

Table 19.2 Particle size measurement instrumentation available from Coulter Electronics Ltd.

Instrument	Type	Particle size range	Analysis time, s	Measurement
Coulter Multisizer	Coulter	0.4-1200 μm (overall) 0.4-336 μm for distributions to be expressed as volume (weight) against size	30 - 90	Counting and sizing
Model ZM (more basic than multisizer)	Coulter	0.4-1200 μm	-	Counting and sizing
Coulter LS Series particle size distribution analyser LS130 LS100	Laser light diffraction	0.1-810 μm	60	Particle sizing
Coulter Series Model N4	Photon correlation spectroscopy (auto correlation spectroscopy)	0.003-3 μm	-	Sub micro particle distribution and molecular weight
Model N45 measures coefficients of particles by laser light scattering and converts them to size or molecular weight.			-	
Model N4MD automated laster based submicron particle size analyser		0.003-3 μm	-	
Model N45D submicron particle size analyser with size distribution analysis capability			-	

Source: Author's own files.

Table 19.3 Suppliers and working ranges of particle-size distribution methods			
Method	Particle size range	Equipment supplier	Model
Dry sieving	63 µm - 63 µm	Fritsch	Analysette 3, (20 mm - 25 mm)
Wet sieving	20 µm - 200 µm	Fritsch	Analysette 18, see Table 7.3 A-C
Microsieving	5 µm - 100 µm	Fritsch	
Sedimentation in gravitational field	0.5 µm - 500 µm	Fritsch	Analysette 20
Laser diffraction	0.1 µm - 1100 µm	Fritsch	Analysette 22
Electrical zone sensing	0.4 µm - 1200 µm	Coulter	Model ZM, Coulter multisizer
Electron microscopy	0.5 µm - 100 µm	-	-
Photocorrelation spectroscopy	0.5 µm - 5 µm	-	-
Sedimentation in centrifugal field	0.5 µm - 10 µm	Fritsch	Analysette 21 (Anderson Pipette centrifuge)
Diffraction spectroscopy	1 µm - 1 mm		
Optical microscale	0.5 µm - 1 mm		
Projection microscopy	0.05 µm - 1 mm		
Image analysis systems	0.8 µm - 150 µm down to 0.5 µm	Joyce Leobl, Leitz Karl Zeiss, Cambridge Instruments	Magiscan and Magiscan P Autoscope P Videoplan II Quantimet 520
Source: Author's own files.			

19.2.3 Photon Correlation Spectroscopy (Autocorrelation Spectroscopy)

The Coulter N4 and its predecessor, the Nano-Sizer, both use autocorrelation spectroscopy of scattered light to determine the time-dependent fluctuations of the scattered light which result from the Brownian motion of the particles in suspension. The technique is also known as photon correlation spectroscopy.

19.2.4 Sedimentation

The use of sedimentation whereby the particle size is determined by the application of Stokes' law is well established. The time taken for fine particles to complete the sedimentation process is far too long for practical utility. However, the development of the wide scanning sedimentometer whereby a suspension of particles that are separated under gravity is scanned by a light beam which is attenuated accordingly has largely overcome this difficulty and a result may be obtained in half an hour by a competent operator using hand calculation methods.

19.2.5 Other Instrumentation

19.2.5.1 Acoustic Spectroscopy

Particle size measurement results obtained from the deconvolution of acoustic attenuation over a broad frequency range is showing increasing potential in the characterisation of emulsions and slurries at process concentrations over particle sizes ranging from 0.01 to 1000 µm [5]. The results of this study reveal that the particle size distribution in both high- and low-density contrast materials can be accurately measured by means of acoustic spectroscopy. This technique is distinct from other size characterisation technologies in three ways: there are no optics involved, since the instrument can measure through opaque high concentration suspensions; it has an extremely wide dynamic range; and it is well suited for on-line applications, being a relatively robust instrument with no dilution required.

19.2.5.2 Capillary Hydrodynamic Fractionation

Venkatesan and Silebi [6] used capillary hydrodynamic fractionation to monitor an emulsion polymerisation of styrene monomer as a model system. A sample taken from the reactor at different time intervals is injected into the capillary hydrodynamic fractionation system to follow the evolution of the particle size distribution of the polymer particles formed in the emulsion polymerisation. After the colloidal particles have been fractionated by capillary hydrodynamic fractionation they pass through a photodiode array detector which measures the turbidity at a number of wavelengths instantaneously, thereby enabling the utilisation of turbidimetric methods to determine the particle size distribution. The particle size measurement is not hindered by the presence of monomer-swollen particles. The shrinkage effect due to the monomer swelling phenomenon is found to be accurately reflected in the particle size measurements.

19.2.5.3 Small-Angle Light Scattering

Boerschig and co-workers [7] mention that light scattering should prove useful for the direct measurement of particle growth during flow-driven coalescence of polystyrene/polymethyl methacrylate blends.

References

1   ASTM D570, *Standard Test Method for Water Absorption of Plastics*, 1998.

2.   DIN 7708-1, *Plastic Moulding Materials: Plastic Products*; Concepts, 1980.

3.   ASTM D543-95, *Standard Practices for Evaluating the Resistance of Plastics to Chemical Reagents*, 2001.

4.   M. Terray, *Informations Chemie*, 1998, **398**, 121.

5.   R. Trottier, J. Szalanski, C. Dobbs and A. Felix in *Proceedings of the ACS Polymeric Materials Science and Engineering Conference*, Orlando, FL, USA, Fall 1996, Volume 75, p.58.

6.   J. Venkatesan and C.A. Silebi in *Proceedings of the ACS Polymeric Materials Science and Engineering Conference*, Orlando, FL, USA, Fall 1996, Volume 75, p.99.

7.   C. Boerschig, B. Fries, W. Gronski, C. Weis and C. Friedrich, *Polymer*, 2000, **41**, 3029.

20 Additive Migration from Packaged Commodities

Plastics are now being used on a large scale for the packaging of foodstuffs and beverages, both non-alcoholic and alcoholic. This is evident for all to see on the supermarket shelf, e.g., margarine packed in polystyrene (PS) tubs, wine and beer in polyvinyl chloride bottles, and bacon in polyolefin shrunk wrap films. As well as at the point of sale, foods are increasingly shipped in bulk in plastic containers. Additionally there is the area of the increasing use of plastic utensils and containers in both the home and during the bulk preparation of food in restaurants, canteens, and food factories. Contact between plastics and packaged substances also occurs in the products of the pharmaceutical and cosmetics industries.

Where direct contact occurs between the packed commodity and the plastic it is likely that some transfer of polymer additives will occur, adventitious impurities such as monomers, catalyst remnants, and residual polymerisation solvents, and low molecular weight (MW) polymer fractions from the plastic into the packaged material with the consequent risk of a toxic hazard to the consumer. The actual hazard arising to the consumer from any extractable present is a function of two properties, namely the intrinsic toxicity of the neat extracted material as evaluated in animal feeding trials (not dealt with in this book) and the amount of the extracted material from the polymer that enters the packed food under service conditions, i.e., during the packaging operation and during the shelf life of the food to the time of consumption.

Most countries now have regulations regarding which polymer additives it is permitted and which it is not permitted to use in plastics that come into contact with food, and indeed which it is permitted to use as a preservative in the food itself. Also, for the permitted additives and other extractable substances present in polymers, regulations exist in many countries for the maximum amounts of these that can migrate into the food under standardised test conditions.

As well as the foods themselves, many authorities have adopted various standard extraction liquids selected to simulate the action of actual foods for use in polymer extraction tests.

The analysis of foods, beverages, or simulant liquids that have been in contact with plastics either in extraction tests or during the shelf life of a packaged commodity presents many fascinating and all too often difficult problems. Thus, the substance to be determined usually occurs at very low concentrations and several extracted substances may occur in the same food or simulant. For example, the extract of a PS may contain an antioxidant, an ultraviolet

(UV) stabiliser, an antistatic agent, mineral oil, and plasticisers. Although it may not be necessary to determine all of these, it is necessary to be aware of any interference that these substances might have on the determination of any particular component. Also, there is the question of additive breakdown either during polymer manufacture or upon contact with the food. Here there are two considerations: possible interference effects of breakdown products on the determination of the additive and the necessity to identify the breakdown products, which themselves must be considered for their toxic effects.

20.1 Polymer Additives

All polymers, in addition to the basic plastic, contain usually several, if not a multiplicity, of non-polymeric components in amounts from less than 1 ppm to several percent. These substances obviously have implications in the suitability or otherwise of the plastic for applications involving contact with foodstuffs. Thus, although the plastic itself, due to its very high MW, might not contaminate the foodstuff, it is apparent that certain of the additives, which are usually of relatively low MW and therefore of higher solubility, will be transferred from the plastic to the foodstuff during storage. This raises questions about the toxicity of the additives, the amounts that transfer, and the possible implications of this, from the toxicity point of view, as far as the food consumer is concerned.

Non-polymeric components are present in plastics either unavoidably as a result of the process of manufacture or as the result of deliberate additions to the plastic in order to improve some aspect of ease of manufacture or final polymer properties. Non-polymeric components can be subdivided into three groups:

1. Polymerisation residues
2. Processing aids
3. End-product additives

Polymerisation residues cover substances whose presence is to a large extent unavoidable, such as low MW polymer oligomers, catalyst remnants, polymerisation solvents, non-polymerisable impurities, and impurities picked up from plant materials. Processing aids include such substances as thermal antioxidants and heat stabilisers added to prevent decomposition of the polymer during moulding and slip additives to facilitate moulding. The third group again are deliberately added to the polymer either during manufacture or subsequently to improve the properties of the final polymer. As discussed next, a very wide range of non-polymeric substances are used in this category ranging from secondary thermal antioxidants to impact improvers, plasticisers, UV stabilisers, antistatic agents, etc.

The situation regarding the presence of non-polymeric low MW additives in polymers can be best illustrated by an actual example concerning a batch of polypropylene (PP), which, upon detailed but by no means complete examination, was shown to contain the constituents listed in **Table 20.1**. In many instances the probable origin of these substances can be determined. Thus, in this single polymer substances falling under the three categories

Table 20.1 Non-polymeric components found in a sample of PP		
Determined	Concentration (ppm)	Origin
Polymerisation residues		
Aluminium	40	Remnants of Ziegler catalysts
Titanium	5	Organoaluminium
Chlorine	230	Titanium halide catalysts
Sodium	30	Catalyst neutralisation
Potassium	18	Alkalis
Silicon	<20	General
Iron	5	Contamination
Manganese	0.1	From plant
Nickel	8	
Tin	0.3	
Vanadium	<0.2	
Zinc	<10	
Chromium	1	
C_6/C_{16} hydrocarbons	900	Unreacted monomer
Dissolved propylene	1 ml/ml polymer	Residual polymerisation solvent
Processing aids		
Calcium	40	Possibly calcium stearate stabiliser for protecting polymer during moulding
Ionol	200	Antioxidant (thermally degraded Ionol also present)
End-product additives		
Barium	1000	Possible barium sulfate filler
Substituted benzopheno-(2-hydroxy-4-*n*-octoxy-benzophenone)		Light stabiliser
Dilaurythiodipropionate	ppm sulfur found	Probable secondary antioxidant
Source: Author's own files		

previously mentioned were found. As shown, these include Ziegler catalyst remnants, neutralising chemicals, residual monomer, and polymerisation solvent (polymerisation residues), a calcium salt and an antioxidant (processing aids), and a further antioxidant, a light stabiliser, and a filler (end-product additives).

20.2 Extraction Tests

The extractability of an additive from a plastic can be determined by contacting the plastic for a specified time and temperature with either the foodstuff or beverage itself or with a range of oily, alcoholic, and aqueous extractants that simulate various types of food.

The specified conditions in the plastics extraction test and the standard extractants prescribed by different authorities differ considerably. As an example, the extractants quoted by the British Plastics Federation [1] and the US Food and Drug Administration [2, 3] are given in **Table 20.2**.

Various workers have discussed the theory governing the rate of migration of extractable components from plastics into liquids [4-11]. Lee and Archer [12] have proposed a mechanism based on shear-induced diffusion to explain the migration of additives to the surface of polymers. Most developed countries in the world now have legislation concerning the use of additives in food-grade plastics. In particular, recent legislation of the European Commission has been discussed [13-20] including reviews of new directives.

Table 20.2 Comparison of specified standard extractants	
British Plastics Federation	**US Food and Drug Administration**
Distilled water	Distilled water
Ethanol:water (1:1)	Heptane
5% aqueous sodium bicarbonate	8% and 50% aqueous ethanol
5% aqueous citric acid	3% aqueous sodium chloride
Olive oil containing 2% oleic acid, or liquid paraffin	20% aqueous sucrose
	3% aqueous sodium bicarbonate
	3% aqueous acetic acid
	Food (vegetable) oil
Source: Author's own files	

Recent work includes extraction studies of aqueous extractants from acrylate coatings and adhesives using liquid chromatography(LC)–mass spectrometry(MS)/LC–MS–MS detection methods [21], extractant studies on elastomeric materials using MS techniques [22], and studies on rubbers using LC–atmospheric pressure chemical ionisation MS [23].

Other investigations involving aqueous and oily extractants are reviewed in **Table 20.3**. The concept of total migration of additives from plastics into foodstuffs and beverages as opposed to specific migration has been discussed by several workers [24–56]. The direct determination of antioxidants in foodstuffs has also been reviewed [57-84].

Occasionally it is found that an additive that has been extracted from a plastic will undergo a chemical change upon reaction with the extract. A case in point is that of the antistatic additive lauric diethanolamide which upon reaction with aqueous extraction liquids is hydrolysed to diethanolamine and lauric acid. In these circumstances it is not only the toxicity of lauric diethanolamide that must be considered but also that of the two hydrolysis products. In fact, it has been ascertained that during a 14 day extraction test on PP carried out at 60 °C some 100% of the extracted diethanolamide is hydrolysed:

$$C_{11}H_{23}CON(CH_2CH_2OH)_2 + H_2O = C_{11}H_{23}COOH + HN(CH_2CH_2OH)_2$$

Hamdani and co-workers [171] carried out an *in vitro* study of the hydrolysis of polymer plasticisers such as poly(1,2-propylene adipate) by digestive fluid liquids. High-performance size-exclusion chromatography indicated that the bulk plasticiser completely disappeared and low MW oligomers were produced within four hours.

Table 20.3 Extractability testing of additives from plastics

Plastic	Extractant	Additive	Method of analysis	Ref.
Polyethylene, polystyrene	Aqueous, alcoholic, oily	Uvitex OB (2,5 *bis*(6-*tert* butylbenzo oxalyl (2) thiophen), UV stabilizer	Ultraviolet spectroscopy with a Moreton and Stubbs correction procedure	a [85]
Polyolefins	Aqueous, alcoholic, oily	Santonox R (4'4 thiobis-6-*tert* bicresol) antioxidant	Ultraviolet spectroscopy	a
Polyolefins	Aqueous, alcoholic, oily	Dilaurythiodipropionate secondary antioxidant	Thin-layer chromatography	a
Polystyrene	Aqueous, alcoholic, oily	Styrene monomer, eumene, ethyl benzene, toluene, total o,m,p xylene, benzene	Gas chromatography	a [86]
Styrene-acrylonitrile copolymers	Aqueous, alcoholic, oily	Acrylonitrile monomer	Polarography	[87, 88]
Polyvinyl chloride	Vinegar, orange juice	Organotin, dioctyltin 5,5' *bis*(isoactyl mercapto acetate)	Spectrophotometrically in catechol-violet complex	[89]
Polyvinyl chloride	Beer	*Bis*(zethylhexyloxy carbonyl) methyl thio dioctyl tin	Spectrophotometrically as catechol violet complex	[90, 91]
Polyvinyl chloride	Vinegar and salad oil		Thin-layer chromatography	[92, 93]

Table 20.3 Continued ...

Plastic	Extractant	Additive	Method of analysis	Ref.
Polyvinyl chloride	Vinegar, orange drink	Mercapto acetate organotin stabilizer	Chloroform extraction then thin-layer chromatography	[92-94]
Polyvinyl chloride, high pressure polyethylene, polystyrene, polymethyl methacrylate	Ethanol (15%, 50% and 96%), heptone, sunflower seed oil	Hydroxybenzophenones, hydroxyphenylbenzo-triazole	Thin-layer chromo-graphically then spectro-photometrically at 290 nm or polargraphically	[96]
Polyvinyl chloride	Milk	Phthalate esters, C_{10}/C_{20} alkonephenyl sulphonates	Spectrophotometric of 284 nm or spectro-photometric at 470 nm	[97-99]
Polyvinyl chloride	Aqueous extractants	Barium, cadmium, lead and zinc organo-metallic stabilizers	Thin-layer chromato-graphy, spots located with aq. sodium rhodizoate	[100]
Miscellaneous	Various	Analyses by determination of trace elements in extraction liquids		
Miscellaneous	Various	Sulphur	Schoniger combustion - potent 10 metric titration with $Ba(ClO_4)_2$	[101, 102]
Miscellaneous	Various	Chlorine, bromine, oriodine	Schoniger combustion - potentiometric titration with Aq NO_3	[103]
Miscellaneous	Various	Phosphorus	Evaluation of phosvanomethyldate coupler at 430 nm	[104]

Table 20.3 Continued...

Plastic	Extractant	Additive	Method of analysis	Ref.
Miscellaneous	Various	Nitrogen	Rjeldahl digestion - spectrophotometry of phenol-indophenol complex at 630 nm	[105]
Miscellaneous	Various	Fluorine	Schoniger combustion spectrophotometric evaluation of alizarin cerous nitrate complex at 610 nm	[106]
Miscellaneous	Various	Boron	Spectrophotometric evaluation of methol borate at 519.5 nm	[107]
Miscellaneous including polystyrene, polyvinyl chloride, polyethylene	Oily extractants including edible oils, liquid paraffin, coconut oil, synthetic tri-glycerides	Miscellaneous	Using radiactive labelled additives e.g. di-*n*-octyl[1-^{14}C] tin-2-ethyl-*n*-hexyl dithioglycollate, 1,3,5 tri ethyl 2,4-6(3,5 *di*-tert butyl, 4-hydroxy benzyl benzene, *n*-butyl stearate and stearic acid	[108-136]
Miscellaneous	Fatty extractants (synthetic tryglyceride)	Antioxidants, ultraviolet stabilizers, heat stabilizers	Using labelled additives	[137]
Miscellaneous	Fatty extractant (synthetic triglycerol-ceride)	Phenolic antioxidants	Visible spectrophoto-metric method	[168]

Table 20.3 Continued...

Plastic	Extractant	Additive	Method of analysis	Ref.
Plastics utensils	Sunflower seed oil	Santonox R antioxidant (4,4'thio-*bis*-6-tert butyl-*m*-cresol)	Thin-layer chromatography	[149, 150]
Polystyrene	Olive oil and aqueous extraction liquids, sunflower oil, linseed oil	Santonox R antioxidant and hydroxybenzo-triazoles	Spectrophotometric and polarographic techniques	[151-153]
Rubber goods, (styrene-butadiene, natural and nitrile rubbers)	Ailimentary fats	2,2'methylene bis-(6-*tert* butyl-4-methyl phenol) antioxidant	Spectrophotometric	[154-156]
Rubber goods	Olive oil, coconut oil	Miscellaneous	Gas chromatography	[157]
Polyvinyl chloride, cellulose triacetate polyethylene, polystyrene	Organic solvents and synthetic triglycerides	Citrate and phthalate plasticizers, buylated hydroxy anisole, organotin stabilizers	Miscellaneous	[158]
Miscellaneous	Butter fat	Phthalate plasticizers	Gas chromatography	[159]
Polyvinyl choride	Synthetic tri-glycerides	Di-*n*-octyltin etc.	Spectrophotometric and radiochemical	[160-165]
High density polyethylene	Olive oil	Benzophenone and benzotriazole UV stabilizers, antioxidant	Study of migration models	[169]

Table 20.3 Continued...

Plastic	Extractant	Additive	Method of analysis	Ref.
Miscellaneous	Aqueous and fatty extractants	*Bis*-2-ethyl hexyl) adipate, *bis*(2-ethyl-hexyl phthalate), octadecyl 3-(3,5 di-*tert* butyl-4-hydroxy phenyl) propionate (Irganox 1076)	Gas chromatography	[170]
Low density polyethylene	Ethanol	Irganox 1010	Gas chromatography	[169]
Miscellaneous	Aqueous and food oils	Monomers, oligomers, additives, modifiers	Gas chromatography	[170]
Miscellaneous	Wine, edible oils	Polydimethyl siloxanes	Proton magnetic resonance spectroscopy	[171]
Plasticized polyvinyl chloride	Aqueous and food oils	*Di-2-n* ethyl hexyl phthalate plasticizer, organotin heat stabilizer, processing aids, external and internal lubricants	Gas chromatography FTIR spectroscopy, atomic absorption spectrometry, differential scanning calorimetry	[172]

a Unpublished work, T.R. Crompton

References

1. *Second Report of the Toxicity Sub-committee of the Main Technical Committee of the British Plastics Federation, with Methods of Analysis of Representative Extractants*, The British Plastics Federation, UK, 1962.

2. *United States Federal Register, Food and Drugs, Chapter 1*, Food and Drug Administration, Department of Health and Education and Welfare. Part 121, Subpart F, Food additives resulting from contact with containers or equipment and food additives otherwise affecting food, Sections 2501 (polypropylene) and 2510 (polyethylene), 1967.

3. *Code of Federal Regulations*, Food and Drug Administration, Washington, DC, USA, 1967, Chapter 1, Part 121, Sections 2514 and 2526.

4. T. Garlanda and H. Masaero, *La Chimica e l'Industria*, 1966, 78, No. 9.

5. T. Garlanda and H. Masaero, *La Chimica e l'Industria*, 1965, 973, No. 47.

6. L. Robinson and K. Becker, *Kunststoffe*, 1965, 55, 233.

7. L. Robinson and K. Becker, *Kunststoffe*, 1965, 55, 234.

8. D. Chung, S.E. Papadakis and K.L. Yam, *Food Additives and Contaminants*, 2002, 19, 611.

9. A. O'Brian and I. Cooper, *Food Additives and Contaminants*, 2002, 19, 63.

10. A. O'Brian, A. Goodson and I. Cooper, *Food Additives and Contaminants*, 1999, 16, 367.

11. M.J. Forrest, *Analysis of Plastics, Rapra Review Report*, No. 149, Rapra Technology Ltd., Shrewsbury, UK, 2002, Volume 13, No.5.

12. H. Lee and L.A. Archer, *Polymer Engineering and Science*, 2002, 42, 1568.

13. J. Leadbitter in *Proceedings of the PUC '96 Conference*, Brighton, UK, 1996, p.315.

14. *European Plastics News*, 2003, 30, 5, 6.

15. A. Coupard, M. Le Huy and H. Khalfoune, *Revue Générale des Caoutchoucs et Plastiques*, 1996, 750, 97.

16. L. Reade, *European Plastics News*, 2003, 30, 6, 33.

17. B. Ashby in *Proceedings of the PIRA International, Plastics and Polymers in Contact with Foodstuffs Conference*, Coventry, UK, 2001, Paper No.12, p.17.

18. C. Gueris in *Proceedings of the PIRA International, Plastics and Polymers in Contact with Foodstuffs Conference*, Coventry, UK, 2002, Paper No.1.

19. B. van Lierop, L. Castle, A. Feigenbaum, K. Ehlert and A. Boenke, *Food Additives and Contaminants*, 1998, **15**, 855.

20. B. Guise, *Packaging*, 1997, **4**, 2.

21. A. Lin, G. Gao, G. Wind and F. Wornick in *Proceedings of the TAPPI PLACE 2002 Conference*, Boston, MA, USA, 2002, Session 13, Paper No.48.

22. J. Sidwell in *Proceedings of the Rapra Technology Rubber Chem 2002 Conference*, Munich, Germany, 2002; Paper No.16, p.117.

23. K.A. Barnes, L. Castle, A.P. Damant, W.A. Read and D.R. Speck, *Food Additives and Contaminants*, 2003, **20**, 196.

24. K. Figge, *Food and Cosmetics Toxicology*, 1973, **11**, 963.

25. K Figge, *Bundesgesundheitsblatt*, 1975, **24**, 27.

26. K. Figge, *Deutsche Lebensmitt-Rundschau*, 1975, **71**, 129

27. K. Figge and J. Koch, *Fette Seifen Anstrichmittel*, 1975, 77, 184.

28. W.G. Aldershoff, *Annali-Istituto Superiore di Sanità*, 1972, 8, 550.

29. E. Baumgartner, *Kunststoffe-Plastics*, 1968, **15**, 3.

30. *Schweizerisches Lebensmittelbuch*, Vorentwurf, V. Auflage, Band II, Kap, 48-Kunststoffe, Federal Health Office, Berne, Switzerland, 1966.

31. K. Figge in *Proceedings of the Gesellschaft Deutscher Chemiker (GDCh), Präparative Radiochemie Conference*, Lindau/Bodensee, Germany, 1968.

32. Italian Ministry of Health, *Ministerial Decree*, April 1966, Gazzetta Ufficiale, May 1961, No. 111, 2295.

33. *United States Code of Federal Regulations Section 121, 2501*.

34. *United States Code of Federal Regulations, Section 121, 2514*.

35. K. Figge and H. Piater, *Deutsche Lebensmitt-Rundschau*, 1972, **68**, 313.

36. *Deutsche Lebensmitt-Rundschau*, 1972, **68**, 37.

37. R. Brugger, University of Berne, 1971. [PhD thesis]

38. J.H. de Wilde, *Deutsche Lebensmitt-Rundschau*, 1966, **61**, 369.

39. E. Fluckiger and C. Rentsch, *Alimenta*, 1968, **7**, 41.

40. L. Robinson-Görnhardt, *Kunststoffe*, 1957, **47**, 265.

41. L. Robinson-Görnhardt, *Kunststoffe*, 1958, **48**, 463

42. *Determination de la Migration des Constituants des Materiaux Destinés à être Mis au Contact des Denrées Alimentaires Ayant un Contact Gras*, Bureaux Internationaux Techniques des Matières Plastiques (BITMP), Brussels, Belgium, 1971.

43. W. Pfab, *Annali-Istituto Superiore di Sanità*, 1972, **8**, 385.

44. D. van Battum and M.A.H. Rijk, *Annali-Istituto Superiore di Sanità*, 1972, **8**, 421.

45. G. Wildrett, K.W. Evers and F. Kiermeier, *Zeitschrift für Lebensmittel-Untersuchung und Forschung*, 1970, **142**, 205.

46. J. Koch, *Deutsche Lebensmitt-Rundschau*, 1972, **68**, 216.

47. K. Figge, *Food Cosmetics and Toxicology*, 1972, **10**, 815.

48. K. Figge, S.R. Eder and H. Paiter, *Deutsche Lebensmitt-Rundschau*, 1972, **68**, 359.

49. K. Figge, *Kunststoffe*, 1971, **61**, 832.

50. K. Figge and H. Paiter, *Deutsche Lebensmitt-Rundschau*, 1971, **67**, 47, 110, 154 and 235.

51. K. Figge and H. Paiter, *Deutsche Lebensmitt-Rundschau*, 1971, **67**, 265.

52. C.G. vom Bruck, K. Figge, H. Piater and V. Wolf, *Deutsche Lebensmitt-Rundschau*, 1971, **67**, 444.

53. K. Figge, *Deutsche Lebensmitt-Rundschau*, 1973, **7**, 253.

54. J. Koch and R. Kröne, *Deutsche Lebensmitt-Rundschau*, 1975, **8**, 291.

55. W. Pfab, *Deutsche Lebensmitt-Rundschau*, 1972, **68**, 350.

56. D. van Battum and M.A.H. Rijk, *Annali-Istituto Superiore di Sanità*, 1972, **8**, 423.

57. A.M. Phillips, *Journal of the American Oil Chemists' Society*, 1973, **50**, 21.

58. K.T. Hartman and L.C. Rose, *Journal of the American Oil Chemists' Society*, 1970, **47**, 7.

59. K. Lemieszek-Chadorowska and A. Snycerski, *Roczn Panst Zakl Hy*, 1969, **20**, 261.

60. Valdehita de Vincente, Maria and Teresa Vincente, *An Bromat*, 1971, **23**, 107.

61. O. Schwein and O.J. Conroy, *Journal of the Association of Official Agricultural Chemists*, 1965, **48**, 489.

62. H-C. Chïang and R-G. Tseng, *Journal of the Pharmaceutical Society*, 1969, **58**, 1552.

63. G. Lehman and M. Moran, *Zeitschrift für Lebensmittel-Untersuchung und Forschung*, 1971, **145**, 344.

64. P.Y. Vigneron and P. Spicht, *Revue Francaise des Corps Gras*, 1970, **17**, 295.

65. A.M.I. Pino, J.V. Leiro and H. Schmidt-Hebbel, *Grasas y Aceites*, 1969, **20**, 129.

66. H.D. McBridge and D.H. Evans, *Analytical Chemistry*, 1973, **45**, 446.

67. U. Köhler, H. Waggon and W.J. Uhde, *Plaste und Kautschuk*, 1968, **15**, 630.

68. D.M. Takahashi, *Journal of the Association of Official Analytical Chemists*, 1970, **53**, 39.

69. D.M. Takahashi, *Analytical Abstracts*, 1969, **17**, 3063.

70. I. Takemura, *Japan Analyst*, 1971, **20**, 61.

71. D. Halot, *Chim Analyt*, 1971, **53**, 776.

72. S.C. Lee, *Chemistry Taipei*, 1968, **43**, 155.

73. S.C. Lee, *Analytical Abstracts*, 1967, **14**, 7864.

74. J.B. Roos, *Fette Seifen Anstrichmittel*, 1958, **60**, 1160.

75. K. W. Beifer and H. Hadorn, *Mitt Gebiete Lebensmittel unters Hygiene* (Berne), 1956, **47**, 445.

76. F. Gander, *Fette Seifen Anstrichmittel*, 1955, **57**, 423.

77. F. Gander, *Fette Seifen Anstrichmittel*, 1956, **58**, 506.

78. F. Brown and K.L. Baxter, *Chemistry and Industry*, 1951, 633.

79. F. Brown and K.L. Baxter, *Biochemical Journal*, 1952, **51**, 237.

80. R. ter Heide, *Fette Seifen Anstrichmittel*, 1958, **60**, 360.

81. J.W.H. Zipp, *Recueil des Travaux Chimiques des Pays-Bas*, 1956, 75, 1053, 1060, 1089, 1129 and 1155.

82. E. Stahl, *Chemiker-Zeitung*, 1958, **82**, 323.

83. A. Seher, *Fette Seifen Anstrichmittel*, 1959, **61**, 345.

84. A. Seher, *Fette Seifen Anstrichmittel*, 1958, **60**, 1144.

85. R.A. Morton and A.L. Stubbs, *Analyst*, 1946, **71**, 348.

86. P. Shapras and G.C. Claver, *Analytical Chemistry*, 1964, **36**, 2282.

87. W.L. Bird and C.H. Hale, *Analytical Chemistry*, 1952, **24**, 58.

88. G.W. Dawes and W. F. Hamner, *Analytical Chemistry*, 1957, **29**, 1035.

89. L.N. Alcock and W.G. Hope, *Analyst*, 1970, **95**, 968.

90. E.J. Newman and P.D. Jones, *Analyst*, 1966, **94**, 406.

91. J. Koch and K. Figge, *Zeitschrift für Lebensmittel-Untersuchung und Forschung*, 1971, **147**, 8.

92. H. Wieczorek, *Deutsche Lebensmittel-Rundschau*, 1970, **62**, 92.

93. H. Wieczorek, *Analytical Abstracts*, 1970, **19**, 348.

94. V. Suk and M. Malet, *Chemist Analyst*, 1956, **45**, 30.

95. W.J. Ross and J.C. White, *Analytical Chemistry*, 1961, **33**, 421.

96. W.J. Uhde, H. Waggon, G. Zydek and U. Koekler, *Deutsche Lebensmittel-Rundschau*, 1969, **65**, 271.

97. G. Wildbrett, K.W. Evers and F. Kiermeier, *Zeitschrift für Lebensmittel-Untersuchung und Forschung*, 1968, **137**, 365.

98. G. Wildbrett, K.W. Evers and F. Kiermeier, *Zeitschrift für Lebensmittel-Untersuchung und Forschung*, 1970, **142**, 205.

99. G. Wildbrett, K.W. Evers and F. Kiermeier, *Fette Seifen Anstrichmittel*, 1969, **71**, 330.

100. L. Slezewska, *Rocan Panst Zakl Hig*, 1972, **23**, 417.

101. A.F. Colson, *Analyst*, 1963, 88, 26.

102. A.F. Colson, *Analyst*, 1963. **88**, 791.

103. A.F. Colson, *Analyst*, 1965, **90**, 35,

104. T. Salvage and J.F. Dixon, *Analyst*, 1965, **90**, 24.

105. L.T. Mann, *Analytical Chemistry*, 1963, **35**, 2179.

106. C.A. Johnson and M.A. Leonard, *Analyst*, 1961, **86**, 101.

107. T. Yoshizaki, *Analytical Chemistry*, 1963, **35**, 2177.

108. H. Waggon and W.D. Uhde, *Ernährungs-forschung*, 1971, **16**, 227.

109. *Gazetta Ufficiale: Regulations for Packages, Wrappings, Containers and Utensils Intended for Contact with Foodstuffs or with Substances for Personal Use*, Italian Ministry of Health, 1963, No. 64, p.18.

110. R. Franck, *Kunststoffe im Lebensmittelverkehr, Empfehlungen der Kunststoff-Kommission des Bundesgesundheitsamtes*, 7th issue, Carl Heymanns Verlag KG, Cologne, Germany, 1967, p.6.

111. *Staatablad van het Konindrijk der Nederlanden: Draft Packaging and Food Utensils Rregulation (Food Law)*, 3rd version, No.143, Netherlands Ministry of Social Affairs and Public Heath, 1968.

112. K. Figge, *Food and Cosmetics Toxicology*, 1972, **10**, 815.

113. K. Figge, *Migration von Hilfsstoffen der Kunststoffnerarbeitung aus Folien, in Nahrungsfette und Fettsimulantein*: Presented at the meeting Aus den Arbeit von Chemischen Forschungslaborotorien, Ortverland, Hamburg, Germany, 1972.

114. K. Figge, *Angewandte Chemie*, 1971, **83**, 901.

115. K. Figge, *Kunststoffe*, 1971, **61**, 832.

116. K. Figge and S.R. Eder and H. Piater, *Deutsche Lebensmittel-Rundschau*, 1972, **68**, 359.

117. K. Figge and H. Piater, *Deutsche Lebensmittel-Rundschau*, 1971, **67**, 9.

118. K. Figge and H. Piater, *Deutsche Lebensmittel-Rundschau*, 1971, **67**, 47.

119. K. Figge and H. Piater, *Deutsche Lebensmittel-Rundschau*, 1971, **67**, 110.

120. K. Figge and H. Piater, *Deutsche Lebensmittel-Rundschau*, 1971, **67**, 154.

121. K. Figge and H. Piater, *Deutsche Lebensmittel-Rundschau*, 1971, **67**, 235.

122. K. Figge and H. Piater, *Deutsche Lebensmittel-Rundschau*, 1972, **68**, 313.

123. K. Figge and J. Schoene, *Deutsche Lebensmittel-Rundschau*, 1970, **66**, 281.

124. H. Piater and K. Rigge, *Migration von Hilssttoffen der Kunststoffverarbeitung aus Folien in Flussige und Feste Fette bzw, Simulation VIII*, Mitteilung Vergleich der Gravimetrische Bestimmsten Rückstände der Extractionslösungen mit den Tatsachlich Extrahierten Additivmengen, 1971.

125. C.G. vom Bruck, K. Figge, H. Piater and V. Wolf, *Deutsche Lebensmittel-Rundschau*, 1971, **67**, 444.

126. C.G. vom Bruck, K. Figge and V. Wolf, *Deutsche Lebensmittel-Rundschau*, 1970, **66**, 253.

127. K. Figge, *Food Cosmeics and Toxicology*, 1972, **10**, 815.

128. K. Figge, *Synthese eines radiokohlenstoff-markierten Organozinn-Stabilisators zur Bestimmung von Migrationsvorgängen*, Presented at the Gesellschaft Deutscher Chemiker (GDCh) meeting Präparative Radiochemie, Lindau/Bodensee, Germany, 1968.

129. K. Figge, *Journal of Labelled Compounds and Radiopharmacueticals*, 1969, **5**, 122.

130. Circulaire relative aux demandes d'autorisation d'emploi de substances chimiques destinées à être introduites dans les aliments ou itilisées dans les matériaux mis au contact des aliments, *Journal Official de la Ministre de la Santé Publique*, France, 1963.

131. R.F. Van der Heide, *Packaging*, 1966, 37, 54.

132. R.F. Van der Heide in *Proceedings of The Safety for Health of Plastics Food Packaging Materials Conference*, Utrecht, The Netherlands, 1964, 32.

133. W. Pfab, *Zeitschrift für Lebensmittel-Untersuchung und Forschung*, 1961, **115**, 428.

134. N.H. Strodtz and R.E. Henry in *Proceedings of the Industrial Methods of the Analysis of Food Additives Conference*, New York, NY, USA, 1961, 85.

135. J. Phillips and G.C. Marks, *British Plastics*, 1961, **34**, 319.

136. ASTM F34, *Standard Practice for Construction of Test Cell for Liquid Extraction of Flexible Barrier Materials*, 2002.

137. T.W. Pfab, *Deutsche Lebensmittel-Rundschau*, 1968, **64**, 281.

138. T.W. Pfab, *Journal of the Association of Agricultural Chemists*, 1964, **47**, 177.

139. E. Fluckiger and H. Henscher, *Deutsche Molkerei – Zeitung*, 1969, **99**, 848.

140. J. Koch and K. Figge, *Deutsche Lebensmittel-Rundschau*, 1975, **71**, 170.

141. K. Figge and J. Koch, *Food and Cosmetics Toxicology*, 1973, **11**, 975.

142. R.F. Van der Heide in *Proceedings of The Safety for Health of Plastics Good Packaging Materials, Principles and Chemical Methods Conference*, Kemink En Zoon NV, Utretch, The Netherlands, 1964.

143. H. Waggon, D. Jehle and W.J. Uhde, *Nahrung*, 1969, **13**, 343,

144. H. Waggon, W.J. Uhde and G. Zydek, *Zeitschrift für Lebensmittel-Untersuchung und Forschung*, 1969, **138**, 169.

145. K. Figge and A. Zeman, *Kunststoffe*, 1973, **63**, 543.

146. I. Phillips and G.C. Marks, *British Plastics*, 1961, **34**, 385.

147. J. Koch, *Deutsche Lebensmittel-Rundschau*, 1972, 68, 401.

148. J. Koch, *Deutsche Lebensmittel-Rundschau*, 1972, 68, 404.

149. W.J. Uhde, H. Waggon and U. Koehler, *Nahrung*, 1968, **12**, 813.

150. W.J. Uhde, H. Waggon and U. Koehler, *Plaste Kautsch*, 1968, **15**, 630. [Analytical Abstracts, 1970, **18**, 327]

151. W. J Uhde and H. Waggon, *Deutsche Lebensmittel-Rundschau*, 1971, 67, 257.

152. W.J. Uhde and H. Waggon, *Nahrung*, 1968, **12**, 825.

153. W.J. Uhde and H. Waggon, *Analytical Abstracts*, 1970, **18**, 2024.

154. A. Sampaolo, L. Rossi, R. Binetti, D. Cesolari and G. Fava, *Raoss Chimo*, 1972, **24**, 3.

155. C.L. Hilton, *Analytical Abstracts*, 1960, 7, 4893.

156. O. Wadelin, *Analytical Abstracts*, 1957, **4**, 982.

157. R. Piacentini, *Industria Gomma*, 1972, **16**, 46.

158. K.G. Bergner and H. Berg, *Deutsche Lebensmittel-Rundschau*, 1972, 68, 282.

159. K. Rohleder and B. von Bruch-hauser, *Deutsche Lebensmittel-Rundschau*, 1972, 68, 180.

160. J. Koch, *Deutsche Lebensmittel-Rundschau*, 1974, **70**, 209.

161. K. Figge, *Kunststoffe*, 1971, **61**, 832.

162. K. Figge and A. Zeman, *Kunststoffe*, 1973, **63**, 543.

163. K. Figge and J. Koch, private communication.

164. K. Figge and W.D. Bieber, private communication.

165. K. Figge, *Verpackungs-Rundschau*, 1975, 8, 59.

166. A. O'Brian, A. Goodson and I. Cooper, *Food Additives and Contaminants*, 1999, **16**. 367.

167. C. Simoneau and P. Hannaert, *Food Additives and Contaminants*, 1999, **16**. 197.

168. I.E. Helmroth, M. Dekker and T. Hankemeier, *Food Additives and Contaminants*, 2002, **19**, 176.

169. I.E. Helmroth, H.A.M. Bekhuis, J.P.H. Linssen and M. Dekker, *Journal of Applied Polymer Science*, 2002, **86**, 3185.

170. J.S. Eberhard and M.M. McCort-Tipton in *Proceedings of the ANTEC 2000 Conference*, Orlando, FL, USA, 2002, Paper No.438.

171. M. Hamdani, L. Thil, G. Gans and A.E. Feigenbaum, *Journal of Applied Polymer Science*, 2002, **83**, 956.

Appendix 1

Instrument Suppliers	
Visible, ultraviolet and infrared spectrometers	Philips Electronics Cecil Instruments Kontron Instruments Perkin Elmer Corp. Gilson International Varian Instruments Foss Electronic
Fourier Transform Infrared Spectroscopy, Near Infrared Fourier and Transform Raman Spectroscopy	Mattson Instruments Ltd. Perkin Elmer Corp. Varian Instruments Philips Electronic Instruments Foss Electronic JOEL Ltd. Applied Photophysics EDT Research
Chemiluminescence analysis	New Brunswick Scientific Amersham Biosciences
Inductively coupled plasma mass spectroscopy	VG Isotopes Ltd. Perkin Elmer Corp. Labtam Ltd.
Inductively coupled plasma optical emission spectrometers	Spectro Inc. Philips Analytical Philips Electronic
Zeeman atomic absorption spectrometry	Perkin Elmer Corp. Varian Instruments
Flame and graphite furnace atomic absorption spectrometry	Thermoelectron Ltd. Perkin Elmer Corp. Varian Instruments GBC Scientific Pty Ltd. Shimadzu Corp. PS Analytical Ltd.

Total elements	Dohrman Instruments now Emerson Process Management
Halogen, sulfur, nitrogen	Sartec Ltd.
Sulfur	Mitsubishi Chemicals Industries
Sulfur, chlorine	EDT Analytical Ltd.
Nitrogen	Foss Tecator AB
Carbon, hydrogen, nitrogen	Perkin Elmer Corp.
Nitrogen, carbon and sulfur	Thermo Electron
Nitrogen	Thermo Electron Mitsubishi Corp. Buchi Labortechnik AG
Spectrofluorimetry	Shimadzu Spectrovision Inc.
Polarography, voltammetry	Metrohm Ltd. EDT Analytical Chemtronics Ltd.
Luminescence and spectrofluorimetry	Perkin Elmer Corp. Hamilton Co. Hamilton Bonaduz AG
Headspace samplers	Perkin Elmer Corp. Shimadzu Corp. Siemens AG Eden Scientific Ltd. Hewlett Packard Inc.
NMR spectroscopy	Varian Instruments Perkin Elmer Corp. Oxford Systems
Energy dispersive and total reflection X-ray fluorescence spectroscopy	Link Analytical Ltd. Oxford Systems Philips Electronic Instruments Richard Seifert & Co. Traor Europa BV
Gas chromatography	Thermo Electron Perkin Elmer Corp. Shimadzu Corp. Dyson Instruments Hnu-Nordion Instruments Co.

Gas chromatography (*Continued*)	Siemens AG Varian Instruments
Pyrolysis – gas chromatography	CDS Instruments Perkin Elmer Corp. Philips Analytical Varian Instruments Foxborough Co.
Gas chromatography – mass spectrometry and mass spectrometry	Thermo Electron Perkin Elmer Corp. Shimadzu Corp. Dyson Instruments Ltd. Varian Instruments HBI Haakon Buchler
Mass spectrometry	Oxford Analytical Ltd. Perkin Elmer Corp. Varian Instruments Thermo Electron Hewlett Packard JOEL Shimadzu Corp. GV Instruments Ltd.
Gas chromatography – Fourier transform infrared spectroscopy	Perkin Elmer Corp. Philips Analytical Shimadzu Europe Varian Instruments
High performance liquid chromatography	Varian Instruments Perkin Elmer Corp. Kontron Instruments Dionex Corp. Amersham Biosciences Shimadzu Corp. Dyson Instruments Hewlett Packard Kratos Analytical Instruments Cecil Instruments Varian AG GV Analytical Ltd.
High performance liquid chromatography – mass spectrometry	Hewlett Packard

Supercritical fluid chromatography	Lee Scientific Inc. Dionex UK Ltd. Pierce Chemical Company
Gel permeation chromatography, size exclusion chromatography	Perkin Elmer Corp. Polymer Laboratories Ltd.
Thin layer chromatography	JT Baker Ltd. Camag Merck Shimadzu Europa Whatman Ltd.
Electrophoresis	Beckman Coulter, Inc. Cole-Parmer Instrument Co.
Thermogravimetric analysis	Perkin Elmer Corp. PL Thermal Sciences Inc. EI DuPont de Nemours TA Instruments Inc.
Thermogravimetric analysis – Fourier transform infrared spectroscopy	Perkin Elmer Corp. PL Thermal Sciences Inc.
Thermogravimetric analysis – mass spectrometry	PL Thermal Sciences Inc.
Differential scanning calorimetry	Perkin Elmer Corp. PL Thermal Sciences Inc. EI DuPont de Nemours TA Instruments Inc.
Differential thermal analysis	Perkin Elmer Corp. PL Thermal Sciences Inc. EI DuPont de Nemours TA Instruments Inc.
Other suppliers of thermal analysis equipment	Varian Instruments Mettler Instrument Corp. Polymer Laboratories Inc.
Differential photocalorimetry	EI DuPont de Nemours TA Instruments Inc.
Thermomechanical analysis	PL Thermal Sciences Inc.

Dynamic mechanical analysis	Perkin Elmer Corp. EI DuPont de Nemours PL Thermal Sciences Inc. TA Instruments Inc.
Dielectric thermal analysis	PL Thermal Sciences Inc. TA Instruments Inc. EI DuPont de Nemours
Viscometry, molecular weight	Brinckmann Instruments Brookfield Engineering Laboratories Contraves Space AG

Thermal Properties of Polymers

Property	Part numbers	Manufacturer
Vicat softening point	10.01004, 10.01009, 10.01003, 10.01010, 10.01041, 10.01042	ATS FAAR UK Supplier: Martin Instrument Co. Ltd.
Melt flow index (or rate)	10.02013, 10.01017	
Low temperature embrittlement	10.12000, 10.12006	
Linear expansion coefficient	10.46000	
Melting point (Fischer-Johns)	0701730	
Minimum filming temperature	10.35000, 10.35001, 10.35002	
Ageing in air	10.69000	

Mechanical Properties of Polymers

Property	Part Numbers	Manufacturer
De Mattia resistance to dynamic fatigue test	10.23000, 10.23001, 10.23003	
Compressive test	10.65000, 10.07000	
Hardness ball indentation	10.48000	
Hardness Rockwell	11.03000	
Hardness IRHD	10.25130	
Hardness shore	11.03100, 11.03101, 11.03102, 11.03110, 11.03120	
Hardness flat	10.25140, 10.25150	
Density	10.06000, 10.14000	
Textile impact strength	16.10050, 16.00121, 16.00122, 16.00126, 16.00127	
Compressive strength	16.00121, 16.00122, 16.00126, 16.00127	ATS FAAR
Flexural strength	16.00121, 16.00122, 16.00126, 16.00127	UK Supplier:
Izod charpy impact strength	16.10000, 16.10500, 16.10600	Martin Instrument Co. Ltd.
Falling dart impact test	16.20003, 16.20004, 16.20001, 16.20002	
Impact resistance at low temperatures	10.65050	
Modulus of elasticity	10.22000, 10.22001	
Modulus of elasticity (bend test)	10.2900	
Cold bend test (cables)	10.65053	
Low temperature brittleness check	10.29000, 10.29005, 10.12010	
Ross flexing test	10.64002	
Measurement of mechanical properties resilience, etc.	16.65100	

Mechanical stability of natural and synthetic latices	10.67000	ATS FAAR UK Supplier: Martin Instrument Co. Ltd.
Gas permeability of elastomers (water vapour, CO_2, O_2)	10.08050, 10.08100	
Ozone resistance	10.47000, 10.47005	
Environmental stress cracking resistance	10.36000	
Hydrogen chloride content of PVC	11.00050	
Carbon black content of polyethylene	11.00001	
Oscillating die rheometer		
Mark II Mooney viscometer		
Mark 8 Mooney viscometer		Wallace Instruments
Gibitre srl		
Weight-density-volume variation		
Hardness measurement, IHRD microdurometer		
Brittleness point TR test		
De Mattia type dynamics flexing testing machine		
Universal tensile testing machine		
Abrasion test machine		
Ozone degradation test, 'Ozonecheck'		
Minimatt, Mark II		Polymer Laboratories
Miniature materials tester		Polymer Laboratories, Inc
Rheological properties of polymers		Thermo Electron

Electrical Properties of Polymers

Property	Part Numbers	Manufacturer
Volume and surface resistivity		
UK Supplier:		
Martin Instrument Co. Ltd.		ATS FAAR
Dielectric constant and dissipation factor		
Tracking resistance		

Optical Properties of Polymers

Property	Part Numbers	Manufacturer
Colour fastness meter		
Stress optical analyser (Minimat Mark II materials tester)		Polymer Laboratories

Physical Testing of Rubbers and Elastomers

Property	Part Numbers	Manufacturer
Measurement of processability and rheological behaviour of raw (unvulcanised) elastomers at variable shear rates and temperatures. THS Rheometer		Wallace Instruments Gibitre srl
Oscillating disc rheometry Rheocheck 100C		
Carbon black dispersion in rubbers, Tangent Electroscanner		

Physical Testing of Polymer Powders

Property	Part Numbers	Manufacturer
Surface area measurement		Thermo Electron Shimadzu Varian Instruments Thass Beckman Coulter Electronics Ltd.
Particle size distribution of powders		Beckman Coulter Electronics Ltd. Fritsch GmbH (UK Agent: Fisher Scientific)

Polymer Flammability Properties

Property	Part Number	Manufacturer
Flammability by glow wire	10.05300	
Rate of burning by rigid specimens exposed to ignition flame at 45°	10.05000	
Oxygen index	10.04070, 10.04050	
Ignition properties of plastics	10.04020	
Incandescence resistance of rigid plastics in horizontal position	10.04000	ATS FAAR UK Supplier: Martin Instrument Co. Ltd.
Smoke density during combustion of materials	10.05470	
Resistance to combustion of materials used inside automotive vehicles	10.05600	
Fire resistance of building materials	10.05200	
Flame resistance of vertical specimens	10.05050	

Fire reaction of specimens exposed to radiant heat	10.05020	
Fire reaction of upholstered furniture	10.05150	
Flammability test	10.05454	ATS FAAR
Flammability test	10.5700	UK Supplier:
Burning rate and flame resistance of rigid insulating materials	10.05400	Martin Instrument Co. Ltd.
Response of plastic to ignition by small flame	10.05500	
Other suppliers of polymer flammability test equipment		Fire Testing Technology Paul N. Gardner Inc.
Calorimeter for oxygen consumption calorimetry		PL Thermal Sciences Mettler Instruments Perkin Elmer Corp. Fire Testing Technology
Electron spin resonance spectrometer		Bruker-Biospin Varian Instruments
Auger spectrometer		Shimadzu Scientific Instrument Co.
Electron spectrometer		Shimadzu Scientific Instrument Co. Perkin Elmer Physical Electronics Division GV Instruments Co.
Electron microprobe microscopy		Philips Analytical GV Instruments Co.
Reflectance equipment		Miltron Roy Analytical Products Division Shimadzu Scientific Instrument Co. Varian Instruments

Secondary ion mass spectrometer		Cameca Perkin Elmer Corp. Shimadzu Scientific Instrument Co.
X-ray photoelectron spectrometer		Shimadzu Scientific Instrument Co. Tracor Europa BV GV Instruments Co. Perkin Elmer Corp.
X-ray analysers and diffusion equipment		Link Analytical Philips Analytical Tracor Europa BV Spectratech Inc.
EDAX		EDAX International Ltd.
Infrared microscopes		Varian Instruments Cambridge Instruments Co. Gallankamp Co. Southern Microinstruments Carl Zeiss Perkin Elmer Corporation Mattson Instruments Ltd.
NMR microimaging spectrometer		Varian Instruments
Electron microprobe microscope		Philips Analytical Japanese Electron Optics Ltd. (JOEL) GV Instruments Co.
Electron scanning microscope		International Equipment Trading Ltd. Philips Analytical
Electron transmission microscope		International Equipment Trading Ltd.

X-ray microprobe		Hilgenberg GmbH Perkin Elmer Corp. Whatman Scientific Co.
Laser spectrometer		Shimadzu Corp. GV Instruments Co.
Photoacoustic spectrometry		Perkin Elmer Corp.

Addresses of Suppliers	
Alpine American Corporation	5 Michigan Drive Natick MA 01760 USA
Applied Chromatography Systems	The Arsenal Heapy Street Macclesfield Cheshire SK1 7JB UK
Applied Photophysics Ltd www.photophysics.com	203/205 Kingston Road Leatherhead Surrey KT22 7PB UK
ATS FAAR www.atsfaar.it	Via Camporiccolserdio 20060 Vignate Italy (UK Agents: Martin Instrument Co.)
Beckman Coulter, Inc www.beckmancoulter.com	4300 N Habor Boulevard PO Box 3100 Fullerton CA 92834-3100 USA Oakley Court Kingsmead Business Park London Road High Wycombe HP11 1 JU UK

Boekel Scientific www.boekelsi.com	855 Pennsylvania Boulevard Feasterville PA 19053 USA
Brinkmann Instruments Inc. www.brinkmann.com	1 Cantiague Road PO Box 1019 Westbury NY 11590 USA
Brookfield Engineering Laboratories Inc. www.brookfieldengineering.com	11 Commerce Boulevard Middleboro MA 02346 USA
Bruker-Biospin www.bruker-biopsin.com	Bruker Biospin Corporation 15 Fortune Drive Manning Park Billerica MA 01821-3991 USA
Buchi Labortechnik AG www.buchi.com	Meierseggstr 40 CH-9230 Flawil 1 Switzerland
Cahn Instruments, part of Thass www.thass.net	THASS Thermal Analysis and Surface Solutions GmbH Pfingstweide 21 61169 Friedberg Germany
Camag www.camag.ch	Sonnenmattstrasse 11 CH-4132 Muttenz Switzerland
Cambridge Instruments www.home.btconnect.com/camsci	12-15 Sedgeway Business Park Witchford Ely CB6 2HY UK
Cambridge Instruments GmbH now Leica Microsystems www.hbu.de	Postfach 1120 Heidebergerstrasse, 17-19 D-6907 Nussloch Germany

Cambridge Mass Spectrometry Ltd. Now Kratos Analytical, part of Shimadzu group www.kratos.com	Wharfside Trafford Wharf Manchester Road M17 1GP UK
Cameca www.cameca.fr	103 Boulevard Saint Denis BP 6 92403 Courbevoie Cedex France
Carl Zeiss Inc. www.zeiss.com	1 Zeiss Drive Thornwood NY 10594 USA
Carlo Erba Strumentazione SpA now ThermoElectron www.ceinstruments.com	Strada Rivoltana 20090 Rodano Milan Italy
CDS Analytical, Inc www.cdsanalytical.com	PO Box 277 465 Limestone Road Oxford PA 19363-0277 USA
Cecil Instruments Ltd www.cecilinstruments.com	Milton Technical Centre Cambridge CB4 4AZ UK
Cole Parmer Instrument Co www.coleparmer.com	625 East Bunker Court Vernon Hills Illinois 60061-1844 USA
Contraves Space AG www.contraves.com	Schaffhauser Strass 580 CH-8052 Zurich Switzerland
Cypress Systems Inc. www.cypresssystems.com	2300 W 31st Street Lawrence KS 66046 USA

Dionex Corp. www.dionex.com	4 Albany Court Camberley Surrey GU16 7QL UK PO Box 3063 1228 Titan Way Sunnyvale CA 94088-3603 USA
Dohrmann Instruments now Emerson Process Management www.emersonprocess.com	Rosemount Analytical Division 1201 North Main Street Orville OH 44667 USA
Dyson Instruments Ltd.	Hetton Lyons Industrial Estate Hetton Houghton le Spring Tyne and Wear DH5 0RH UK
EDAX International Inc. www.edax.com	915 McKee Drive Mahwah NJ 07430 USA
EDT Research Ltd. www.edt.bham.ac.uk	Department of, Electronic, Electrical and Computing Engineering University Birmingham, Edgbaston Birmingham B15 2TT UK
EI DuPont de Nemours Inc. www.dupont.com	Concord Plaza Wilmington Delaware 19898 USA
Finnigan MAT Now Thermo Electron corporation www.thermo.com	1 Saint George's Court Hanover Business Park Altrincham WA14 5TP UK 1601 Cherry Street, Suite 1200 Philadelphia PA 19102 USA

Fisher Scientific UK Ltd. www.fisher.co.uk	Whitbrook Way Stakehill Industrial Park Middleton Manchester M24 2RH UK
Fisons Instruments now GV Instruments www.gvinstruments.co.uk	GV Instruments Crewe Road Wythenshawe Manchester M23 9BE UK
Foss Electronic www.foss.dk	NIR Systems Ltd. 7703 Montpellier Road Laurel MD 723 USA
Foss Tecator AB www.foss.tecator.se	Box 70 S263-21 Honagas Sweden
Fritsch GmbH www.fritsche.de	Industriestrasse 8 D-55743 Idar Oberstein Germany UK Agent: Fisher Scientific UK Ltd.
Gallenkamp Ltd. www.gallenkamp.co.uk	Units 37-38 The Technology Centre Epinel Way Loughborough LE11 3GE UK
GBC Scientific Pty Ltd www.gbcsci.com	Monterey Road Dandenong Victoria Australia 3175
Gibitre srl www.gibitre.it	Via Mercii I 24035 Curro (BG) Italy

Gilson International www.gilson.com	Box 27 3000 W Beltine Highway PO Box 620027 Middleton Wisconsin 53562-0027 USA
Gilson SAS www.gilson.com	19 Avenue de Entrepreneurs Villiers-le-Bel BP 45 F-95400 France
GV Instruments	Ion Path, Road Three Winsford Cheshire UK
Hamilton Bonadzu AG www.hamilton.ch	Via Crush 8 PO Box 26 CH-7402 Bonadzu Switzerland
Hamilton Company www.hamiltoncompany.com	PO Box 10030 Reno Nevada 89520-0012 USA
Hewlett Packard www.hpl.hp.com	1501 Page Mill Road Palo Alto CA 94303-0890 USA 150 Route du Nant-d'Avril CH-1217 Meyrin 2 Geneva Switzerland
Hewlett-Packardstrasse	Post 1180 D-7517 Waldbron Germany
Hilgenberg GmbH www.hilgenberg-gmbh.de	Struachgraben 2 PO Box 9 D-34323 Malsfeld Germany

Hnu-Nordion Ltd Oy www.hnunordion.fi	Atomitie 5 A 3 PO Box 1 SF 00370 Helsinki Finland
Horiba Jobin Yvan, Inc. www.jobinyvon.com	3880 Park Avenue Edison NJ 08820-3021 USA 2 Dalston Gardens Stanmore Middlesex HA7 1 BQ
HPLC Technology Ltd. www.hplc.co.uk	3 Little Mundells Mundells Industrial Centre Welwyn Garden City AL7 1EW UK
International Equipment Trading Ltd. www.ietild.com	960 Woodlands Parkway Vernon Hills IL 60061 USA
Japanese Electron Optics Ltd. (JEOL) www.jeol.com	1-2 Musashino 3-chome Aikshima Tokyo 196-8558 Japan
JT Baker Inc. Division of Mallinckrodt Baker www.imaging.mallinckrodt.com	222 Red School Lane Philipsburg New Jersey 08865 USA
Kratos Analytical Inc. www.Kratos.com	100 red Schollhouse Road Building A Chesnut Ridge NY 10977 USA
Labtam Ltd. www.labtam.com.au	33 Malcomb Road Braeside Victoria Australia 3195

Martin Instruments Co. Ltd.	6 Windsor Drive Market Drayton Shropshire TF9 1HX UK
Merck K GaA www.merck.de	Frankfurterstrasse 250 Postfach 4119 D-64293 Darmstadt Germany
Metrohm AG www.metrohm.com	68 Obersdorf Strasse CH-9101 Herisau AR Switzerland
Mettler Electronics Corp. www.mettlerelectronics.com	1333 S Claudina Street Anaheim CA 92805 USA
Mettler-Toledo Ltd www.mt.com	64 Boston Road Beaumont Leicestershire LE4 1AW UK
Mitsubishi Chemical Industries Ltd. www.mitsubishi.com	Instruments Department Mitsubishi Building 5-2 Marunouchi-2-Chrome Chiyoda-ku Tokyo 100 Japan
New Brunswick Scientific (UK) Ltd. www.nbsc.com	17 Alban Park Hatfield Road Hertfordshire AL 4 0JJ UK
Oxford Scientific Instruments Ltd. www.oxisci.com	Culham Innovation centre D5 Culham Science Centre Abingdon Oxon OX14 3DB UK
Paul N Gardner Co. Inc. www.gardco.com	316 Northeast First Street Pompano Beach FL 33060 USA

Perkin Elmer Corpoartion www.perkinelmer.com www.de.instruments.perkinelmer.com www.las.perkinelmer.co.uk	Chalfont Road Seer Green Beaconsfield Buckinghamshire HP9 2FX UK Life and Analytical Sciences 549 Albany Street Boston MA 02118-2512 USA Perkin Elmer (LAS) GmbH Ferdinand Porsche Ring 17 D-63110 Rodgau Germany
Pharmacia LKB Biotechnology Now Amersham Biosciences www/amershambiosciences.com	Amresham Bioscineces Europe GmbH Filial Sverige Bjorkgaten 30 75125 Uppsala Sweden Pollards Wood Nightengales Lane Chalfont St Giles Bucks HP8 4 SP
Philips Electronic Instruments www.analytical.philips.com	PANalytical EMEA Office Twentepoort Oost 26 7609 RG ALMELO The Netherlands York Street Cambridge CB1 2QU United Kingdom PANalytical Inc. 12 Michigan Drive Natick MA 01760 USA
Pierce Chemical Company www.piercenet.com	PO Box 117 Rockford IL 61105 USA

PL Thermal Sciences	Surrey Business Park Kiln Lane Epsom Surrey KT1 7JF UK
	300 Washington Boulevard Mundelein IL 60060 USA
	Polymer Laboratories Kurfuersten Anlage 9 6900 Heidelberg Germany
Polymer Laboratories www.polymerlabs.com	Essex Road Church Stretton Shropshire SY6 6AX UK
	Amherst Fields Research Park 160 Old Farm Road Amherst MA 01002 USA
	Polymer Laboratories BV Sourethweg 1 6422 PC Heerlen The Netherlands
	Polymer Laboratories GmbH PEKA Park T5 (001) Otto-Hesse Straße 19 D-64293 Darmstadt Germany
	Polymer Laboratories SARL GVIO Parc de Marseille Sud Impasse du Paradou Bâtiment D5 BP 159 13276 Marseille Cedex 09 France

PS Analytical Ltd. www.psanalytical.com	Arthur House Crayfields Industrial Estate Main Road Orpington Kent BR5 3HP UK
Sartec Ltd. www.sartec.co.uk	Century Farm Reading Street Tenterden TN30 7HS UK
Shimadzu Deutschland GmbH www.eu.schmadzu.de	Albert-Hahn Strasse, 6-10 D-47269 Duisburg Germany
Shimadzu UK www.shimadzu.com	Unit 1A Mill Court Featherstone Road Wolverton Mill South Milton Keynes MK12 5RD International Marketing Division Shinjuku Mitsui Buildings 1-1 Nishi Shinjuku 2 Chrome Shinjuku ku Tokyo 163 Japan
Siemens AG www.siemens.com	Siemens AG Wittelsbacherplatz 2 80312 Munich Germany Siemens plc Siemens House Oldbury Bracknell RG12 8FZ
Southern Instruments www.southernmicro.com	Southern Micro Instruments, Inc. 1700 Enterprise Way Suite 112 Marietta GA 30067-9219 USA

Spectratech Inc.	652 Glenbrook Road Stamford Connecticut CT 06906 USA
Spectro Analytical UK Ltd. www.spectro.com	Fountain House Great Cornbow Halesowen West Midlands B63 3BL UK
Spectro Anaylytical Instruments, Inc. www.spectro-ai.com	450 Donald Lynch Boulevard Marlborough MA 01752 USA
Stanton Redcroft Ltd. Now Fire Testintg Technology Ltd www.fire-testing.com	Fire Testing Technology Ltd Charlwoods Road East Grinstead West Sussex RH19 2HL
TA Instruments www.tainstruments.com	109 Luckens Drive New Castle Delaware 19720 USA
Teledyne Isco Inc. www.isco.com	4700 Superior Street Lincoln NE 68504 USA
Thermoelectron Ltd.	830 Birchwood Boulevard Birchwood Warrington Cheshire WA3 7QT UK 590 Lincoln Street Waltham MA 02254 USA
Tracor Europe BV www.tracor-europe.tripod.com	PO Box 333 3720 311 Bilthoven The Netherlands

Valco Instruments www.vici.com	Parkstrasse 2 CH-6214 Schenkon Switzerland 7806 Bobbitt Houston TX 77055 USA
Varian AG www.varianinc.com	Steinhauserstrasse CH-6300 Zug Switzerland
Varian Instruments www.varianinc.com	Varian Limited 6 Mead Road Oxford Industrial Park Yarnton Oxford OX5 1QU UK Varian Inc., Corporate Headquarters 3120 Hansen Way Palo Alto CA 94304-1030 USA
Varian Techtron Pty Ltd.	679 Springvale Road Mulgrove Victoria Australia 3170
Wallace Instruments www.hwwallace.co.uk	Unit 4, St Georges Industrial Estate Richmond Road Kingston KT2 5BQ UK
Whatman Inc www.whatman.com	200 Park Avenue Suite 210 Florham Park NJ 07932 USA
Whatman International Ltd. www.whatman.com	Springfield Mill James Whatman Way Maidstone Kent ME14 2LE UK

Abbreviations and Acronyms

AA	Atomic absorption
AAS	Atomic absorption spectrometry/spectrometers
ABS	Acrylonitrile–butadiene–styrene terpolymers
ADC	Analogue-to-digital converter
AES	Auger electron spectroscopy
AFM	Atomic force microscopy
AMMS	Anionic micromembrane suppressor
amu	Atomic mass unit
aPP	Atactic polypropylene
ASTM	American Society for Testing and Materials
ATR	Attenuated total infra-red internal reflectance
BDEK	Benzyldiethylketal
BMDK	Benzyldimethylketal
BS	British Standards
CI	Chemical ionisation
CID	Collision induced dissociation
CIMS	Chemical ionisation mass spectrometry
CMMS	Cationic micromembrane suppressor
CP	Constant potential
CVAAS	Cold vapour atomic absorption spectrometry
DC	Direct current
DETA	Dielectric thermal analysis
DIN	Deutsches Institüt für Normung
DLS	Dynamic light scattering
DMA	Dynamic mechanical analysis/analyser(s)
DMTA	Dynamic mechanical thermal analysis
DPC	Differential photocalorimetry
DR	Dichroic ratio(s)

DRS	Diffuse reflectance spectroscopy
DSC	Differential scanning calorimetry/calorimeter(s) calorimetric
DSIMS	Dynamic SIMS
DTA	Differential thermal analysis
DTGS	Deuterated triglycine sulfate
DTUL	Deflection temperature under load(s)
DWI	Falling dart impact method
ECD	Electron capture detector
EDAX	Energy dispersive analysis using X-rays
EDXRF	Energy dispersive X-ray fluorescence
EGA	Evolved gas analysis
EI	Electron impact
EPDM	Ethylene–propylene–diene
ESCA	X-ray photoelectron spectroscopy
ESI	Electrospray ionisation
ESR	Electron spin resonance spectroscopy
FAA	Federal Aviation Authority
FAB-MS	Fast atom bombardment-mass spectrometry
FAR	Federal Air Regulations
FHA	Federal Highway Administration
FI	Field ionisation
FID	Flame ionisation detector/detection
FIT	Flash ignition temperatures
FLS	Digital filtering and least squares
FMVSS	Federal Motor Vehicle Safety Standard
FPD	Flame photometric detection
FT	Fourier transform
FTC	Federal Trade Commission
FT-ICR-MS	Fourier transform ion cyclotron mass spectrometry
FT-IR	Fourier transform infrared spectroscopy
FT-MS	Fourier transform - mass spectrometry
GC	Gas chromatography/chromatographic/chromatograph
GFAAS	Graphite furnace atomic absorption
GPC	Gel permeation chromatography

HDDA	Hexanediol-diacrylate
HDPE	High-density polyethylene(s)
HPLC	High-performance liquid chromatography
HRMS	High-resolution MS
ICP	Inductively coupled plasma
ICP-AES	Inductively coupled plasma - atomic emission spectrometry
ICP-MS	Inductively coupled plasma - mass spectrometry/spectrometer
ICP-OES	Inductively coupled plasma - optical emission spectrometer/ spectrometry
id	Internal diameter
IGC	Inverse gas chromatography
IMCO	Inter Governmental Maritime Consulting Organisation
iPP	Isotactic polypropylene
IR	Infrared
IRHD	International Rubber Hardness Degrees
ISO	International Organisation for Standardisation
IUPAC	International Union of Pure and Applied chemistry
LC	Liquid chromatography/chromatographs
LC-UV-MS-MS	Liquid chromatography–ultraviolet spectroscopy–mass spectrometry–mass spectrometry
LDPE	Low-density polyethylene(s)
LIMA	Laser ionisation mass analysis
LIPI	Laser-induced photoelectron ionisation
LVN	Limiting viscosity number
MALDI	Matrix-assisted laser desorption/ionisation
MALDI-MS	Matrix-assisted laser desorption/ionisation mass spectrometry
MALDI-CID	Matrix-assisted laser desorption/ionisation–collision-induced dissociation
MAS	Magic angle spinning
MBET	Bis(2-hydroxyethyl) terephthalate
MFI	Melt flow index
MHET	Mono(2-hydroxyethyl) terephthalate
MMA	Methyl methacrylate(s)
M_n	Number average molecular weights

mp	Melting point
MS	Mass spectrometry/spectrometer
MS - MS	Tandem mass spectrometry
MW	Molecular weight
M_w	Weight-average molecular weights
MWD	Molecular weight distribution(s)
NAA	Neutron activation analysis
NBS	National Bureau of Standards
NIST	National Institute of Science and Technology
NMR	Nuclear magnetic resonance spectroscopy
Oa-TOF	Orthogonal acceleration - time-of-flight
OIT	Oxidative induction time
OIT	Oxygen index test
PA	Polyamide(s)
PALS	Positron annihilation lifetime spectroscopy
PAR	Pyridyl azoresorcinol
PC	Polycarbonates(s)
PDSC	Pressure differential scanning calorimetry
PE	Polyethylene(s)
PEEK	Polyether ether ketone
PEG	Polyethylene glycol
PES	Polyethersulfone
PET	Polyethylene terephthalate
PGC	Pyrolysis - gas chromatography/chromatograms
PID	Photoionisation detector
PMMA	Polymethyl methacrylate(s)
PMR	Proton magnetic resonance
PMS	Pyrolysis–mass spectrometry
PP	Polypropylene(s)
ppm	Parts per million
ppt	Parts per trillion
PQC	Piezoelectric quartz crystal
PS	Polystyrene(s)
PSA	British Property Service Agency

PTFE	Polytetrafluoroethylene
PU	Polyurethane(s)
PVC	Polyvinyl chloride(s)
PVOH	Polyvinyl alcohol
RF	Radio frequency
rpm	Revolutions per minute
SAW	Surface acoustic wave
SAXS	Small-angle X-ray scattering
SBS	Styrene–butadiene–styrene
SCOT	Solid capillary open tubular column
SEC	Size exclusion chromatography
SECM	Scanning electrochemical microscopy
SEM	Scanning electron microscopy
SFC	Supercritical fluid chromatography/chromatographic/chromatographs
SIMS	Secondary ion mass spectrometry
SIT	Self-ignition temperature
sPS	Syndiotactic polystyrene
ss	Sunshine
SSIMS	Static SIMS
TCD	Thermal conductivity detector
TEM	Transmission electron microscopy
T_g	Glass transition temperature(s)
TGA	Thermogravimetric analysis
THF	Tetrahydrofuran
TLC	Thin-layer chromatography
T_m	Melting temperature
TMA	Thermal mechanical analysis
TMBPA-PC	Tetramethyl bisphenol-A–PC
TNO	Dutch packaging research organisation
TOC	Total organic carbon
TOF	Time-of-flight
ToF-MS	Time-of-flight - mass spectrometry/specrometer
ToFS	Time-of-flight spectrometer
ToFSIMS	Time-of-flight secondary ion mass spectrometry

TRXRF	Total reflection XRF
TTL	Transistor-transistor logic
TVA	Thermal volatilisation analysis
UL	Underwriters' Laboratory
UV	Ultraviolet
WAXD	Wide-angle X-ray diffraction
WAXS	Wide-angle X-ray scattering
WDXRF	Wavelength dispersive XRF
WLF	Williams-Landel-Ferry
XPS	X-ray photoelectron spectroscopy
XRF	X-ray fluorescence
XRFS	X-ray fluorescence spectrometry
ZAAS	Zeeman atomic absorption spectrometry

Index

Page numbers in italic, e.g. *30*, refer to figures. Page numbers in bold, e.g. **293**, signify entries in tables.

A

W

X

Y

Z

Printed in the United States
106880LV00001B/20/A

9 781859 575260